LIFE IN MUD AND SAND

S. K. ELTRINGHAM
B.SC., PH.D., M.I.BIOL.

Director, Nuffield Unit of Tropical Animal Ecology, Uganda;
formerly Lecturer in Zoology,
King's College, University of London

Distributed in the United States by
CRANE, RUSSAK & COMPANY, INC.
52 Vanderbilt Avenue
New York, New York 10017

CRANE, RUSSAK & COMPANY, INC.
NEW YORK

Published in the United States by
Crane, Russak & Company, Inc.
52 Vanderbilt Avenue
New York, N.Y. 10017

ISBN 0-8448-0007-4
Library of Congress Catalog Number 72-79497

Printed and bound in Great Britain

FOREWORD

Dr. S. K. Eltringham, the author of this book, has a wide experience in ecological research ranging from the gribbles of Southampton Water and the wild ducks at Slimbridge to the elephants of the Queen Elizabeth National Park in Africa. His approach to the diversity of problems which he has investigated has always been meticulous and thorough and his present work on life in mud and sand follows in the same mould.

He has attempted with very considerable success to treat muddy and sandy shores as an ecological unit and to illustrate the general principles underlying the ecology of such shores. Recognizing the paucity of data, Dr. Eltringham has indicated either directly or by implication the sort of problems that need investigating. Many of these problems may serve to test the intellect of investigators but doubtless many would be useful guide lines in understanding such problems as pollution of shores and estuaries. Hitherto the amount of work done on sands and muds, when compared with other types of shore line, has not been large mainly from lack of guidance and from an incorrect assumption that they are usually barren and uninteresting. Dr. Eltringham's present contribution should do much to correct these misapprehensions, particularly as it is so clearly expounded and well illustrated.

Don. R. Arthur

PREFACE

During my time as a university teacher responsible for a course in marine biology, I was frequently conscious of the lack of a book on the ecology of the depositing shores which could be recommended to students. A variety of books on marine ecology exists, covering such topics as the plankton, rocky shores, the shallow sea and the identification of shore species but none has been devoted solely to intertidal sand and mud. The subject is dealt with in chapters of more general books but often in a form which is either too brief or too selective to quite fill the gap. It is not suggested that the gap has been adequately closed with the publication of this book, but I hope it is at least a step in the right direction. I had intended to produce a detailed survey of the literature which would have appealed to research workers but shortly after I had begun the task, I left Britain to take up an appointment in Africa where, for various reasons, my original plan proved to be impracticable and consequently I decided to confine myself to an exposition aimed primarily at undergraduates specialising in marine ecology. In some ways, this is a difficult audience; one does not wish to patronize but on the other hand, I know from experience that one can over-estimate the background knowledge of the tyro zoologist. However, I have assumed in the reader a familiarity with the vocabulary of a modern biologist and with the main groups of animals and plants. These modest attainments should be within the capacity of the average sixth-former and I hope, therefore, that the book will be of some value in schools.

I have tried to treat the shore as an ecological unit and have looked for features which distinguish it from other marine habitats. Consequently, this is not an account of the biology of marine animals and still less a guide to their identification. Where I have felt it necessary, I have gone into the biology of a particular species in some detail but only to illustrate general principles of shore ecology. Many of these examples are chosen from the British fauna but the conclusions should apply to all temperate seas. Most of the references are to readily accessible literature in the English language. To some extent, this restriction was imposed upon me as I have been largely dependent on the limited resources of my own library but it has also been deliberate since it is too much to hope that the average undergraduate, or the above-average one for that matter, has the time or inclination to track down obscure references in specialist libraries.

Any value this book may have lies not so much in the information it contains as in the extent to which it reveals our ignorance of the ecology of the depositing shore. This is not to decry the quality of the work which has been carried out on the shore, for the research has been as good as any pursued in other habitats, but it is sadly deficient in quantity. It is hoped that the publication of this text may in some small way stimulate an interest in the young biologist for the depositing shore. Physically, a mud flat is uninviting, if not repellant, but it has much to offer the intellect.

I am deeply grateful to Professor Don. R. Arthur who encouraged me to write this book and who has given me the great benefit of his criticism of the finished product although it goes without saying that I accept full responsibility for any errors or defects in the work. I would also like to thank the following publishers and proprietors for permission to reproduce figures from the journals listed below:

Archivio di Oceanografia e Limnologia. Fig. 35.

Biological Bulletin. Fig. 30.

The British Ecological Society (*Journal of Animal Ecology* published by Blackwell Scientific Publications). Figs 14, 15, 17, 18 and 24.

The Cambridge University Press (*Biological Reviews*). Figs 16 and 17.

The Company of Biologists Limited (*Journal of Experimental Biology* published by Cambridge University Press). Figs 7 and 10.

Council of the Marine Biological Association of the United Kingdom (*Journal of the Marine Biological Association of the United Kingdom* published by Cambridge University Press). Figs 11, 19, 20, 21, 22, 23, 25, 27, 28, 36, 39 and 40.

The Ecological Society of America (*Ecology*). Figs 44 and 45.

The Game Conservancy (*Transactions of the International Union of Game Biologists*). Figs 32 and 33.

Journal of the Fisheries Research Board of Canada. Fig. 31.

Macmillan (Journals) Limited (*Nature*). Fig. 9.

Taylor and Francis Limited (*Annals and Magazine of Natural History*). Figs 8, 38 and 47.

Vie et Milieu. Fig. 29.

The Zoological Society of London (*Proceedings of the Zoological Society of London*). Figs 6 and 37.

The following publishers have granted permission for the reproduction of figures from their books: Holt, Rinehart and Winston (Fig. 43); Oliver & Boyd (Fig. 13); Pergamon Press (Fig. 41).

The original authors of the illustrations are acknowledged in the relevant text figures.

S. K. E.
July, 1969

CONTENTS

	Foreword	iii
	Preface	iv
Chapter 1	**The physical background**	1
Chapter 2	**Categorising the shore fauna**	15
Chapter 3	**The natural history of the shore fauna**	19
Chapter 4	**The spatial distribution of the fauna**	64
Chapter 5	**Seasons on the shore**	83
Chapter 6	**The shore environment and its limiting factors**	95
Chapter 7	**The special case of estuaries**	147
Chapter 8	**The plants of the shore**	164
Chapter 9	**The shore as an ecosystem**	179
	References	204
	Index	211

CHAPTER ONE
THE PHYSICAL BACKGROUND

The sea shore is a term in common usage which is difficult to define precisely. Certainly, the bold lines drawn on a map have little meaning for the marine biologist. The shore line of a lake is more easily described, being simply the region where the water meets the land, but the limits of the sea shore alter as the tide advances and recedes over a tidal cycle. Any definition of the shore must be arbitrary but we will confine ours to describe that region between extreme high water and extreme low water of spring tides. We will exclude the area above high water mark even though it may be considered to be 'shore' by the layman and indeed, may have the same physical characteristics. For example, a sandy shore may grade imperceptibly into dunes but the fauna with which we are concerned, will change more abruptly. On rocky shores, the region above high water mark forms the splash zone and properly falls into the province of the marine biologist but rough seas, a necessary precursor of splash, are not a feature of sandy or muddy shores for reasons which will shortly be apparent. Even so, our limits for the shore are not clear cut and the lower one is particularly indecisive with identical species living on either side of it. However, the fauna between the tide marks shows a number of common features which together, make up a biological unit similar, in many respects, to a community in the sense of the biological component of an ecosystem, but whether the shore does in fact constitute an ecosystem is a point which will be discussed later. The shore on our definition is synonymous with the littoral zone or the intertidal region although these terms may be used in different senses by some authors. Fig. 1. is a diagram of the shore showing tidal levels and some of the terms in current use in describing these levels.

Shores can usually be divided into four main types: rock, pebble, sand and mud although mixed shores of sand or mud with rocky outcrops are very common. These types of shore are distinct not only in appearance but also in their fauna. Of course, the different types of shore will grade into one another, sand into mud, rock into pebble, and their faunas will also intergrade. Nevertheless, they are sufficiently distinct for these broad categories to be retained and the animals found in each show distinct modifications which enable a sample to be readily assigned to its habitat. It is perhaps true to say that the rocky shore has been studied more than the others and certainly the animals and plants are more apparent. The ecology of rocky shores has been extensively reviewed by Lewis (1964), but it is doubtful whether sufficient data exist for a similar account to be written about the ecology of mud and sand. The habitat is difficult to study for the daily activities of the animals are

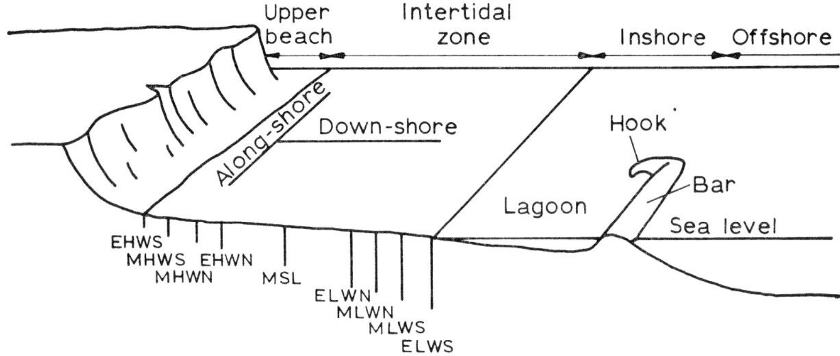

Fig. 1 A typical depositing shore with an explanation of the more common terms. Some of these are imprecise e.g. there is no acceptable diagnostic difference between the inshore and offshore regions. 'Along-shore' and 'down-shore' are used to describe the distribution of animals. The 'vertical' orientation follows the direction of any of the lines indicating the tidal levels.

EHWS	— extreme	high	water	springs	tidal	level
MHWS	— mean	,,	,,	,,	,,	,,
MHWN	— ,,	,,	,,	neaps	,,	,,
EHWN	— extreme	,,	,,	,,	,,	,,
M.S.L.	— mean sea level					
ELWN	— extreme	low	water	neaps	,,	,,
MLWN	— mean	,,	,,	,,	,,	,,
MLWS	— ,,	,,	,,	springs	,,	,,
ELWS	— extreme	,,	,,	,,	,,	,,

The maximum tidal range is between EHWS and ELWS and the minimum between EHWN and ELWN

carried out under a veil of sand which hides them from the observer. Changes in physical factors and their influences on the fauna are clear cut on rocky shores but are blurred and not at all obvious on the beach. Technical problems are greater; simply to move about on a mud flat is frequently difficult and often hazardous while the separation of animals from such a substratum is tedious and difficult, particularly with those species whose size approaches that of the particles themselves. For these and other reasons, good quantitive ecological research into the fauna of mud and sand, especially of the microscopic forms, is rarer than is the case for rocky shore animals. Nonetheless, there are plenty of published studies which show, for example, that zonation and seasonal changes occur in mud and sand as much as on rock, although it is doubtful whether there are sufficient data to permit an evaluation of the geographical variations such as has been done with the rocky shore fauna. A number of general reviews has been written on muddy and sandy shores (Yonge 1949, 1953, Stephen 1953, Russell and Yonge 1963) but these have mainly been concerned with the larger animals, or macrofauna, while the microscopic forms have been neglected. There have been good reasons for this neglect for the microfauna is of less interest to the non-specialist and in any case, not much is known about it, but an attempt to deal with the whole biology of sand and mud—all animals both large and small and, in as far as they exist, the plants—seems justified.

Before considering the environments of sandy and muddy shores it is expedient to consider the formation of shores as a knowledge of this aspect is essential to an understanding of shore ecology.

It is important to realise that the coastline is not static but is being constantly worn away by the sea. Coast erosion has been a problem for centuries and many instances are known of erstwhile seaside villages or forests which are now under water. If coast erosion continued unchecked by any balancing phenomenon, all the land would eventually disappear under the waves and the whole earth would become covered with a shallow sea. That this does not occur is due to the fact that as fast as material is eroded in one area, it is being deposited elsewhere. Thus, to set beside the drowned villages, we have examples of stranded ports such as Rye in Sussex which is now a couple of miles inland.

There are, therefore, two fundamentally different types of shores; the erosion shore and the depositing shore.

Erosion Shores

Erosion shores are created by wave action. Waves are regular oscillations in the surface of the sea caused by wind—usually storm winds of high velocity. The energy of the wind is transferred to the wave whose motion continues until its energy is expended or until some solid object stops or deflects it. Waves which arrive on exposed shores as swells have a lot of energy stored within them because they have long wave lengths. The wave length, the distance between successive troughs or crests, is the product of the velocity and the period, which is the time taken for one vibration of the wave. Hence long waves have greater velocities and, therefore, store more energy than short ones. In order to understand what happens when a wave reaches the shore and breaks, it is necesary to know the structure of a wave (Fig. 2). The water particles do not themselves move with the wave to any extent but they do describe a circular motion the diameter of which is equal to the wave height. Thus wave action is essentially a surface phenomenon with little effect existing below a depth equal to the wave length. As the wave approaches the shore, its action becomes modified because once the shallow areas are reached, friction with the sea bed causes the water particles to move in an elipse rather than a circle. With decreasing depth, this movement becomes exaggerated and when the depth becomes equal to about half of the wavelength, the particles near the sea floor are moving more or less backwards and forwards. As a result of the friction, this deeper water tends to lag behind so that the crest of the wave tends to overhang the trough in front of it until it eventually topples over and the wave breaks. It now becomes a wave of translation with only a crest and no trough.

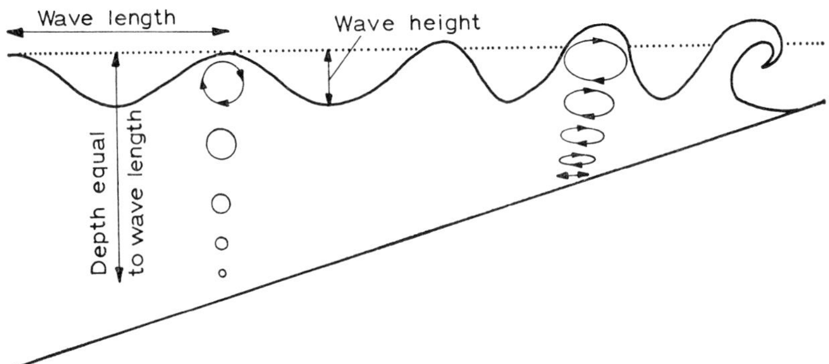

Fig. 2 Wave structure. In deep water (left), the water particles move in a circle the diameter of which decreases with depth until at a depth equal to the wave length there is no movement at all. As the wave approaches the shore (right), the deeper water becomes retarded by friction with the sea bed and at a depth equal to about half the wave length the waves become steeper and closer together. The water particles now move in an elipse rather than a circle while those near the bottom are moving backwards and forwards. Eventually the top of the wave becomes so far in advance of the bottom that it topples over and breaks to become a wave of translation

The water particles in such a wave are not oscillating but move in the direction of the wave striking any object in their path with considerable force. It is this lateral movement which causes erosion, but not through the water particles themselves for apart from limited dissolving action, water has no effect on rock which can only be broken down by objects harder than itself. However, the water mass rushing forward with a breaking wave picks up pebbles and chunks of rock and hurls them at the shore with great force to cause the actual erosion. One has only to watch a stormy sea pounding the defences of a seaside town to appreciate the tremendous forces involved. Wave pressures as high as 29 700 kg per square metre have been recorded on the shore during storms. The particles which are broken off tend to be carried out to sea by the undertow currents moving over the sea floor. The currents caused by the breaking waves are more powerful than the undertow so that sorting of the eroded particles occurs. The large heavy lumps of rock remain close to the

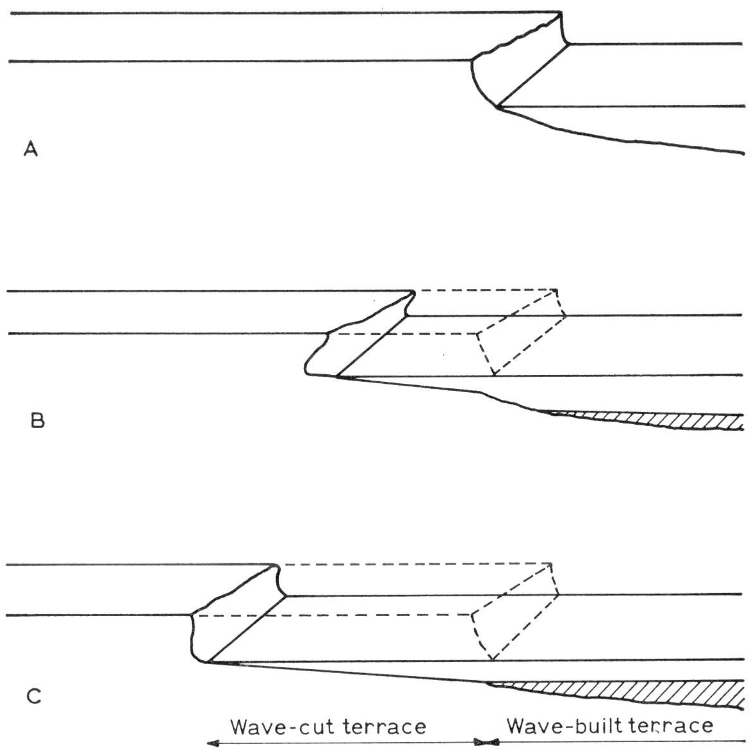

Fig. 3 Sequence of events leading to the formation of a depositing shore.
A Early stage of erosion shore with little or no beach.
B Cliff becomes progressively undercut by wave action. A flattened wave-cut terrace is formed while the eroded material (cross hatched) is deposited offshore.
C The process continues until a wide expense of wave-cut terrace is formed. The material deposited offshore accumulates to form a wave-built terrace. The waves now break well away from the cliff which is, consequently, no longer eroded. The terraces develop into a depositing shore as material is brought to the area in water currents.
The dotted lines show the original extent of the cliffs

shore and are broken up further by being rolled over one another and also by being thrown at the shore again until they are sufficiently ground down to be taken out to sea and deposited. Hence, there is a gradation of particles on an erosion shore from large, heavy pebbles near high water mark to fine sand offshore.

It should be clear from the above that erosion shores are found in regions subject to heavy surf where hard rock rises quite abruptly from the sea. The great majority of rocky shores are erosion shores although erosion may, for one reason or another, no longer be taking place. The sequence of events leading to the formation of an erosion shore is shown in Fig. 3. Here we see how wave action undercuts the base of a rocky cliff until the overhang collapses, the process is repeated over many years so that the cliff gradually recedes leaving a flat area known as a wave-cut terrace in its place. The finer particles cut from the cliff are taken out to sea in the undertow and deposited in deeper water. Over a long period of time, these particles accumulate to form a platform continuous with the wave-cut terrace and called, from its origin, a wave-built terrace. The situation has now changed from one of a cliff rising steeply out of relatively deep water to one of a cliff fringed by shallow water over a wide terrace. Because the sea is shallow, the waves now break much further out from the shore and most of their energy has been dissipated long before they reach the cliff. Hence, the waves have lost much of their destructive power so that erosion is slowed down and eventually ceases. Conditions are now ripe for the terraces, both wave-cut and wave-built, to become covered with material brought from elsewhere. The sequence is complete for the erosion shore has become a depositing one.

Depositing Shores

Depositing shores are characteristically formed from particles which have been carried in water currents from other areas. Much of the material has probably been derived from erosion shores but a lot of particulate matter, especially mud or silt, is brought to the shore in suspension by rivers. Depositing shores are often called beaches, although the term is not very appropriate in the case of mud. The areas in which beaches are formed are dependent upon the velocity of the water currents because the particles are carried along in suspension and fall out only when the current velocity drops below a certain level. The first particles to be deposited are the large pebbles followed by the various grades of shingle and sand until there remains only mud and finally silt which does not fall out before the water is quite still. Pebble beaches are found, therefore, in relatively exposed areas where wave action is too great for the deposition of finer particles while sandy beaches occur in more sheltered areas. Muddy shores are confined to very quiet waters and hence are typical of estuaries and upper reaches of semi-enclosed bays. In many protected sandy bays, the sand is found on the upper shore but gives way to mud as the low water mark is approached, a fact familiar to anyone who has, for instance, ever spent a holiday at a Bristol Channel resort. The reason for this distribution is that the sand is deposited first from the undertow currents while the mud is carried further out to sea.

Other water movements besides the undertow currents are important to the formation of depositing shores. The most significant, perhaps, are the longshore currents which run parallel to the shore. These are formed when more water is brought ashore by waves than can escape in the undertow. This leads to a piling up of water which eventually escapes by running away parallel with the beach. Longshore currents occur particularly in regions where the waves strike the shore at an angle. Normally, waves break parallel with the coast whatever their original orientation, because the wave approaching at an angle is retarded by friction at the shallower end so that the deeper end swings round and catches up. However, the action is not always completed before the wave breaks and under such conditions, longshore currents are prevalent.

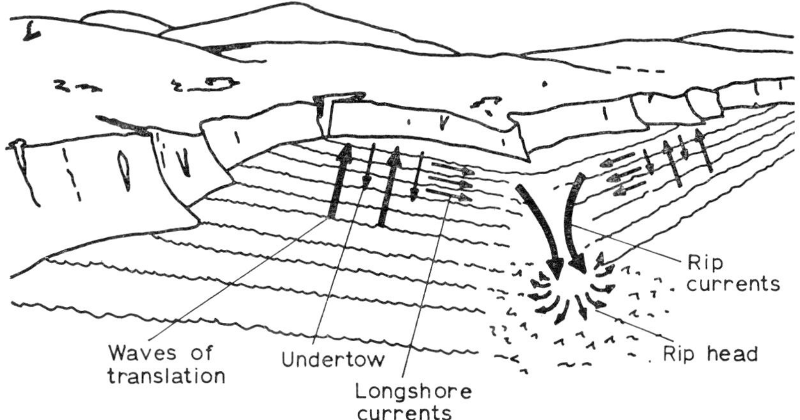

Fig. 4 The principal current system close inshore. Longshore currents are most common where the waves approach the shore at an angle. They result from more water being brought ashore than can escape in the undertow. The water eventually finds its way back to the open sea in high velocity rip currents which may become R.I.P. currents for the unwary swimmer. Note the rough water of the rip head

These currents do not, of course, continue indefinitely and eventually they turn out to sea, often through a break in a sand bar, to form rip currents which disperse off-shore in areas of turbulence known as ripheads. Rip currents may reach speeds of four or five knots and are extremely hazardous to swimmers as, indeed, are longshore and undertow currents. The relationship of some of the currents found near the shore are illustrated in Fig. 4.

It should be clear by now that depositing shores are not necessarily stable for their integrity depends upon the persistence of the currents leading to their formation. The velocity of these currents will vary, often quite considerably, as storms give way to calm periods and *vice versa.* Most depositing shores are dynamic with material being removed in undertow and longshore currents as fast as it is being deposited by other water movements. In most cases, the shores are stable with the rate of deposition equalling that of removal but sometimes even long established beaches can be radically altered by changes in the current system. Thus excavation off the South Devon coast for material with which to build the Plymouth Breakwater had such profound effects that the village of Hallsands near Start Point was seriously undermined and eventually destroyed by a great storm in 1917. Less spectacular but significant changes in the composition of beaches often follow in the wake of a storm and it is not for nothing that seaside resorts conserve their sandy assets with breakwaters and the like.

Material removed from the beach is usually deposited offshore and may form a bar (Fig. 1). Very often this breaks through to the surface either as a continuous bar or as a series of sandy islands which may or may not be covered at high tide. Sometimes a bar will stretch from one side of a bay to the other and enclose a lagoon which soon becomes fresh as at Slapton Ley, Devon, or more strikingly at Chessil Beach near Weymouth. Under natural conditions, such bars are continously breached by the sea and then reformed, but human engineering is capable of stabilising them. Similar bars are found at the mouth of rivers, or in any area where currents are deflected seawards, for on meeting deeper water, these currents slow down and deposit material previously held in suspension. Eventually, a bar is formed which characteristically has a landward facing hook caused by tidal currents. Such bars are extremely unstable and maritime charts of only a few years' standing are notoriously unreliable in the positioning of such features.

Types of Depositing Shores

This book is concerned with the biology of mud and sand so we will leave the rocky shores and turn to the types of depositing shore. Essentially, there are three types classified from the materials from which they are formed—shingle, sand or mud. Although some shores can be firmly placed in one or other of these categories, there is an infinite number of gradations from shingle to silt for which descriptive terms are quite inadequate. One man's muddy sand is another's sandy mud and the only certain way of defining the nature of a beach is to make a particle size analysis and quote the median value or some other measurable parameter. Even so, it is not uncommon to find such a gradation of particle size that it is impossible to classify the beach. It may for example be muddy at one point and sandy only a short distance away, this in addition to the phenomenon of sand grading into mud down the shore. It is even possible to find what appears to be erosion and depositing shores together for many rocky coasts have sandy or even muddy bays. In most cases such rocks are no longer being eroded and we have in fact a stage in the formation of depositing shores where the wave cut platform is being covered with sand. Even the stickiest mud flat can have some of the features of a rocky shore in the form of shells or scattered pebbles which, with such human artefacts as groynes, breakwaters and concrete anchor points, provide hard surfaces for the attachment of typical rocky shore animals. Many mud flats or sandy beaches are, in effect, backed by a rocky shore in the shape of concrete sea walls. Hence the rocky shore is never far away and an empty bottle discarded on the beach soon becomes thickly encrusted with barnacles or mussels and other rocky shore creatures. It is this mixture of shore characteristics that gives rise to such curious associations as a sessile coelenterate (*Cereus*) with a muddy shore. If dug out, the anenome will be found to be attached to a pebble or shell fragment a few inches under the surface.

Apart from the size of the particles, there are other structural differences between the three types of shore. As we have seen, shingle beaches are formed in areas where wave action is considerable and the piling up of material leads

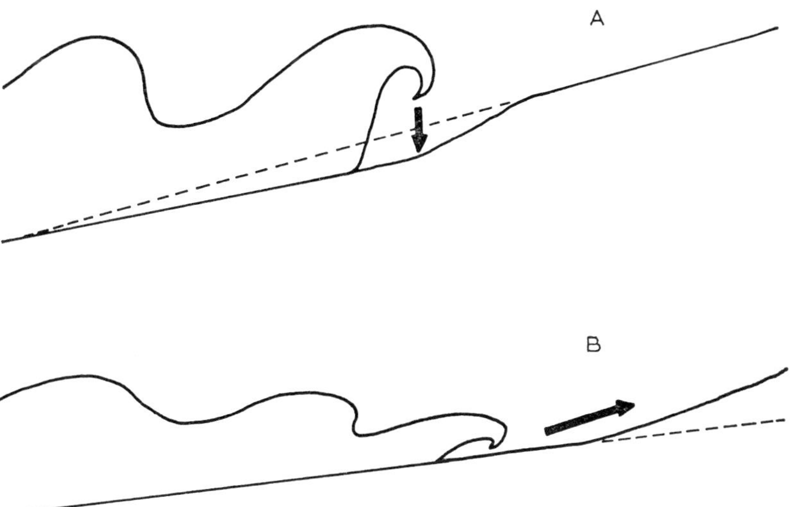

Fig. 5 Plunging or destructive waves (A) cannot run far up the beach after they break because of the steep slope. Consequently, most of their force is directed downwards (large arrow) to remove the surface particles which are carried out to sea in the undertow. On more gentle beaches (B), the breaker becomes a wave of translation and a body of water runs up the shore. Such waves are constructive as they build up the beach. Dotted lines show the original beach levels. Destructive waves are higher with a shorter wave length than constructive waves

to a steep slope for the beach. On such a shore the waves are of the plunging type, that is to say that when they break, most of the water plunges downward with considerable force. This is due to the steep slope because the water cannot run far up the shore when the wave breaks and the plunging action further increases the angle of the beach (Fig. 5). One consequence of biological interest is that such beaches dry out rather quickly through drainage.

Owing to the extreme mobility of shingle beaches, it is difficult for any animal to live in them. A large organism would soon be crushed as the stones rolled over one another. It is this continual motion which causes pebbles to be ground smooth and round whatever their original shape when dislodged from the parent rock. This grinding process is exemplified by the beautifully tinted pebbles which prove on closer examination to be the fragments of a broken beer bottle which have been shaped to the typical rounded form. Hence, life, in shingle is possible only for very tiny animals and even then only in relatively stable beaches. A few archiannelids may be found clinging to the water film around each pebble but for the most part, shingle beaches are devoid of life and are of little biological interest.

The sand grains of beaches are usually quartz particles which have been broken off from the original rock during the erosion process. Movements of the sand particle is appreciable, particularly at the surface, but each grain is surrounded by a film of moisture which acts as a cushion preventing mutual abrasion with its neighbours. Sand grains are not round, therefore, as are pebbles but may be seen under the microscope to have a variety of irregular shapes. Desert sands, on the other hand, lack the protective water film and are worn smooth by wind action. Thus marine and desert sands can be distinguished easily, a fact which is sometimes of value in revealing the marine origin of terrestrial deposits. Most sands include a high proportion of shell fragments. In some cases, the whole beach is formed almost entirely of shells which are transported by currents in the same way as are eroded particles and which usually pile up in an area remote from that of their origin. Being softer than sand grains and also considerably more soluble, they are quite rapidly worn away.

Sandy shores are much more gradual in their slope than are shingle beaches and this is a reflection of the gentler wave action. As a consequence, they dry out more slowly and with their greater stability, are better suited for animal life. Animals cannot live, however, in the absence of organic matter for they must eat and pure sand, if it existed, would be quite sterile. All sands, however, contain a certain amount of detritus derived from decaying seaweed or from the faeces and dead remains of animals, while much organic matter is blown onto the beach from the shore. Organic particles are very light and most settle out only in quiet waters so that there is an inverse relationship between the organic content of a beach and the turbulence of the overlying water. Detritus tends to clog the interstices between the grains and to bind them together until the deposit takes on the characteristics of muddy sand. The interstices between the particles are filled with water which drains away when the beach is exposed—quickly in the case of clean sands but more slowly as the organic content increases. As the water trickles down through the substratum it tends to lose its oxygen content partly as a result of the respiration of micro-organisms and partly through the oxidation of chemical substances within the sand. Eventually, all the oxygen is used up at a depth which may vary from a few millimetres to nearly a metre depending on the organic content. The position where conditions become near anaerobic is marked by a black boundary known as the sulphide layer because the dark coloration is due to ferrous sulphide. Conditions are not necessarily anaerobic in the upper layers of the black sand but at least the amount of oxygen present is insufficient to oxidise the ferrous sulphide which is very unstable, particularly when wet. The ferrous sulphide is produced indirectly by the activity of saprophytic bacteria and hence decaying organic matter is an important factor in the establishment of the sulphide layer since it provides food for the bacteria. The bacteria produce hydrogen sulphide but this reacts with iron salts in the sand to form the ferrous sulphide which is normally immediately oxidised to ferric oxide. This is yellow or brownish-yellow and it helps to preserve the light colour of clean sand. It is only when the supply of oxygen is insufficient to oxidise all the ferrous sulphide that the black contamination appears.

One of the most interesting features of black sand is the presence of chemosynthetic bacteria within it. These bacteria can exist anaerobically and are chemosynthetic because they produce complex organic matter from simple inorganic compounds in the manner of green plants but without the necessity of sunlight. Instead, they utilise the energy released during the chemical reaction, which they themselves bring about, whereby hydrogen sulphide, thiosulphates and elementary sulphur are converted to sulphates. One genus, *Thiospira*, carries the reaction only to an intermediate stage and assimilates free sulphur into the cell body but others, such as the pseudomonad *Thiobacillus thioxidans*, complete the oxidation of the sulphur to sulphates. Sulphuric acid is produced as a by-product and with free hydrogen sulphide is responsible for the generally low pH of the black layer. The free sulphur produced is also reduced to hydrogen sulphide by the action of bacteria and this reaction is the basis of the alternative name of reducing layer which is sometimes given to the black sand. Sulphates are also reduced to hydrogen sulphide in the absence of oxygen. This excess hydrogen sulphide is responsible for the offensive smell of black muds familiar to anyone who has delved too deeply while building sand castles.

Conditions within the sulphide layer have been discussed by Perkins (1957). The boundary between the clean and black sand is not flat but is thrown into a series of villiform projections which reflect the differential de-oxygenation of the water as it percolates through the sand. However, the depth varies only within narrow limits at any one place and time, but there are significant variations at different times of the year with an inverse relationship between depth and the environmental temperature. At Whitstable the average depth was found to vary from about 8 mm in February to less than 2 mm in June, July and August. It is likely that the causes of this seasonal change are complex and relate to the poorer oxygenation of the water during the summer as well as to the seasonal fluctuation in the abundance of the bacteria responsible for the blackening of the sand.

The water content of sand is another factor of great ecological significance for the ease with which burrowing animals can progress through the sand depends upon the amount of water present (Chapman 1949). All seaside visitors must have noticed the whitening of wet sand under the foot. This is due to water being driven from the interstices by pressure until the sand becomes dry and hard packed. Such sands are called dilatant and are obviously difficult to penetrate for the application of pressure causes them to harden. Other sands behave in a reverse manner. When pressure is applied they become softer and easier to penetrate. This type of sand is called thixotropic. Thixotropy is a general phenomenon and describes a system which becomes less viscous upon agitation but sets solid when undisturbed. The well known non-drip paints are based on this principle; these flow easily when sheering forces are applied but solidify when the brushing stops. The most notorious examples of thixotropic soils are, of course, quicksands which liquefy when pressure is applied. As far as burrowing animals are concerned, thixotropic soils are the more desirable because of the greater ease of passage. The difference between the two types depends, to some extent, on the size of the soil particles but this is not the whole story. In general, thixotropic sands have smaller particles than do the dilatant ones or, at least, they have a higher proportion of clay particles. However, the water content is also significant as it is found to vary inversely with the hardness of the sand irrespective of the particle size. Freundlich and Roder (1938) found that with quartz particles 1–5μ in diameter, dilatancy occurs only when the volume of solid matter is within the range of 42–45%; below this range the mixture remains fluid, while above it stays solid.

Muddy shores have many features that distinguish them from sandy beaches. In the first place, the slope is so gradual as to be barely perceptible, a property which is reflected in their general appellation of mud flats. Because they are so flat, the tide goes out a long way yet they remain wet long after the sea has receded as a result of the absence of slope and the capillary attraction of closely packed particles. Mud is a much more complex substance than sand and contains a high proportion of organic matter. Yet even pure soft mud has some inorganic components which appear under the microscope as miniature sand grains of hard, irregular fragments of quartz. All mud, therefore, contains some sand but by far the more important constituents are silt and clay. Silt consists of very fine inorganic particles which are usually held in suspension at the surface of the mud if there is the slightest water movement. Clay is basically hydrated aluminium silicate but normally it contains iron and other impurities. The particles are so small that they are frequently in a colloidal state. A mud flat which is composed largely of clay becomes very compacted and hard and, ecologically, takes on some of the properties of rock with a sessile fauna and even rock borers such as *Pholas* within it. Perhaps the rock-boring habit evolved by this path.

Mud is more than inorganic particles, however, for it has a very high organic content relative to other marine soils. The organic matter is in the form of detritus—tiny particles in the process of being broken down by bacteria which are themselves of great significance ecologically (Newell 1965). In addition to detritus of marine origin, intertidal muds are heavily enriched by organic matter brought down to the sea in rivers and such material of course, can settle out only in very still waters. As with sandy beaches, much organic matter reaches the mud flat in the wind blowing from the land. Because of the high organic content, water within the mud is soon depleted of oxygen so that the reducing layer is very close to the surface, sometimes no more than a centimetre down. This has important ecological concomitants as we shall see later.

Mud is much more thixotropic than sand, a fact readily apparent to anyone who has attempted to walk over it. In some cases, research workers have found it necessary to don mud skis (George 1963) to prevent themselves joining their objects of study, but whatever difficulties it presents to human progression, the thixotropic nature makes it much easier for animals to burrow through mud than through sand. We shall shortly discuss the influence this has had on the shape of burrowing animals and on their way of life.

The above account of the formation of depositing shores and the nature of the deposits has necessarily been abbreviated and perhaps oversimplified but it should form a basis to enable one to understand the biology of the animals inhabiting them. We might conclude this chapter by summarising the differences between the three broad divisions into which depositing shores have been divided, viz. shingle, sand and mud. It should be remembered that in every particular, there is a continuous gradation from one type to another and that the arbitrary division into types is made purely for convenience in describing the salient features.

1 Wave action

Wave action is greatest on shingle beaches and becomes progressively less on sand and mud. It is the degree of water movement which determines the nature of the deposit but the topography of the shore has some influence on the wave action; for example, the energy of waves is quickly dissipated over shallow level substrata.

2 Slope of beach

This bears a complex relationship with wave action. A flat beach reduces wave action while plunging waves increase the slope of a steep one. However, shingle beaches are steepest and muddy shores flattest.

3 Area of beach

It follows from 2 that the intertidal zone is narrowest on shingle beaches and widest on muddy shores.

4 Particle size

The particles comprising the beach vary from large pebbles several centimetres in diameter in shingle to minute particles less than a micron across in mud.

5 Interstitial space

The interstitial volume (the space between particles) decreases with the size of the particle and is greatest, therefore, in shingle beaches.

6 Thixotropy

The term is hardly applicable in the case of shingle, but thixotropy increases from sand to mud.

7 Water content

The water content bears an intricate relationship with the particle size and thixotropy of the soil. It also varies with the drainage. Hence the slope of the shore and the capillarity, which depends on interstitial volume, are important. Shingle dries out most and mud least when the tide is out.

8 Organic content

This is very negligible in shingle and becomes increasingly high as sand gives way to mud. Perhaps the most important factor controlling the distribution of organic matter is the degree of water movement.

9 Depth of reducing layer

In the absence of organic matter, there can be no reducing layer in shingle. The depth of the layer decreases progressively from sand to mud.

10 Surface stability

The movement of surface particles varies with the severity of the wave action. The pebbles of a shingle beach are constantly rolled over one another but the mobility of the particles decreases with their size.

11 Flora

The flora of beaches will be discussed later but the instability of the surface makes settlement of rooted plants between the tide marks impossible on shingle. Rooted plants occur on sand particularly if some mud is present but higher plants are not a feature of beaches. Plants of the saltmarsh can withstand occasional flooding but *Zostera*, the eel grass, is one of the few flowering plants at home in the intertidal zone. The flora of mud banks consist largely of diatoms although such green algae as *Enteromorpha* and *Ulva* are often found growing on the surface.

12 Fauna

The fauna of sand and mud show many important differences, the consideration of which forms the bulk of this book.

CHAPTER TWO

CATEGORISING THE SHORE FAUNA

It is a human characteristic, some would say failing, to impose order on nature by placing natural phenomena into categories. Scientific progress would be difficult or impossible without this tendency but one should never lose sight of the fact that all classifications are artificial for few natural phehomena are distinct. Hence, whatever classification is adopted, there will always be exceptions and borderline cases but this should not deter one from the attempt. In this chapter, we are concerned with the broad grouping of animals according to their habitats.

It is fairly easy to distinguish between the faunas of rocky and depositing shores for the high proportion of species with modifications for clinging to exposed surfaces readily identifies the former. We will not concern ourselves further with rocky shores but there are several ways of classifying the animals of depositing shores. One is on the basis of the shore deposit, i.e. a division into pebble, sand and mud fauna. This is perhaps the least satisfactory as the gradations from one type of deposit to another is reflected in the gradation shown by the fauna. There is, for example, a typical mud fauna and a typical sand fauna—animals which are found only in thick mud or in clean sand—but there are many other types which occur in intermediate deposits or even in all types of substratum. The latter distribution is particularly characteristic of animals which live on the surface because the nature of the deposit is of less importance to these forms than it is with those which are in more intimate contact with the beach material. As was mentioned in the last chapter, one may even find animals of the rocky shore on firm mud while such species as the shore crab or shrimps are equally at home on rock, sand or mud.

Another method of separating the animals is based on the region in which they live. Some are found on the surface while others burrow in the substratum and modifications necessitated by these habits impose a certain uniformity upon these animals regardless of their taxonomic grouping. Thus we distinguish between the *epifauna* which lives on the surface of the mud or sand and the *infauna* which burrows through or into the material. Infauna, however is too imprecise a term for it covers organisms from the 30 cm-long gaper clam

(*Mya*) to the invisible protozoans. This size difference has long been recognised by the familiar terms *macrofauna* and *microfauna* based loosely on those animals that can be seen and those that cannot, or slightly more accurately, between those that are retained on a half inch (1·3 cm) sieve and those that are not. There is, however, an intermediate type of organism—too small to be considered as macrofauna yet too large to fit conveniently with the microfauna. To this group the name *meiofauna* has been given, a term first used by Mare in 1942. The meiofauna can be distinguished only on the basis of size and is usually defined as those animals which can pass through a 500μ (0·5 mm) mesh sieve but which are retained by a 50μ mesh. Most of these organisms are metazoans but a few of the larger protozoans also qualify for membership. Anything below this size range belongs to the microfauna which consists exclusively of protozoans. The size range is not very precise and in fact, separation by sieving is not very reliable owing to the elongated worm-like shape of most members of the meiofauna. These will be retained if stranded broadside on but may easily pass through if they meet the sieve at right angles. A better and biologically more significant definition of meiofauna is reflected in the alternative name of interstitial fauna. The interstices of the beach are the spaces between the sand grains and the meiofauna can be well defined as those animals which live in the interstitial space through which they move without dislodging the sand grains. Such animals must obviously be very small. This definition of meiofauna succeeds in the case of sand-dwelling species but in very fine mud, the interstitial space is so reduced that it is doubtful whether even the smallest animal can progress through it without moving aside the particles.

In the case of sand, Boaden (1964) has delineated three habitats. The animals living on the surface, the epifauna, have as their habitat the epipsammon (Gk. psammos—sand). Within the substrata itself, the endo- and meso-psammon can be distinguished; these being respectively, the habitats of the macrofauna and meiofauna. The distinction between these faunas is given above, i.e. it depends on whether the animal moves through the sand by displacing or swallowing the grains or whether it moves through the pore spaces without disturbing the particles.

The infauna, both macro- and micro-, are usually, but not always, permanent inhabitants of the beach. An example of a temporary member is the common prawn which joins the infauna when it buries itself during the day but belongs to the epifauna during its nightly foraging for food. All infauna species must, at some point in the life cycle, leave the deposit if dispersal of the species is to be ensured. In some cases, this is achieved by means of planktonic larvae but it is doubtful whether active dispersal occurs in the case of the majority of the meiofauna. It is more likely that these species are distributed passively by transport of either the eggs or adults in water currents when the sand or mud is washed away. An alternative method is in the mud adhering to the feet of wading birds. On the whole, however, the infauna forms one or more homogenous communities whose major habitat is the mud or sand itself. This is not true of the epifauna, for these species may not spend all or even the majority of their time on the beach. The habitat which we are considering is the intertidal zone and this means that for half the time it is covered with water and for the other half it is exposed to the air. When the tide is in, the epipsammon is available to animals such as fish which are normally found below the tide mark; when the tide is out the region is accessible to terrestrial mamals and to birds. In addition, we have epifaunal species which spend most of their time within the intertidal zone irrespective of the state of the tide. Very often, these animals are members of the epifauna when the tide is in and of the infauna when it is out, for burrowing is an easy answer to problems of survival for a marine animal when it finds itself marooned on a soft substratum.

Thus the epifauna can be subdivided into two major categories; the permanent and the temporary with the latter further divided into terrestrial and marine species. The problem of the epifaunal species which regularly burrow into the substratum can be best resolved by ignoring etymological niceties and defining the endofauna more precisely as those species which spend all or most of their time within the mud or sand, and the permanent epifauna as those species which spend much of their time on the surface but which may bury themselves within the substratum at regular or irregular intervals.

Beanland (1940) made an attempt to classify the shore fauna into a 'community system' of recognisable animal communities related to the bottom deposits. Although his results show a soil/fauna association similar to the classical divisions made by Petersen (1918), the area in which he worked was too small and his data too restricted for general conclusions to be inferred. In the light of subsequent work, it seems unlikely that the same pattern of distribution should occur, since the shore now seems to be a distinct habitat separate from the sub-littoral regions and one would not expect to find similar organisations within the animal communities.

If this chapter has by now thoroughly confused the reader, its major object will have been achieved. For it should now be quite clear that any attempt to pigeon-hole beach faunas into neat self-contained categories will not succeed. However, even a defective classification is better than none at all and the following is offered mainly to impose some sort of order upon the remaining pages of this book.

The fauna of mud and sand

Name	Definition	Examples
1 PERMANENT MEMBERS		
1a Epifauna (= Epipsammic spp.)	Species which spend long periods moving on the surface of the soil but which may at times bury themselves.	*Crangon* (sand) *Hydrobia* (mud)

Name	Definition	Examples
1b Infauna	Species which spend all or the vast majority of their time under the surface.	
1b i Macrofauna (= Endopsammic spp.)	Species which move or occupy space by displacing or swallowing beach particles.	*Nepthys* *Tellina* (sand) *Mya*, *Nereis* (mud) *Arenicola* (muddy sand)
1b ii Meiofauna (= Mesopsammic spp. interstitial fauna)	Species which live within the interstitial spaces or which are within the size range of 50–500μ.	Gastrotrichs, Nematodes, Harpacticoid copepods.
1b iii Microfauna	Protozoans less than 50μ in length.	
2 Temporary members		
2a Marine	Species which visit the beach from below low tide mark when the tide is in.	Many fish, seals, plankton.
2b Terrestrial	Species which visit the beach from above high tide mark when the tide is out.	
2b i Facultative spp.	Species which occur in other habitats and which are not dependent on the beach.	Many mammals (rats, otters). Some birds (rooks)
2b ii Obligate spp.	Species which are dependent on the intertidal zone for some essential resource, usually food.	Many seabirds (shelduck)

CHAPTER THREE

THE NATURAL HISTORY OF THE SHORE FAUNA

Biology lacks a term to describe the way in which animals exist, the sum total of physiology, behaviour, ecology and anatomy which together make up the life of the organism. No doubt one could be invented and indeed, such clumsy expressions as mass physiology or environmental biology have been offered but, quite rightly, they have not gained general currency. In the absence of anything better, we will fall back on the much abused but time-honoured term of Natural History. If you wish to annoy an ecologist, you should describe his work as natural history; but good natural history is not to be despised for it is the summation of ecology, physiology, ethology and all the other disciplines. It is in this sense that we are concerned with natural history here and we will consider in turn the various categories of shore fauna that were used in the classification given in the last chapter. No attempt will be made to describe the anatomy of any species nor will the discussion be at all comprehensive, but it is hoped that sufficient examples will be given to illustrate the general biological principles which direct life on the shore.

1 Permanent Members

1a The epifauna

The surface of mud or sand is not an easy environment for animals and, as is characteristic of such habitats, we find a paucity of species but a great number of individuals. When the tide is out, very little is seen on the surface except for *Hydrobia*, a tiny gastropod mollusc, which may be present in enormous numbers on mud while closer inspection may reveal the ubiquitous amphipod *Corophium* in its muddy tube or on the surface. A sandy beach appears entirely deserted and few visitors to the seaside would suspect that any life at all exists in the wide stretches of yellow sand on which they park their deck chairs. When the tide is in, it is a different matter. All sorts of organisms appear on the surface and move about their perpetual business of finding food and mates. Some inkling of this traffic is apparent from an inspection of what would be rock pools on another type of shore. Most muddy shores have shallow depressions which remain water-filled when the tide recedes and these often contain a crab or shrimp. However, there is very little one can say about the epifauna of intertidal mud and sand since so little of it exists. This should not be taken to imply criticism of the use of such a term for were this book about sub-littoral deposits, one could expand an account of the epifauna over several chapters. We should, nevertheless, spend some time in considering *Hydrobia*, for although it burrows into mud for long periods it is one of the very few intertidal species with a claim to be ranked as epifauna. *Hydrobia* looks very like a small winkle. There are a number of species, some freshwater, but the one most commonly found on the shore is *Hydrobia ulvae* which is also known as *Peringia ulvae*. The species has received attention from Newell (1962, 1964, 1965). *Hydrobia* has a relatively simple pattern of behaviour although extraneous factors of topography and tide complicate matters.

An important aspect of the natural history of a species is its feeding behaviour and *Hydrobia* illustrates very well the adaptations required by a species which lives on mud. It was mentioned earlier that *Hydrobia* exists in millions all over the mud but this requires qualification for the density of the species is far from uniform and Newell (1965) has shown that numbers can vary greatly even between areas a few hundred yards apart. However, there is a very definite correlation between numbers and grade of deposit, with high densities on fine mud and lower densities on deposits with an increasing sand content. Muddy deposits contain more organic matter than do sandy shores and it is possible that some factor related to this could be responsible for the observed distribution. For instance, the three most likely attractions for the snails are the higher carbon content, the increased nitrogen supply and perhaps the particle size itself. Newell investigated each of these factors and concluded that it is the third, the particle size, which is important. He found that it is not so much the amount of organic detritus that is significant as the greater surface area which is presented by the fine grained deposits and which provides a larger habitat for micro-organisms such as bacteria on which *Hydrobia* was shown to feed. Muddy deposits, therefore, have a higher proportion of edible organic matter because they support a larger population of micro-organisms than do coarser deposits. The presence of so many molluscs will inevitably increase the organic matter in that much faeces will be produced, but Newell showed that although *Hydrobia* will feed on faeces, both their own and from other species, they derive no nourishment from them. Their purpose in eating the faeces is, presumably, to digest the bacteria adhering to them. Thus the carbon content of the faeces remains constant during their second passage through the gut but the nitrogen content falls dramatically showing clearly that it is the bacterial component and not the faeces which is digested. This shows conclusively that it is the particle size, not the amount of organic matter available, which is responsible for the high nitrogen values in muddy

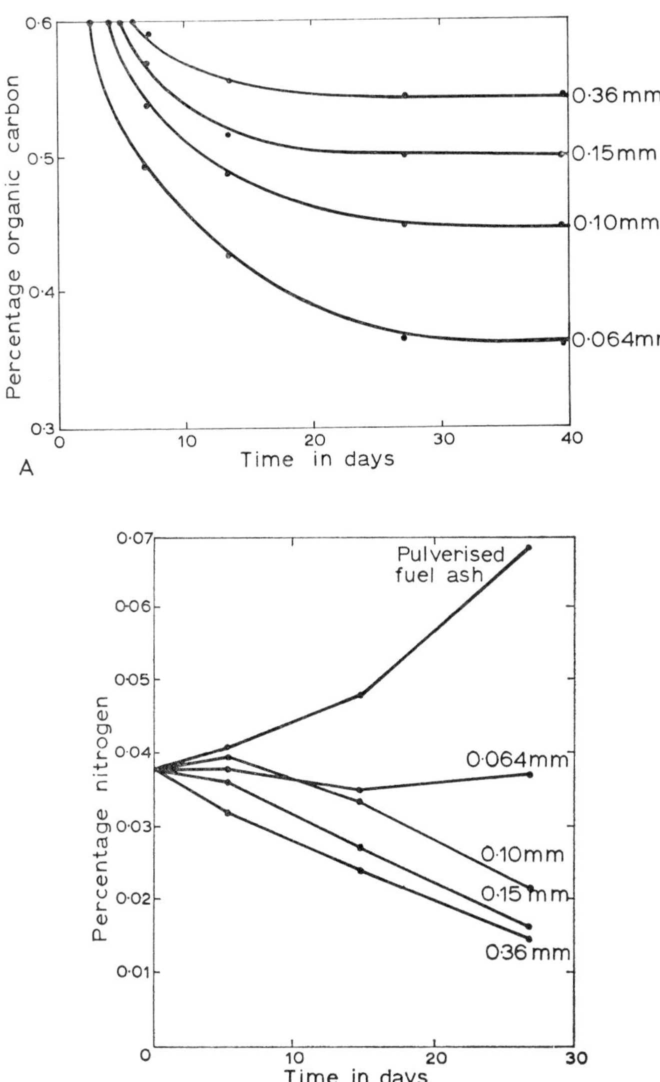

Fig. 6 (A) Decrease in the amount of organic carbon in samples to which 0·65% powdered *Fucus* had been added. The grade of deposit is shown on the right hand side.
(B) Changes on the percentage of nitrogen in samples to which 0·65% of powdered *Fucus* had been added.
Note that the carbon content falls most in coarse deposits and that the nitrogen content increases in fine deposits. Each point is the mean of duplicated estimations on four samples and all deposits were kept at 19°C. (after Newell 1965)

deposits. These micro-organisms are not photosynthetic and they derive their energy by breaking down the complex organic matter in the mud. The nitrogen necessary to build up the new protoplasm may be taken from the nitrogenous compounds in the sea water or perhaps from the air; if so, we have a maritime nitrogen fixation comparable with the more familiar leguminous form. The evidence for these conclusions is summarised in Fig. 6 which shows what happens when the same amount of powdered, dried seaweed is added to each grade of an artificially composed range of deposits. It can be seen that in fine grades, the carbon content falls due to its being broken down by bacterial flora whose increasing numbers are responsible for the rise in nitrogen. On the other hand, the carbon content hardly changes in coarse deposits, unsuitable for bacterial life, and there is no build up of protein; the nitrogen originally present (in the seaweed) is lost.

Some important conclusions derive from this work, which incidentally was carried out in parallel with the bivalve *Macoma balthica* with similar results. It seems that these animals eat organic particles, known as 'detritus', but do not digest them. Instead they digest the bacterial film adhering to the particles. This is probably true not only of *Hydrobia* and *Macoma* but of many other 'detritus feeders' as well. Since bacterial films are found on inorganic particles, the animals can just as well eat them too. This requires a reconsideration of the widespread term 'detritus feeder' since it is really the bacterial film not the detritus which is digested. Newell suggests the adoption of the synonymous term 'organic debris' for 'detritus'. He points out that the organic nitrogen content reveals the level of the whole micro-organism population while the organic carbon value reflects only the organic debris component since the carbon content of bacteria is low compared with that of the deposits on which they are found. Organic nitrogen estimations are, therefore, the more significant ecologically.

It was mentioned earlier that *Hydrobia* is one of the very few species which has any claim to the rank of epifauna of mud or sand. This absence of epifauna is due simply to the fact that the animals present can burrow into the soft substratum which is more hospitable than the surface. Rocky shore animals cannot burrow and consequently, they have, for the most part, to endure the rigours of a fluctuating environment whenever the tide ebbs or flows and the skies rain or shine. Although *Hydrobia* burrows at times, it spends considerable periods on the surface when the tide is out. Has it any particular modifications to cope with the extremes of intertidal life? The most common factors which fluctuate widely in the intertidal zone are salinity and temperature. Salinity varies for a number of reasons—when it rains, when a river is in flood or when water left behind as a pool on the shore begins to evaporate under a hot Sun. Temperature varies with the diurnal fluctuations when the shore is exposed, and frost can be just as serious a problem as heat. The position has been investigated by Todd (1964) who studied the osmotic concentration of body fluids in *Hydrobia ulvae* under different conditions of temperature and salinity. She found that the urine (taken to be isosmotic with the blood) was at about the same osmotic pressure as the external medium in animals collected during the summer and kept in solutions ranging from 100% seawater (32‰) to 50% dilution. Below this, estuarine *Hydrobia* were able to maintain their urine considerably hyperosmotic to (higher than) the medium i.e. they were able to osmoregulate. This was at 5°C: at a higher temperature (15°C) the osmoregulatory mechanism did not come into operation until the concentration of the outside medium fell to 25% seawater. In the winter, there was no difference between the concentrations at either temperature. These results show that temperature and salinity cannot be considered independently for the effect of one is modified by the other. This is a well known phenomenon which will be considered later. The point that concerns us here is that *Hydrobia* appears to osmoregulate at low salinities and so can survive periods when the water is fresh or almost fresh. However, under such conditions, the animals are inactive and withdraw into their shell. The survival value of this is not clear for no physical barrier exists between the blood and seawater under these conditions. Experiments with phenol red have shown that the dye penetrates the tissues in fresh water even when the snail is withdrawn. This is surprising as *Hydrobia* has a thick operculum but it is possible that the operculum opens for respiratory purposes and allows the dye to enter. Contradictory data were reported by Avens (1965) who found a difference between the blood concentration of active animals and of those which remained in their shells. In the former case, blood was isosmotic throughout the salinity range while the concentration of the blood from inactive animals was intermediate between the concentration of the medium and that of the water in which the snails had been keept previously. In other words, her results suggest that *Hydrobia* does not osmoregulate but merely tolerates low salinity. The apparent hyperosmoticity, which Todd found, is presumably due to the body fluids being separated from the medium by the imperfect but partially effective mechanical barrier of the shell and operculum. Thus *Hydrobia* copes with low or high salinities (which are usually of short duration in nature) by closing up and waiting for conditions to improve. However, Avens found that there were no weight changes on transferring *Hydrobia* from one medium to another, i.e. there was no passage of water between the animal and the surrounding medium. A likely explanation is that the snail looses salts from its body in hyposmotic solutions thus maintaining equality between the osmotic pressures of the blood and external media. Such transfer of salts is held to take place so quickly that there is no time for swelling or shrinkage such as occurs when the larger, related winkle (*Littorina*) is used (Avens and Sleigh 1965). Avens points out that the British estuarine snails are of small size and conjectures that the smaller species have better volume control than the larger marine forms and that it is this faculty which enables them to colonise the estuarine regions.

Avens (1965) also studied the blood of *Hydrobia* when it was exposed to the air for long periods. She found that the blood concentration rose rapidly on exposure under dry conditions at room temperature but that no harm ensued, e.g. the blood concentration rose by about 90% within 4 to 5 days in one experiment without deaths occurring. On reimmersion in seawater, there was very rapid dilution of the blood. These experimental conditions are much more severe than the organisms would experience in nature.

Hydrobia ulvae is obviously a very euryhaline animal, more so than other gastropods studied, and this is a property which makes it well suited to life in estuarine regions. In addition, it has behavioural characteristics which ensure that it misses the worst of environmental extremes e.g. it will burrow or seek favourable microclimates such as that afforded by the moist roots and stems of aquatic plants. The latter are not features of muddy shores but *Hydrobia* does occur in saltmarshes where plants abound and there the problem is one of dehydration and exposure rather than dilution. Perhaps *Hydrobia* is really an estuarine form not particularly adapted anatomically for life on mud. However, mud is such a demanding habitat that few animals can colonise it and those that can survive are faced with a wealth of untapped resources and are able to spread prodigiously. There are many examples away from the shore of few species but many individuals living in exacting environments such as polar or desert wastes.

Surprisingly for such a common animal, reproduction in *Hydrobia* is imperfectly known. The seasons of peak breeding activity have been recorded in a number of areas and generally spawning extends over several months, although the time of the year is by no means constant from place to place. In common with many other British prosobranch molluscs of the shore, *Hydrobia* produces a small number of eggs contained in separate egg capsules which in turn are embedded in a mucilaginous base. The gelatinous mass with the egg capsules is known as spawn. Primitively the pattern of reproduction in prosobranch molluscs is probably the production of planktonic eggs which hatch into free swimming planktonic larvae known as veligers. After an extended life in the plankton, during which it feeds, the veliger larva settles on a suitable substratum and undergoes metamorphosis into the juvenile adult form. One of the principal functions of the planktonic phase is to effect dispersal of the species. As the hazards of life in the plankton are great, including predation and translocation to unsuitable areas, the number of eggs produced is large to compensate for the inevitable losses. British prosobranch molluscs which show many of these features in their reproduction are the two species of winkle, *Littorina littorea* and *L. neritodes.* These produce capsules which are released into the plankton where they hatch out into veliger larvae. Each capsule of *L. neritoides* contains one egg, while that of *L. littorea* has from one to five. The amount of albumen around each egg is small and it is not an important food source for the developing larva which feeds on its own account in the plankton.

The spawn of *Hydrobia ulvae* consists of a number of capsules; up to 22 have been recorded, and it is attached to the shells of other *Hydrobia* irrespective of sex. The eggs, which vary from 3 to 25 per capsule, float in a copious quantity of albuminous fluid within the capsule. It is not too certain whether or not the eggs hatch into a planktonic form. A veliger larva has been described for this species but it is doubtful that it spends anything but the most transient period in the plankton and almost certainly it does not feed there. The abundance of albuminous fluid infers that this is the major food source of the larvae during their first few days.

In its reproduction, *Hydrobia* provides an interesting link between the two species of *Littorina* which have planktonic veligers and the other two species of British winkles, *L. littoralis* and *L. saxatilis* (= *L. rudis*). *L. littoralis* produces a spawn mass containing 90–150 eggs, not in separate capsules but embedded in a mass of hard jelly. Each egg is supplied with a large quantity of albumen sufficient to nourish the developing embryo during the veliger stage which takes place *inside* the mass of jelly. At the end of this stage the jelly absorbs water and becomes so soft that the young, which are now miniature adults, can bore their way out to the outside world. *L. saxatilis* takes the process one step further. This species does not produce any spawn but retains the eggs within the brood pouch inside the female where development proceeds. Each egg is surrounded by a quantity of albumen greater than that found in any of the other three *Littorina*. The veliger stage occurs in the egg through which the fully formed snail bores its way to freedom in the oviduct of the female, from whence it passes to the exterior through the genital aperture.

There is thus a gradation from a fully planktonic veliger to a viviparous juvenile which suppresses the pelagic phase altogether. Some other correlations are apparent. Those species which release eggs into the plankton provide them with a tough covering and almost literally do not put all their eggs in one basket. By contrast, those which deposit spawn on the substratum leave many more eggs in one place. This suggests that the latter procedure is safer. Planktonic eggs have a better chance of survival if they are spread out rather than left in a clump which might be swallowed in one gulp by a passing fish, but presumably egg masses on the ground are more difficult for predators to find. The penalty paid by those species which dispense with the planktonic phase is, of course, a reduction in the efficiency of dispersal. An increase in the amount of albumen provided for the eggs is also noticeable as one passes from those species with planktonic larvae to those without. This is simply a reflection of the fact that the planktonic forms feed in the plankton and are not dependent on a stored source of food. It is also the reason why a great many more eggs can be produced if they are planktonic; the parent has only a finite amount of food material to spare and it can either go into many small eggs or into a few large ones.

There is one other prominent species that has some claim to rank as an epifaunal species, although its best known characteristic is that of burrowing. This is *Corophium* which is an amphipod genus with many species but the one commonly found on mud is *C. volutator*. Its general behaviour has been described by Meadows and Reid (1966). The larger specimens are the ones most usually seen on the surface as the smaller forms remain permanently within the burrows. When the tide is out, *Corophium* crawls with a characteristic looping movement achieved by pulling itself along with its long antennae and pushing with its rear appendages. Under water, it crawls more slowly using its walking legs (pereiopods) or alternatively, it darts over the surface with rapid strokes of its pleopods, the swimming appendages. Normally, it does not show the looping movement under water unless the substratum is particularly soft and fine. *Corophium* is not a good swimmer and rarely swims away on leaving its burrow. Social contact between the members of a population of *Corophium* must be frequent for the large animals do not remain permanently within their burrows but emerge after one or more days to re-explore the surface of the mud. They will then either re-bury themselves or enter another burrow if one is encountered. Should this be empty, the intruder will remain but if the burrow is already occupied there is a fight which is usually won by the original occupant irrespective of size. The fact that it is the large (i.e. adult) animals which periodically leave the burrow is not fortuitous for reproduction is through copulation and contact between the sexes is essential for propagation of the species. Reproduction in many sedentary or tubiculous animals is achieved by the shedding of gametes into the water where fertilisation takes place and dispersal is achieved by the planktonic larvae. In *Corophium*, however, the eggs are carried in the brood pouch of the female until hatching when the young proceed to tunnel from or near to the parent burrow. There is, therefore, no question of the young effecting dispersal but this is probably achieved by the adults or by translocation with the substratum itself.

While out of the burrow, *Corophium* shows a series of tactic reactions which keeps it within the intertidal zone. On damp mud, *Corophium* is negatively phototactic and positively geotactic, i.e. it crawls away from light and down a slope, reactions which would tend to keep it away from the upper shore. When the mud is covered with water, however, the light reaction is reversed and it will crawl or swim towards light, i.e. away from the darker sublittoral regions.

The distribution of all species of *Corophium* examined is apparently controlled by the nature of the substratum for examination and selection of the mud when crawling over the surface are prominent features of the animal's behaviour (Meadows 1964a, b). If given a choice, a specimen removed from its burrow will always select mud from its original habitat. Thus the mud-dwelling *C. volutator* will select a substratum which has been maintained under anareobic conditions while *C. arenarium*, which occurs in sand, will choose an aerobic one. Experimental conditioning of animals to an atypical substratum has no effect upon subsequent preferences when a choice is again available. Neither the level or illumination nor the colour of the surface is of any significance, but not surprisingly more animals burrow when the mud is illuminated than when it is not, a tendency which is more marked in large species than in small (Meadows 1967). Destruction of the primary film around the sand grains renders the substratum unattractive, but the situation is not a simple one since removal of the bacteria alone does not seem to be detectable by *Corophium* while attempts to restore the attractive properties to sand stripped of its primary film by acid-cleaning or other treatment met with failure.

The depth of the substratum was found to be significant. If given no choice, *C. volutator* will burrow in mud of any depth but normally it avoids deposits shallower than 0·5 cm. Presumably it must first attempt to burrow in these shallower muds in order to discover the depth. *Corophium* will settle in shallow deposits but only if the density of animals in deeper levels reaches a high enough value (Table 1). At high population densities (>0·1 animals per square cm), Meadows (1964b) found some evidence of territoriality while at low densities there was a tendency towards aggregation. Size is important in

TABLE 1

Depth selection by *Corophium volutator* at high and low densities. In more crowded conditions a greater proportion of animals selected the least suitable 0·5 cm depth (after Meadows 1964b)

Depth of mud (cm)	% of *Corophium* High density (190 animals)	Low density (17 animals*)
0·5	12	3
2·0	22	29
5·0	27	33
9·0	39	34

*mean of 7 expts.

the colonisation of deep muds and the larger animals are found at the greatest depths (Fig. 7). Under natural conditions, therefore, there is a topographical separation of the age classes with the younger near the surface and the older further down. This does not necessarily mean that the large animals drive away the smaller but more probably that the bigger animals are physically unable to construct a large enough tunnel in shallow deposits. It is the size of the tunnel, not that of the animal, which controls the depth at which *Corophium* can survive. Its burrow is semi-circular or U-shaped, open at both ends, with the lowest point at a depth equal to at least the distance between the openings. From an examination of the relation between body size and the distance between the openings, Meadows (1964b) concluded that *Corophium* at depths of 0·5 cm average about 1·15 mm in length and those from 2·0 cm down average 6·9 mm. Because few animals are as small as 1·15 mm, few can survive in muds of 0·5 cm depth and this conclusion is reflected in the experimental results (Table 1). Some *Corophium* exceed 6·9 mm in length, however, and require depths greater than 2·0 cm. This may be the explanation for the somewhat higher proportions found in the deeper experimental muds.

The burrowing behaviour of *Corophium* has been studied by Ingle (1966) who chose to work with *C. arenarium*, which is a burrower into sand and consequently easier to work with than *C. volutator*. Differences between the

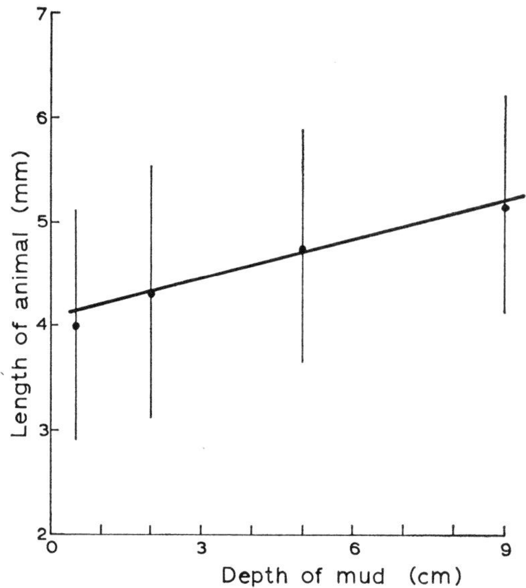

Fig. 7 Relationship between the size of *Corophium* and the depth at which they are found. The vertical lines show the standard deviation. The increase in length with depth is highly significant statistically ($P < 0{\cdot}001$) (after Meadows 1964b)

two species may be assessed by reference to the description given by Meadows and Reid (1966) of the rather less complicated burrowing procedure in *C. volutator*.

Ingle found that there were five stages in the burrowing sequence, namely (1) exploration of the substratum, (2) excavation of the sand, (3) levering action of appendages, (4) orientation within the burrow and (5) excavation of a secondary opening within the burrow. It is worth examining this sequence in some detail as an example of the way in which a species adapts to life on the shore. The various stages are depicted in Fig. 8.

The exploratory phase, which may take from a few seconds to ten minutes, consists of alternate insertions of the antennae into the sand. When the animal begins to burrow, the swimming legs start to beat more quickly and the current so produced carries away sand particles excavated by the anterior appendages and piles them up in a small heap behind the animal. After the

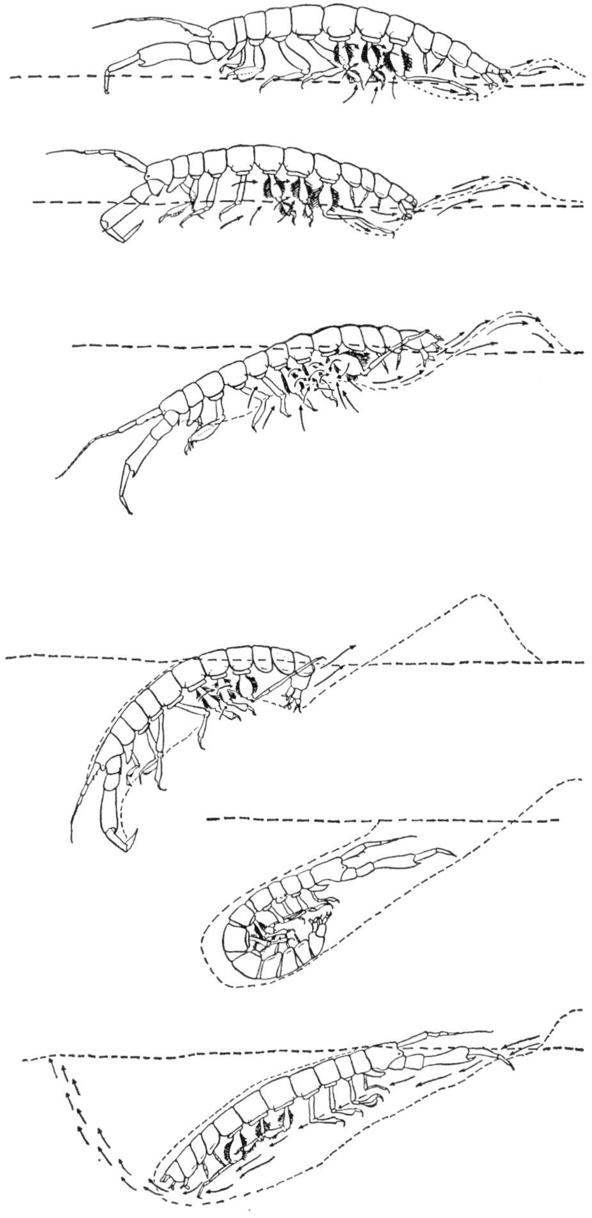

Fig. 8 Burrowing sequence of *Corophium*. Broken line indicates the level of the substratum. Arrows show the direction of water currents (after Ingle 1966)

anterior appendages have got a good grip, the frequency of the beating of the pleopods is increased still more, while the posterior part of the body oscillates from side to side. The resulting high velocity water current removes more and more of the sand and the animal quickly sinks into the substratum until the uropods, the posterior appendages, can grip the wall of the burrow. It is now that the levering action of the appendages comes into play; the posterior part of the body is alternately flexed and extended while the uropods loosen the sand which is thrown clear by the movements of the body. In a short time, the animal disappears from view. Then follows stage (4), the orientation within the burrow, during which the animal turns over and bends back upon itself so that the head comes to the burrow opening and the ends of the antennae are rested on the surface. After a short pause, the pleopods being to beat again and draw water into the tube. On reaching the base of the burrow, the water current is deflected upwards, for reasons which are not altogether clear, and loosens the sand particles in its path. The animal then turns over, doubles back upon itself and pushes its way through the loosened sand to form a secondary burrow. The horse-shoe shaped burrow is now complete and the animal alternates from one end of the tube to the other spending only a short time at each. While resting at either end, a slow water current is drawn into the tube and it is from this current that food material is filtered off.

We may have lingered too long in describing the way in which two organisms, one a mollusc and the other a crustacean, live on the shore but from the details we can detect some general principles. First, *Hydrobia* feeds by swallowing particles picked up from the surface, i.e. it is a *deposit feeder*. *Corophium*, on the other hand, filters its food from a water current of its own making, i.e. it is a *filter feeder*. These are the two fundamental methods of feeding in the fauna of mud or sand. Many of the seemingly bizarre anatomical modifications seen in shore animals are often no more than adaptations to improve the efficiency of the filter feeding mechanism. Deposit feeding is a more simple process and morphological adaptations in these types are not so extreme. The two methods of feeding are not mutually exclusive. In fact *Corophium* itself is probably more of a deposit feeder than a filter feeder as it sifts quantities of mud with its gnathopods (head appendages) and conveys the larger particles to its mouth where they are crushed by the mandibles. Smaller particles are strained from the water currents by a fringe of fine hairs or setae on the gnathopods.

Both filter and deposit feeders may choose the food they eat. If so, they are known as *selective* feeders. Selective filter feeders may select food on the basis of size as do many lamellibranch molluscs (on the evolutionary assumption that particles of a given size are usually food), or they may reject particles after examining them chemically—tasting them—at the mouth. This is a less common method of rejection although it is not easy to demonstrate whether selection at the mouth is gustatory or on the basis of size. Deposit feeding may be *selective* or *non-selective*. Non-selective feeding is seen in such worms as

Arenicola which burrow through muddy sand by swallowing it and digest the organic matter contained therein. Other deposit feeders are more particular and ingest only those particles which are nutritious. An example is the mud-dwelling bivalve *Scrobicularia.*

It is relatively easy to guess whether a museum specimen of a bivalve was a deposit or suspension feeder. If the inhalent and exhalent siphons are long and separate, the chances are that it is a deposit feeder but if the siphons are joined, the animal was probably a filter feeder. The dichotomy is not complete, however, for the separate-syphoned *Macoma balthica,* normally a deposit feeder, seems to act as a filter feeder on occasions (Brafield and Newell 1961). The important point, however, is that deposit feeders must be able to pick things up and a long flexible 'trunk' is one of the most effective mechanisms. Normally the deposit feeder is more or less sedentary and clears only the area within reach. A long syphon is, therefore, important and a 5 cm *Scrobicularia plana* will extrude a 15 cm tube if left undisturbed in a dish of sea water.

The type of feeding method has other concomitants important to an understanding of the ecology of depositing shores. Filter feeders sieve off their food from water currents and it is essential that large quantities of water should be filtered in order to obtain sufficient food. It is desirable that food particles should not be too small; otherwise it would be necessary to have such a fine net that it would be easily clogged. Hence, filter feeders tend to be found in coarser grained deposits or else in water so still that little particulate matter is in suspension. The latter conditions are fulfilled within the burrow of the tubiculous polychaete *Chaetopterus* which draws particles into its burrow and entangles them on strings of mucus or on a mucous bag which it secretes and later swallows along with the food.

Desposit feeders, on the other hand, are more typical of fine grain shores. Many of the organisms are small and would be unable to cope with large particles. If, as seems likely, deposit feeders are subsisting on the bacterial flora around the mud particles, a fine grain deposit with its larger surface area will provide the most food.

A further generalisation, based on the same logic, is that filter feeders will tend to be found higher up on the shore than will deposit feeders because the grade of deposit decreases as one moves towards the sea (p. 6). It must be stressed, however, that food is not the only factor involved and the clogging or otherwise of respiratory surfaces is a potent factor in determining distribution. For example, *Arenicola,* the lug worm, is very sensitive to particle size for these reasons and the grade of deposit which limits its distribution is similar in widely spaced geographical regions.

Some general conclusions concerning the osmoregulatory powers of intertidal invertebrates emerge from the earlier discussion of *Hydrobia.* It is clear that the species is remarkably euryhaline and can tolerate wide ranges of salinities both high and low. To a large extent this is probably true of all shore animals, but it is particularly marked in *Hydrobia ulvae* because it is basically an estuarine animal. Estuarine animals are not uncommonly found on the open shore and when they do occur there, they tend to be restricted to the higher levels of the beach. The reason for this may be that these regions, while not estuarine, have in common with estuaries a greater range in the fluctuations of the environmental conditions than is the case lower down the shore. The mechanism by which euryhaline organisms control their body fluids will be dealt with in more detail in a later chapter, but it is worth recalling the *Hydrobia* is able to shut itself from the environment to some extent by closing the shell opening with the operculum. This is an example of an avoidance reaction which is one of the most typical responses shown by animals of the depositing shore when faced with adverse conditions. This contrasts with the behaviour of rocky shore animals which in the main have to endure environmental extremes.

The tolerance shown by *Hydrobia* extends beyond salinity extremes and includes fluctuations in temperature and exposure as well as most other environmental factors. Again, this is a feature of the majority of intertidal animals. Indeed, it is hardly surprising since organisms without this faculty would not be able to survive for long in a habitat where conditions vary so widely.

The description of reproduction in *Hydrobia* and *Corophium* has thrown some light on the breeding biology of shore animals. In contrast to invertebrates of the open sea, which produce millions of gametes and scatter them blindly into the water, shore animals take some care over their reproduction. In the first place, copulation occurs. This ensures that the spermatozoa are placed in the most advantageous position for fertilising the ova. The significance of reducing the risk of non-fertilised ova lies in the considerable investment in food reserves given to each egg. A small number of yolky eggs seems to be characteristic of shore animals. Because the number is small, a greater proportion must survive than would be necessary if many were produced. Consequently one finds examples of parental care which is developed to a high degree in *Corophium* whose larvae develop within the brood pouch of the female. In many other ways the eggs and young are protected from untimely death, for example by the development of protected egg capsules in *Hydrobia* or by the suppression of the vulnerable planktonic larva.

Why should the shore fauna show the traits summarised above? The reasons for developing a tolerance to environmental extremes are obvious enough and specialised feeding techniques are quite understandable, but why has the reproduction of the shore fauna departed from the more typical and presumably primitive method of planktonic development shown by most marine animals? The penalty, which has been pointed out more than once in these pages, is a reduction in the powers of dispersal. A clue to the possible answer can be found in a comparison of the reproductive habits of all marine animals in relation to latitude. It has been found that marine invertebrates of the Arctic, whether littoral or not, show the same features of a reduction in the number of eggs, a suppression of planktonic larvae and a development of brood pouches and other forms of parental care. This tendency is progressively lost as lower latitudes are approached until in tropical seas a planktonic development with many eggs is the rule (See Table 2).

TABLE 2

Percentage of gastropod molluscs having planktonic larvae in relation to latitude (after Thorson 1940)

Area	Approximate Latitude	With planktonic larva
East Greenland	70–80° N	0
East Iceland	65° N	10
South Iceland	63° N	30
Faeroe Islands	62° N	40
British Isles	50–60° N	64
Danish Waters	54–58° N	65
Iranian Gulf	25–30° N	75
New Caledonia	20° S	57

The probable reason for the correlation between this reproductive trend and latitude is that conditions become more difficult as one moves from the balmy tropics to the harsh conditions of the polar regions. The conditions are harsh to planktonic larvae not so much for the low temperature as from the short period during the summer when feeding is possible in the polar seas. If planktonic development were the rule, all species would need to release their eggs into the plankton at one time instead of spacing the breeding seasons over many months, as occurs in lower latitudes. The resulting overcrowding and competition would be to the advantage of none. Even if the planktonic phase were successful, the outlook for the newly settled animal would be bleak indeed, particularly if it were a filter feeder, faced as it would be with the approaching winter and the consequent cessation of planktonic production. Far better to evolve a large yolky egg which can develop to an advanced stage before it has to face the rigours of an independent existence.

The reasons for the trend towards the suppression of planktonic eggs are not the same in polar regions as on the shore, but the two environments have this in common: they both impose severe conditions on a would-be coloniser. The precise nature of these conditions is considered elsewhere in this book, but there is no doubt that they are severe. The severity of the environment is likely to bear more hardly upon the young than upon the adults, and it is not surprising that whenever conditions become difficult, greater care is taken of the young forms. Many examples of this tendency can be found in terrestrial environments. An added hazard to an intertidal animal with planktonic larvae is the risk that the larvae will be carried out to sea with the tide and deposited in unsuitable areas. This risk has to be balanced against the advantages of dispersal—some larvae would certainly find themselves in suitable circumstances—but on an evolutionary scale the advantages have apparently not outweighed the disadvantages.

The loss or reduction in the powers of dispersal is a serious matter for should conditions worsen in its home area, as well they might in such a fickle habitat as the sea shore, the species faces a real danger of extinction. Animals vary in their ability to disperse and a poor performance in this aspect of their biology may well be a contributory factor in the natural extinction of species. The shore fauna, however, shows no tendency towards general extinction and although populations may disappear from one area, they as readily appear in another. Shore populations, in fact, have all the hallmarks of species with good dispersal powers. However, dispersal is accomplished not through planktonic larvae but through the agency of the current systems described in the first chapter. Such dispersal is largely passive and, in the case of the macrofauna, involves only the younger forms since the adults are often too big or deeply buried to be moved. However, whole populations of all age classes are translocated in the case of the meio- and microfaunas. Dispersal by these means is not as blind as it may appear since the physical sorting of particles through the differential velocities of the currents ensures, for example, that the mud-dwelling meiofauna is not deposited until still water is reached. These conditions are usually associated with a mud flat or other suitable habitats. Further methods of dispersal include transport in the mud adhering to the feet of wading birds which feed in the intertidal zone. Again, this method is selective since the birds tend to visit similar areas where conditions are suitable for intertidal invertebrates.

1b The infauna

1bi *The Macrofauna*

The division between a discussion of the epifauna and one of the infauna is about as blurred as the difference between the concepts themselves. However, it is true that the infauna as defined in the last chapter, makes up the great majority of the fauna of the beaches. The first essential in any consideration of a fauna is to know what types of animals compose it. This can be done quite simply by listing the species found. Table 3 shows a list of species collected from three shores in South Devon by a party of students on a field course. Species lists, though essential in ecology, do not make vital reading and the data are presented here in terms of the taxonomic group to which each species belongs. These are not complete lists of all the animals which could have been found by a more intensive search, but they serve as comparable samples since they were all collected by the same group of people. A list of rocky shore species is given for comparison. This last list is slightly different from the others as it comprises the species collected from a transect running down the shore while the other lists are of animals gathered during a general search of the whole beach. The muddy shore is Wareham Point near the Salstone in the northern half of the Kingsbridge Estuary while the sandy shore is Mill Bay, also in the Kingsbridge Estuary but to the south of Salcombe on the opposite shore. The rocky shore is Sharpers Head in Lannancombe Bay, South Devon.

TABLE 3

Comparison of fauna from muddy, sandy and rocky shores

Taxonomic group	Mud (Wareham Point)		Sand (Mill Bay)		Total for depositing shores		Rock (Sharpers Head)	
Porifera	0		0		0		3	
Cnidaria	2		2		2		5	
Turbellaria	1		0		1		0	
Nemertini	1		1		2		0	
Annelida	9		7		14		2	
Sipunculoidea	1		0		1		0	
Crustacea								
Cirripedea	0	4	0	6	0	8	4	14
Isopoda	0		0		0		3	
Amphipoda	1		0		1		4	
Decapoda	3		6		7		3	
Pycnogonida	1		1		1		0	
Mollusca								
Polyplacophora	0	15	0	6	0	21	1	15
Gastropoda	8		2		10		13	
Lamellibranchia	7		4		11		1	
Polyzoa	0		0		0		1	
Echinodermata	0		1		1		1	
Protochordata	3		2		3		0	
Total species	37		26		54		41	

The first fact that is apparent from this analysis is the relatively high number of molluscs on all three types of shore and the equally high proportion of annelids which, however, are confined to the depositing shores. This is a reflection of the fact that while molluscs include species which both move on the surface and burrow in the substratum, most annelids are burrowers. It will be noted that on the rocky shore, most of the molluscs are gastropods which, typically, are forms which crawl about on the surface and rarely burrow. Gastropods are scarce on sandy shores (the two species recorded here were *Nassarius brachiata* and *Onchidoris muricata* which is a burrowing opisthobranch species). They are more numerous on mud flats since the firmer substratum permits winkles and topshells (e.g. *Calliostoma zizyphinum* and *Gibbula umbilicalis*) to establish themselves, but it is unlikely that they would be found on a mud flat unless there was a rocky shore nearby as is the case at

Wareham Point. Therefore, the large proportion of gastropods found at Wareham may be considered atypical. Much more typical, however, is the large percentage of lamellibranch species compared with the rocky shore where the only one found was *Mytilus edulis* one of the few bivalves which is attached to the substratum. The presence of so many bivalve species on the depositing shore follows from the widespread burrowing habit displayed by the group.

Most of the crustaceans found on the depositing shores were decapods. The isopods and in particular the amphipods are probably under-represented for more than one species should have been collected. *Carcinus maenus*, the shore crab is equally at home on all three habitats and indeed, few intertidal decapods are exclusively confined to one type of beach. Exceptions include *Corystes*, the masked crab, and the curious burrowing shrimps, *Upogebia* in sand and *Callianassa* in muddy sand. The presence of two species of Cnidaria on the muddy shore may seem unusual and to some extent it is for this group includes the sea anemones which are normally associated with the hard substratum of the rocky shore. However, some species are adapted to life on soft deposits. One is *Cereus pedunculatus* which buries itself in mud with only the rossette of tentacles showing above the surface. If the animal is dug up, it will be seen that it has retained the ancestral habit of adherence to a firm structure, for the bottom of the stalk-like body is attached to a stone or a fragment of shell about three inches below the surface. A sea anenome which has departed still further from orthodox ways is *Peachia hastata* which is a truly burrowing animal as it is usually found several inches below the surface and comes up only to feed. The base of the column is bulbous and not flat as in most sea anenomes. Although this species occurs in the Kingsbridge Estuary, it was not found on this particular occasion. The other cnidarian found was *Halliclystus auricula*, the stalked jelly fish, which attaches itself to the leaves of seaweeds or, in this case, to *Zostera*, the eel grass. A number of other species collected at Wareham Point have not been recorded in Table 3 since they occurred on boulders or stones lying on the mud. They include three other cnidarians, four species of sponge, a limpet, a barnacle, a sea urchin and a brittle-star all of which are essentially rocky shore animals and owe their presence at Wareham to the nearby Salstone. They serve to illustrate the fact that a species list of muddy shore animals can be surprisingly cosmopolitan unless one eliminates animals living on boulders or other hard objects. It may be argued that these are ubiquitous features of muddy shores and their inhabitants should, therefore, be included but to safeguard ourselves, we will consider the list to be of animals from mud and sand rather than from muddy and sandy shores. The single echinoderm species of the Mill Bay sand was *Leptosynapta inhaerens*, a burrowing sea-cucumber, but other species might well have been found. The tide was not low enough to allow the heart urchin, *Echinocardium cordatum*, to be collected. This burrows to a depth of several cm and is quite common at the low water mark and below. A brittle star typical of sand, is *Acronida brachiata* which occurs in the Zostera beds of Mill Bay although it was not found there by the students.

In terms of species diversity, there is not much difference between the rocky and muddy shores but the number of species on the sandy shore is well below that of the other types. The poor showing of the sandy shore is due to the unstable nature of the habitat as well as to the low level of organic matter relative to mud or rock.

A similar analysis of species from mud and sand but in this case from the same area has been made with data provided by Holme (1949). Holme published a list of species taken from a 300 metre transect near the mouth of the Exe Estuary. The transect ran from sand high up this shore through mud to sand again. Samples were taken by sieving a quarter of a square metre of soil through a 1 mm mesh, a method which retains only the macrofauna. The distribution of the species according to the grade of deposit is shown in Table 4. In this example, the number of species within each substratum is of little

TABLE 4

Number and density (per 0·25 m^2) of species on a downshore transect running from sand through mud to sand. Mouth of the Exe Estuary 1947 (after Holme 1949)

Group	Mud		Upper sand		Lower sand	
	No. of spp	Density	No. of spp	Density	No. of spp	Density
Nemertinea	?	1·9	?	4·9	?	1·2
Annelida	14	62·9	16	11·4	14	2·4
Crustacea						
Isopoda	1	0·6	1	25·9	2	0·2
Amphipoda	4	4·1	5	35·5	7	10·9
Mysidacea	1	0·7	3	0·9	1	0·9
Decapoda	2	24·1	2	2·8	2	3·4
Insecta						
Dipterous larvae	1	0·4	1	0·8	1	0·2
Mollusca						
Polyplacophora	1	0·2	—	—	—	—
Lamellibranchia	5	16·8	6	3·7	4	10·8
Gastropoda	3	651·5	2	28·4	1	52·4

moment since the number of stations sampled was not the same in each case, but the densities of the species are of significance. It can be seen at once that, with the exception of isopods and amphipods, the density of the fauna is much greater in mud than it is in sand. Again, this is a reflection of the better food supply in mud. The densities in the upper and lower reaches of sand do not vary much except for that of the isopod fauna which is high in the upper sand because of a concentration of *Eurydice pulchra*. The annelids in mud were dominated by *Ampharete grubei* a species which is rarely found in other areas. It is apparently a mud-dwelling species which cannot tolerate diluted sea water and as most mudflats are restricted to estuarine areas, it has few opportunities to be common. An equally rare worm, *Pygospio elegans*, was the most numerous polychaete in the upper sand. The high density of bivalves in mud is due to the abundance of *Scrobicularia plana* and *Cardium edule* while *Hydrobia ulvae* is mainly responsible for putting the density of gastropods in mud way ahead of their density in sand.

A further idea of the taxonomic structure of shore species may be obtained from an analysis of the list given in the Plymouth Marine Fauna (Marine Biological Association, 1957) of the burrowing animals from both mud and sand in Salcombe Harbour. A total of 78 species is listed of which no fewer than 29 (38 %) are polychaetes and 27 (35 %) are lamellibranchs, i.e. nearly three quarters of the animals belong to one or other of these two groups. The next most numerous group is the coelenterates, with six species, followed by echinoderms with five, and decapod crustaceans with four.

From these comparisons, it is clear that the infauna is compounded largely of lamellibranchs and polychaetes although representatives of almost all phyla may occur. The success of these two groups suggests that they have developed the burrowing habit to a high degree of efficiency and we should next look at the methods by which they burrow. Rather surprisingly, there is little significant difference between the burrowing mechanisms in the two groups. Furthermore, the method of penetrating a soft substratum is basically similar in all soft bodied animals. Trueman (1968) has summarised the method and has introduced new terms to describe the two types of anchor which are the basis of the mechanism. The first is the 'penetration anchor' which is formed by bulbous extension of the body wall some distance behind the organ which is actually penetrating the soil. The purpose of this anchor is to prevent the body from being pushed backwards when the burrowing 'probe' is extended. The second anchor, the 'terminal anchor', is brought into play after the penetration thrust is completed. This anchor is formed by the distension of the terminal end of the burrowing organ and its purpose is to lock the front of the animal in the soil so that when the longitudinal muscles contract, the body is pulled towards the tip rather than the tip towards the body. The animal proceeds, therefore, in a series of jerks with the following sequence: (a) formation of penetration anchor, (b) extension of penetrating organ by contraction of circular muscle, (c) formation of terminal anchor and (d) contraction of longitudinal muscles resulting in forward (or downward, depending on the orientation) movement of the body. While the animal is lying on the surface, the penetration anchor cannot be applied and it is only the weight of the body which prevents the organism from being lifted into the air and enables the penetrating organ to enter the ground. The mechanism, which is shown diagrammatically in Fig. 9, is obviously a hydrostatic one with the coelom (in the case of annelids) or haemocoele (in the case of bivalve molluscs) serving as the hydraulic organs. However, this method of burrowing is not confined to coelomate animals for it is seen even in the coelenterates (Fig. 9). Here, the coelenteron or gut forms the hydraulic organ but there is one major disadvantage in mixing the hydrostatic and alimentary systems i.e. the mouth must be

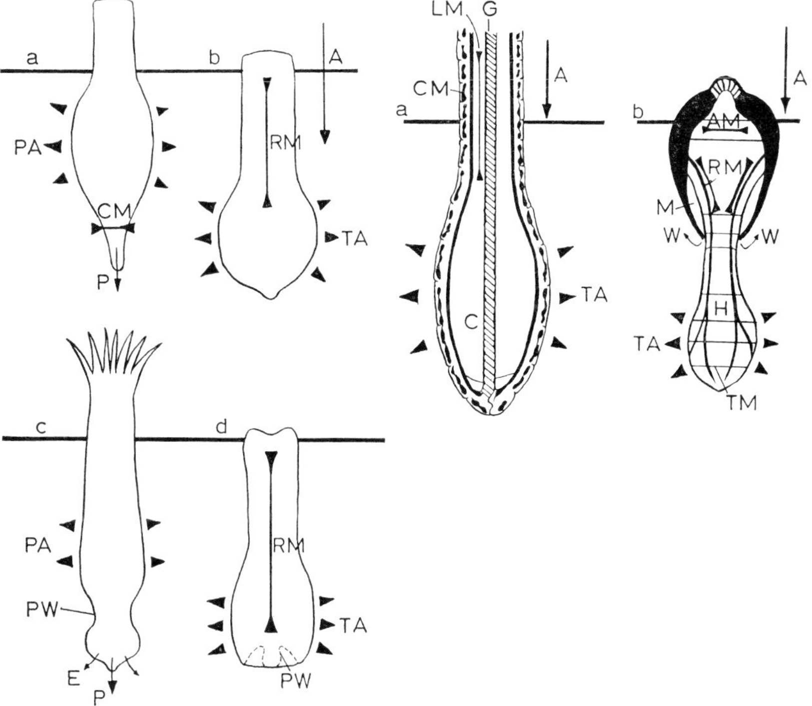

Fig. 9 (left) The two principal stages in the burrowing process of a generalised soft-bodied animal (a and b) and of an anenome (c and d).
(a) formation of penetration anchor by contraction of circular muscles.
(b) formation of terminal anchor which allows body to be pulled into the substratum when longitudinal muscles contract.
(c) formation of penetration anchor after passage of peristaltic wave and achievement of further penetration by eversion of the base of the column.
(d) formation of terminal anchor. Body drawn into sand by contraction of longitudinal muscles after tentacles are infolded and mouth closed.
(right) (a) *Arenicola* in longitudinal section and (b) a bivalve mollusc in transverse section. The mechanism of burrowing in *Arenicola* is close to that described for the generalised soft-bodied animal. In the bivalve molluscs, closure of the shell by the adducter muscles causes jets of water to issue from the mantle cavity and loosen the sand. This facilitates penetration.
A—direction of movement of body. AM—adducter muscles, C—coelom. CM—circular muscle. E—eversion of base of column. G—gut. H—pedal haemocoels. LM—longitudinal muscle. M—mantle cavity. P—direction of penetration. PA—penetration anchor. PW—peristaltic wave. RM—retractor muscle. TA—terminal anchor. TM—transverse pedal muscles. W—water jet (after Trueman 1968)

closed during changes of shape, and it is not surprising that a separate hydraulic system has evolved in other animals. This view supposes that the coelom has evolved specifically to facilitate the burrowing habit. It is significant that the coelom or haemocoele is most strongly developed in those animals which burrow and there can be no doubt that its greatest importance to these groups lies in its hydrostatic functions.

A constant feature of the penetrating organ, whether it be coelenterate disc or annelid head, is that it tends to be circular in cross-section. Such a shape allows maximum contact with the substrata on all sides and permits the development of powerful longitudinal muscles which are essential for pulling the organism forwards. Trueman (1968) suggests that it is this factor which is responsible for the radial symmetry in coelenterates and not the sessile habit which is more commonly invoked. The bivalve's foot is an exception for it is flattened (the alternative name for the group is Pelycipoda which means axe-foot) and consequently pressure is exerted over a wide area. The foot of the Tellinacea is particularly flattened forming a broad triangle. This may have value in rapid burrowing through a dilatant sand/water mixture for when compressed, the sand grains pack hard and provide rigid walls against which the broad foot may press for anchorage. Other molluscs, such as *Ensis*, have a bulbous terminal anchor similar to that described for the schematic burrower. The latter is probably a more advanced condition, since the force which pulls the shell downwards is 800 g in *Ensis arcuatus* and only 5–6 g in *Mercenaria mercinaria*, a member of the Veneridae, which has an axe-shaped foot. *Donax vittatus* with an intermediate type of foot produces a pull of 50 g (Ansell and Trueman 1967).

Although the mechanism of burrowing is essentially the same in all groups, there are, of course, many differences depending on the body shape of the animal. A worm-like organism has few difficulties for where the terminal anchor has been, the body can surely follow, but the mollusc has a further problem in that the foot, which is the penetrating organ, has to drag a large shelled body after it. This is more difficult in hard sand than in soft mud and, consequently, we tend to find that bivalves from sandy shores are slim—sometimes wafer thin—while those from mud are broader. However, even amongst sand-dwelling forms, there are differences in shape. Slim shelled species such as *Donax* and *Tellina* have no problems in slicing through the sand but the stouter *Cardium* and *Macoma* find the process more difficult and take longer to enter when on the surface. Those with bulging shells spend more time in probing with the foot and effect a see-saw motion to facilitate penetration.

The rapidity with which *Donax* can move through the sand is an essential factor in its ecology. Wade (1964) showed that *D. denticulatus* in Jamaica is specially adapted to life on a steep, wave-swept beach. As the tide rises and falls, *Donax* migrates up and down the shore in order to keep within the splash zone. Only an exceptionally fast-moving mollusc would be able to do this.

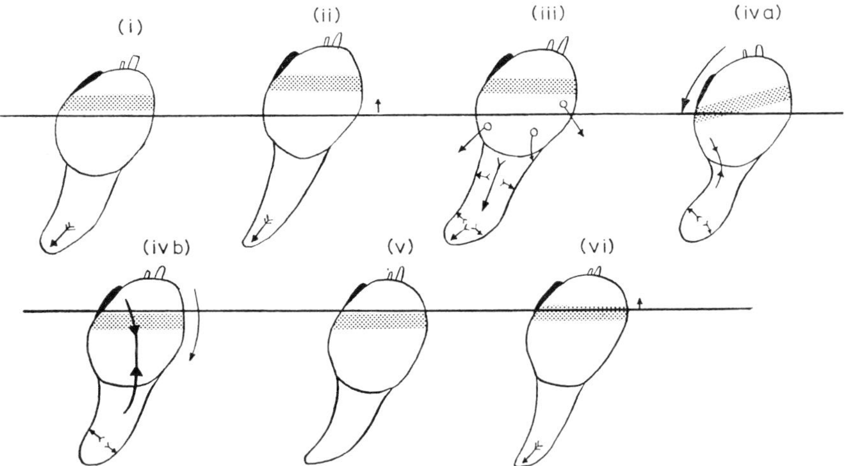

Fig. 10 Burrowing sequence in a generalised bivalve. The stippled line on the shell is intended as a reference level relative to the surface of the sand which is indicated by the horizontal line.
(i) Foot probing downwards. (ii) siphons close while foot still probing, shell moves upwards slightly. (iii) adduction (A) of valves, water ejected from mantle cavity. (iva) Siphons open, formation of terminal anchor, anterior retractor contracts and shell moves downwards at left. (ivb) posterior retractors contract and shell moves downwards at right (v) adductors relaxed, valves gape, no anchorage. (vi) Foot commences to probe for next phase, shell pushed slightly upwards. Normally, a penetration anchor is formed at (i) and (vi) to prevent the shell being pushed upwards.
← movement of shell. ←o ejection of water from mouth cavity ←< probing of foot by pressure changes due to intrinsic muscles. ←—< hydrostatic pressure derived primarily from adductor muscles and to lesser extent, from retractor muscles →—← contraction of retractor muscles. (after Trueman *et al.* 1966)

The bivalve shows a further adaptation for moving through the substratum. After the foot has made a major downward thrust and becomes locked in the substratum by the terminal anchor, the siphons close preventing water leaving the mantle cavity during the next stage which is a rapid adduction of the valves. This causes an increase of pressure in the mantle cavity and a consequent ejection of water through the lower margins of the valves. This jet of water on both sides of the body loosens the sand particles and further facilitates penetration of the soil.

Trueman *et al.* (1966) have divided the sequence of events into six stages in *Tellina tenuis*. The cycle should be read in conjunction with Fig. 10 for full understanding.

Stage (i) the foot makes a thrust downwards (which can raise the body if penetration is difficult) and becomes anchored at the terminal point.

Stage (ii) The siphons close.

Stage (iii) The valves close rapidly (within 0·1 s) and force the water from the ventral margins of the shell. During this stage the foot shows maximum dilation.

Stage (iv) The pedal retractor muscles contract, the anterior being slightly ahead of the posterior, but as the foot is anchored in the sand, it is not withdrawn. Instead, the body is pulled through the sand in a see-saw action. Downward movement follows from the shell settling passively into the cavity formed by the water jet from the valves in the previous stage as well as as a result of the contraction of the pedal muscles. During this period the siphons open and at the end of the stage, pedal dilation is at its minimum.

Stage (v) The adductor muscles relax and the shell valves consequently gape open.

Stage (vi) There follows a static period until the next cycle commences at (i). During this period, repetitive protrusion of the foot at intervals of 0·5 to 1 second and slight lifting of the body occur.

Specific variations on this general theme are described by Trueman *et al.* The rocking movement in stage (iv), caused by the sequential contraction of the retractor muscles, is more marked in *Cardium* and *Macoma* than in *Tellina* while it is almost absent from *Donax*. The greatest specific difference is found in stage (vi), the static period. The length of this period depends on a number of factors including the efficiency of the foot in penetrating the sand to effect an anchorage. The ability of the foot to anchor the animal is itself directly related to the shape of the shell and the ease with which the shell can pass through the sand during stage (iv). Hence shell shape, which is a specific character, has an influence upon the static period.

The siphons of the bivalve molluscs studied by Trueman *et al.* remain in contact with the surface of the sand during the burrowing process. As the animal penetrates further into the substratum, so the siphons elongate to keep pace. This is an adumbration, perhaps of evolutionary significance, of the condition found in some bivalve molluscs which burrow into deep, thick mud. In such forms, exemplified by *Mya arenaria* the soft shelled clam, the animal grows as it tunnels so that the whole burrow becomes filled by the body which may be up to 30 cm in length. The fleshy, upper portion of the clam consists of the fused and enormously enlarged siphons which the mollusc is quite incapable of withdrawing into its shell. This is rarely, if ever, attempted under natural conditions but if the mollusc is dug out the siphons can contract somewhat although the fleshy mass protrudes from the gaping shell and gives the animal its alternative name of gaper clam. The foot is almost vestigal once the organism is fully grown and if removed from the mud, the animal can rebury itself only with the very greatest difficulty.

One of the principal differences between sand and mud as far as burrowers are concerned is that in the former, permanent burrows are rarely possible since the relatively large particle size and the characteristically strong wave action soon cause the collapse of any that are formed. In the case of mud, however, permanent burrows are quite feasible and members of the infauna may spend most of their lives in the same place. A good example is *Chaetopterus* which, as was mentioned above, weaves a net within its burrow to trap food. *Chaetopterus* lines its U-shaped burrow with a parchment-like secretion and lies at the bottom drawing in water by rhythmically beating the flap-like extensions of three segments in the middle of its body. A curious feature of the worm is that it produces a luminous mucus but it is difficult to appreciate the value of luminescence in such a creature.

On sandy shores in relatively sheltered areas, many of the sedentary polychaete worms have succeeded in producing permanent homes by lining their burrows. Unlike *Chaetopterus*, which fits loosely into its tunnel, these worms are tightly enclosed in a tube which may be sandy (e.g. Sabellariidae) or membranous though often encrusted with sand grains (e.g. Terebellidae) or mucous (e.g. many Sabellidae). Perhaps the most delicate sandy tube is built by *Pectinaria* (Amphictenidae). The tusk-like tube, which is often washed up empty on the shore, may be strongly curved or almost straight depending on the species but it is very fragile being made up of a single layer of fine sand grains cemented together. *Pectinaria koreni* shows the finest workmanship with the sand grains placed edge to edge forming a perfect fit. The worm is buried upside down in the sand with the narrow end protruding above the surface. The excavated sand is conveniently removed by discharging it through the narrow end of the tube. It feeds by constructing a conical depression in the sand with its head at the bottom of the funnels. Particles of food drop into the depression and are picked up by the feeding tentacles. *Pectinaria* is unusual amongst 'sedentary' polychaetes in being mobile, a condition which is probably forced upon it by the limited amount of food which is likely to fall down the funnel.

The tube of the sand mason, *Lanice*, a Terebellid, is a familiar feature on the lower reaches of sandy shores and one of the few surface indicators of life below. The tubes, which can be locally abundant, project from the surface of the sand as ragged tufts. When the tide is in, the tentacles of the head protrude and move sinuously over the surface seeking food for this is a deposit feeder. If danger threatens it can quickly shoot back into its tube which is considerably longer than the worm itself.

The burrowing habit of *Arenicola*, the lug worm, which was thoroughly worked out by Wells several decades ago, is perhaps, too well known to bear redescription but it cannot be ignored. The presence of *Arenicola* is very obvious from the worm-like castings found all over the shore. It prefers a beach with enough mud to provide detritus for food but with sufficient sand to prevent its respiratory organs from being clogged. Little depressions in the sand are further evidence of *Arenicola*. These are formed at the head end of the U-shaped burrow which can be divided into three sections, the head shaft, gallery and tail shaft (Fig. 11). The gallery is a permanent structure lined with mucous and is the part in which the worm spends most of its time. The vertical tail shaft is used only when the worm backs up it to defaecate at the entrance. The head shaft which is also vertical, is invariably filled with sand but its position can always be detected from its yellow colour which contrasts strongly with the black sand of the reducing layer in which it is normally located. The

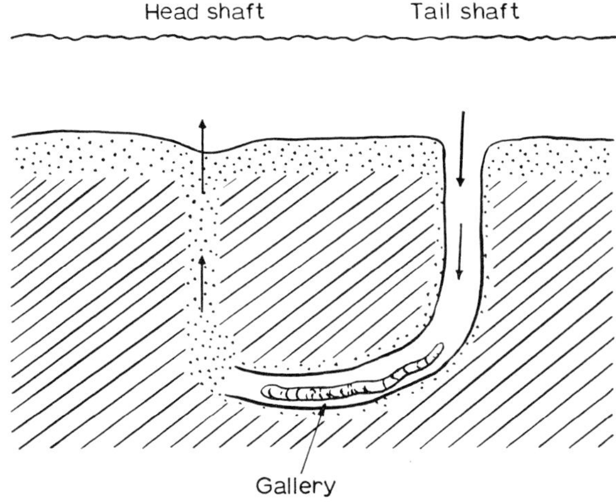

Fig. 11 *Arenicola* in its burrow. The open burrow is L-shaped and consists of a tail shaft, which extends down to the sulphide layer and may become choked with sand, and a gallery lined with mucus. The worm spends all its time in the gallery but pushes its tail through to the surface in order to defaecate. The head shaft is filled with yellow sand which is eaten by the worm. As the sand is consumed, more falls down from the surface and leaves the typical slight depression. Water for respiratory purposes is drawn in through the tail shaft and passed up the head shaft by waves of contraction which pass forward over the worm from the tail end. This current may be partly responsible for the yellow (oxidised) sand in the head shaft (adapted from Wells 1945)

stippled — yellow sand.
striated — black sand (sulphide layer)
White — water.
arrows — direction of water current

colour would suggest that the sand of the head shaft is aerated and this is indeed the case. It is brought about by a sequence of actions beginning with the worm swallowing the sand at the foot of the head shaft. This sand is replaced by surface material falling into the tunnel and causing the typical depression, a process which is aided by the worm loosening the sand at the base of the shaft with its head. The aeration itself is achieved by a current of water produced by the worm's movements which flows in at the tail shaft and up through the head shaft. This water current is essential to the worm as it carries oxygen into the burrow. When the tide is out, water movement is impossible but asphyxiation is prevented by the strong affinity between oxygen and the haemoglobin in the blood and by the extraction of oxygen from a bubble of air which can be drawn into the burrow by the tail. The worm derives its nourishment from organic matter within the sand and the constant replacement of the surface by the ebb and flow of the tide ensures a continuous food supply which permits the animal to remain in one place throughout its life.

So far we have considered only the sedentary polychaetes but there are a large number of species which move actively within the mud or sand. These errant polychaetes are largely carnivorous and hence the ability to move is important in the capture of prey which is often seized by formidable jaws. The jaws, which are sclerotinised hooks of taxonomic significance, are everted with the pharynx through the mouth. The most familiar species, perhaps, is *Nereis diversicolor*, one of the rag-worms. This is a worm typical of muddy rather than sandy shores and one which is distinctly estuarine in its distribution. Unlike most errant forms, *Nereis* may have a more or less permanent burrow which is lined with slime. It is perhaps as much scavenger as predator and will feed on any organic matter that it can find. Sandy shore forms include species of *Nepthys*, e.g. *N. hombergii*, the cat-worm and *Glycera convoluta*, a transparent worm which coils into a spiral when handled.

The mechanism of burrowing in the errant polychaete has not been fully worked out but it is thought that the everted proboscis plays a part. In many soils, little specialised technique is necessary for the worm can progress simply by pushing aside the sand particles.

Before leaving the macrofauna, mention must be made of the truly intertidal fish which may be considered either as epi- or infauna. Fish which are genuinely adapted to shore life are more common on rocky coasts than on depositing shores probably because the presence of rock pools in the former habitat provides an aquatic medium when the tide goes out. Fish from sandy shores need to develop more radical modifications in order to survive. Even so, rocky shore fish such as sea-scorpions and blennies can survive out of water for long periods and it is perhaps surprising that more species have not colonised sand or mud. However, a true sandy shore species is *Gobus minutus*, the sand goby, which is between 6 and 11 cm in length. It often occurs in large numbers over sand but it is difficult to see because of its cryptic coloration. Another sand dweller is *Trachinus vipera* the lesser weever or sting fish. The fish is well named for although it is small (10–15 cm) its dorsal and opercular spines produce an irritant poison which can cause severe pain. It is easy to step on since it lies buried in sand with only its eyes and the top of its head showing above the surface. Normally it feeds on shrimps. Perhaps the best known fish from depositing shore is the sand eel, *Ammodytes*, of which there are two common British species, the greater (*A. lanceotatus*) and the lesser (*A. tobianus*). They are not related to the common eel but they have eel-like bodies which probably help them to burrow into the sand in which they hide from their enemies. The actual digging is effected with the spatulate lower jaw. When the tide is in, sand eels swim about in large shoals but cannot easily be caught with a hand net even in shallow water.

Shore fish differ from their deep water relatives in a number of ways. One of these is in their shape. All are teleosts, the great division of bony fish whose typical shape is close to that of a goldfish, but on the shore, they come in a variety of shapes and sizes from the narrow tubular pipe fish and snake-like sand eels to the squat toad-like lumpsuckers and scorpion fish. A very common feature is an elongated dorsal fin. These odd shapes are a consequence of the fish leaving their ancestral home in the open sea and becoming adapted to life under boulders or in sand. Another common feature of shore fish is their pattern of reproduction. Life for eggs and larvae is hard on the shore; free floating eggs or planktonic young would easily be washed out to sea or be stranded on the beach. Consequently, the shore fish have abandoned the ancestral habit of shedding the gametes into the sea and allowing the eggs and young to fend for themselves. Instead, they lay a small number of eggs, to each of which the mother can afford a large yolk supply, and the clutch is guarded by the parent, usually the male, until the eggs hatch. We are familiar with the courting and nest-building habits of fresh-water sticklebacks and the shore fish behave in much the same way. Fertilization is ensured by a form of copulation which follows a much more elaborate courtship display than is usual in fish. The courting behaviour of shore fish is no doubt responsible for their bright colours compared with the drab steel grey of many pelagic fish. The cuckoo wrasse and dragonet are excellent examples of this trend and the colours are most intense during the breeding season when they may also have a territorial significance.

A consideration of an analysis made by Craig & Jones (1966) provides a suitable conclusion to this discussion of the macrofauna. These authors were interested in the relationship between the grade of deposit and the benthic fauna from a palaeoecological point of view. As a contribution to this problem, they analysed the feeding methods of both infauna and epifauna from different types of deposits. The data used refer to animals from the central part of the northern Irish Sea and hence are not closely relevant to the intertidal situation, but the principles involved are not so very different. Four types of deposit—mud, muddy sand, find sand and gravel/shell—were considered by Craig & Jones, but gravel will be omitted from this discussion since it supports radically different faunas above and below the tide mark. Some of the richest communities of the sea bed are found on gravel, whereas shingle beaches are almost devoid of life in intertidal regions. For the purpose of this analysis the animals were divided into four types according to their feeding habits. These were:

1. Suspension feeders
2. Deposit feeders

3. Carnivores
4. Uncertain feeding habits

A clear distinction between the epi- and infaunas could not always be made, but the percentage of each type found on the various grades of deposit is shown in Table 5. This shows that there is a marked increase in the proportion of suspension feeders as the grade of deposit coarsens; this is particularly marked in the data for the infauna. There is no great change in the proportion of deposit feeders although the numbers tend to decrease from mud to

TABLE 5

The percentage of benthic species found on each grade of deposit according to their feeding behaviour (after Craig & Jones 1966)

% on each type of deposit

Feeding Method	Infauna			Epifauna			Infauna and Epifauna combined		
	mud	muddy sand	fine sand	mud	muddy sand	fine sand	mud	muddy sand	fine sand
1 Suspension	0	20·5	36·5	0	0	0	0	17·5	12·5
2 Deposit	62·5	61·5	54·5	0	20	33·4	65·5	52	54·5
3 Carnivores	37·5	9	9	100	80	66·6	43·5	30·5	33
4 Uncertain	0	9	0	0	0	0	—	—	—
Total no. of species	8	24	11	1	5	3	23	127	68

sand and fall off markedly on shell gravel (not shown here). The percentage of carnivores tends to show a decline with an increased grade of deposit but it is difficult to see why this should be so unless the carnivores prey more on the deposit feeders which show a similar pattern.

The relationship between the type of species and the grade of deposit was also studied, and in general it was found that the proportion of epifaunal species in the community decreases from gravel through sand to mud. Fine grain deposits, therefore, favour the development of infauna, consisting predominantly of deposit feeders, while coarser deposits support both infauna and epifauna with a larger proportion of suspension feeders. On all types of deposit except gravel, the infauna species outnumbered those of the epifauna.

These conclusions can be applied to the shore only with caution, but the general trends are similar. Perhaps the most obvious difference is in the proportion of carnivorous species which are not as numerous on the depositing shore as they are under water. On the other hand, the greater tendency toward filter feeding on coarse deposits is also true on the shore and has, indeed, already been pointed out above.

1bii *The Meiofauna*

The alternative name of interstitial fauna for this group of animals is useful in emphasizing the habitat in which they occur for sand or mud particles are not jammed so closely together that no spaces or interstices occur between them. It is only within the last forty years or so that the most elementary interest at the taxonomic level has been given to this fauna but nevertheless a considerable corpus of knowledge has been accumulated on its biology. It is difficult to make a distinction between the intertidal and sub-littoral fauna since in either case conditions are much the same and it is likely that ecological problems are similar to both. An interstitial fauna is not confined to marine environments for it is found in fresh water in the bottoms of both lakes and rivers. Even the waterlogged subsoil on land has a similar fauna. Within the marine environment itself, differences in faunal types may be recognised, related apparently to the size and mineral composition of the deposits. Such differences are additional to the more obvious ones found between sand and mud. Thus, a siliceous deposit will support a fauna distinct from a shell sand region and both will differ from an estuarine community. Almost all animal phyla are represented in the meiofauna and some groups are practically confined to it, e.g. one of the two orders (Macrodasyoidea) of the Gastrotricha comes into this category. The first zoologist to give serious attention to the meiofauna was Remane whose book, published in German in 1940, reviewed the data then available. A more recent book in French is by Delamere-Deboutteville (1960) while a modern summary in the English language has been provided by Swedmark (1964) from which the following account has largely been taken.

It will be remembered that the definition of the meiofauna is a group of animals which moves through the substratum without displacing the particles. This is possible only if the animal is sufficiently small. This does not mean that the environment is open only to groups of organisms which are primitively small for it has just been stated that all phyla are represented. Small size must therefore have been evolved to enable some phyla to adapt to interstitial conditions. On the other hand, representatives of other phyla have had to evolve a large size; thus, most interstitial protozoans are bigger than their relatives from other habitats e.g. *Helicoprorodon maximum* is 4 mm long. The principle, of course, is that an optimum size is essential for any mode of life and in the case of interstitial sand, the optimum is between about 50µ to 3 mm. Examples of small metazoans from sand with large relatives include the hydroid *Psammohydra nanna* (1 mm), the archiannalid *Diurodrilus minimus* (350 µ), the prosobranch mollusc *Caecum glabrum* (2 mm) and the sea cucumber *Leptosynapta minuta* (2 mm). Animals from interstitial mud are smaller still. The harpacticoid order of copepods are very largely confined to interstitial mud and rarely reach 1 mm in length.

Swedmark has pointed out some of the problems involved in regressive evolution. One of the consequences of small size is that the number of cells available for the make up of organs is considerably reduced. Interstitial animals, therefore, are simpler than their large relatives but it is a secondary simplicity, e.g. some gastrotrichs have only one testis although related forms have the normal two. Even whole structures may be missing in the smaller forms. Swedmark has compared the only two species of the polychaete family Psammodrilidae, *Psammodrilus balanoglossoides* and *Psammodriloides fauveli*, the former having a volume some 25 times greater than the latter. *Psammodriloides* lacks a number of structures including the pharyngeal apparatus, reversible diaphragm sacs and nephridia, all of which are present in *Psammodrilus*. In fact, *Psammodriloides* is similar to a juvenile *Psammodrilus* in structure and also in size. This observation suggests that regressive evolution has been due to neoteny, the retention of juvenile features of the ancestor in the adult form of the descendant.

Fig. 12 shows a selection of species from interstitial sand. The common feature which is most obvious is the shape of the animals all of which are elongated or vermiform. This shape is found equally in species which belong to groups which are not normally worm-like (e.g. molluscs), as to those which are (e.g. annelids). The tendency towards vermiformity in the meiofauna has been analysed for some phyla and shown to be statistically significant. The reason is perhaps, not difficult to find. An organism which has to move through spaces between irregularly packed sand grains will find the task much easier if it is long and thing. Not all meiofauna are of this shape, however, for flat, broad organisms are not uncommon.

A third feature which distinguishes the meiofauna is the body wall. The sand in which these organisms live is constantly moving and one of the main dangers of living in such a habitat is that of being crushed. Some protection is afforded by the strong body wall which may be covered by a well developed cuticle as in nematodes and crustaceans. Spines and scales are a feature of the whole group of gastrotrichs and these must give some protection against crushing. Many other interstitial animals have an internal skeleton in the form of calcareous spicules which are found in such diverse groups as the Protozoa e.g. *Remanella*, a ciliate, and the opisthobranch molluscs e.g. *Rhodope* and *Hedyopsis*.

A different mechanism for avoiding mechanical damage is the ability to contract, a facility which is strongly developed in many species including ciliates, turbellarians, gastrotrichs and opisthobranchs. Many of these taxa do not normally contain members with this feature which must, therefore, be considered as an adaptation to interstitial life.

Adhesive organs are a further major characteristic of the meiofauna. The adaptive significance of these structures is presumably related to the dynamic nature of the environment. Adhesive organs are mostly glandular and very often concentrated in certain areas of the body but in some groups, such as arthropods, they are in the form of hooks or claws.

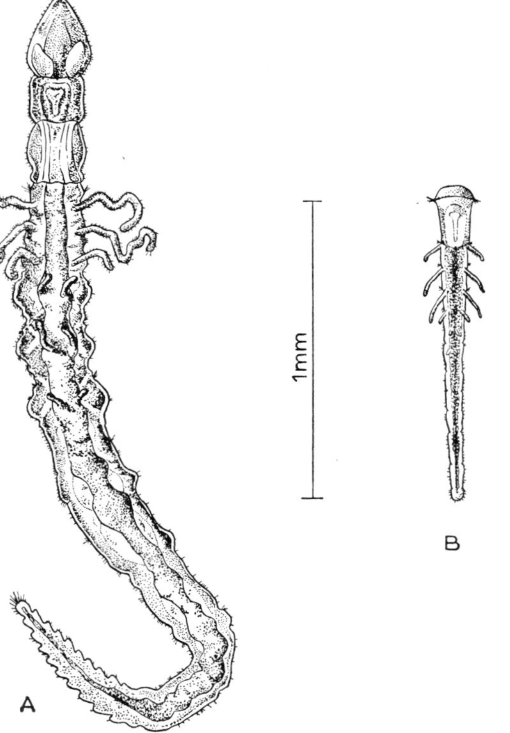
A
1mm
B
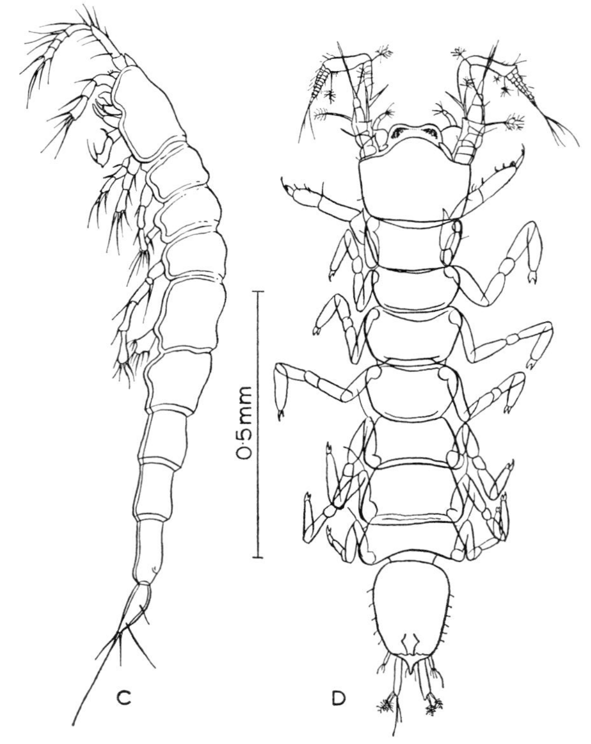
0·5mm
C
D

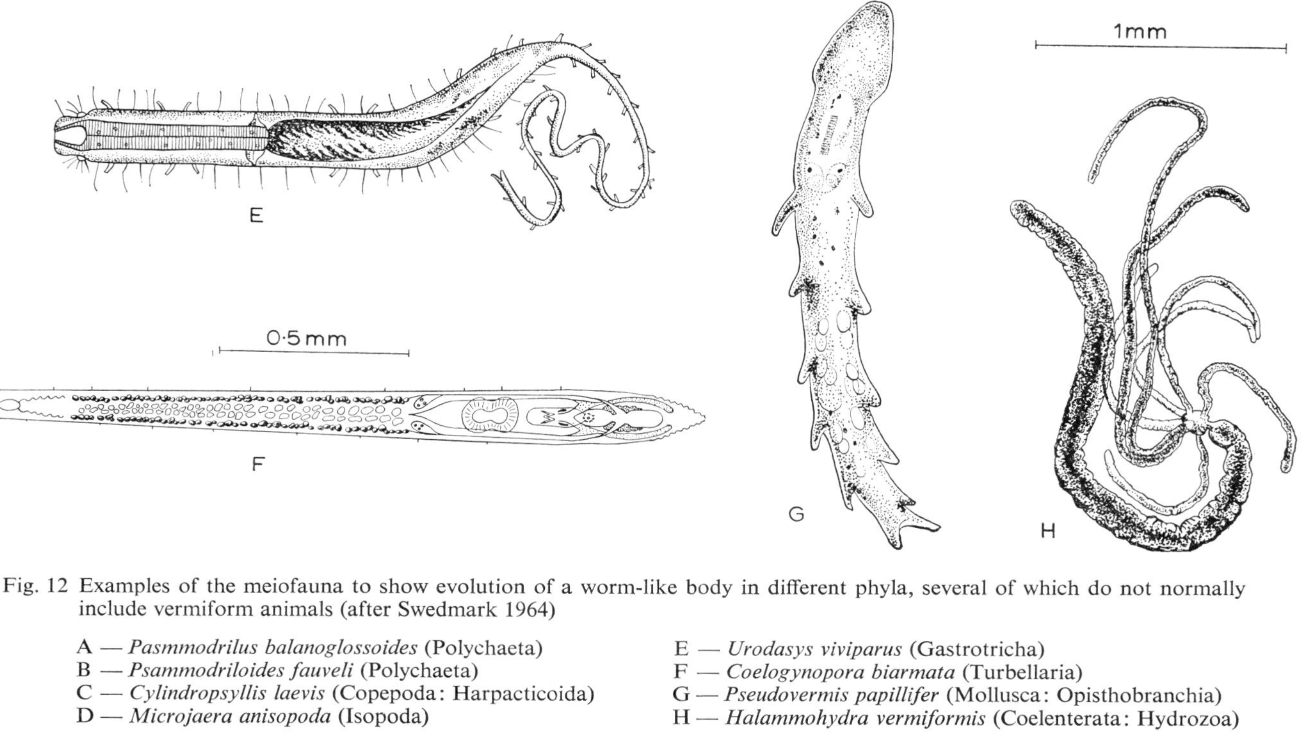

Fig. 12 Examples of the meiofauna to show evolution of a worm-like body in different phyla, several of which do not normally include vermiform animals (after Swedmark 1964)

A — *Pasmmodrilus balanoglossoides* (Polychaeta)
B — *Psammodriloides fauveli* (Polychaeta)
C — *Cylindropsyllis laevis* (Copepoda: Harpacticoida)
D — *Microjaera anisopoda* (Isopoda)
E — *Urodasys viviparus* (Gastrotricha)
F — *Coelogynopora biarmata* (Turbellaria)
G — *Pseudovermis papillifer* (Mollusca: Opisthobranchia)
H — *Halammohydra vermiformis* (Coelenterata: Hydrozoa)

A major problem for the interstitial animal in the absence of visual clues, is to know which way up it is and it is not surprising to find that statocysts are well developed in many groups.

All these morphological adaptations for interstitial life have resulted in a considerable degree of convergent evolution, so much so that widely unrelated animals tend to look alike and it is not difficult to recognise a member of the meiofauna from its morphology even if it cannot be identified.

The general biology of the meiofauna shows similar convergence. Locomotion in many members is by ciliary gliding. It is seen in such diverse groups as ciliates, coelenterates, turbellarians, archiannelids, polychaetes and molluscs. In several of these taxa, ciliary gliding is the usual method of progression in other habitats but in some, polychaetes for instance, it must be a specific adaptation to interstitial life.

Another method of locomotion is by writhing. A collection of living interstitial animals constitutes a wriggling mass and if examined closely, many of the specimens will be seen lashing from side to side. In a glass Petri-dish, this gets them nowhere but in mud and sand, it is a different matter. Presumably the body pushes against the sand grains in such a way that the animal is propelled forward but no clear demonstration of the mechanics involved seems to have been made and it is a field which would repay further investigation.

Some of the meiofauna have legs and crawl through the interstices but not all limbed forms do this. Thus, the harpacticoid copepods, which have legs, usually progress by the writhing movements described above. However, movement within the soil is of little significance in the distribution of the interstitial fauna because the principal agent in their dispersion is undoubtedly wave action or the current systems described in Chapter 1. These carry the actual shore deposits with them and must inevitably transport the meiofauna since many of the organisms cling to individual sand grains with their adhesive organs while their eggs are often sticky and adhere to the particles. A further possibility is that the meiofauna are dispersed on sand that sticks to the feet of shore birds for certain species have been isolated from brushings taken from the feet of mallard and other ducks (Boaden 1964).

Nutrition is another aspect in which the biology of the meiofauna shows specialisations. Swedmark divides the fauna into four broad nutritional groups viz. (a) predators, (b) diatom and epigrowth feeders (c) detritus eaters and (d) suspension feeders.

The predators include coelenterates e.g. *Halammohydra* and many turbellarians, groups which are normally carnivorous in other environments. The diatoms and other epiphytes on which the second group feeds, grow in a film surrounding each sand grain. The film is basically bacterial in origin and besides diatoms, the plants include peridians and members of the Chlorophycae and Cyanophycae all of which are autotrophic and, therefore, limited to the upper sand layers where there is some light. Bacteria are particularly abundant in the encrusting film and concentrations of 50 000 to 5 000 000 per ml of sand have been reported (Pearse, Humm and Wharton 1942). A wide variety of meiofauna feeds on the interstitial flora including nematodes, copepods, ostracods, rotifers and archiannelids. Some feed by scraping the plants off the grains by a process described as 'browsing' by Swedmark and as 'grazing' by Boaden but neither are very good terms since they have precise meanings in large mammal ecology. Other groups feed by sucking in the film, usually through pharyngeal action. These include gastrotrichs, some nematodes and turbellarians. Specialised suckers are found in the Tardigrades, the 'bear animalicules' which taxonomists find difficult to place. These have a tube-like mouth containing two stylets which pierce the cell walls of plants while the muscular pharynx sucks out the cell contents. Tardigrades are more typical of fresh water or damp terrestrial environments rather than the sea, but some species occur in sand and their specialised feeding habits require a separate category to accommodate them.

Some authors include the sand-lickers as a type of meiofauna. These comprise certain amphipods and cumaceans amongst the Crustacea. However, these organisms tend to manipulate the sand grains while removing the organic film and hence, do not qualify as meiofauna which, it will be remembered, is defined on the fact that the animals do not disturb the deposits in which they live. Sandlickers are, therefore, members of the macrofauna. They may also eat the smaller interstitial animals themselves and their eggs which are often adhering to the grains. This is not properly a case of predation any more than a cow preys on the caterpillar it inadvertently swallows with its food, but whether intentional or not, the result is the same for the animal consumed.

Detritus eaters are represented by some gastrotrichs, nematodes and archiannelids and presumably, the detritus is eaten for its bacterial content as in the case of *Hydrobia* discussed earlier. Suspension feeding is unusual in the interstitial environment—not a surprising fact in view of the mobile nature of the habitat—but some suspension feeders do occur e.g. *Monobryozoan*, a bryozoan, and the brachiopod *Gwynia* as well as some ascidians. *Monobryozoan* is a peculiar animal for its class as it is both solitary and mobile.

The reproduction of meiofauna shows some obvious adaptations for interstitial life. The number of gametes is small, because of the reduced total number of cells, and clutches of single eggs are very common. Ten is usually the maximum number. Pelagic development is rare because of the high mortality in such a mode of life which the unprolific species could not afford. About 98% of the meiofauna lack a pelagic phase in the life history and those species which belong to groups in which this type of development is normal, show unmistakable evidence of its evolutionary suppression. The few interstitial forms with pelagic larvae are the ones which tend to produce a relatively large number of eggs of 100 or more.

The fact that so few eggs are produced demands a greater degree of parental care than is usual amongst marine invertebrates. The hazardous egg phase is often eliminated altogether through the evolution of viviparity. In others, the egg may be incubated in a brood pouch and the young not released until they are fully independent. One of the more bizarre adaptations is seen in the nerillid archiannelids in which the developing embryos remain attached to the rear end of the mother and undergo development whilst being towed about. The eggs themselves contain much yolk, except in pelagic types, and consequently the young are able to develop to an advanced stage before an external food supply is required.

In some meiofauna, egg production is enhanced by the production of cocoons which are usually sticky and adhere to sand grains. Those eggs which are laid singly without protection are also sticky and become attached to the grains.

Many of these adaptations towards the protection of eggs would not have been possible but for the effective exclusion of a dispersal phase from the life history. This, in turn, has been possible only through the extreme mobility of the habitat which either moves the animal itself or, which amounts to the same thing, brings a new environment to it.

The low number of gametes produced by the meiofauna precludes the common method of fertilisation in marine invertebrates—the shedding of the genital products into the water. Instead, a variety of behaviour patterns has been developed to ensure the union of the male and female gametes. Copulation is obviously the most efficient and it is found in many groups particularly those in which it is the norm in other habitats, e.g. copepods. The production of spermatophores, packets of spermatozoa, is another insurance of efficient fertilisation. In the opisthobranch mollusc *Microhedyle lactea*, a spermatophore is attached to the body of the sexual partner and the sperm pass through the epidermis into the body cavity from where they find their way to the developing egg. Many interstitial forms are hermaphrodite which increases the efficiency of copulation in that two 'females' instead of one are fertilised at each mating.*

1b iii *The Microfauna.*

This group concludes the catalogue of permanent members of the shore outlined in Chapter 2. As in all environments, micro-organisms abound in the interstitial habitat but full treatment of the microfauna would take us into the field of microbiology which is outside the scope of this book. There is indeed,

*While this book was in production, an important review of the meiofauna was published by McIntyre (1969). Some of his conclusions imply that the meiofauna is not of great ecological significance. Thus, it is responsible for only a small proportion of the total respiratory exchange of the deposit as a whole and only a small part of the meiofauna is utilised by animals of higher trophic levels. These conclusions apply to the meiofauna as a whole and not only to that of the shore. The discussion in the last chapter of this book concerning the energy relationships of the shore has suggested, however, that a rather more significant role should be attached to the shore meiofauna.

very little which can be written here. Most papers on marine microbiology are concerned with bacteria and principally with the bacteria of the open sea. In this chapter, we are concerned mainly with the fauna and hence we should now consider the small protozoans of the shore. It would be possible to compile a list of such species as have been recorded on the beach but it is doubtful that the effort would be worthwhile since practically nothing is known of the ecology of such forms. A recent account of marine microbiology is given by Kriss (1963) although his work is confined to the heterotrophs—bacteria, fungi and the like—of the deep sea. Again, one can only point to a gap in our knowledge of the ecology of shore fauna in the hope that some future research worker will make a mental note before passing on to the next section.

2 Temporary Members

2a Marine

The temporary marine members of the shore fauna are those that visit the shore from the sea when the tide is in. The category includes mainly fish and plankton but swimming invertebrates are also included. 'Visit' is perhaps an unfortunate term for a distinction must be made between those animals which enter the beach deliberately and those which are simply carried there passively with the flow of the tide. On the whole, fish come into the former category and plankton into the latter. One of the few reasons which brings fish ashore is food and another is sex. Several species of fish feed on the rich concentrations of animals on the shore many of which are either on the surface or protruding from the sand when the tide is in. There are, of course, species of fish which are genuinely intertidal and rank as permanent members of the shore fauna. These have already been described.

It is not easy to give an account of the temporary visitors from the sea since most of the inshore fish are found in the intertidal zone at one time or another. However, a line must be drawn somewhere and we will deal here principally with those fish which interact in a direct way with the permanent members. The most usual point of contact is through predation but it must be remembered that any organism which spends a high proportion of its time in an ecosystem must inevitably through defaecation, excretion and respiration, as well as through predation, interact with the other organisms there. The temporary fauna less often acts as food for the permanent members.

The most significant fish as far as the permanent shore fauna is concerned are bottom-dwelling or demersal fish. The flat fish are an order of the Teleosta (Heterosomata) which are very common inshore. The most striking feature of the group is the positioning of both the eyes on one side of the body. 'Side' becomes 'top' but some are right-sided and others are left-sided. There is no general rule although related forms are similar. The order includes many of the fish which appear at the table such as the dab, lemon sole, Dover sole, plaice, turbot and halibut. Steven (1930) has given an account of the feeding behaviour of some of these species. It may come as a surprise to some to realise that these apparently awkward fish are active carnivores feeding principally on the sedentary polychaete worms in the sand. The lemon sole, (*Microstomus kitt*) feeds almost entirely on polychaetes which it catches visually. When hunting, the fish rests in a characteristic posture with the front part of the body raised well off the ground which is scanned with the extremely mobile eyes. Should the tentacles of a worm emerge from the sand, the dab leaps onto it arching its body so that its head descends almost vertically onto the prey. In the absence of food the fish swims off a short distance to try its luck elsewhere. Incidentally, the worm often suffers no more than the loss of its tentacular crown which can be regenerated, a form of renewable resource with few parallells in other predator/prey relationships. The closely related dab (*Limanda limanda*) and plaice (*Pleuronectes platessa*) also hunt by sight but with a less exaggerated raising of the forebody. The dab feeds on a greater variety of prey than does the lemon sole but it is not as successful in catching polychaetes. The plaice is less active and feeds more randomly taking molluscs and some crustaceans as well as errant polychaetes. The Dover sole (*Solea solea*) is a step further along the line from active hunters to opportunist feeders. It feeds mostly at night when sight is of little advantage and moves slowly over the sandy bottom grubbing in the sand in places where its 'whiskers' have detected food.

Not all flat fish are equally likely to be found intertidally. The plaice is found close inshore when young but it migrates progressively further out to sea as it grows older so that the adults occur only in deep water. One would be most unlikely to find a halibut (*Hippoglossus hippoglossus*), the giant of the flat fish world, in any but very deep water, but the turbot (*Scophthalmus maximus*) and flounder (*Platichtys flesus*) are shallow water forms and may visit the shore. Other species, such as the Dover sole, show annual migrations being found only in deep water during the winter but close inshore in the summer.

The other broad grouping of teleosts, the round fish, do not represent a taxonomic division as do the flat fish. The roundness refers to their shape in cross-section and they are, in fact, the typical animals which the layman thinks of as 'fish'. Most of these fish, whether pelagic or demersal, spend most of their time at sea but a number of species occur close inshore and these are quite likely to find their way into the littoral zone when the tide is in. One has not to wait very long at the end of a pier in the clear waters of south-west England before one sees a shoal of mullet or even whiting. To some extent, their presence is fortuitous for they do not necessarily seek out the intertidal fauna as food and they rarely make any impact on the shore ecosystem. Like astronauts circling the moon, they hover over an alien world and are soon gone but unlike astronauts, they occur in large numbers.

The cod family, Gadidae is one of the most important groups of food fish and one whose members are essentially demersal in their feeding habits. The family includes such familiar table fish as cod, pollack, haddock, whiting, ling and coal fish although the latter is usually eaten innocently as cod in fish and chip shops. The family also includes the rocklings which are true shore fish although, as their name suggests, inhabitants of rocky rather than depositing shores. A fish closely related to the cods is the hake (*Merluccius merluccius*).

The food of gadoids tends to be invertebrate particularly when the fish are young. The adult cod is a predator feeding largely on herring, mackerel and any other fish it can catch including the young of its own species. However, adult cod rarely come close to the shore and we need not consider them further here. The whiting, (*Gadus merlangus*) is more likely to be found in shallow water. This looks like a small cod, rarely exceeding 40 cm in length, but it is more silvery with a very small barbule on the chin. A feature which distinguishes it from closely related forms, is a black spot at the root of the pectoral fins. It is an active predator taking fish smaller than itself but it also feeds on shrimps and other invertebrates. It is, however, not in the same class as the haddock (*Gadus aeglefinus*) as an invertebrate feeder. This fish takes all sorts of invertebrate bottom dwellers for which it searches assiduously by rooting in the mud. They are also very partial to the egg masses of other fish particularly herring spawn. They do not, however, impinge on the shore fauna as they are primarily creatures of the open sea and we must, therefore pass on.

The reproductive behaviour of the gadoids emphasises their essential involvement with the open sea rather than the shore. Unlike the littoral fish, mating is a casual affair with huge numbers of eggs and sperm (milt) being discharged simultaneously into the sea. The cod holds the record in its family with a production of up to four million eggs at one time. As very few need to survive to replace adult mortality, it is obvious that the losses among the young fish are prodigious. Parental care is non-existent; another feature which sets them apart from the shore fish. The eggs themselves differ, e.g. they hatch soon after laying, they have very little yolk and the young, known as fry, have to fend for themselves. The eggs float and with the fry constitute an important sector of the plankton.

Not all temporary visitors from the open sea are fish. Occasionally, lobsters and spider crabs are found near low tide mark and less frequently, shoals of octopus may come ashore. A recent instance occurred in 1950–51 when *Octopus vulgaris* became extremely abundant on the English Channel coast. A previous 'plague' year was 1899–1900. *Octopus* does not breed in our waters but originates in the triangle of sea defined by the Cherbourg Peninsula, the Channel Islands and Ushant where the larvae may be very abundant. The causal factor of octopus plagues seems to be a migration of adults to the French coast resulting in a building up of the breeding population. Provided conditions are not too severe—a warm winter is probably important—numbers will continue to increase. The appearance of the adults on the English coast appears to be the result of a northward migration across the channel caused primarily by food shortages (Rees and Lumby 1954). Normally, *Octopus* does not have any significant effect on the shore ecology but during plague years, they are very destructive to crabs and lobsters which they take as food.

The definition we have adopted of the marine members of the temporary fauna is those species which visit the shore from the sea when the tide is in. So far, these have been organisms which live under the surface of the sea and respire dissolved oxygen. However, the definition needs to be extended to include the air-breathing, marine animals. We must exclude the whales which come ashore only very occasionally and then to die, but the other major group, the Pinnipedia or seals, deserves further mention. Around the British Isles there are two indigenous species which spend most of their time at sea but they come ashore for breeding and moulting or in the intervals between fishing in inshore waters. The larger of the British species, the grey or Atlantic seal (*Halichoerus grypus*), tends to come ashore on rocky coasts to breed but sometimes a sandy beach or sandbank is chosen. The nature of the shore is probably less important than its isolation. Haul-out areas used between fishing forays may be of any shore type depending on the geography of the fishing area. The other species in Britain is *Phoca vitulina*, the common seal, which ecologically is similar to the grey seal. This species is found principally off the west coast of Scotland, the Hebrides, Orkneys and Shetland, yet it may also occur almost anywhere in Britain, e.g. it is quite abundant on the Wash and small groups occur in the Bristol Channel. The distribution of the grey seal overlaps that of the common seal, but the former species is more frequently found on the west coast of England, a fact to which it owes its alternative vernacular name of Atlantic seal. It is also found in western Scotland with major breeding colonies in the outer and inner Hebrides, North Rona, Orkneys and Shetlands. It is not confined to the Atlantic for the large colony on the Farne Islands consists of this species. Other favoured areas are the Pembrokeshire coast and, in Ireland, the shores of Kerry and Cork in the south west and Clare Island, Co. Mayo in the north. Grey seals tend to breed on the beach but in exposed areas, such as the western coast of the Outer Hebrides, the seals pass over the beach to the grassland beyond where they can rear their pups in safety. In Cornwall, some colonies have taken to using caves which can be entered only from the sea. However, neither species is of much ecological significance to the shore since they do not feed there and are present for such relatively short periods during the year.

2b Temporary members—terrestrial

We come now to the last major category of shore fauna in the classification drawn up in Chapter 2. It will be remembered that distinction was made between those species which may visit the beach from time to time but are not dependent upon it for any major resources and those which are. The first were termed facultative and the second obligate.

2b i. *Facultative Species*

The species which feed on the shore when the opportunity arises include a wide range of animals. Amongst the mammals, the brown rat is perhaps the most ubiquitous and may almost be a full-time member of the shore fauna, taking most of its food in the habitat and retreating up the beach or out of reach of the water when the tide comes in. However, it is never an obligate member, for should it for any reason be translocated to another environment it can survive perfectly adequately. On the beach, rats feed mainly on organic rubbish in the strand line but on rocky shores they will take shellfish and any other prey they can catch or break open. They are also suspected of taking the eggs of shore-nesting waders or ducks but there is little direct evidence of this. Most shore birds of any size are usually quite capable of defending their nests against rats. Voles and field mice, which are vegetarian, occur on shores with a background of vegetation such as a saltmarsh. Their burrows are close to the high tide mark and are often flooded during high tides. Shrews can be abundant, while larger carnivores include the otter which takes chiefly crabs and fish. It is, however, found more often on rocky shore where food is more plentiful.

The avian sector of the facultative shore fauna contains such common species as the rook, raven, swans and a variety of ducks while almost any common bird near the coast may be found on the shore at one time or another. Gulls belong here too, for although most common on the shore they can certainly thrive away from it and feed inland to a large extent. The attraction for all these is, of course, food. The rook (*Corvus frugilegus*) is well known for its habit of taking mussels, whose shells it breaks with a powerful blow of the bill. It is naturally more abundant on rocky shores but sufficient artifacts occur on other types of shore to support mussels and attract rooks. Mute swans (*Cygnus olor*) are feeders on aquatic vegetation which they normally take in fresh water but in suitable areas, they readily feed in the shallow sea. They come ashore to rest and preen and contribute very little to the shore ecosystem apart from the donation of some nutrients in their droppings. In the north of Britain, the whooper swan (*Cygnus cygnus*) is addicted to the shore although it often breeds inland. Certain ducks feed so intensively on the shore fauna that one might consider them as obligate inhabitants were it not for the fact that others of their kind never come near salt water. Examples are the mallard (*Anas platyrhinchos*), which may feed extensively on saltmarsh plants and shore crustacenas yet in inland areas, the dominant food consists of the seeds of terrestrial plants and freshwater organisms (Olney 1964). A similar situation is found in the goldeneye *Bucephala clangula* (Olney 1963). Other species of birds may be facultative members of the shore fauna by virtue of their seasonal appearances there. We shall see later that some species of waders spend the summer inland and the winter by the sea. Whether these should be considered as temporary obligate species or as facultative is a moot point.

We reach similar semantic problems with the terrestrial invertebrate visitors to the shore. Many of these live in the strand line which has the properties of both a marine and a terrestrial environment with the lower wrack beds containing a higher proportion of maritime species than the upper which may be inhabited by solely terrestrial species. The latter are really facultative members of the shore fauna since they would probably do very well if they were transported to the leaf litter on a woodland floor, yet they normally spend their whole life on the shore and have some claim to obligate status. These organisms are considered more fully in a later chapter in which the different habitats within the shore are considered. Incidentally, the large number of wrack flies on the shore attract a further group of facultative species—the insect-eating birds such as swallows.

2b ii. *Obligate Species.*

The majority of terrestrial obligate species on the shore consists of birds. There are no mammals, reptiles or amphibia in this category in British waters although presumably, one would include crocodiles and marine iguanas in a world-wide context. Few invertebrates qualify, because those which have of necessity to spend their whole time on the beach have become so well adapted to marine life that they can no longer be considered terrestrial. Thus, it would be stretching the definition to breaking point to include the marine insects *Anurida maritima* and *Petrobius maritimus* as obligate terrestrial forms. Certain of the wrack flies presumably require the strand line in which to breed and in this sense may be considered as obligate species.

The number and variety of marine birds are astonishing and range from the host of waders to cormorants, gannets, skuas, kittiwakes, puffins, razorbills, guillemots, shearwaters, shelducks, scoters, eiders and even geese. However, many of these are truly marine birds and come ashore only to breed or roost for the night, often on high cliffs or other habitats remote from the beach. Many of them, on the other hand, feed exclusively, or nearly so, in the intertidal zone and these are the ones relevant to this book. Of the waders, a term which describes a heterogenous collection of shore birds not necessarily closely related, only two qualify in Britain as truly obligate shore species. These are the oyster-catcher (*Haematopus ostralagus*) and the ringed plover, (*Charadrius hiaticula*) both of which breed principally on the shore itself and take most of their food there. Even so, they may be found inland but only very rarely. Many of the other waders—sandpipers, dunlin, sanderlings, avocets, curlews and many more besides—are predominantly shore birds dependent on the intertidal zone for food during most of their lives although several may breed inland. The same is true of the shelduck (*Tadorna tadorna*) which only rarely is found away from the shore except during the breeding season when it seeks out an old rabbit burrow or a hole in a hay-stack in which to rear its young.

True geese are found on the shore. In Britain, there are two species, the barnacle goose (*Branta leucopsis*) and the brent (*Branta bernicula*). The latter is the more genuinely intertidal since its preferred food is *Zostera*, the eel grass, which grows near the low water mark of sandy and muddy shores. The country-wide decline in the *Zostera* beds had a most adverse effect on the brent geese wintering in England but their numbers are now well on the way to recovery. The barnacle goose spends much of its time feeding in grass and cultivated fields and so demonstrates its independence of the shore. It is, however, never far from the coast.

The eider duck (*Somateria mollissima*) is an obligate species as it breeds on the shore and takes its food from the mollusc beds in the intertidal zone as well as in deeper water. Indeed, it is one of the few birds which is more or less confined to sand rather than mud. The largest colonies in Britain can be found in the Farne Islands, at the mouth of the Tay and in the Ythan Estuary north of Aberdeen. The bird is a year-round resident and undertakes only very limited local movements.

We find even a passerine with excellent credentials for the status of an obligate shore species although the shore is rocky rather than sandy. This is the rock pippit, *Anthus spinoletta*, which is very similar to its meadow and tree relatives. It nests on the shore or rocky crevices and feeds on marine invertebrates in the tidal reaches.

3 The origin of the shore fauna

It should be clear by now that the permanent shore fauna constitutes a distinct biological entity and does not simply consist of animals from the nearby land or sea which have extended their range onto the shore. The shore is an ancient habitat which has existed since the land first arose from the sea, but it is not, as was once thought, the most probable region in which life appeared on earth. It is extremely unlikely that the first primaeval delicate organisms could have survived under the exacting conditions of the intertidal zone and it is much more likely that life originated in the shallow sea. The shore, therefore, must have been colonised by organisms which evolved in other habitats but this colonisation need not have taken place early in geological time. Aeons may well have passed before the level of physiological complexity had evolved to such an extent as to permit the invasion of the intertidal regions. Nor need the colonisation have been a once and for all event. Species have, no doubt, spread on to the shore on a number of occassions and may yet still be doing so.

Assuming that the shore was colonised after the evolution of most of the invertebrate phyla, there are three likely environments from which the invaders could have come. These are the sea, the land and other types of shore. A fourth possibility, is by way of colonisation from fresh water through the estuaries, but this would probably be of minor importance because of the small number of fresh water species relative to those in other environments.

Some of the estuarine species, e.g. cyprid ostracods, almost certainly originated in fresh water. The reverse movement has certainly occurred and within historical times the zebra mussel (*Dreissensia polymorpha*) and the hydroid *Cordylophora lacustris* have changed from marine to fresh water species as a result of penetrating estuaries. Such abrupt changes in the preferred habitat are probably the result of genetic changes and some of the movements onto the shore may have been equally sudden.

Yonge (1953) discussed the origins of the fauna of muddy shores and his explanatory diagram is reproduced here (Fig. 13). He considers that for the most part, the mud fauna spread on to mud from other types of shore and

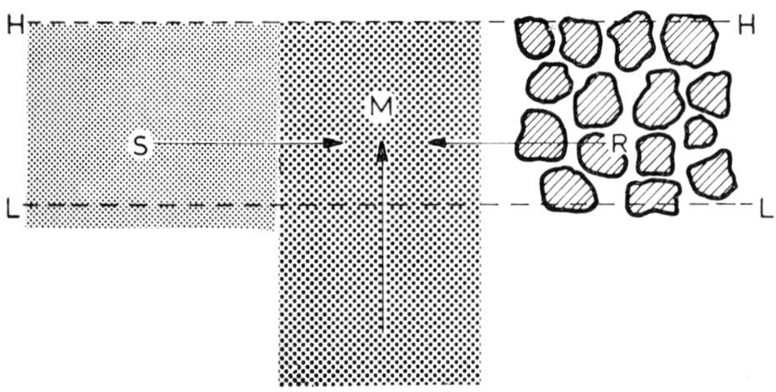

Fig. 13 Diagram indicating the possible routes whereby muddy shores were colonised. Most of the epifauna is thought to have arrived by horizontal movement from other types of shore while the more specialised infauna probably moved up from sub-littoral mud. H—high-tide level, L—low-tide level, M—mud, R—rock, S—sand (after Yonge 1953)

from sub-littoral muds. In the former case, the animals had already become adapted to intertidal life but were not necessarily modified for existence in mud. As a result, most species of this type belong to the epifauna which has a less intimate contact with the substratum than has the infauna. Examples include *Littorina littorea*, *Carcinus maenas* and *Mytilus edulis*, which are considered to have migrated from rocky shores and *Cardium edule* and *Crangon vulgaris* which are believed to have originated in sand. Although *Cardium edule* is more often thought of as an infaunal species of sand, Yonge considers it can rank as part of the epifauna of mud as it spends so much of its time there moving over the surface. None of these species shows any particular modifications for life in mud although all are well adapted physiologically for intertidal existence.

By contract, Yonge infers that most of the infauna species have arrived from sublittoral muds. These animals would obviously be adapted to muddy conditions but would need to evolve tolerance of intertidal factors before moving onto the shore. This was probably the path by which most of the polychaetes and many lamellibranch molluscs reached the shore. The more highly specialised the organism is to life in mud, the more likely it is to have originated in this manner.

The penetration of estuaries has undoubtedly been overwhelmingly from the sea rather than from fresh water, although some species belong to primarily fresh water groups. These will be mentioned in a later chapter. Most of the intertidal fauna of estuarine muds have moved there from the coast rather than from sublittoral regions.

The origins of the temporary fauna, whether terrestrial or marine, are so obvious that they need not detain us further here. It should be noted, however, that the invertebrate terrestrial component of the shore fauna has not been successful in penetrating very far down the beach but is restricted to the region around high tide mark or to semi-terrestrial habitats such as salt marshes. The reason, no doubt, is that most ecological niches suitable for shore invertebrates were already filled by marine forms before an invasion from the land was possible.

CHAPTER FOUR

THE SPATIAL DISTRIBUTION OF THE FAUNA

The zonation of marine life on a rocky shore is obvious to the most casual observer. Most of the work on the zonation of shore fauna has been carried out on rocky coasts to the neglect of other types but it should not be thought that the phenomenon does not exist on mud and sand. Zonation, however, is not so marked because the environmental factors, which presumably control the distribution of animals, change less abruptly on a beach than on rock. For example, the gentle slope of a mud flat results in relatively little drainage between tides so that desiccation is unlikely to be much of a problem to the animals particularly as they are buried within the damp substratum and not exposed on the surface as is the case with the rocky shore fauna. There is, indeed, no *a priori* reason why zonation should occur at all in mud or sand but nevertheless, it is almost as distinct a feature as in any other environment. As we shall see, the limiting factors controlling the distributions are more subtle and in some cases still not fully understood.

The spatial distribution of any shore fauna is more complicated than that of, for example, a grassy field because of the physical gradation up and down the shore. This has been recognised by distinguishing between the horizontal distribution, i.e. along the shore and the vertical distribution, i.e. up and down the shore. The latter is not a good term for apart from the fact that the gradation is not literally vertical, the expression is ambiguous in the case of animals which can move in a further dimension—vertically upwards and downwards—when burrowing in the soil. In this book, the word 'vertical' will be confined to this type of distribution and 'along-shore' and 'down-shore' will refer to the distribution parallel and at right angles, respectively, to the shore. The meanings are made clear in Fig. 1. The term 'horizontal distribution' is also avoided as it, too, is ambiguous as both along-shore and down-shore distributions can be said to be horizontal.

In this chapter, we will select one or two examples of each type to demonstrate the validity of spatial distribution in mud and sand. It should be borne in mind that the demonstration of zonation in these types of habitat is statistical and, therefore, vulnerable to sampling error. One cannot take a photograph to demonstrate the zonation as one can so eloquently do in the case of rocky shores. What has first to be done is to demonstrate that the distribution of animals is not random and then it is necessary to correlate the observed distribution with the supposed variable factor, in our case, the slope or depth of the substratum. As in all such exercises, there is a real danger of false correlations and ecological observations of this type should not be accepted unless there is at least some experimental evidence in their favour. It should be remembered that with the same population, any sample will show a continuous distribution if it is small enough or an aggregated one if it is sufficiently large while intermediate size samples may show that the same population is randomly distributed. These pitfalls are well known to the terrestrial ecologist but the ecologist reared exclusively on rocky shores tends sometimes to overlook them. The ecology of mud and sand has much in common with terrestrial soil ecology and many of the techniques are interchangeable but the marine science has the added complication, and interest, of tidal ebb and flow.

With these reservations in mind, let us turn to some examples which prove beyond doubt the reality of zonation in mud and sand. One of the most extensive studies from a geographical point of view is that of Dahl (1952), who investigated the zonation of crustaceans on six widely separated sandy shores in northern Norway, south-western Sweden, Venezuela, central and southern Chile and in the Magellan Straits. His work is of particular interest in that it shows a pattern for the distribution of malacostracan crustaceans which is similar over a wide geographical area, and which mirrors a similar general pattern in the distribution of animals on rocky shores. Just as the rocky shore fauna can be divided into a sub-terrestrial (*Littorina*) zone, a mid-littoral (balanoid) zone and a sub-littoral zone, so can the fauna of a sandy shore be divided into three comparable regions. It is always encouraging to be able to extract ecological principles from what is sometimes thought to be an amorphous science and it is worth examining Dahl's conclusions in some detail.

Dahl's equivalent of the rocky shore sub-terrestrial zone is populated by two main types of crustaceans, the ghost-crabs (*Oxypode*) in the tropics and the talitrid amphipods, of which the sand hopper *Talitrus saltator* is a common example, in temperate zones. It seems likely that each form replaces the other between the tropics and temperate regions. This suggests that both groups fill a similar ecological niche and indeed, they have much in common ecologically. Both types are scavengers on animal or vegetable debris thrown up by the tide and both are believed to be predacious to some extent. They show many other similar features, both are largely nocturnal and all species have good eyesight. All forms live in burrows which are made by the animals themselves. These are permanent for the crab but temporary for the isopod. The talitrids of temperate shores have more in common with the tropical ghost crabs than they have with the tropical talitrids for the latter tend to be found in damp wrack on the strand line or on the shores of fresh water ponds and lakes or they may even be terrestrial and live on the damp forest floor. Where they do occur as a sand form, they are found further down the beach. The differences can probably be related to factors of desiccation. Because of their thinner integument and habit of brooding the eggs in pouches, the talitrids are sensitive to humidity conditions and need a damp environment. They would be unable to survive on the hot tropical beaches. The crab, however, is less affected owing to its greater resistance to desiccation and its possession of an aquatic larva.

The middle regions of sandy shores are dominated by isopods of the subfamily Cirolaninae. This is a cosmopolitan group whose members are usually scavengers or ectoparasites on fish but they can also be active predators. They are most active when covered by the tide and hide in the sand during low water. They do not, however, dig deeply and perhaps for this reason, they are never found in sand which dries out thoroughly. Thus in tideless seas, they are confined to sand below the water level e.g. *Eurydice pulchra* is a sub-littoral form on the west coast of Sweden. The universal occurrence of the cirolanids and their ecological similarities justifies the recognition of a 'cirolana belt' on the mid-shore of all sandy beaches.

Lower down the shore towards the sub-littoral zone, the number of ecological niches increases and as a result, a greater number of species is found but again, one can detect ecological similarities which suggest that a third zone can be recognized. One distinctive category of animals includes those which regularly leave the sand to feed in the water. In temperate seas, this ecological type is represented by amphipod crustaceans and on tropical shores, by a group of anomuran crabs belonging to the Hippidea. However, in the more congenial tropical environment, greater ecological variation is found, even between taxonomically closely related species. For example *Emerita* is a filter feeder, straining particles from the water with its setose antennae, while other species such as *Blepharipoda occidentalis* are scavengers. The temperate equivalents appear to be the amphipods of two families; the Haustoriidae represented by such species as *Bathyporeia elegans*, *Haustorius arenarius*, *Urothoë grimaldi*, and the family Oedicerotidae including especially *Pontocrates arenarius*. Dahl emphasises particularly the ecological similarities between the tropical *Emerita* and the temperate *Haustorius*. Both forms are nocturnal and both are found in the region where waves are breaking. Both are said to be filter feeders although other reports maintain that *Haustorius* is a sand-licker. They even look alike, a feature not uncommon in animals displaying convergent evolution.

The possible reasons for this down-shore zonation of the fauna into three belts will be left until a later chapter.

It is useful to turn to a specific example to see how far Dahl's principles apply in individual cases. Exhaustive sampling of the macrofauna of beaches is not often made but there are some records in the literature including that of Watkin (1942) from Kames Bay, Millport. He took samples at intervals of

4.5m from high water mark to low water mark of spring tides. Watkin's results (Table 6) are generally in agreement with Dahl's views. No crustaceans were found until Station 6 and none of these was a talitrid so that Dahl's upper community is unrepresented. However, the mid shore representative, *Eurydice pulchra*, is present and the lower regions are well populated with amphipods of the family Haustoriidae. Note, however, that one of these amphipods, *Bathyporeia pilosa*, is found well up the shore.

Apart from illustrating Dahl's principles, Watkin's results are of interest in showing the very real zonation of species within a sandy beach. Such zonation is not confined to crustaceans for it is very obvious in Watkin's data on the polychaetes (Table 7). In both groups, the absence of representatives from the upper 27 m of the beach emphasises the harsh conditions which exist for marine organisms near the high water mark. *Nereis diversicolor* is very characteristic of the region around high tide mark and on more muddy beaches this worm can be very common indeed. It is really an estuarine species and its presence high up the shore can probably be related to the lower salinities obtaining there.

The molluscs of Kames Bay are also zoned. Stephen (1929) found that the dominant lamellibranch species was *Tellina tenuis* which is commonest in the lower regions of the intertidal zone although it extends half way up the beach. *Tellina fabula*, a sub-littoral form, was sometimes found above low watermark while another lamellibranch, *Donax vittatus*, was also concentrated towards the lower parts of the shore.

Other beaches studied by Stephen had an admixture of mud in the upper reaches. This contradicts the theoretical principle that the grade of deposit decreases in a sea-ward direction but presumably, conditions had changed and the beaches were evolving from one type to another. However, the gradation from muddy sand to sand is clearly reflected in the distribution of the fauna. Near low water mark in clear sand, *Tellina tenuis* is dominant but at about half way up the beach it is replaced by *Cardium edule*, the familiar cockle. *Cardium* extends throughout the upper shore but it becomes increasingly scarce towards high tide mark. Here the dominant lamellibranch is *Macoma balthica* which, however, extends well down the shore to overlap the distribution of *Cardium*.

Zonation is equally characteristic of really muddy shores although the wide expanse of mud flats results in an attenuated distribution with very gradual changes. The polychaetes show a very clear zonation. The upper reaches of muddy shores can be very sparsely populated unless the flow from a nearby stream renders conditions somewhat estuarine and encourages the settlement of *Nereis diversicolor*. If the mud has sand in it, *Arenicola* may be found from near high water mark to well down the beach—below low water mark of neap tides but rarely much lower. A characteristic polychaete of the mid-tidal reaches is *Scoloplos armiger* while lower down, *Phyllodoce maculata*, *Nephthys caeca* and *Audouinia tentaculata* are common.

TABLE 6

Downshore distribution of crustaceans at Kames Bay, Millport. Amphipods of the family Haustoriidae are marked with an asterisk. The stations are placed at intervals of 4·5 metres from high water mark of spring tides (1) to low water mark of spring tides (42). No crustaceans were found above station 6. The midshore distribution of *Eurydice pulchra* and the preponderance of haustoriid amphipods in the lower regions are in accordance with Dahl's classification of shore communities (after Watkin 1942)

Station No.	*Bathyporeia pilosa**	*Haustorius arenarius**	*Eurydice pulcha* (Amphipoda)	*Urothoe brevicornis**	*Pontocrates norvegicus* (Amphipoda)	*Bathyporeia pelagica**	*Cumopsis goodsiri* (Cumacea)	*Pseudocuma cercaria* (Cumacea	*Bathyporeia elegans**	*Bathyporeia guilliamsoniana**	*Pontocrates arenarius* (Amphipoda)
6	3										
7	21										
8	29										
9	311										
10	602	1									
11	884	1									
12	1673	21									
13	1584	6	1								
14	109	2									
15	68		35								
16	19		8								
17	13		13	4							
18	8		5	3							
19	3		8	1	3						
20	4		3	13	11						
21	4		10	22	29						
22			16	12	30						
23		1	3	24	71						
24	2		4	14	52	1					
25		1	2	11	78						
26	1		3	20	86	5					
27			3	26	96	4					
28		1	7	32	90	15	7				
29		1	2	24	128	34	4				
30			6	33	141	48	4				
31				22	188	289	9				
32			5	40	221	160	2				
33			1	26	201	299	14	4			
34		6	1	17	105	323	8				
35			1	18	112	293	4	2	8		
36		1		30	84	135	5	7	22	2	
37			1	25	36	115	5		37	3	3
38				25	59	35	2	5	47	3	3
39				31	36	3	3	2	47		2
40				42	24	5	4	17	53	3	37
41				19	29		2	22	120	3	17
42		1		6	15		2	34	103	8	43

TABLE 7

Downshore distribution of polychaetes at Kames Bay, Millport. The stations are the same as those shown in Table 6. Most species are sharply zoned but some, e.g. *Scolelepis ciliata* and *Nepthys caeca* extend over a wide area (after Watkin 1942)

Station number	*Nereis diversicolor*	*Nerine cirratulus*	*Ophelia cluthenis*	*Arenicola marina*	*Scolelepis ciliata*	*Eteone flava*	Species of Capitellidae	*Nephthys caeca*	*Phyllodoce groenlandica*
8	1								
9									
10	4								
11	17								
12	11	2							
13	21	43	1						
14	8	43	13	3	3				
15	1	9	85	1	4	1	2		
16		11	150	2	11		6		
17	1	7	184		12	2	8	1	
18		8	145	2	17	1	6	1	1
19	1	6	101	2	8	4	9		1
20		1	42	3	10	3	2	1	
21		4	16	1	6	3	1	1	1
22		1	6	1	6	1	1		
23				1	22				2
24		1	2		27		2	1	3
25					7	1		1	1
26		1		1	22	2			1
27				1	12			3	4
28					22	1		2	1
29					37				2
30					30			2	2
31					21			1	
32					43			1	2
33		1		1	22	1		1	2
34			1	1	12			5	1
35					20			3	2
36					23			4	1
37					33	1		4	2
38		1			20	1		8	2
39		1			23			5	3
40					51		1	3	2
41					21	1		3	2
42					23			5	2

The molluscan fauna of mud is much the same as that of muddy sand discussed above. *Tellina tenuis* does not persist in pure mud but its place is taken by *Macoma balthica* or perhaps *Scrobicularia plana* although both species live rather higher up on the shore than does *Tellina*. The typical molluscan inhabitant of the lower shore is *Mya arenaria*, the edible soft-shelled clam.

A specific example of zonation on the muddy shore is given by Brady (1943) from a mudflat at Tynemouth descriptively named Black Middens. The distribution of the macrofauna is shown in Fig. 14. *Nereis diversicolor* is seen in its typical position high up on the shore. *Scoloplos armiger* is abundant throughout the entire tidal range although at its maximum lower down the

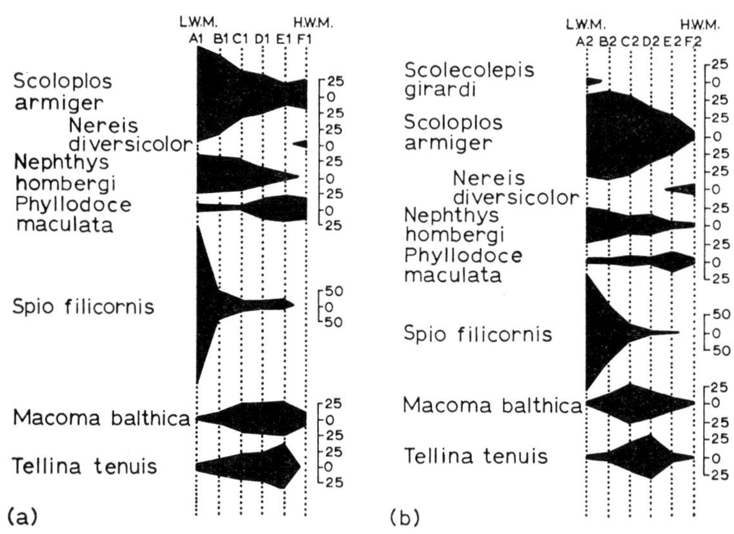

Fig. 14 Distribution and relative density of the fauna of the Black Middens: (a) distribution in March 1931; (b) distribution in April 1931. The two distributions do not reflect seasonal changes as they are from different parts of the shore. The scale on the right shows the number of animals (after Brady 1943)

shore. *Nephthys hombergi* and *Phyllodoce maculata* appear to be complementary, for when one increases the other declines, but in general *Phyllodoce* occurs further up the shore. *Spio filicornis* is the dominant annelid, particularly on the lower shore. Higher up, it is replaced by *Macoma balthica* and *Tellina tenuis* which tend to replace each other. The shore was obviously sufficiently sandy to support *Tellina* but not *Arenicola*, which cannot exist in too muddy a habitat.

Brady also studied a shore, intermediate between mud and sand, at Budle Bay on the Northumberland coast. Although a down-shore transect was taken, it ran from near a fresh water stream to high water water mark at a sandy spit which encloses the bay and did not, therefore, represent a typical down-shore gradation. However, the distribution of the fauna (Fig. 15) again showed *Scoloplos armiger* at its greatest abundance at the lowest station which was, however, affected by the stream. *Macoma balthica* and *Tellina tenuis* were most abundant at the same station at about mid-tide level. Usually each species tends to replace the other, but in Budle Bay the distribution of *Macoma* (and *Tellina*) appears to be related to that of *Scoloplos* with their numbers varying inversely. Both the molluscs and the polychaete have similar ecological requirements and feed on the same food—detritus on

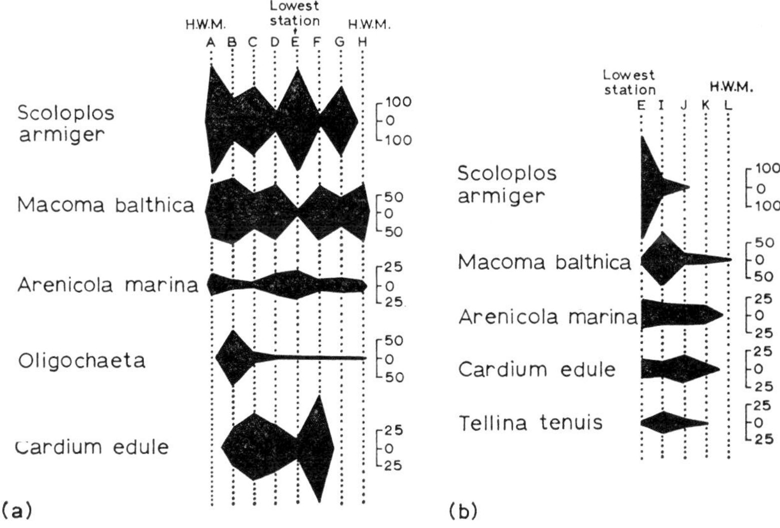

Fig. 15 Distribution and relative density of the fauna of Budle Bay along two transects, one passing straight across the centre of the bay (a) and the other from the lowest station of the first transect to the sandy spit at the mouth of the bay (b), June 1931 (after Brady 1943)

the mud surface—which *Scoloplos* takes up by means of its extrusible gullet and *Macoma* through its siphons. *Macoma* can also act as a filter feeder and hence has the edge over *Scoloplos* under conditions where they might compete. *Macoma*, however, needs firmer sand than *Scoloplos* which in turn prefers a lower salinity and a higher organic content. The preference of *Arenicola marina* for muddy sand is shown by its abundance at Budle Bay compared with its absence from the mud at Tynemouth and its rarity at Cullercoats, an area of clean sand also studied by Brady.

Brady's data exemplify the well-established principle found on all types of depositing shore that the number of species declines as the shore is ascended

(Table 8). The reasons for this distribution are complex and will be considered more fully in the next chapter. However, as conditions become less marine towards the upper shore, they must necessarily become more difficult for animals which are primarily marine. The number of species able to contend with these harsher conditions is, not surprisingly, few but those which can do so may be very abundant. This is yet another of many examples in ecology of

TABLE 8

The number of species recorded at various levels on intertidal mud and sand on the Northumberland coast. The decline in the number of species from low to upper shore is most marked (after Brady 1943)

Cullercoats (clean sand)		Budle Bay (muddy sand)		Black Middens (mud)	
Station	Average No. of species	Station	No. of species	Station	Average No. of species
A (low shore)	3·5	D (low shore)	11	A (low shore)	10
B	5	C	9	B	9·5
C	5·5	B	7	C	9
D	5	A (upper shore)	3	D	9
E	4	E (low shore)	10	E	8·5
F	4·5	F	7	F (upper shore)	7
G	4	G	4		
H	1	H (upper shore)	3		
I (upper shore)	0	E (low shore)	10		
		I	7		
		J	7		
		K	6		
		L (upper shore)	1		

exacting conditions supporting a few species but many individuals. Thus, at Cullercoats, the density was the same throughout the transect at 90 individuals of all species per square metre.

On the whole, the distribution found by Brady agrees with the general remarks made earlier, although there are some discrepancies, e.g. *Phyllodoce maculata* is more common in the upper than in the lower regions of Black Middens. Such differences are inevitable for individual examples will rarely conform in every particular to general principles.

Although less attention has been paid to zonation on muddy shores compared with other types, it is clear from this brief summary that zonation is as real a phenomenon here as anywhere else. It is, however, slightly different in kind for whereas down-shore zonation on rock or sand tends to reflect a change from marine to terrestrial conditions, the gradation on mud flats is rather from marine to estuarine. We have seen that on both rocky and sandy shores, three broad yet distinct regions can be distinguished viz. a sub-terrestrial, a mid-littoral and a sub-littoral but a similar classification of muddy shore does not seem to have been made. However, three distinct zones are equally apparent on mud flats. The sub-terrestrial zone can be defined rather negatively by the absence of organisms or, if the conditions are right, by the presence of the estuarine rag worm *Nereis diversicolor*. The mid-tidal zone is characterised by a number of molluscs principally *Scrobicularia plana*, *Cardium edule* and *Macoma balthica*. In sandy mud *Arenicola marina* is a good indicator of this region. The sub-littoral zone can be defined by the presence of *Mya arenaria* with, again on sandy mud, *Tellina fabula*. A number of polychaetes are found in this region which might almost be designated the 'polychaete zone' to distinguish it from the 'lamellibranch zone' of the mid shore. The muddy shores of the world would well repay a comparative study along the lines of Dahl's analysis of sandy shores to see how far these suggestions are parochial or whether they are of general application.

So far we have considered only the macrofauna, chiefly because this is the group of animals upon which most work on zonation has been conducted. There are, however, some data which show that zonation is equally characteristic of the meiofauna or interstitial fauna. Barnett (1958) studied the harpacticoid copepods of a mud flat in Southampton Water and found a distinct zonation. He took regular quantitative core samples from three equally spaced sites in the intertidal zone. Two species of *Platychelipus* were found in the mud, with *P. littoralis* being most abundant in the upper levels and the other, *P. laophontoides*, in the lower. Both species were found at the mid-tidal station in roughly equal numbers. Other species of harpacticoid copepod were most numerous in certain regions of the shore, e.g. *Stenhelia palustris* was most frequently found high up on the shore while *Harpacticus flexus* and *Canuella furcigera* had their centres of abundance in the lower levels.

Boaden (1963) lists over 100 species of interstitial animals from mostly sandy beaches in North Wales and indicates the tidal levels at which each occurs. Although precise quantitative data are not given, it is clear that many of the species, which belong to all taxonomic groups, have preferred regions of the shore and a quantitative survey would surely reveal zonation.

We must not confine ourselves solely to the variation in numbers of an animal in relation to the shore level because resort to a ruler shows that other differences occur. If individuals of the same species are measured at various levels on the shore, it may be found that there is a significant gradation in size. In some species, the largest specimens are high up the shore while in others they occur lower down. In *Tellina tenuis*, for example, Stephen (1928) found the biggest specimens in the uppermost regions of its range and the smallest near the low watermark. The reverse distribution is found with *Cardium edule* whose largest specimens are found low down on the shore (Stephen, 1932). The reasons for these differences have not been satisfactorily explained. One possibility is that the largest specimens migrate to their preferred area although the reason why it should be so attractive to them remains to be explained and there is, in fact, no evidence for a migration of this nature.

A more likely explanation is that the growth rate is better at some levels than at others. There is good evidence that this is the case because the size of the same age groups increase or decreases, as the case may be, with the changing levels of the shore. Many lamellibranch molluscs lend themselves readily to analysis of this nature for their ages can be assessed by counting the rings on their shells which are similar in principle to those found in cross-sections of tree trunks, The ring itself is laid down in the winter when growth is retarded and the space between the rings represents the more rapid summer growth.

The most obvious factor causing a rapid growth rate would seem to be food but why should one part of the shore favour the growth rate of *Tellina* and another part that of *Cardium*? The answer may lie in the different feeding methods adopted by these organisms for *Tellina* is a deposit feeder while *Cardium* is a filter feeder. A filter feeder can take in food only while covered with water and an animal further down the beach would be submerged by the tide for a longer period than one higher up and might well benefit from the extra food available and grow more quickly. The deposit feeding *Tellina* is less dependent on the tide and can feed when the tide is out but there is little evidence that the supply of food is greater around mid-tide level where the largest specimens are found than it is around the low water mark. It may be that one should look to another factor for an explanation and population density seems to be a significant one. The shore zone occupied by a species is not uniformly inhabited and in the case of *Tellina*, the densely packed population around low tide mark peters out in the middle regions of the shore. With fewer animals to exploit the available food supply, the mid-tide animals should be better fed and hence grow more quickly. There is always the risk of a fallacy in this type of argument as we are assuming that no other species are present to compete with the well-fed *Tellina*. However, it is certainly true that the growth of *Tellina* varies inversely with the population density (Stephen 1930) but which is the cause and which the effect remains a moot point.

The possible causes of zonation are discussed in Chapter 6 but it may help to realise that zonation is not entirely an intertidal phenomenon. Thus Clutter (1967) has revealed a pattern of zonation in nine species of sublittoral mysids, which are primitive shrimp-like crustaceans. This observation is of particular interest in that five of the species are pelagic, swimming freely in the water, and could quite easily move from one zone to another. We will return to these mysids when we consider the causes and functions of zonation later on.

We have said very little about the 'horizontal' or along-shore distribution of shore animals. One of the reasons for this is that not much information is available in the literature, for the preoccupation with down-shore zonation has led to neglect of the techniques of spatial sampling which is one of the notable tools of terrestrial ecology. It can safely be said, however, that organisms are not uniformly distributed along the beach at a particular tidal level but more work needs to be done before one can draw any conclusions.

The nature of the substratum is undoubtedly one of the most significant factors determining the distribution of a species. Beaches are rarely uniform in texture and changes are very often obvious from the most casual inspection. A change in lateral distribution may well be a reflection of increasing estuarine conditions and we will consider this special case in a later chapter.

Before concluding this consideration of the spatial distribution of the shore fauna, some mention should be made of the vertical distribution. On the shore, this type of distribution is possible only in mud and sand apart from rocks which are attacked by borers. It is, however, necessary to distinguish between those animals in the mud which remain in contact with the surface and those which live within the mud itself. The distinction here is really between the macrofauna and the meio- and microfaunas. Although the body of the gaper clam *Mya arenaria* may be 30 cm or more below the surface of the mud, it is really a surface animal for its long trunk-like siphons connect it to the surface and it respires surface water and feeds on surface food. It would certainly be unable to survive if wholly dependent on the anaerobic, sulphide affected environment in which the bulk of its body is found. This is true of all macrofauna found in mud. The sulphide layer is much lower in sand and hence the burrowing animals there are usually living physiologically in the habitat which surrounds them. This is particularly true of the smaller crustaceans but some of the infauna of sand belong ecologically to the surface environment, e.g. *Corystes*, the masked crab, remains in contact with the surface through its enormously elongated tube-like antennae, although the body of the crab itself is buried several inches in the sand. Most of the molluscs which burrow in sand also remain in contact with the surface water.

The situation is different with the meiofauna or microfauna. These are indeed intimately part of the environment in which they live. They also exhibit vertical zonation for if a core sample taken from the beach is sectioned and the animals contained in each are extracted, it will be found that most of them are in the top one or two segments. This is particularly true of soft mud in which the meiofauna exists only in the upper four or five millimetres. Barnett (1968) reports that 95% of the harpacticoid copepods at Hamble Spit, Southampton Water, were within 5 mm. of the surface. Very few, if any, interstitial animals occur below 0·75 cm in pure mud. The reasons for this can be found in the oxygen content of the interstitial water which falls off very rapidly below the surface of mud. Barnett found that the copepods were so reluctant to enter low oxygen conditions that they would not move downward into the mud even when subjected to reduced salinities at the surface. In muddy sand or in clean sand, life extends to a greater depth but the sulphide layer forms an impenetrable ecological barrier to further downward movement. It has yet to be established whether or not specific differences in vertical zonation occur but it remains a distinct possibility, particularly in sandy beaches.

Swedmark (1964) gives a diagram, reproduced here as Figure 16, of the vertical distribution of the meiofauna of a sandy beach in the Bahamas. The data, which were collected by a French worker, J. Renaud-Debyser, show that the animals are well down at depths between 50 and 70 cm during the winter (January) but that they come much nearer to the surface in the summer. Such seasonal movements are probably a reflection of the differing temperature conditions.

This discussion may have given the impression that the position of an animal on the beach is immutable but it is not necessarily so and in fact, some organisms move considerable distances up and down the shore every day.

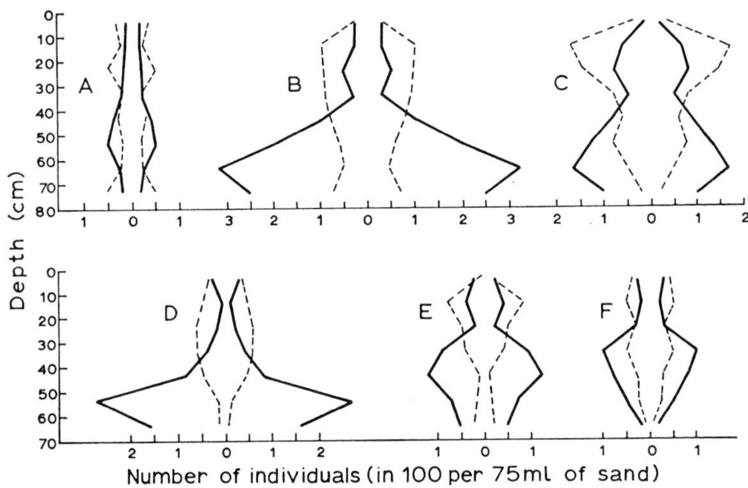

Fig. 16 Vertical distribution of the meiofauna of a sandy beach in Archachon, Bahamas. Note that the fauna moves downwards in winter. A—high water mark, B—E—intermediate positions, F—low water mark. Solid line—winter (January 1955), dotted line—summer (August 1955) (from Swedmark 1964 after Renaud—Debyser 1963)

This is true of the well known *Hydrobia ulvae* which we have met several times already. This is a creature capable of a speed of perhaps a few metres per hour yet it is one which undertakes a journey of a hundred metres or more twice a day. The tidal movements provide an explanation of this seeming paradox. When seen on the shore at low tide, the *Hydrobia* are most probably feeding which they do while crawling over the surface and their tracks in the mud are very evident. They then burrow into the mud but reappear when the tide comes in and float on the water surface by means of a mucous raft which also acts as a food web or net. Thus, the animals are feeding during this part of the tidal cycle as well as when they are on the mud and perhaps it is this ability to obtain food from two quite different environments that has made these molluscs so successful and abundant. When the tide recedes, the snails are carried back down the beach and are stranded in regions where there is a concavity in the slope of the shore. Newell (1962) has shown that this point coincides with the upper limit of the muddy-sand zone and consequently, the animals are returned to much the same place as that from which they started.

This is only one of a number of instances of animals making use of the tide to move up and down the beach. Such a facility is rarely employed by the inhabitants of rocky shores but the reason is probably a nutritive one. Food is perhaps more abundant and certainly more readily available on the rocky shore than it is in mud or sand. Very many rocky shore animals are suspension feeders whose food is brought to them in the water thus eliminating the need to move. Those animals which have to forage for food are either relatively large and able to move quickly or are surrounded by a superabundance of algal food. Hence, there is little need for rocky shore animals to wander far from their zone. Conditions are rather different on the depositing shore. The scarcity of food requires the inhabitants to forage widely, but their small size and the difficult substratum, which is not the easiest to move over, create real problems for a vagrant species. Hence the powerful selective pressure towards adopting the tide as an aid to locomotion. The practice is not confined to organisms the size of *Hydrobia*; it is also found in such large and sophisticated animals as the shelduck (*Tadorna tadorna*). The family to which the wildfowl belong, Anatidae, is peculiar in that it is one of the few groups of birds whose members suffer a simultaneous loss of flight feathers at the annual moult and as a result, become flightless. Shelduck frequent mud flats where they prey heavily on *Hydrobia*. The birds feed on the ebb tide which they follow down to the regions where the *Hydrobia* are stranded. In order to repeat the feeding cycle they must transport themselves back to the high tide mark. This can normally be accomplished by a few flaps of the wings but it would seem that a flightless bird would either have to walk or swim. The shelduck does neither but merely floats in with the incoming tide.

Corophium volutator is another species which makes use of the tide and, in this case, something is known of the mechanism by which it co-ordinates its activities with the various phases of the tidal cycle. Morgan (1965) showed that *Corophium* reacted to reduced hydrostatic pressure by swimming (Fig. 17). (It has been discovered only relatively recently that marine invertebrates are sensitive to hydrostatic pressure although how this can be so in the absence of a gas filled sensor is not at all clear—see however, Digby 1967.) *Corophium* reacts to the incoming tide, which causes the hydrostatic pressure to increase, by leaving its burrow although it does not swim. Even so, the animals tend to be carried shorewards and, if they did nothing more about it, they would be deposited further up the beach than their usual haunts. However, as the tide begins to ebb the pressure falls off and because of the reaction mentioned above, the *Corophium* begin to swim. In this way they are carried back with the tide and are deposited at the level from which they started. A similar reaction is displayed by *Nymphon*, one of the sea spiders, which swims most when the pressure decreases as the tide ebbs (Morgan, Nelson-Smith and Knight-Jones 1964). A more complicated behaviour pattern is seen in a group of mysids investigated by Rice (1961). The species investigated, *Schistomysis spiritus*, *Siriella armata*, *Praunus flexuosa* and *Praunus neglectus*, show a complex reaction to light and pressure. When the hydrostatic pressure is increased, the mysids swim towards the light or, under conditions of darkness, upwards. In technical language, they become positively phototactic and negatively geotactic when the pressure is increased. When the pressure is decreased, the opposite reactions occur and the animals swim away from the

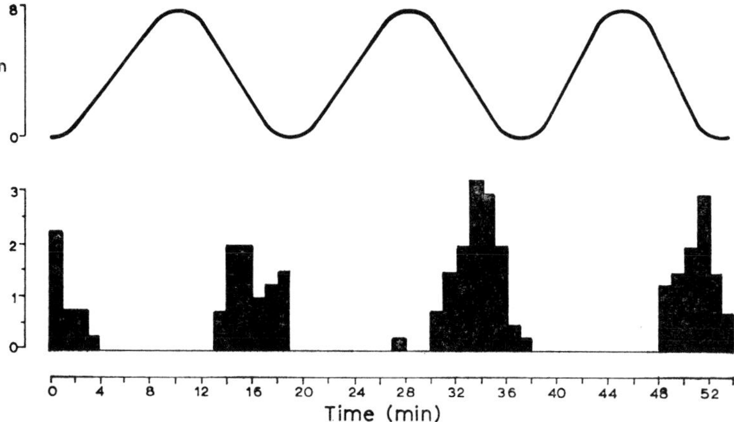

Fig. 17 Effect of hydrostatic pressure changes on *Corophium* from the intertidal zone. The amphipods were subjected to pressure changes (solid line) equivalent to a change in tidal level of 8 metres and with a period of about 18 minutes. The histograms show the mean number (out of 20) of *Corophium* swimming above half depth at one minute intervals. Note that the *Corophium* tend to rise, i.e. swim, as the pressure falls (after Morgan 1965)

light or downwards if it is dark. If these laboratory results are translated into field reactions, we can see that when the tide comes in, the animals will rise to the surface, through their light reactions by day or their gravity reaction at night, and will be carried up the shore. They are returned to their original position on the shore when the tide recedes. Presumably, the reactions to a falling pressure ensure that they do not rise into the water when the tide is receding as they would then be swept out to sea.

Not all shore animals make such epic voyages up and down the beach but few stay in one spot all their lives. Many need to move in order to graze as on rocky shores where the foraging activities of the limpet coupled with its remarkable homing ability are well known. Similar movements occur on mud or sand. Although the winkle, *Littorina*, is considered a rocky shore animal, it is also abundant on muddy sand and much of our knowledge of the orientation of winkles comes from studies of sand-dwelling populations. Newell (1958 a, b) studied *Littorina littorea* in some detail and was able to show by marking them with a spot of paint that winkles, although mobile, tend to remain in approximately the same position for many weeks and that they use the Sun to orientate themselves. He found that as the tide recedes, winkles tend to move either towards or away from the Sun as they forage but that after a time, each animal reverses its direction of travel, a reaction which brings it back to about the place from which it started. Whether they move towards or away from the source of light appears to depend on the history of the individual, dark adapted winkles being positively phototactic. On a cloudy day, the winkles continue to show this orientation and it is likely that they react to the brightest point in the sky; there was no evidence that they orientate to the plane of polarized light in the sky, as bees appear to do. For most of the time, the winkles remain stationary whether submerged or exposed to the air and the stimulus which activates them is probably the increased wave activity as the tide rises or falls. The time spent in feeding, therefore, is relatively short.

The winkle is capable of more complex movements than those described above as it can also react to gravity. On vertical surfaces, such as groynes or other man-made structures with a vertical face, the winkle maintains its U-shaped feeding path but the reaction is principally to gravity, reinforced by a positive phototaxis. Under such conditions, the animal first crawls downwards, then horizontally then finally upwards before settling head uppermost just above the waterline. On rocky shores, the reactions are more complicated as the animal shows reversal of taxes when it is inverted and this ensures that it is not ethologically trapped when it moves into a horizontal crevice. This aspect is dealt with in some detail by Newell (1958 b). It is of interest to note that winkles taken from a flat shore do not react to gravity when experimentally placed on a vertical surface until they have been exposed to the new conditions for some time which can be as much as ten days. On most 'natural' depositing shores, the animals would be light reactors only, because they are unaffected by gravitational stimuli below a slope of 10–20 degrees which is much steeper than that found on muddy or sandy shores.

Decapod crustaceans show similar movements up and down the shore. Efford (1965) in a study of aggregation in *Emerita*, a sand crab from the Californian coast, found that the aggregations moved up and down the beach in phase with the tidal cycle (Fig. 18). The aggregations move up with the rising tide but do not follow the retreating tide all the way. Instead, they burrow into the sand just before the water reaches its lowest ebb and remain there until disturbed by the rising tide at the next cycle. These reactions are presumably geared to their feeding behaviour which takes place only at the water's edge. The tropical ghost crabs (*Ocypode*) show similar movements which are correlated with the tidal cycle (Barras 1963).

Nearer home, Edwards (1958) studied the individual movements of the shore crab (*Carcinus maenas*) in Southampton Water. This species show a seasonal zonation as it migrates from the intertidal region during the winter to deep water offshore. In the summer, they are found throughout the littoral zone at all levels of the beach but it is thought that the individual crabs maintain more or less the same situation on the shore and do not, as was once supposed, wander far from their chosen position. Edwards, however, found from

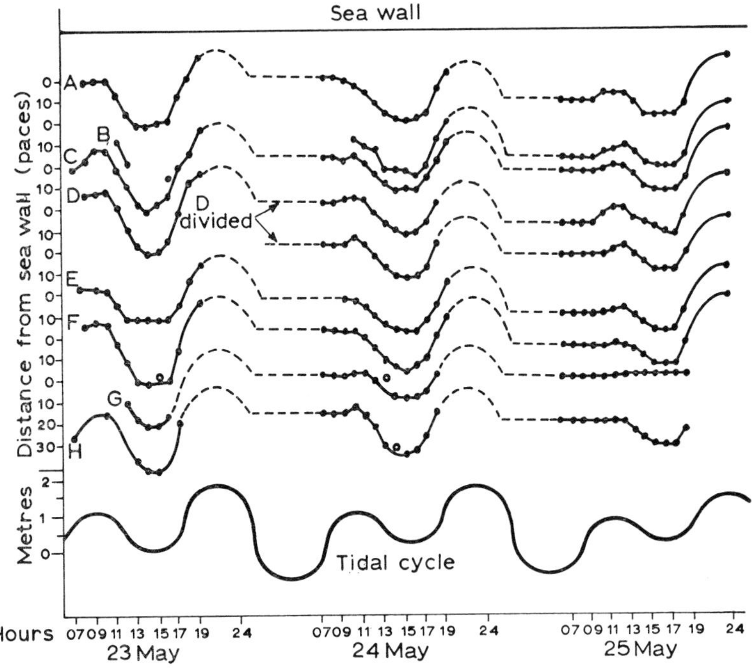

Fig. 18 Movement of eight aggregations (A—H) of the sand crab, *Emerita analoga*. Note how the crabs move up and down the shore in phase with the tidal cycle (lower line). Solid lines—observed movements, dotted lines—projected movements based on knowledge of subsequent conditions (after Efford 1965)

marking experiments that there was considerable interchange between the littoral and sub-littoral zones involving the majority of the crabs although some remained at about the same level throughout.

Littorina is an obviously mobile animal which can be seen to move about at fair speed but the lamellibranch molluscs are usually considered to be sedentary. It is, therefore, of great interest that some have been shown to display regular movements very similar to those demonstrated in *Littorina*. *Macoma balthica*, is an example of such a lamellibranch. *Macoma* is principally a non-selective deposit feeder and is soon able, with its inhalant siphons,

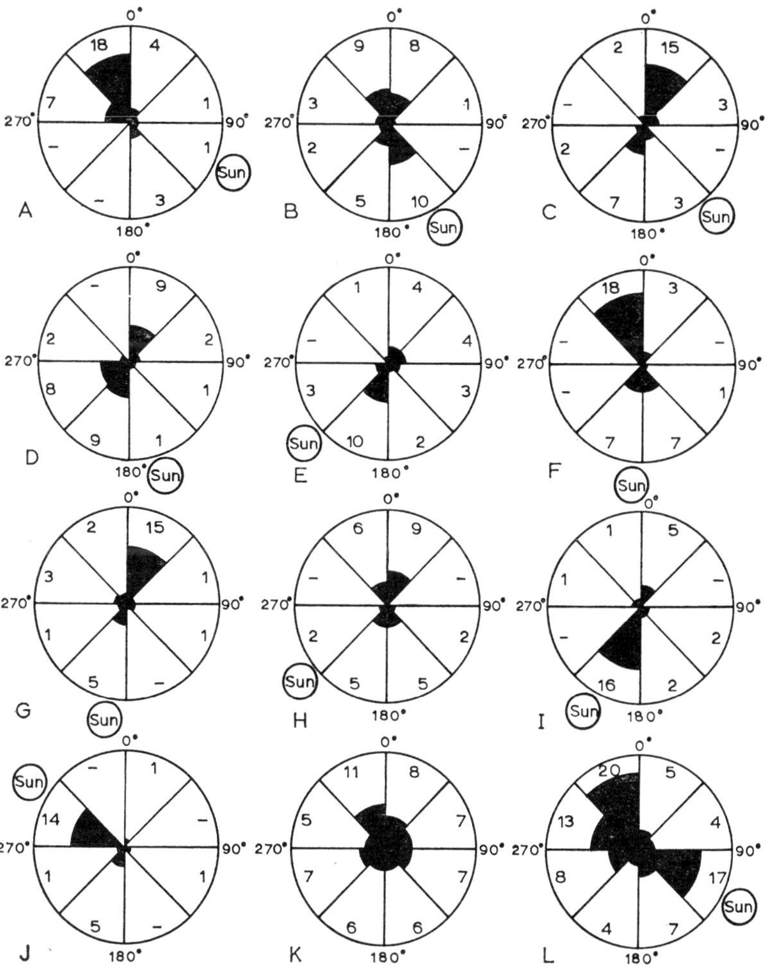

Fig. 19 The compass directions of tracks made by *Macoma balthica*. The position of the Sun is shown except in K which was recorded at a time of heavy cloud cover with non-directional light. Figures indicate the number of tracks lying in each octant. Note that the tracks are strongly orientated whenever the Sun was visible and that for the most part, the tracks point towards is away from the Sun (after Brafield and Newell 1961)

to clear a circular area which may be as much as 4 cm across. It would obviously be of advantage to this animal if it were able to move to other areas in order to trap further food supplies. Brafield & Newell (1961) investigated this possibility and found that a shore densely populated by *Macoma* is covered with a network of tracks obviously produced by the molluscs although it was only rarely that an individual was seen in the act of moving. The mechanism of movement is as follows: the foot is extended anteriorly at regular intervals and pushes the animal through the sand while it is lying partly on its side. The resulting disturbance traces a furrow about 5 mm wide. One specimen was observed to extend its foot at 15 second intervals and to travel 5 cm in four minutes. It was found that many of the furrows were straight although some were curved or even circular. The compass directions of the straight tracks were measured and the results grouped in octants and plotted as polar diagrams. The results, which are reproduced in Fig. 19, show a strong tendency for the tracks to be oriented towards or away from the Sun. On one occasion when the sky was obscured by heavy cloud, the tracks were randomly distributed showing that the Sun is necessary for orientation. Some of the examples given, e.g. C and D may not be too impressive but it is possible that the tracks were formed some time before and, therefore, under a different Sun regime than at the time when the observations were made. What is quite clear is that the tracks are strongly orientated and, as the authors point out, there are no other environmental factors which could so readily be used as clues in orientation. The mechanism offered as an explanation of these movements is that the animal becomes photopositive or photonegative for a time and then reverses the sign of its response and moves in the opposite direction. The fact that so many straight lines were recorded can be explained on the basis that the animal either exactly retraced its 'steps' or that it had not completed its out and back journey at the time of the observation. Alternatively the first formed tracks may have been obliterated by the incoming tide before the investigator arrived on the scene. However, some U-shaped and circular (closed loops) tracks were recorded and indicate the possibility of such movements.

There is plenty of evidence to show that the type of out and back movement described above is not an aberrant behaviour pattern confined to one or a few species, but that it is widely distributed throughout the shore fauna and it is probable that it is a universal behaviour mechanism which ensures that zonation is maintained.

CHAPTER FIVE
SEASONS ON THE SHORE

The seasonal changes on land are very obvious and we are accustomed to seeing the trees lose their leaves in the autumn and to the appearance of lambs in the spring, but the sea presents a constant appearance throughout the year. Nevertheless, considerable changes are taking place. The plankton, which may clog a net in May, is sparse or non-existent in November, while fish eggs or the larvae of sea creatures are not to be found at all times of the year. Such seasonal variations are also a feature of the shore. This should not be surprising for the annual changes in environmental factors are themselves very marked, with perhaps light and temperature being the most significant. Seasonal changes in the fauna may be manifest in a number of ways. In some cases, a species may be present only at one time of the year, or it may be present all the year round but vary in numbers or alter in weight, size or some other parameter. Other aspects of the biology which can be seasonal include the reproduction and dispersal behaviour. Very often, the seasonal cycle we observe is really a reflection of the life cycle of the animal concerned, for the life span of many of the shore fauna probably does not exceed a twelve month period and in the case of interstitial fauna, may be very much less. However, it is difficult to generalise on these matters for little attention has been paid to seasonal changes in depositing shores and very little published data are available. There is no reason to believe that seasonal variation in the inhabitants of sand and mud is radically different from that of the rocky shore fauna and one might expect to find similar principles involved. Perhaps the most important difference between the two habitats lies in the degree of protection afforded to the fauna. As we have seen, the animals in mud and sand are usually buried under the surface for at least some of the time and can escape the worst effects of desiccation or temperature fluctuations, but the rocky shore fauna has for, the most part, to endure environmental changes. The precise degree to which the surface changes are cushioned by the layer of mud or sand is a subject that would repay further study. We might, therefore, expect seasonal changes in the fauna of depositing shores to be less abrupt than those on eroding shores and it is possible that some phenomena which are seasonal in other habitats, are continuous in sand or mud.

A convenient way of evaluating seasonal changes on the shore is to examine the life cycle of a hypothetical organism whose life span is one year and to see how actual animals fit into the pattern. The life cycle of the individual obviously begins at fertilisation and this predicates the existence of an earlier sexual reproduction. If breeding in our hypothetical animal is seasonal, all subsequent events in the life cycle will necessarily be seasonal as well. The fertilised egg hatches—again a seasonal phenomenon—and, if our model has a planktonic larvae, a seasonal production of young forms takes place in the overlying water. After further development in the plankton, the larvae settle on the mud or sand in a seasonal cloud. They then metamorphose and proceed to grow, taking several months to reach maturity. Their size at any particular moment will depend upon the time of the year i.e. the season. Thus, we find in our hypothetical animal a number of seasonal phenomena: first there is the breeding season, then the hatching season followed by the settling season, the metamorphosis season and the growing season. However, there is really only the one primary seasonal phenomenon and that is reproduction. Hence, if we wish to investigate the factors which determine the seasonal processes on the shore it is usually the breeding season to which we should turn. The term 'usually' is used advisedly for it is theoretically possible that reproduction is not the seasonal pace-maker. All marine animals have a dispersal phase in the life cycle. In sedentary species, dispersal must inevitably be effected by the young forms, usually larvae but possibly eggs. Dispersal in such cases is dependent upon reproduction and cannot be the primary seasonal phenomenon. On the other hand, the dispersal phase can be centred in adults and if reproduction follows and is dependent upon dispersal, then dispersion and not reproduction, is the primary seasonal phenomenon. A well authenticated example of such a species is the common gribble, a marine wood-boring crustacean, which moves to fresh wood in a 'migratory' season which is temperature controlled. Reproduction follows dispersal although it cannot be dependent upon it for non-migratory animals show a breeding season. In this case, we find two basic seasons with dispersal being independent of reproduction. There is probably an example of such a species in the shore fauna although, as far as I know, one has not yet been found.

It can be accepted that for the majority of the shore fauna at least, the onset of the reproductive season controls the timing of the remaining seasonal processes and, therefore, the factors which control reproduction are responsible for seasons on the shore. Before proceeding further with a consideration of such factors it is as well to establish that seasons do, in fact, occur on the shore. This is not easy to do for published data based on soil samples are sketchy in the extreme. Most evidence comes indirectly from the seasonal abundance of larval forms in the plankton. The technique of placing test panels in the sea to collect the settled organisms has proved invaluable in studying the seasonal settlement of rocky shore and ship-fouling organisms but the method is not available to the student of the depositing shore as his organisms will not remain on such uninviting material. There is, however, abundant evidence that reproduction in marine soil animals is seasonal. This is well known to fishermen connected with the shellfish industry whose livelihood may depend upon the size of the 'spat fall' in a particular year. Thaι this is not constant is well illustrated by Stephen's observations (1932) on the number of spat from *Tellina tenuis* in Kames Bay, Scotland. The density varied from 2620 per square metre in 1926 to 232 in 1929. The existence of a seasonal spat fall in commercially valuable species points to seasonal reproduction in the shore fauna as a whole.

It is perhaps time one turned from theoretical conjecture to a consideration of an actual investigation on the shore. One of the few detailed studies of a muddy shore organism is that by George (1964a) who investigated the life

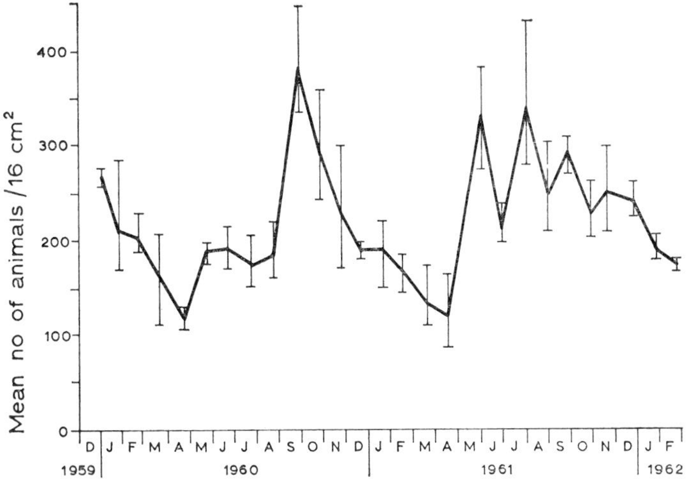

Fig. 20 Seasonal fluctuations in the total population of *Cirriformia tentaculata*. The vertical lines show the extreme values around the mean (from George 1964a).

history of a cirratulid worm, *Cirriformia tentaculata*, in Hamble Spit, an intertidal mud flat in Southampton Water. *C. tentaculata*, perhaps better known by its alternative generic name of *Audounia* is one of the commonest sedentary polychaetes in Great Britain. It occurs in a variety of habitats; in gravel and sand as well as in finer deposits, provided there is a considerable admixture of mud. It often forms a slimy tube under the stones and may even be found on rocky shores. It occurs high up the beach at Southampton but in other areas tends to be found lower down. In all, it is a most adaptable animal.

George took core samples at monthly intervals for a little over two years and recorded the following data: (a) the number of animals present in the core, (b) the length of the worms, (c) the presence or absence of developing gametes in the coelom and the shape and size of the oocytes. It soon became clear that numbers were not constant throughout the year for there were sharp increases in number during the summer months (Fig. 20). This is strong *prima facie* evidence of seasonal reproduction with the extra worms being recruits from a breeding season. The decline during the autumn and winter presumable represents a mortality factor. Analysis of the reproductive data confirmed that a breeding season occurs for, as Fig. 21 shows, the percentage of

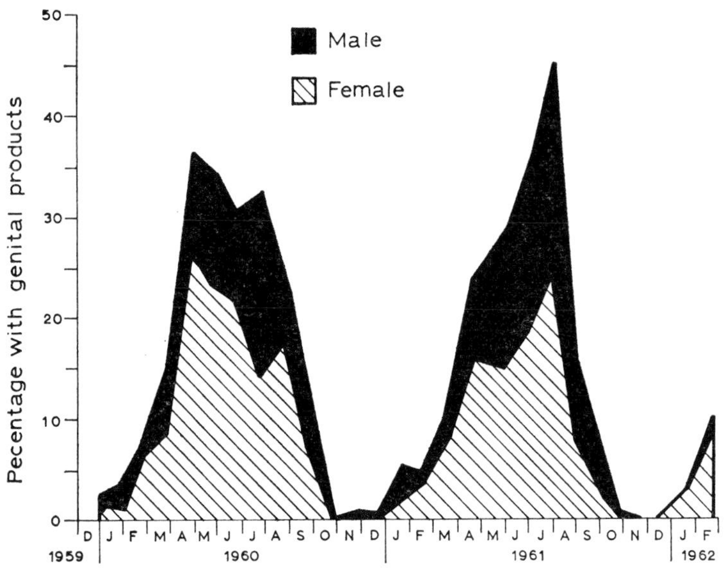

Fig. 21 Season variation in the percentage of sexually mature *Cirriformia tentaculata* containing genital products (from George 1964a)

sexually mature animals also increases during the summer months. From these observations, it seems that the period of maximum reproduction takes place between March and September with a cessation of breeding activity in November and December. Confirmatory evidence comes from an analysis of the size of the oocytes in the females. These were smallest in January but by April, all sizes of oocytes could be found. Fluctuations in the large oocytes suggests that bursts of spawning occurred, a supposition supported by a similar series of sudden increases in the number of small sized worms.

Such a life cycle is probably typical of most of the inhabitants of the shore. We see a definite breeding season but all the reproduction does not occur simultaneously. Rather, reproduction is a continuous process with eggs and young at all stages of development being present within the breeding season.

The evidence for spawning bursts, however, suggests that most of the females breed at the beginning of the season and at regular intervals throughout the breeding period. Because reproduction is seasonal, the fluctuations in the total population is seasonal as is the appearance of young forms. Thus, *Cirriformia* displays many of the characteristics of our hypothetical animal, but what of the dispersal phase? George looked for planktonic larvae in the overlying sea water but, curiously, was unable to find any at the time when they might be expected to be present. A little later, the reason became clear (George 1967). *Cirriformia* proved to be polymorphic in a cryptic way. In some areas the worm was found to possess both free swimming planktonic larvae and bottom living demersal forms. This is probably the normal condition but in Southampton, all the larvae are demersal. The point to make, of course, is that the larvae, whether planktonic or demersal, make a seasonal appearance which is dependent on the underlying rhythm of reproduction.

Such a species may be taken as typical of many mud and sand dwelling animals as far as the seasonal pattern is concerned. *Cirriformia* is a summer breeder but many other species breed at other times of the year.

In general, marine animals from temperate zones can be divided into those which are at the southern limits of their range, those which are at the northern limit with a third category of those living in about the middle of the range. In the northern hemisphere, those animals of the first category will tend to breed during the winter when the weather is coldest while the warm water species will reproduce in the summer. Intermediate species are most likely to be found breeding in the spring although the breeding season may extend over a much longer period. These differences in species may be noticeable only in respect of their reproduction for there may be no other differences in their reactions to temperature. It is usual to classify animals on the basis of their tolerance to temperature into two broad classes; first, the eurythermal forms which can tolerate a wide range of temperatures and second, the more numerous stenotherms—those with a narrow temperature tolerance. The latter can be separated into those which are adapted to cold and those which prefer warm conditions. Thus there are really three classes; the eurythermal, cold-stenothermal or oligothermal and warm-stenothermal or polythermal. However, the matter is not quite so simple. It has been found in a large number of marine species that the organism may be eurythermal in respect of its general activities but be stenothermal as far as its reproductive habits are concerned. This situation is so common that the concept of reproductive stenothermy has gained wide acceptance. Reproductive stenothermy is most marked in young forms particularly embryos in early cleavage stages which are much less tolerant of temperature fluctuations than are larval forms.

It would appear from the above comments that temperature is the most significant factor controlling the onset of reproduction provided that other environmental factors lie within tolerable limits, and, therefore, temperature is responsible for all the seasonal phenomena exhibited by shore animals.

There are few environmental factors apart from temperature, that exhibit a seasonal variation. Day length, for example, is a seasonal variant but it is difficult to see its significance to animals buried in sand or mud. It is inconceivable that marine organisms are sensitive to tidal forces, which vary annually as well as monthly, or that the animals have an inherent rhythm independent of external clues. There is adequate evidence of inherent rhythms in shore animals but usually such 'clocks' need to be 'set' by environmental stimuli. Very often the stimulus is the tidal cycle but the factor which governs the seasonal activities of most shore animals is undoubtedly the annual fluctuation in the environmental temperature. It appears as if each species has a particular temperature value which triggers off the reproductive processes when it is reached, either as the temperature is rising in the spring or when it is declining in the autumn. This is saying, in so many words, that the species starts to breed when the temperature conditions are most suitable. The period of reproductive stenothermy does not necessarily coincide with the production of gametes or young for it may occur during the earlier development of the gonads.

Recent studies of the common lugworm, *Arenicola marina,* have shed some light on these problems. *Arenicola* is probably a boreal, or northern species, which might, therefore, be expected to breed during the colder months or at least not during the summer. In fact, the breeding season in Europe is usually during the autumn although there have been reports of spring spawning. Howie (1959) suggests that the difference in the breeding seasons can be related to the position which the population occupies on the shore. These have been termed laminarian if they occur below the level of low water at neap tides and littoral if they are found above. In the case of the autumn breeders of *A. marina,* Howie found that the spawning period at St Andrews coincided with a number of marked environmental changes. Of significance in the present discussion, was the fact that the beginning of spawning took place at the time of a sharp fall in the minimum air temperature; usually during the first really cold snap of the year. There were, however, other significant correlations. The breeding season commenced towards the end of a period of neap tides while peak spawning occurred during the following full moon spring tides (Fig. 22). Other weather factors appear to be of no significance. This situation was unusual in that Newell (1948) and Duncan (1960) had shown quite conclusively that in most areas the reverse occurs i.e. the worms begin to spawn during spring tides and reach their period of peak reproduction during the neaps. These results are interesting, not so much for the contradictions, but in so far as they reveal that the tidal cycle has some control over the spawning activity. It must be remembered that air temperature is significant to an intertidal animal only when the tide is out; hence, one would expect the sudden drop in temperature to be most effective if it occurred when the animals were exposed to the air temperatures as would tend to be the case during a period of spring tides. The effect would be the more marked since at the turn of the year, the difference between the air and the water temperatures is greatest. Because the intertidal animals are exposed directly to the air temperatures, one would expect the seasonal changes to affect them earlier than they would affect the fauna below the low tide mark and this seems to be the case. Some of the species low down on the shore extend into the sublittoral where the onset of the breeding season tends to lag behind that in the intertidal zone.

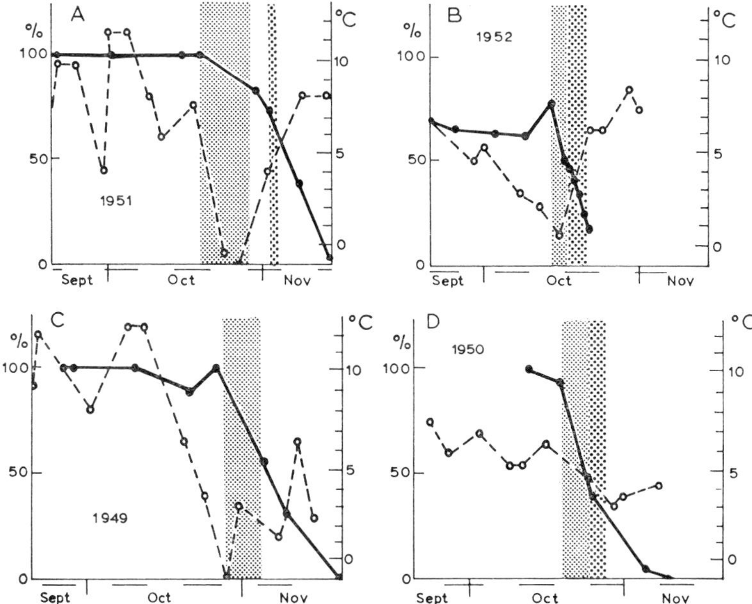

Fig. 22 The breeding season of *Arenicola marina* at St Andrews related to minimum air temperatures and tidal conditions. Open circles—minimum air temperatures in °C. Closed circles—percentage of worms containing genital products. Coarse stippling—spawning period. Fine stippling period of peak spawning. A double line on the abscissa indicates a period of spring tide (after Howie 1959)

Thorson (1946) made an exhaustive study of reproduction in Danish bottom invertebrates and found that the reproductive seasons of various species began progressively later in the year as the depth of water increased. The most likely explanation is that reproduction is temperature dependent and is delayed in deep water because of the slower temperature changes which occur in large bodies of water.

It is possible that the factor which triggers the spawning period is not temperature *per se* but some other stimulus which is effective only after the significant temperature regime has been reached. Temperature is most likely the ultimate factor but there is evidence that the Moon may be the proximate one. The peculiar dependence upon the phases of the Moon of the Pacific palolo worm (*Eunice viridis*) is well known and provides a precedent for this type of behaviour. Battle (1932) recorded a lunar rhythm in the spawning of *Macoma balthica* and *Mya arenaria*, while Brafield and Chapman (1967) found that the spawning period of *Nereis virens* at Southend coincided with the appearance of the new Moon. However, a lunar rhythm is associated with a tidal rhythm and, indeed, it seems that the tide is the important factor in the case of the *Macoma* and *Mya* populations at least for it was found that spawning occurred during periods of the spring tides. But in the area concerned, low water of spring tides occurs in the morning and evening so that the mud flats on which the animals live are exposed during the day to the heat of the Sun and it is, therefore, most likely that temperature is the operative factor. This example should counsel one against drawing false correlations although it is not inconceivable that the tidal movements themselves can influence reproduction, for it has been shown (see p. 77) that some intertidal animals at least are sensitive to pressure and can detect intertidal changes. It is even possible that the tidal forces themselves can be detected although this has not yet been demonstrated. One wonders how else the palolo worm achieves its remarkable timing. Other possible factors which could stimulate reproduction once the suitable temperatures are reached, include light, food, the release of pheromones by the opposite sex or by other species and perhaps salinity changes. The stimulation of one sex by the other is essential for animals which release their gametes into the water; in such cases, it is usually the male which spawns before the female. Very often, the presence of sperm is essential for the spawning of the female. It should not be assumed that the temperature value associated with the production of gametes or young is necessarily significant for it is quite likely that an earlier stage in gametogenesis is the one which is temperature dependent. It is even possible that each of the various developmental stages has its own range of temperature tolerance. It has been found, for example, that in some lamellibranchs, gametogenesis will begin at a particular environmental temperature but will not proceed to the production of gametes if the temperature is experimentally prevented from rising further. *Venus mercenaria* under such conditions, appears to be able to hold the gametes in cold storage and to release them when conditions become suitable, but in other species the genital products may be reabsorbed.

The duration of the breeding season, or of any other seasonal phenomenon, shows a wide range of values amongst the various species on the shore. Spawning itself may be restricted to a relatively short period as in *Arenicola*, whose breeding season is rarely more than two or three weeks, or it can extend over the greater part of the year. In the latter case, it is usual for peaks of increased intensity to occur. A typical pattern is that described earlier for *Cirriformia*. In this species, sexually mature individuals of both sexes can be found in most months of the year but those with ripe genital products occur only between April and September and within these months, a series of spawning bursts takes place.

Brafield and Chapman's study (1967) of *Nereis virens* has revealed a single spawning period of no more than two weeks per annum although the process leading to the spawning is prolonged and occupies more than one year. The stage of the gametogenesis was determined in the female by measuring the diameter of the oocytes in the coelemic fluid and in the male, by the separation

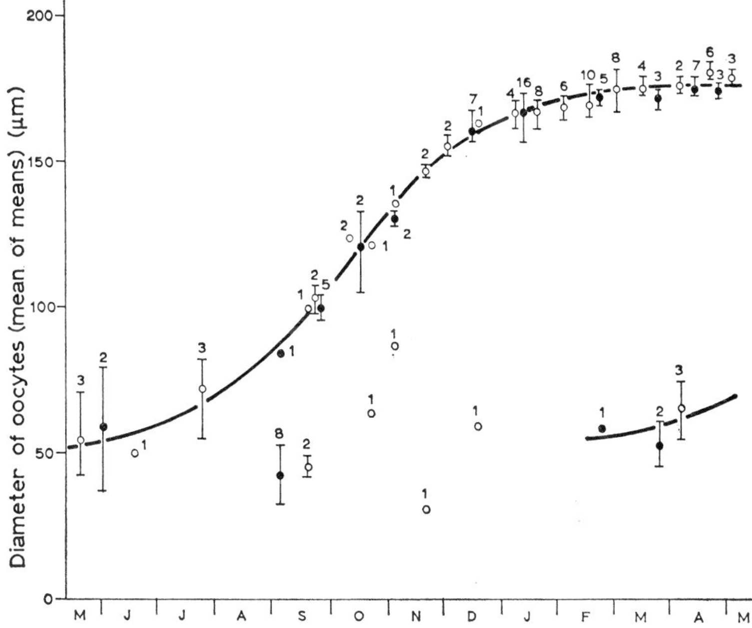

Fig. 23 The mean diameter of oocytes taken from *Nereis virens* at different times of the year. The abrupt disappearance of large oocytes at the beginning of May indicates spawning. Note that some small oocytes may be found throughout the year. These have probably failed to develop properly but the small oocytes found between February and March were probably from young worms and would have developed normally in the ensuing autumn.
Open circles—September 1963—August 1964
Closed circles—September 1964—August 1965
(after Brafield and Chapman 1967)

and mobility of the developing sperms. Fig. 23 shows that a few worms with small oocytes can be found throughout the year but most grow rapidly between September and December, after which little increase in size occurs. The large oocytes disappear abruptly in early May presumably at the time of spawning. The spermatozoa take about eight months to develop to maturity from coelomic sperm plates and again, disappear from the animals sampled in

May. It is likely that the worms die after spawning, for many dead or dying animals are washed up on the shore at this time. All those examined proved to be males so it is possible that only the males actually leave their burrows and swarm. An analysis of the weight distribution of samples of the worms before and several months after spawning suggests that *Nereis virens* normally takes two years to mature and spawn, after which it dies.

A number of patterns emerge from the review made in this chapter. On the one extreme, we have a brief burst of spawning confined to a period of a week or two and on the other, a prolonged period of reproduction with one or more peaks of increased reproductive intensity within it. The most potent controlling factor is the environmental temperature which may either act directly on the production of gametes or it may set in train the process of gametogenesis so that the moment of spawning will depend upon the duration of the maturation of the genital products and not upon the environmental temperatures obtaining at the time of their release. Under such conditions, the date of spawning each year can be very constant irrespective of whether the weather is unusually cold or warm. However, other factors may have a more immediate effect upon the onset of spawning and in several cases, a strong correlation with the lunar and tidal cycles has been demonstrated. It is of course quite likely that more than one factor is concerned with the immediate cause of spawning but the ultimate, over-riding significance of environmental temperature seems indisputable.

A form of seasonality rather different from that so far discussed in this chapter was found by Brady (1943) who showed that the pattern of zonation on a sandy beach varied with the season. Brady worked in Cullercoats Harbour, on the Northumberland coast, where the shore is composed of clean, firm sand. The distribution of some of the species for November and March is shown in Fig. 24. All are polychaetes with the exception of *Enchytraeus albidus*, an oligochaete, and the isopod *Eurydice pulchra*. The general

TABLE 9

The number of species at 9 stations along a transect on a sandy beach at Cullercoats from about 5 metres above MLWS (station A) to just above MHWN (station I). Each station was 11 metres apart (after Brady 1943)

Station	A	B	C	D	F		G	H	I
	number of species								
November 1930	3	4	5	5	5	6	6	2	0
March 1931	4	6	6	5	3	3	2	0	0
November 1930		12			16			8	
March 1931		16			11			2	

impression from this figure is that the animals are further down the beach in March than in November and an analysis of the number of species at each station confirms this (Table 9). The seaward trend is particularly clear when

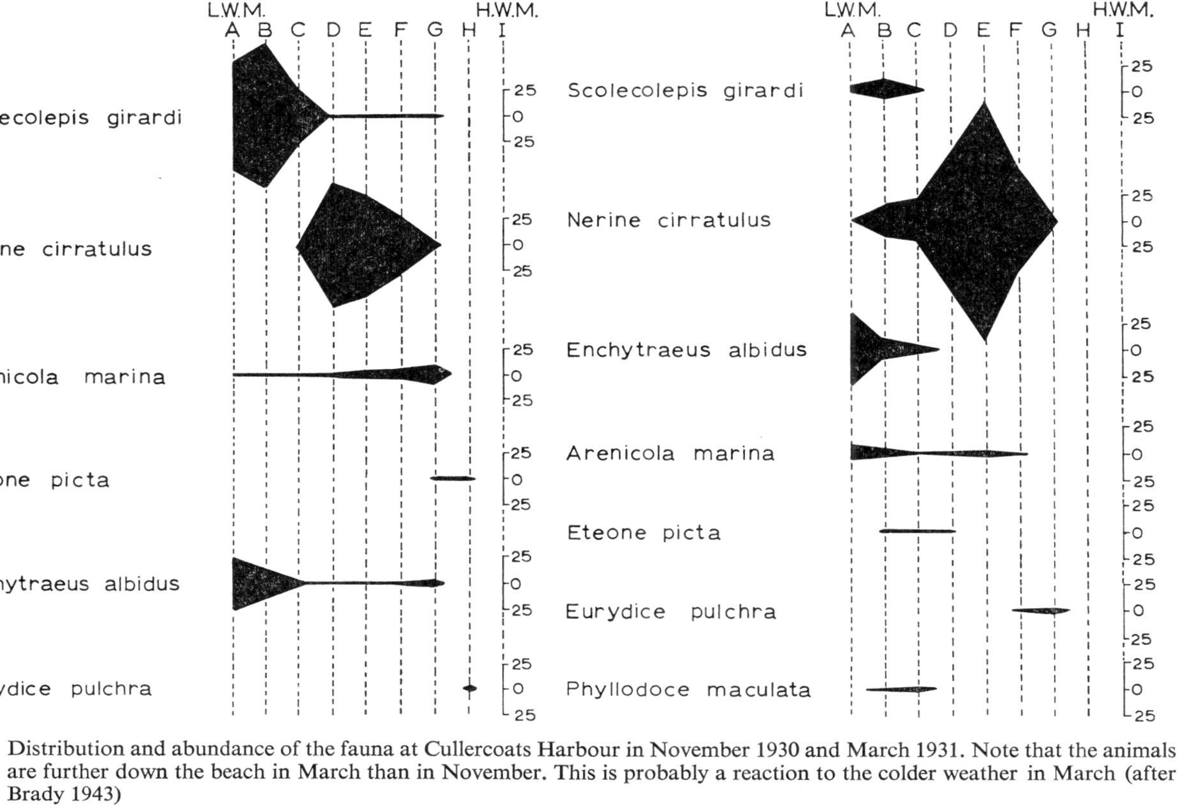

Fig. 24 Distribution and abundance of the fauna at Cullercoats Harbour in November 1930 and March 1931. Note that the animals are further down the beach in March than in November. This is probably a reaction to the colder weather in March (after Brady 1943)

the data are grouped into the upper, middle and lower thirds. The seasonal change is not identical in each species. Thus in *Scolecolepis gerardi* the centre of gravity of the population has not changed although the numbers are much fewer in March and no individuals at all are found in the middle regions of the shore at this time of the year. In the case of *Nerine cirratulus*, there is a definite shift of the centre of gravity towards the sea as well as a seaward extension of the range. *Eteone picta* and *Enchytraeus albidus* are alike in moving further down the beach during the winter. The probable causal factor for this seaward migration may be found in the low winter air temperatures. Except during the harshest winters, the sea does not freeze in Britain and its temperature rarely falls below + 2°C, but air temperatures well below zero are commonplace in January and February. Consequently, intertidal animals which move towards the lower tide mark during these months are moving into a warmer climate since the period during which they are covered by the relatively warm sea is substantially increased. There may have been an interaction with salinity at Cullercoats where the salinity at Station A (low water) was fully marine at 35·1% compared with 24·0‰ at the high water mark. As will be seen in the next chapter, the ability of marine invertebrates to tolerate reduced salinities declines as the temperature conditions become extreme. It is possible, therefore, that the downward movement of the fauna at Cullercoats was due partly to a need to avoid low salinities during cold weather. The salinities will in any case be lower during the winter owing to the increased rainfall and reduced solar radiation.

This pattern of seasonal change in zonation is typical of the beach fauna and becomes much more marked in colder regions of the world where the sea freezes over regularly.

CHAPTER SIX

THE SHORE ENVIRONMENT AND ITS LIMITING FACTORS

The environment consists of all the factors of an area which impinge upon the animals and plants living there. The environment is peculiar to an individual organism which is itself part of its own environment because it interacts with its surroundings and were it to disappear, the environment would change. The environment is not equivalent to the habitat which is simply the type of country—shore, mountain, river or desert—in which the organism lives and which would continue to exist even if all the animals and plants therein were to be wiped out by some catastrophe. The habitat forms part of the environment as do all the animate and inanimate factors within it. In fact, the environment of an organism can be broken down into its component factors, all of which must remain within tolerable limits if the organism is to survive. If any one factor should vary beyond the tolerated range, the species would be eliminated even though all the other factors remained suitable. Such a factor is said to be limiting and it is the factors which are capable of becoming limiting which are important in controlling the distribution of animals in any ecosystem. In this chapter, we will consider how far these factors may become limiting to the animals living on the shore and to what extent they may account for the spatial and temporal variations in shore populations which have been discussed in earlier chapters. Limiting factors may operate either in space or in time; for example, a population may be prevented from spreading into an adjacent area either because some essential resource is entirely missing, or because it is absent only at certain times of the year.

It may help to clarify this discussion if we first consider the major components of the shore environment and pinpoint those aspects which may become limiting. We need to distinguish between limiting factors and regulatory factors in the environment. Factors which control the size of a population without destroying it are regulatory, e.g. disease or predation may prevent a population of animals from expanding and yet not prevent its continuous existence. At the same time, regulatory factors can, under certain circumstances, become limiting. Thus predation does not normally do more than control the numbers in a population but if the balance between predator and prey is tipped in one direction, the predator could eliminate the prey population or prevent its spread. This is most likely under conditions in which the predator is already supported by a prey population when a new potential prey species attempts to extend its range. The predator is, therefore, a regulatory factor in the ecology of the first prey species and a limiting factor in the case of the second. This brings out the point that limiting factors are relative to a particular ecosystem and a factor which is limiting under one set of environmental conditions may be of no significance whatever in different circumstances. An obvious example which will be discussed later is oxygen. This is always in plentiful supply in the water overlying a muddy shore yet a few centimetres below the surface its presence or absence is of vital importance to the animals living there.

The concept of limiting factors as weak links in the ecological chain is of long standing and was probably first formalised by the German agriculturist Liebig in the first half of the nineteenth century. The concept is sometimes known as 'Liebig's Law of Minimum' which states that any essential element which is in short supply, whether or not the total amount required is large or small, will retard or prevent the growth of a population. Leibig was concerned with the growth of crop plants but the law can be extended to cover other plant and animal communities. It can also be extended to include the concept of a tolerable range of factors so that too much of an essential element can be limiting as well as too little. Thus, temperature can be limiting by being too high as well as by being too low. The question of interaction between factors must also be considered; a resource at a certain level may not be limiting if a second resource lies within an appropriate range but should the second factor change, then the first may become limiting even though its level remains the same. The most significant interaction between factors on the shore is that between temperature and salinity. This realisation allows us to establish a second generalisation on limiting factors; the level at which a factor becomes limiting in a particular ecosystem will vary with the value of other environmental factors.

A thorough understanding of the concept of limiting factors is important both in the efficient exploitation of natural resources and in the elucidation of the mechanisms of the ecosystem. Farmers and agriculturists apply the principles of limiting factors in their daily work whether they realise it or not. Fertilisers are applied to the land to provide a suitable environment for the growing plant but the farmer must first establish which elements are likely to be in short supply, i.e. limiting. It would be of little use adding vast quantities of nitrogenous fertiliser if an essential trace element such as barium was absent altogether. For the scientific study of ecosystems, the concept of limiting factors is of great value as it permits the application of the experimental method to ecology. Even the simplest natural ecosystem is so complicated that it is difficult by observational analysis alone to establish which factors are of significance in the ecology of a population or community of organisms. It is possible to carry out field experiments such as the reduction of a predator or the exclusion of a herbivore to see what effect this has on the rest of the ecosystem but the results are rarely clear-cut because so many other factors are changing throughout the experiment. A simpler technique is to bring the animal or plant into the laboratory and rear it under controlled conditions in which only one factor is altered. Such experiments will give more precise results and show the physiological tolerance of the organisms but they do not take into consideration the interaction between factors whose significance may not be apparent to the experimenter. Nevertheless, the method has yielded data of the highest importance for an understanding of the ecosystem. An important point in this respect is the realisation that animals and plants are not necessarily living under conditions which are physiologically the best. Thus a saltmarsh plant may not be where it is because it 'likes' salt, in fact it may grow infinitely better in a salt-free environment. The reason it is there may be that it can tolerate high salinity better than other plants and hence is able to exploit other resources such as space, which are denied to its less tolerant relations. The existence of tubificids in near anaerobic conditions provides another obvious example. *Tubifex* would find life much easier if the oxygen supply were unlimited but it would also find that it would have to share other desirable resources with many other organisms which at present are excluded from the habitat through the low oxygen conditions. It is a general ecological principle that under harsh conditions, the number of available niches is strictly limited but any organism which is able to tolerate the near limiting conditions will be able to expand its range and numbers in the absence of competition from other organisms. This situation is found in the polar regions where there are few species but enormous numbers of such animals that can exist there. The depositing shore is a harsh environment in many ways and the same situation of a few species and many individuals is found there.

The last chapter has shown that a factor at a certain level of intensity may be limiting to an individual at one stage of its life cycle but not at another. We saw, for example, that many species are reproductive stenotherms although in other respects they are eurythermal. Thus, it may happen that a species is able to live in an area but be unable to breed there because the temperature is too low. Alternatively, an animal may find the temperature conditions tolerable during the summer but lethal during the winter. In either case the continued existence of the population is dependent upon regular immigration from other areas outside.

Before we turn to a detailed consideration of limiting factors on the shore, it would be as well to divide the environment into its component parts to see which are likely to be significant on the shore. There have been many attempts to do this but for convenience we will take that of Andrewartha and Birch (1954) who categorised the environment into four major divisions. These are

(i) weather (ii) food (iii) other animals and organisms causing disease, and (iv) a place in which to live. This is a more useful classification than the simpler one of biotic and abiotic factors but there is obviously some overlapping in the categories, for example a prey species could be put in either class (ii) or (iii). However, it is a classification of the environment of a single animal and if each category is mutually exclusive, there need be no confusion. The term weather has a terrestrial connotation but it includes such factors as temperature, light, salinity and exposure in the marine context. A place in which to live may seem a surprising division to some, but on reflection, it is seen to be important. It is rather more precise than the similar term of habitat which implies a certain uniformity over a large area. A place in which to live is, however, more intimate. Andrewartha and Birch quote the example of a rabbit which is able to exist in a wood because the wood contains a hollow log of sufficiently large size to allow it to enter but sufficiently small to prevent predators from following. In the context of the muddy shore, we could cite the existence of scattered boulders which provide living space for barnacles and other sedentary organisms.

In the following discussion, the various factors which could be limiting to the animals on a depositing shore will be considered under the several categories devised by Andrewartha and Birch. Not only will this procedure help to maintain some order in the chapter but it is also useful to apply an ecological classification which is biased by terrestrial overtones to the marine environment. It is inevitable that writers on ecology should be specialised in one particular field or another so that their books on general ecology tend to be overweighted with examples from one area, whether insects, the soil fauna or some other speciality. It is likely, although it has yet to be established, that the ecology of all animals is similar and that it is the methodology, not the material, which has divided ecology into a number of subsidiary disciplines.

Weather

Weather on the shore has a somewhat different meaning from that on land. As terrestrial beings, we tend to judge weather from the atmospheric point of view while an inhabitant of the sea bed would think, if it could, of weather from the standpoint of changes in the aquatic medium. The animals of the sea shore, however, experience both types of weather and it might be as well to maintain this distinction in the following account.

Atmospheric weather

When the tide is out, the littoral habitat is exposed to the atmospheric weather although this may not affect the inhabitants of depositing shores whose covering of mud or sand mitigates the worst of the climatic extremes. Nevertheless, it is true that any organism on the shore is exposed to such factors as rain, wind, Sun or frost which are of no significance to organisms only a metre or so below the low tide mark. With few exceptions, the rocky shore fauna has to endure the atmospheric weather while the animals of the depositing shore avoid it through their burrowing habits. It is possible for non-burrowers to avoid the worst of the atmospheric weather by seeking a less severe micro-climate such as might exist underneath a stone or a frond of seaweed. Such possibilities are of course open to rocky shore animals or indeed to any animal living in a difficult habitat, e.g. many desert animals can exist only because they are small enough to take advantage of microclimates.

Temperature

Temperature on the shore can be limiting if it rises too high when the Sun is shining or falls below freezing during periods of frost. The significance of the effects of exposure to atmospheric temperature varies from one area to another depending on the time of day during which low water occurs. Although there are daily variations in the tidal rhythm, the occurrence of low and high water in any one place tends to fall within a range of a few hours at a particular time of day. If low water occurs during the morning or evening, the worst effects of the hot Sun at noon or frost at night are avoided, but should it fall around midday or midnight, then the animals are exposed to the full rigours of climatic extremes. The severe British winter of 1962/3 showed very clearly that cold can kill off whole populations in the intertidal community (Crisp 1964) but it should be remembered that such conditions, while abnormal in British waters, are by no means unusual on polar shores where species similar to those in Britain occur. Although such species have become adapted to extreme cold, they are not immune and many avoid the worst of the low temperature by migrating below the tide mark in winter. Such adaptations are obviously impossible in sedentary forms and high mortalities from cold can ensue. Kinne (1963), in a review of the effects of low temperature, quotes the case of a sandy beach in Denmark which suffered two months of unusually low temperatures. The entire populations of *Scoloplos armiger*, *Mytilus edulis*, *Scrobicularia plana* and *Littorina littorea* were completely wiped out while severe mortalities were suffered by *Arenicola marina* (90 %), *Cardium edule*, *Mya arenaria* (both 80 %), *Nereis diversicolor* (70 %) and *Macoma balthica* (33 %). The main hazards from low temperature include cell damage from ice formation in intra-cellular fluids as well as gross damage from ice crystals in the coelomic or blood vascular systems. Normally full strength sea water freezes somewhere in the region of −1·8°C and most marine invertebrates have body fluids which are not very different in concentration from the external medium. Hence, the organism is not seriously inconvenienced until its body temperature falls to that level. This is unlikely in marine creatures for the large volume and high thermal capacity of the sea makes it very unlikely that it will freeze under British conditions other than in such notable exceptions as those of 1963. Intertidal forms, however, are exposed to air temperatures for which a fall to −1·8°C is not exceptional. The method whereby the intertidal forms resist freezing are not fully understood but it is not unlikely that they use a form of anti-freeze. Certain fish and insects are known to add organic solvents, such as glycerol, to the body fluids and the same may be true of marine forms. It is certainly true that some marine animals can increase the osmotic pressure of the blood and hence lower its freezing point. Under conditions of low temperature, this is done simply by increasing the concentration of salts and not by the addition of exotic substances to the blood. Other mechanisms are outlined by Kinne (1963) whereby marine animals avoid the effects of subzero temperatures. These include supercooling, which is most likely to occur if the solution is contained in capillary spaces as in the invertebrate body, and the dehydration and gelation of frozen extra-cellular fluids. The latter process results in an increased intra-cellular concentration which delays freezing of the more vital cell fluids. This is analogous to the behaviour of freezing sea water. The first ice to form is salt-free so that the remaining water becomes more saline and its freezing point is correspondingly reduced. Small cells of brine form within the ice and are not frozen until very low temperatures are reached. Something very similar appears to happen to animals which are subjected to freezing temperatures. It was found, for example, that in intertidal animals kept at −15°C, only 55–66% of the body water was actually frozen.

Death from high temperatures is perhaps rarer. It is difficult to be sure that dead animals found during hot weather have succumbed to temperature *per se* and not to oxygen lack or some other temperature-dependent factor. Nevertheless, the upper lethal temperature is usually found to be surprisingly close to the environmental maximum. One needs to be very careful, however, in the definition of lethal temperature. Most estimations with marine organisms result from laboratory experiments, but the criteria of death used by the experimenters as well as the histories of the experimental animals can vary widely and lead to radically different conclusions. Survival experiments are useful for comparing populations or species but their results can be transferred to the field only with the greatest caution. Death through high temperatures can arise from a variety of causes; the oxygen supply may be affected or excessive desiccation may take place. Eventually the proteins become coagulated but death normally precedes this because of extensive interference with the biochemistry of the cell.

Southward (1958) has considered the temperature tolerances of marine animals and although his material came from the rocky shore, his results have relevance here. Southward separated three ranges in the temperature tolerance of an animal. The first is the irritability range within which the animal carries out its normal activities. This is a wide range which the animal can endure without being killed although towards the extremes, certain activities may cease but the point at which reaction to stimuli is lost and coma ensues marks the limits of this irritability range. The survival range extends into temperatures which cause temporary coma but a point is finally reached when the conditions become irreversible and the animal dies. These concepts are shown diagrammatically in Fig. 25. The third, the activity range, is the ecologically important one and in the long run, is the range which controls the distribution of the species since populations which spend most of the time in a coma would be unlikely to become well established. Nevertheless, the considerable

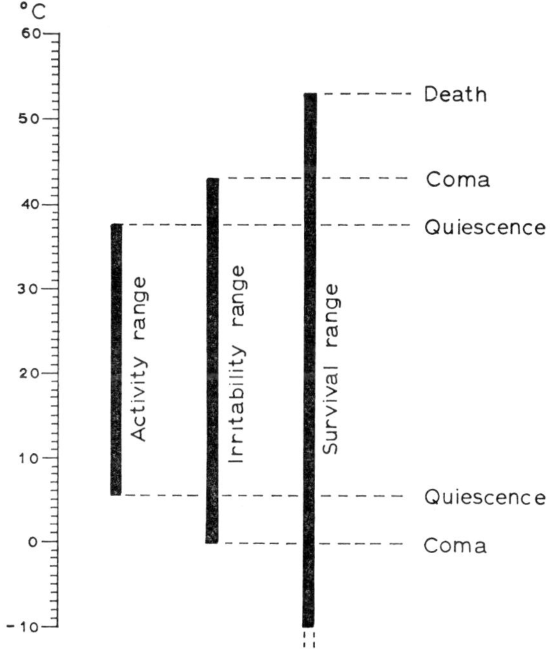

Fig. 25 Concept of temperature tolerance of an animal. The data were derived from experiments with the barnacle *Chthamalus stellatus* for which the lower death limit was not ascertained (after Southward 1958)

potential extension of the viable range which most animals possess provides a valuable safeguard against extinction during unseasonable weather conditions.

If Southward's results are of general applicability, they provide some interesting pointers to the probable situation on depositing shores. As might be expected, he found that sedentary barnacles were more tolerant of temperature extremes than were the mobile topshells and that animals highest up the shore were more tolerant than those lower down. Southward's data for molluscs are similar to those given by Evans (1948) for the same and other species some of which occur on depositing shores. The upper temperature causing 50% mortality in experiments—often considered by experimenters to be the lethal temperature—fell for the most part in the lower forties (Centigrade) for *Littorina* and *Patella*, with high values in the high thirties for most topshells. These are ecologically very high temperatures which are unlikely to be reached on the shore. Southward found that the temperatures of inanimate bodies did not rise above 29°C on the beach with averages varying between 7·4°C in November and 28·4°C in August. He also found that the temperature of the bodies of the animals themselves as measured with thermocouples, did not always correspond with the temperature of these inanimate bodies, as might be supposed, but were usually higher. During exposure to cold air temperatures at low tide, the animals of the shore maintain enough of the warmth from the sea to keep their temperature above zero while they readily absorb heat from the Sun, however fleeting its appearance. It is only when the sea temperatures fall below freezing and there is little Sun, that widespread mortality is likely to occur.

Similar findings were reported by Lewis (1963) for a barnacle and two species of gastropod molluscs found on a tropical rocky shore. He found that the tissue temperatures of the animals were always higher than the ambient air temperature but below those of a black body, an inanimate object and the rock surface. The differences were greater on sunny than on cloudy days. The chief interest in these results from our point of view, is that living animals do not absorb solar radiation to the extent which non-living bodies do. Cooling seems to be effected by evaporation and the molluscs' shells were seen to be raised from the rock surface with the result that evaporative loss was possible from the moist surfaces of the foot. In the barnacle, in which there is little opportunity for evaporation since moist surfaces are not exposed during low tide, it is significant that the tissue temperatures were closer to those of the inanimate and black bodies than was the case with the molluscs. The price for such temperature regulation is, of course, water loss but the measured quantity of water lost was negligible being only about 0·1% after six hours on a cloudy day and no more than 0·4% after two hours on a sunny day. It was found that the species best able to control its body temperature lived high up on the shore while the other two occurred lower down at mid-tide level. Lewis concludes that the ability of intertidal forms to regulate their body temperatures may have a bearing on their zonation, at least in the tropics. It is quite likely, however, that the same hypothesis is true of the fauna of temperate, depositing shores.

The conclusions of Southward and Lewis refer to the rocky shore where the temperature conditions are more severe than is the case on depositing shores. It is extremely unlikely, therefore, that burrowing animals on depositing shores are ever seriously troubled by temperature extremes under normal conditions but the records of wholesale mortality amongst such protected animals as *Arenicola* and *Cardium* during extremely severe winters indicates that the sand and mud do get sufficiently chilled to make the medium uninhabitable. Such a lethal limiting effect is abnormal and the more usual manner in which temperature becomes a controlling factor is through its non-lethal influence on reproduction, activity, interspecific competition or similar biological phenomena. Thus, Ansell (1961) found that spawning in *Venus striatula* took place during May and June when the sea temperatures were generally above 10°C. Laboratory experiments upon the effect of temperature on the development of the veliger larvae suggest that the species is controlled by temperature in this aspect of its biology.

Salinity

Salinity is a 'weather' factor peculiar to the marine ecosystem if one excludes such rarities as saline lakes. Salt water is the natural milieu of very many invertebrates and it is only when sea water is diluted for any reason that salinity becomes a problem. Salinities higher than those of natural sea water are less common although they do occur in rock pools and in the interstitial water of mud flats.

In general, marine invertebrates react to salinity changes in one of three ways (see Fig. 26). First, there are the conformers whose blood is almost isosmotic (i.e. similar in concentration) with the external medium throughout the salinity range tolerated by the species. Secondly, there are those forms whose blood is hyperosmotic (higher in concentration) to the medium when the animal is in diluted sea water, but isosmotic at concentrated salinities. Finally, there are those whose blood is hyposmotic (lower in concentration) to the medium at concentrated salinities and hyperosmotic at reduced salinities. In this context, high and low salinities are relative to that at which the animal normally lives and to which it is acclimatized. Animals in the latter two categories are called regulators since they can regulate, to a greater or lesser extent, the concentration of salts in their blood. Such organisms are on a physiologically higher level than conformers, with those in the last group on the highest level of all.

Conformers are also termed poikilosmotic, with the same connotations as poikilotherms in a temperature context, while regulators are homoiosmotic. The first category of homoiosmotic animals are the salinity equivalents of homoiotherms which can keep warm when the temperature drops but cannot keep cool when it rises. The second group of regulators are equivalent to the mammals and birds which can regulate both sides of the normal level. Further nomenclature follows that used in temperature studies with euryhaline (or eurysaline) referring to animals with wide salinity tolerances and stenohaline (or stenosaline) to those with narrow tolerances. Stenohaline forms can be further divided into those acclimatized to high salinities as found in the open sea (polystenohaline) and those which are adapted to low salinities or to fresh water (oligostenohaline). The degree of tolerance to salinity can, however, vary widely with a form of reproductive stenohalinity analogous to the reproductive stenothermy noted earlier.

Most open sea, primarily marine invertebrates are conformers. Such organisms never encounter much environmental change in salinity and the selective pressure for the evolution of osmoregulation is absent. They are usually stenohaline. Conformers may, however, be met on the shore although they are not common. Examples are the burrowing 'prawns' *Callinasasa* and *Upogebia*, which burrow in muddy sand and sand respectively, but these tend to be sublittoral or occur well down the shore where salinity changes are rare.

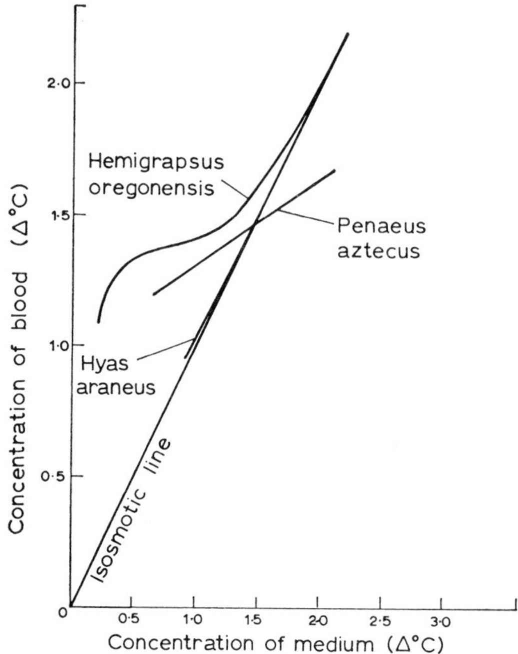

Fig. 26 The three categories of reaction to salinity changes in marine invertebrates (a) Poikilosmotic (conformers) e.g. *Hyas arenarius*, (b) Hyperosmotic in reduced salinities and poikilosmotic in high, e.g. *Hemigrapsus oregonensis*, (c) Hyperosmotic in reduced salinities and hyposmotic in high, e.g. *Penaeus aztecus*. Note: subsequent work has shown that *H. oregonensis* belongs to category (c) provided acclimation is slow (after Lockwood 1962)

Animals which experience wide changes in environmental salinity can usually regulate the blood concentration to some extent. These include particularly the estuarine forms but many other coastal organisms have the facility. There is a wide variability in the degree to which osmoregulation takes place. A few animals, termed holeuryhaline, can tolerate all conditions from fresh water to full strength sea water. Examples on the shore include the turbellarian *Macrostonum appendiculatum* and the oligochaete *Lumbricillus lineatus*. In the great majority of cases, however, the ability of the euryhaline forms to tolerate salinity changes is limited, with wide variations, between species. Hyperosmotic animals on the shore include *Carcinus maenas* which has considerable osmoregulatory powers in diluted sea water but is a conformer in concentrated media. Another shore crustacean, *Crangon crangon*, the common shrimp, is an example of the third type which can regulate in concentrated sea water. Why there should be such a difference between shore animals is not always clear, for example neither *Carcinus* nor *Crangon* is more likely than the other to experience concentrated sea waters, but in other cases, the adaptive significance can be inferred. Thus, many of the animals in mangrove swamps are hyposmotic regulators as are the animals which migrate from fresh water to the sea. *Artemia salina*, the brine shrimp, comes, not surprisingly, within the third category.

It must not be supposed that animals which can control the osmotic pressure of the blood maintain their body fluids at all times at a concentration equal to that of full strength sea water. In fact, most of them allow the blood concentration to fall considerably on contact with low salinity water but it is never permitted to fall below a certain level. The difference between the external and internal concentrations increases as the salinity of the water declines. This limited tolerance of low salinities by the tissues has obvious advantages. Since the blood concentration is allowed to fall, the osmotic pressure difference between the body fluids and the surrounding dilute medium becomes less than it would be were the original blood concentration maintained. A consequence of this state of affairs is a marked reduction in the amount of energy necessary for osmoregulatory activity.

Marine animals have alternative means of reacting to salinity changes other than by osmoregulation. An obvious example is the ability to move away from the area, a behaviour pattern which usually involves a seaward migration, but other devices may be brought into play such as shutting the organism off from the environment by plugging burrows or closing shells with impervious substances, e.g. the operculum of *Hydrobia* which was mentioned on p. 23. Even a layer of mucus around the body will reduce exchange between the body fluids and the surrounding water. On the shore, salinity changes are usually of short duration e.g. the salinity on the beach may drop sharply after a heavy storm during low tide but within six hours at the most, full salinity values will be restored. Hence, the ability to close down metabolic operations for a short time is all that is necessary to resist salinity changes. Some intertidal animals, such as barnacles or clams with impervious shells, can remain shut off from the environment for remarkably long periods provided there are sufficient intervals of suitable salinity for the essential functions of feeding and waste elimination to take place. Such an ability may well be a factor in determining the zonation of the species. Animals living at the extreme high water of spring tides for example, must have adaptations along these lines.

Osmoregulation is a more complex affair than the simple removal of water which enters by osmosis. Water removal is, however, necessary for volume control in soft-bodied animals to prevent rupturing of the body wall; bursting of the body is one of the more catastrophic symptoms of a breakdown in osmotic control. More important, however, is ionic regulation. The concentration of ions in the cell fluids is often different from that of the surrounding medium which, in turn, may differ from that of the environmental water. The major ions in sea water are sodium (475·4 millimoles per kilogramme at 20°C and at a salinity of 34·33‰) and chlorine (554·4 mM/kg) but other ions, such as Ca^{++}, K^{+}, Mg^{++} and SO_4^{--}, are essential to life although they are present in much smaller amounts. If this chapter were about fresh-water animals, it would be necessary to consider these ions as limiting factors but the proportion of salts in the sea is remarkably constant irrespective of the concentration and no element is missing. Trace elements are another matter and, as we shall see later, infinitesimal quantities of certain chemicals are essential for the existence of some species. However, if the composition of the blood of marine animals is studied, it will be found that the proportions of ions present are different from those in the sea even though the overall concentration may be the same. Table 10 shows the ratios of various ions between the blood and external medium medium of some animals from the shore.

TABLE 10

Ratio of blood/external media in respect of various ions (adapted from Prosser & Brown 1961)

Animal	Na^{+}	K^{+}	Ca^{++}	Mg^{++}	Cl^{-}	SO_4^{--}
Phascolosoma	1·04	1·10	1·04	0·69	0·98	0·91
Aphrodite	0·99	1·26	1·00	0·99	1·00	0·99
Arenicola	1·00	1·03	0·99	1·00	0·99	0·92
Carcinus	1·02	0·95	1·23	0·42	0·98	—
,,	1·11	1·21	1·27	0·36	0·99	0·57
Palaemon	0·85	0·85	1·05	0·20	0·85	0·10
Uca	0·83	1·22	1·33	0·52	0·93	0·91
Ligia	1·18	1·30	3·32	0·37	1·03	0·17

No general conclusions are immediately apparent from this analysis. It is clear that wide variations can occur between different types of animals, for example *Ligia* accumulates Na^{+}, Ka^{+}, Ca^{+} and excretes Mg^{++} and SO_4^{--} while *Arenicola* shows relatively little change from the proportions found in sea water. This is typical of the groups to which each belongs, for crustaceans, on the whole, are good ionic regulators while polychaetes are poor in this respect—a further demonstration, perhaps, of the high level of organisation shown by arthropods.

A further important constituent of blood is protein. Not only can protein bind inorganic ions to itself and effectively remove their osmotic influence, it can itself exert an osmotic effect. A very well known example of a non-ionic substance with osmotic significance is the urea retained in the blood of elasmobranch fish but similar substances are found in some marine invertebrates which have considerable quantities of protein in the plasma, e.g. 80 g per litre in decapod crustaceans and over 100 g per litre in cephalopods. Quantities as great as these must have important osmotic effects and have to be taken into consideration.

This brief excursion into the physiology of osmoregulation has merely served to show how it is possible for marine invertebrates to tolerate variations in the environmental salinity. It is now, however, necessary to see to what extent animals on the shore are subjected to salinity variations. The region where salinity changes are most common is, of course, the estuary which will be considered in more detail later, but can salinity changes of any magnitude exist on the non-estuarine shore? There is evidence that they can, although there are problems of definition such as, for example, the case of a small fresh-water stream running across a sandy beach and creating estuarine conditions locally. Nor can the existence of the water table be forgotten. Because fresh water floats on top of sea water, the interstitial water of a beach may be quite fresh. This is particularly true of a shingle bar which separates a fresh water lagoon from the sea but even on a sandy beach, fresh water seepage can occur and give estuarine conditions in an apparent marine habitat. A well known example is Kames Bay at Millport where a flourishing population of the brackhish water polychaete *Nereis diversicolor* occurs on a beach covered with water of 32·9‰ mean salinity (Smith 1955). In this case, there is no evidence of a fresh-water table resting on more saline water but it seems that geological conditions are such that fresh-water drainage from the land flows into the sand with sufficient force to maintain low salinity conditions even at high tide. Fig. 27 shows this clearly, and gives a very obvious indication of the correlation between low salinity and the distribution of *Nereis* and, at the same time, the decreasing tolerance of *Arenicola* and *Nepthys* to dilution of the water. The spatial distribution of *Nereis* is shown in Fig. 28 which is a map of Kames Bay with the transects studied by Smith superimposed upon it. The presence of *Nereis* along the western transect is associated with a small fresh-water stream flowing over the beach while the wide band towards the upper shore on the eastern transect coincides with a strip of sand which never dries out even at low water. This strip is wetted by a seepage of fresh water. Whereas the eastern transect had, at least in the area of seepage, salt water overlying a more dilute interstitial water, the position was reversed along the western transect. It was found here that the fresh water running over the sand had little effect on the salinity of the underlying, interstitial water, except near the high tide mark. At 100 metres down the shore the flowing surface stream had a salinity of 4·3‰ yet the salinity of the underlying water remained at 29‰, only a little below the value of the sea water in the bay. This situation appears to be typical of most beaches and it is unlikely that the interstitial salinity is much affected by the introduction of fresh water to the surface as happens during heavy storms. It follows, therefore, that salinity is unlikely to be a

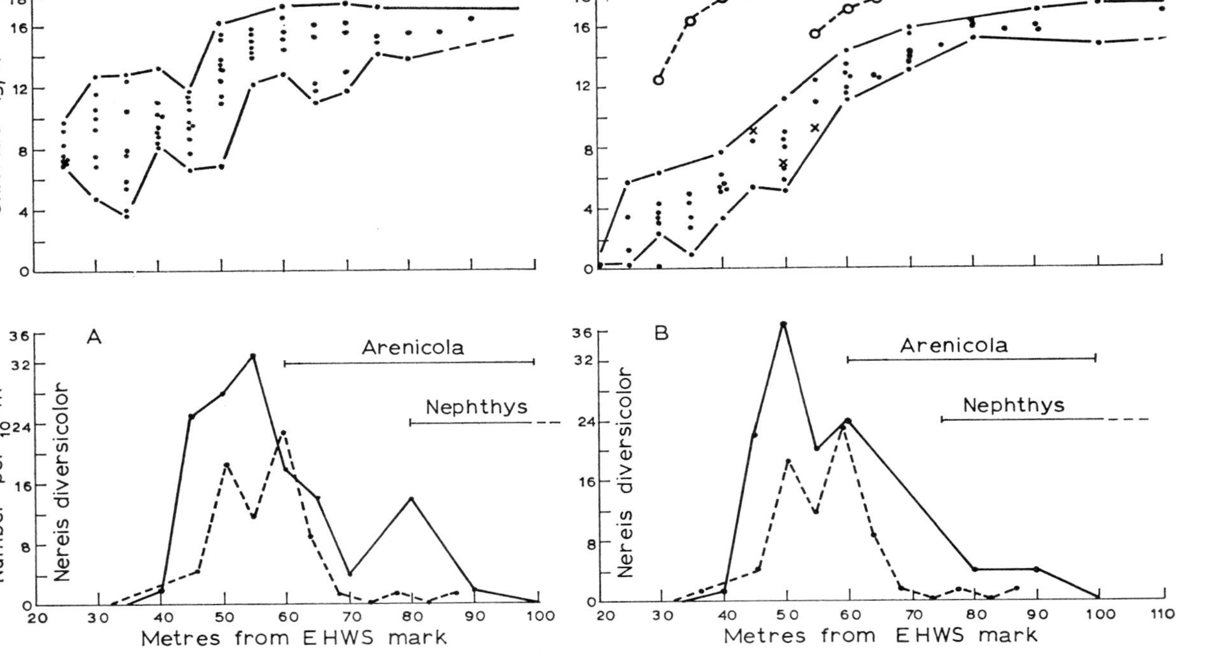

Fig. 27 The two upper graphs show the salinity (expressed as grammes per litre (‰) of chloride) of the interstitial water of Kames Bay, Millport in relation to tidal level. The details are of no particular relevance, but A includes data for one rising and two falling tides, 31 October–2 November 1953 and B the same for one rising and one falling tide, 12–13 February. The open circles show the salinity of the overlying sea water. Note that the interstitial water high up the beach is very much more dilute than the overlying sea water. This is due to seepage of fresh water under the beach with sufficient pressure to maintain low salinity conditions. The lower graphs show the distribution of *Nereis diversicolor* (dotted and unbroken lines represent separate studies). Note how *Nereis* predominates in sand with dilute interstitial water. The horizontal bars show the range only of *Arenicola* and *Nepthys* which are obviously less tolerant of low salinity (after Smith 1958)

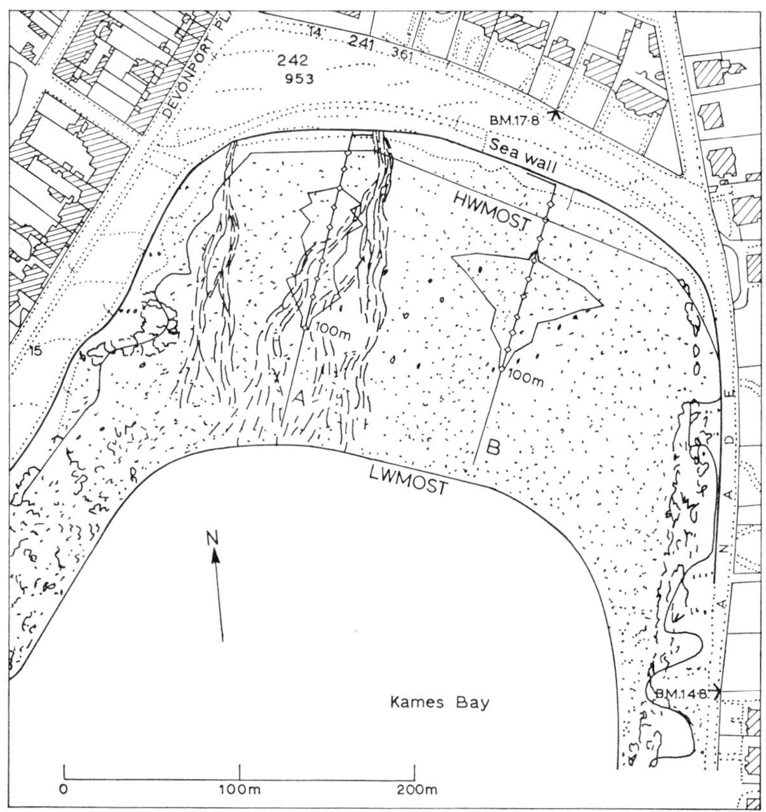

Fig. 28 Map of Kames Bay, Millport. A and B are transects on which the relative densities of *Nereis diversicolor* are shown. A fresh-water stream flows over A but has little effect on the interstitial salinity. Transect B is unusual in that fresh-water seepage occurs in the area at which *Nereis* is abundant and the interstitial salinity is much lower than that of the overlying water. (after Smith 1955)

limiting factor on the shore for not only are salinity changes slight, but the duration of exposure to them can never be more than the period between two high tides, i.e. 12 hours at the most. The work at Kames Bay, however, shows that under certain unusual conditions, salinity can exert a limiting effect on the shore.

A distinction must be drawn between the macro- and meiofauna as far as salinity is concerned. Although the ions present in sea water are normally remarkably constant in their proportions, this is not necessarily so in the case of interstitial water. The reason for this is that particles of decaying organic matter may produce locally huge concentrations of one or more ions. Thus, the water around the decaying seaweed is rich in iodine and chlorine ions, greatly in excess of their concentrations in the open sea. Such breakdown products are quickly diluted in the sea while on the shore their concentrations are too small or too localised to be of any significance to large animals, but to a small organism living in the intestitial water, they represent an important part of its environment. Caution is necessary, therefore, in carrying out salinity experiments with interstitial animals and discrepancies between laboratory results and field observations may be due, in part, to neglect of these considerations. For the same reasons, it is not reliable to use a chemical method for measuring the salinity of interstitial water. Normally, only one ion, the chloride, is measured and the salinity is calculated on the assumption that this ion bears a constant ratio to the others but, for the above reasons, this assumption does not hold with interstitial water. The most reliable way of deriving the salinity of such water is to measure the conductivity which is directly related to the number of ions irrespective of their nature.

Aquatic weather

Temperature and salinity have been treated as facets of the atmospheric weather as changes in their properties depend largely on atmospheric factors namely Sun and frost in the case of temperature, and rain in the case of salinity. In most cases, however, the consequences of changes in the atmospheric weather are effected through the aquatic environment and the distinction adopted here between the two types of weather should not be maintained too rigidly. Some weather factors, however, are more firmly aquatic and in this category we will consider oxygen and other gases, hydrostatic pressure, light, water currents, salts and organic constituents and desiccation. The penultimate factor is not entirely aquatic for we shall see that the attractive constituents of sand or mud may be associated with the particles themselves rather than the overlying or interstitial water. Such constituents are, however, part of the inanimate environment and are most properly considered here. Several of these factors do not change significantly, apart from a regular cycling, and are, therefore, unlikely to become limiting to established populations, although they might well be limiting in the sense of determining whether or not life in a particular area is possible in the first place. It will also be necessary to consider the interaction of factors. The most obvious examples are temperature, salinity and oxygen concentration. Oxygen and temperature are inversely proportional to each other in their concentrations as is salinity with either factor. The severity of temperature or salinity extremes is increased if the other factor is approaching tolerance limits. On the whole, marine organisms are able to endure lower salinities at benign temperatures than at either extreme and the same principle holds if the roles of temperature and salinity are reversed. This is but an example of a wider ecological fact that the level at which factors become limiting varies with other environmental conditions. We are familiar with the situation in our own lives; cold weather can be invigorating if we are dry yet miserable if we are soaked to the skin.

Each component of the aquatic weather will be considered separately followed by a discussion of the interaction between these factors.

Oxygen

The concentration of oxygen in the water, or oxygen tension as it is sometimes loosely called, is a factor which is normally of greater significance to fresh-water organisms than to those in the sea although its solubility in sea water is only some 80% of its value in fresh water. This is because bodies of fresh water are usually smaller and more liable to stagnation than is the case with sea water. Sea water 'ponds' occur, of course, as in the case of rock pools which are high up the shore and which are replenished only around the period of high water springs. Otherwise, oxygen is rarely a significant factor to the rocky shore fauna whose surrounding water is well oxygenated during the breaking of the waves. The position of the depositing shore is not so simple. Normally breaking waves are not a feature of such shores particularly of mud flats where the rising tide is no more than a gradual creep up the shore. In summer, the mud exposed at low water becomes heated by the Sun and the heat is transferred to the water further lowering its oxygen content. Under some meteorological conditions, the oxygen content of the overlying water on a mud bank can fall quite considerably particularly if there is much decaying matter on the surface such as occurs when the aftermath of a storm brings a floating mass of detached seaweeds onto the shore. Seriously polluted waters provide another instance of low oxygen concentrations in a marine environment and, as many holiday-makers have discovered to their cost, such situations are regrettably not unknown around our shores. However, it is only in landlocked bays with little tidal movement that oxygen concentrations in sea water are likely to fall anywhere near limiting values. Animals have remarkable resources for dealing with low oxygen conditions including a form of suspended animation which is maintained until more suitable conditions return. In some species, the oxygen capacity of the blood is increased by various mechanisms. The blood of most marine invertebrates has a low oxygen capacity with solubilities varying from 1–5%, depending on the taxonomic group concerned. These values may not be very different from the concentration of a simple oxygen solution but some organisms, of which the red-blooded vertebrates provide the most notable examples, have haemoglobin or other pigments in the blood which increase the oxygen carrying capacity. One might expect such adaptations to be present in organisms normally living under low oxygen conditions and this tends to be the case. The degree to which an organism is able to cope with low oxygen conditions is shown by the dissociation curve, which is a plot of the percentage oxygen saturation of the blood against oxygen tension of its environment. An animal adapted to near anaerobic conditions will retain a high level of oxygen in the blood until the external oxygen tension has fallen to a very low level indeed, at which point the oxygen is released to the tissues at a time when they most need it. Conventionally, the point at which oxygen is released to the tissues, the 'unloading tension', is taken to be that at which the percentage concentration of the blood has fallen to 50% and the 'loading tension' to be that at which the blood becomes 95% saturated with oxygen. Both values will be lower in animals adapted to low oxygen conditions than in those which normally experience no oxygen deprivation. In fact, the values of the loading and unloading tensions give one an insight into the ecology of the animal concerned. Thus, in a marine mammal (for example a porpoise) with an ample supply of oxygen, the unloading tension is as high as 31 mm of mercury while the loading tension is 105 mm Hg. By contrast, the corresponding values for *Arenicola*, which live in stagnant mud, are 1·8 mm and about 7 mm Hg. respectively. This is made possible by the high affinity which the haemoglobin of *Arenicola* has for oxygen.

We may take *Arenicola* as an example of a beach inhabitant which is adapted to low oxygen conditions. Normally, *Arenicola* respires by circulating water through its burrow as described on p. 44, but at low tide it is unable to do this and the oxygen concentration within the burrow may fall to a very low level. Up to a point, *Arenicola* can adjust to these conditions by the oxygen affinity mechanism of its blood but eventually the worm resorts to drawing bubbles of air into its burrow to cover the gills. Should the water drain completely from the burrow, *Arenicola* is able to respire atmospheric air provided the surface of its body remains wet (Wells 1945). Other mud-dwelling polychaetes are able to utilise atmospheric oxygen in a similar way.

Other avenues are open to invertebrates exposed to anaerobic conditions. One is simply to reduce the metabolic rate to such a low level that the amount of oxygen required becomes negligible. This is perhaps the most important method developed by many intertidal polychaetes which can survive for surprisingly long anaerobic periods. Dales (1958) reviewed some published accounts which reported survival under anaerobic conditions for 9 days in *Arenicola*, 21 days in *Owenia*, 1 day in *Amphitrite* and 1 day in *Nereis*. Measurements of the glycogen content in some cases revealed high concentrations which suggest that glycolysis takes place, that is the glycogen is broken down to give energy without the intervention of oxygen. In mammals, such glycolysis results in the accumulation of lactic acid in the tissues. This builds up an oxygen debt which has to be paid off by an increased consumption of oxygen when normal conditions return. The fact that such an increase in oxygen consumption does not take place in *Arenicola* shows that glycolysis does not lead to the production of lactic acid or that, if it does, that the lactic acid is continually excreted. Dales' work (1958) makes it clear that the former possibility occurs and that glycolysis results in the production of other acids at present undetermined. The absence of an oxygen debt mechanism from these worms may be a reflection of an inability to ventilate the respiratory surfaces adequately—a necessary requisite for clearing the accumulated acid. Although *Arenicola* is able to utilise glycogen during anaerobic conditions, Dales concludes that this is not a significant feature in the normal life of the animal but that the ability to remain quiescent, coupled with the aerial respiration mentioned earlier, is sufficient to (literally) tide it over adverse conditions. There was no evidence that *Owenia fulsiformis*, a tubiculous worm from low down on sandy shores, is able to utilise glycogen although the concentration of this substance in its body is high, perhaps as much as 5% by weight. During the experiments, these worms remained curled up in the centre of their tubes showing no signs of activity. There was no evidence of the metabolism of oil in either species to produce energy under anaerobic conditions. If these results are typical, they suggest that shore animals contend with anaerobic conditions, which in nature will not endure for more than a few hours, by becoming quiescent and so reducing energy requirements to a bare minimum.

In the context of oxygen as a limiting factor, it is necessary to re-emphasise the distinction between the macrofauna and the meiofauna since the two types respire in radically different environments. The macrofauna, although buried in the mud, usually maintains contact with the overlying water either by the construction of burrows, through which the water circulates, or by drawing in the water through siphons or similar devices. The meiofauna, on the other hand, has to subsist on the oxygen dissolved in the interstitial water which is oxygenated only to a shallow depth which may be no more than 5 mm in thick mud. To the macrofauna, therefore, oxygen is rarely a limiting factor except in badly polluted areas but to the small forms, it certainly can be. It could be said that the oxygen requirements limit the meiofauna to the upper layers of the mud but other factors such as high pH or toxic constituents would in any case prevent deeper penetration by these animals. The factors are, of course, inter-related; the sulphide layer is present because there is insufficient oxidation of the organic matter. The depth to which oxygen can penetrate depends upon the speed with which oxygenated water can percolate through the deposit and upon the amount of organic matter present.

The actual oxygen content of the interstitial water has not received much attention due, no doubt, to the difficult technical problems involved, but Brafield (1964) has devised a method for its measurement. The major difficulties in the technique are the prevention of contamination of the water sample with atmospheric oxygen and the determination of the sources of error arising from chemical pollution of the beach which, unfortunately, is all too likely around our coasts although even badly drained 'natural' beaches are likely to be contaminated. The first problem requires the construction of a suitable apparatus and the one devised by Brafield avoids contact with the atmospheric oxygen as well as being robust and simple to use in the field. A description of his sampler is out of place here but full details are given in the published account. The problem of pollutants demands chemical pre-treatment of the water sample, techniques which again are not relevant here since we are principally interested in the results.

The main conclusion to be drawn from Brafield's work is that interstitial water is generally very poorly oxygenated. The most significant factor influencing the oxygen concentration is the drainage of the beach which, in turn, is controlled by the slope and the particle size. Poor drainage is characterised by standing pools on the surface while the percentage of fine sand (particles less than 0·25 mm in diameter) in the deposit is an excellent indicator of drainage condition. If the percentage of fine sand exceeds 10%, the oxygen concentration of the interstitial water cannot rise above 20% of the air saturated value. There is, however, no limit to the oxygen level should the proportion of fine sand be below 10%. As far as drainage is concerned, there is a marked improvement once the percentage of fine sand falls below 20%. Table 11 shows the results of analyses made below mid-tide level at ten beaches on the Isles of Scilly. Certain conclusions emerge from these comparisons. There is a very definite positive correlation between low oxygen concentrations and poor drainage as indicated by the presence of surface water. There is also an inverse relationship between the oxygen level and the percentage of fine sand (Fig. 29) particularly below values of 5%. Above this figure, the correlation is not impressive. Station B is anomalous in that the percentage of fine sand is very high (94%), yet the oxygen content is also high (16·5%) and the sand is clean. The explanation appears to lie in the presence at this site of rivulets of highly oxygenated surface water which must, therefore, be able to influence the interstitial water to a depth of at least 5 cm, the depth at which the water samples were taken. Similar streams were present at site H and are, no doubt, responsible for the very high values recorded for the oxygen content. Other correlations are obvious; the oxygen level rises with the cleanliness of the sand, the depth of the reducing layer and the slope of the beach. These conclusions would be expected on theoretical grounds outlined in the first chapter.

Brafield made a more detailed study of the interstitial water on the flats at Whitstable, Kent. His results there revealed similar correlations as were found in the Scilly Isles and brought out the further point that although the oxygen content of the surface water left behind by the receding tide increased during the period of immersion, there was no appreciable change at a depth of 5 cm. The surface increase is the result of photosynthetic activity of diatoms which migrate to the surface at low tide, but it is clear that they have no influence on the deeper interstitial water although at 2 cm a slight increase in oxygen levels was detected.

From the standpoint of limiting factors, Brafield's results support our earlier conclusions that oxygen lack limits the meiofauna to the sand above the reducing layer. Brafield did not consider the interstitial animals but he noted the presence of some of the larger forms which apparently show clear correlations with oxygen conditions, for example *Scoloplos* is abundant only

TABLE 11

Relationship between the oxygen concentration of the interstitial water and various physical factors of the beach. Water and soil samples were taken from a depth of 5 cm except at site J (3·5 cm) where there was insufficient depth (abridged from Brafield 1964)

Site	Mean oxygen level % of air saturation	Colour and depth of reducing layer		Surface water	Drainage slope of beach	% fine sand
A	0·0	Grey	2 cm	Pools persisted	Flat	98·2
E	3·0	Black	1 cm	,,	,,	7·2
C	3·5	,,	1 cm	,,	,,	71·7
J	8·5	,,	1–2 cm	,,	,,	37·1
G	11·5	Grey	4 cm	Pools drained later	,,	18·5
B	16·5	Clean		Thixotropic at first dilatant later	Sloping	94.0
D	21·5	,,		No pools	,,	3·4
F	33·5	,,		,, ,,	,,	3·7
I	39·5	,,		Very little quickly drained	Flat	5·0
H	81·5	,,		No pools	Sloping	2·6

between mean percentage saturation values of 3 and 8·5% while *Arenicola* is not found in sand with oxygen values above 3·5%. However, such correlations are, in all probability, spurious since the macrofauna is mostly independent of the interstitial oxygen conditions and the operative factor is more likely to be the presence of food or the absence of competition under such unfavourable conditions.

One can sum up the position of oxygen on the shore by stating that under natural conditions, i.e. in the absence of gross pollution, the water covering

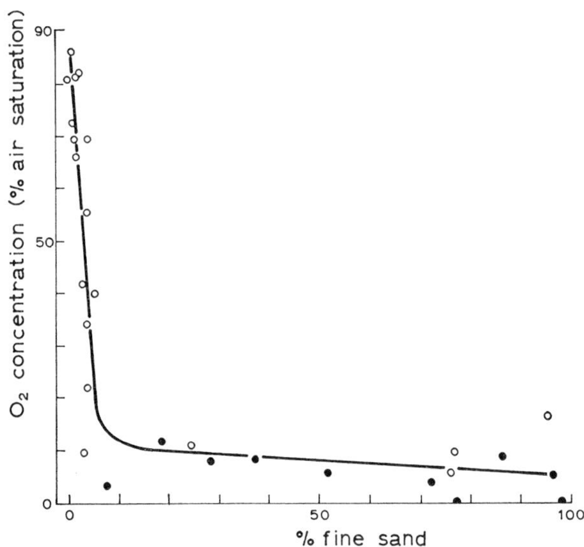

Fig. 29 The relationship between the oxygen concentration of the interstitial water and the percentage of fine sand in the substrate. Solid circles indicate beaches where the sand was blackened. Samples taken in the Isles of Scilly (10), Whitstable (4), Jersey (3) and Banjuls (French Mediterranean) (11) (after Brafield 1965)

the shore is sufficiently oxygenated to support the macrofauna buried under the surface with the exception of some tubiculous forms, the water in whose tubes may become stagnant. The meiofauna and microfauna on the other hand can encounter oxygen lack and, for this reason, are restricted to the upper layers of the deposit.

Other gases

The most common gases, other than oxygen, in the air are nitrogen and carbon dioxide. A certain amount of nitrogen from the air is dissolved in sea water but being an inert gas, it is in no way a limiting factor. The carbon dioxide content of the sea is very high relative to its concentration in fresh water as its solubility is some fifty times greater in sea water. Nevertheless, this is still quite small (c. 1·6% of all gases dissolved in sea water). It is also very constant due to the buffering action of the carbonate and bicarbonate ions which are normally associated with the excess base elements in the sea. This huge alkali reserve ensures that marine plants are never short of carbon dioxide when they need it and it prevents the concentration of the gas from building up during respiration.

Apart from carbon dioxide in true solution, the gas is present as the dissociated carbonic acid radical HCO_3^-, as carbonates and bicarbonates and even, in minute amounts, as undissociated carbonic acid. Because of the presence of bound CO_2 the carbon dioxide content of the sea is remarkably high being, when saturated, about 100 times the concentration of carbon dioxide in an equal volume of atmospheric air. The quantity in saturated fresh water is somewhat greater than, but of the same order as, that in an equal volume of air. Consequently, marine plants have a superabundance of carbon dioxide available for photosynthesis and this gas can never become limiting to life in the sea through its being in short supply. Nor, except under the most unfavourable conditions, can it become limiting through being present in excessive quantities since excess gas is quickly taken into combination and neutralised.

Hydrogen sulphide, which is the only other gas of any significance on the shore, may be formed by the reduction of free sulphur by bacteria present in the blackened reducing layer of sand or mud. It is a toxic gas and in solution, can limit the downward penetration of those animals which cannot maintain contact with the overlying water. Free hydrogen sulphide does not occur independently but is usually associated with a high organic content and anaerobic conditions which themselves may be limiting. We have here yet another example of the interdependence of environmental factors.

Before leaving the subject of gases, it is worth recording that the interstitial fauna is much more susceptible to micro-climatic conditions than similar sized organisms in the open sea. Small particles of rotting seaweeds, for example, can generate local concentrations of gases which might hardly be measurable and yet could dominate the environment of a nematode or copepod living in the interstitial spaces.

Hydrostatic Pressure

Until comparatively recently, it was not thought that hydrostatic pressure could be of any significance to marine invertebrates since there appeared to be no method by which they could detect it. Most instruments for measuring pressure depend on the compressibility of a gas. Liquids are incompressible for all practical purposes and the pressure within a liquid is the same at similar levels. It seems impossible, therefore, than an animal consisting of a series of solutions, which is one way of describing a marine invertebrate, can detect pressure changes within another solution—sea water. It is understandable in a fish with a gas filled air bladder or a marine animal with lungs or other gas filled cavities which are capable of being compressed. However, recent research has proved beyond doubt that marine invertebrates can react to pressure changes and must therefore be able to detect them. Furthermore, the pressure response mechanism is very sensitive and can detect such small pressure differences as those between high and low water. We saw on p. 77 how this reaction was made use of by *Corophium* and others in maintaining their position on the shore. There have been a number of suggestions as to the mechanism of pressure detection but no firm conclusions can be drawn and the problem remains unsolved. Digby (1967) has advanced an ingenious theory based on electrical principles to which the interested reader is referred for further information.

In the context of this chapter, we need proceed no further with a discussion of hydrostatic pressure for it is certainly never a limiting factor on the shore however important it may be for other reasons, e.g. in the mechanics of zonation. Perhaps the position is different in the abyss where high pressure seems to be necessary for the well-being of certain organisms but that is another story.

Light

As a weather factor, light is neither more nor less aquatic than atmospheric.

Illumination of the shore originates from two sources—the Sun and the Moon—but the intensity of full moonlight is so much less than that of the Sun (about one five thousandth) that for most purposes it can be ignored. Light can affect the inhabitants of the shore in a number of ways. It is, of course, essential for photosynthesis although the primary producers are not obvious on the shore. Nonetheless, diatoms do occur in large numbers and contribute towards the energy supply of the ecosystem.

As far as animals are concerned, light tends to be a deleterious factor. The ultra-violet and infra-red rays in sunlight can be lethal to some organisms although these rays, known as actinic, are very quickly absorbed by sea water. It may be partly for these reasons that the majority of shore animals tend to remain hidden under the mud or sand when the tide is out and emerge only when covered by a protective layer of water. There are other reasons of course; it is more difficult for soft bodied animals to move without the support of water and respiratory needs have to be considered, but more difficult problems than these have been surmounted during evolution.

It is unlikely that animals buried in mud or sand ever become exposed to light and shore animals must spend their lives in continual darkness apart from the temporary epifauna. Many species, however, possess eyes and even those with none are often sensitive to light. This reaction is important for it prevents the animal from wandering into a lighted area where it might be harmed by actinic rays. Boaden (1963b) showed that the archiannelid *Trilobodrilus heideri* is photophobic and this is probably true of most interstitial forms. As early as 1912, Goodrich described a similar light reaction in the archiannelid *Nerilla antennata* although, unlike *Trilobodrilus*, it possesses a true phototaxis and moves along a light gradient. Such reactions tend to keep the animals below a depth of about 10–15 mm, the extent to which Pennak (1951) found that light could penetrate into sand.

Although light can be dismissed as an important limiting factor on the shore it can be of great significance in controlling the movements of shore animals. The role of the Sun in the orientation of winkles was discussed at some length in Chapter 4 and many of these light-controlled reactions of shore animals contribute to their ability to survive and to maintain zonation. Light, however, can hardly become a limiting factor as it is one of the most unchanging factors on the shore with a regular and constant winter/summer variation.

Water currents

Currents in the sea are the analogue of winds on the land but ecologically, currents and winds are quite distinct. Whereas wind can cause serious desiccation problems to a terrestrial organism, there is no similar deleterious effect with currents. True, currents can wash away settled organisms in a similar fashion to the wind blowing away an exposed terrestrial animal, but currents differ from winds in being far more constant and predictable. While a bird may succeed in building a nest in an exposed area and in rearing its young, there is always a chance that a gale will blow up and send the nest crashing to the ground. In the sea, on the other hand, the prevailing regime of currents will either permit the establishment of sedentary organisms or it will not. It is true that storms will vary the velocity of currents, but the difference between water movements under calm and storm conditions is proportionally much less than is the case with winds.

The various types of currents and their causes were discussed in the first chapter, where we found that sandy and muddy shores tend to be formed in areas where water currents are slight. Accordingly currents are likely to be of little significance in limiting the distribution of the shore fauna, and movement produced by wave action is more important. Big waves are exceptional on mud flats but large breakers may occur on sandy shores particularly on those exposed to ocean swells such as the beaches of northern Cornwall. Plunging waves stir up the surface layers of the sand extensively, so that the resulting thick suspension is sufficient to coat the hair and swimming costume of anyone bathing in the surf. However, only a narrow area of beach is affected at any moment and the period of disturbance is short, because the tide soon moves on. Some animals are adapted to these conditions and even require wave action to survive. Nevertheless, the disturbance is sufficient to dislodge or uncover any animals buried near the surface and this is probably one of the main reasons why organisms tend to burrow deeper into sand than into mud. Active animals such as *Donax variabilis* are adept at reburying themselves, but excessive wave action constitutes a limiting factor in the distribution of the less mobile sand burrowers and Ansell (1961) suggests that this is the reason why *Venus striatula* does not occur intertidally in Kames Bay, Millport, although it is found there in the sublittoral. In other areas, this species may extend above low water of spring tides. Shores which are constantly subjected to heavy wave action are usually poorer in fauna than sheltered beaches.

Water movements are important in ensuring the distribution of the animals inhabiting the shore. Some species have planktonic larvae which represent the dispersal phase of the species and if such a stage was not present in the life cycle, a population of sedentary animals would soon outgrow its resources. Much of the observed distribution of shore animals is seemingly capricious. A thriving population may be present on one side of a headland while on the other, with apparently identical conditions, not one individual is to be found. In most cases, the explanation is simply that the colony has been founded by larvae or adults brought in currents which do not touch the barren shore. In this sense, water movement is essential for the well-being of a species for under completely motionless conditions, adequate dispersal would be impossible.

Boaden (1963b) discovered some interesting facts concerning the significance of water currents in the life of interstitial animals. The currents with which he was concerned were very gentle, resulting from the percolation of fresh-water drainage and ground water seepage over and through sand. Boaden found that the archiannelid *Trilobodrilus* was rarely found in the current itself but was abundant in pools produced by small boulders damming the current. Laboratory experiments showed that the worm is positively rheotactic and will swim against any current whose velocity is between 0·37 and 2·4 cm per second. In nature, it would, therefore, progress up the shore. However, the water flowing down the beach is fresh and if the reaction were continued indefinitely, the animal would move into conditions of unsuitably low salinity. However, Boaden showed that the rheotactic reaction is lost as the salinity of the water falls, with the result that the animal is carried back down the shore in the current. These simple reactions account for the observed zonation of *Trilobodrilus* within a restricted region of the shore. Animals higher up are in water of low salinity and are carried down passively while those low on the shore, swim up against the current which in this region is of high salinity.

Nutrients and Organic Constituents

The justification for including nutrients and organic constituents as part of the aquatic weather rests on the fact that they are carried to the habitat in water currents. Their significance as limiting factors lies in the degree to which they render the substratum attractive to settling larvae. In many of the investigated cases, planktonic larvae will not necessarily remain in the area, where they first settle from the plankton. Indeed, most larvae have been found to be highly selective in choosing an area in which to settle and they are able to delay metamorphosis until they find suitable conditions. The classic work of Wilson (e.g. 1953) has shown that this ability is widespread amongst polychaetes.

Larvae may not settle on a particular substratum either because it lacks an attractive factor or because it contains a repellent one. There is evidence that both aspects apply but for the most part, larvae settle for positive rather than for negative reasons. Experiments have shown that some sands neither attract nor repel larvae and such substrata are classed as neutral by Wilson (1953). It is relatively simple to show that a specific attractive factor, other than the physical composition of the sand grains, must be associated with preferred substrata because the attractiveness can be removed by washing the sand in hot, concentrated acid. The sand is, of course, well rinsed before using it in experiments. The precise nature of the attractive factor can be determined by adding various substances to the acid-washed sand, and the most effective way to restore its attractive nature is to soak in natural sea water. This introduces micro-organisms which flourish on the surfaces of the sand grains and suggests that it is the presence of micro-organisms, which presumably are used as food, that attract the settling larvae. In the case of *Ophelia bicornis* larvae, a non-intertidal marine worm, it seems that diatoms alone are insufficient to restore the attractiveness of acid-washed sand but the presence of flagellates is slightly effective (Wilson 1954). The most important factor is probably the bacterial film adhering to the sand particles. However Wilson does not think that bacteria are used as food by *Ophelia.* A number of subsequent studies have supported the belief that the bacterial film is responsible for the attractive nature of sand. Meadows (1964a) found that sand in which this film was altered in any way, lost its attractiveness to *Corophium arenarium* and *Corophium volutator.* Gray (1966) found a similar situation with an interstitial form, the archiannelid *Protodrilus symbioticus.* The attractiveness of sand to adults of this species was completely destroyed by acid cleaning, heating or drying. Gray attempted to discover whether one bacterium species was more significant than any other in increasing the attractiveness of the sand and, if so, whether the bacteria from the home beach were alone responsible or whether another species could make sand equally attractive. This was achieved by inoculating samples of acid-cleaned sand with four species of marine bacteria. Bacteria of mixed species from the beach and from garden soil were also tested. The results are shown in Table 12. The statistical methods used to obtain valid comparisons need not concern us here, but we can see from the transformed means that natural sand bacteria are the most attractive followed by sterile sand treated with soil bacteria. Sand treated with species of *Pseudomonas* was quite attractive, being statistically not inferior to soil bacteria, but the other bacteria species, which were all of similar action, were much less able to confer attractiveness. Even so, they were greatly superior to the sterile controls. The results show, therefore, that while natural sand bacteria are the most suitable agents for restoring attractiveness to sterile sand, any bacterium

TABLE 12

The influence of bacteria in rendering acid-cleaned sand attractive to the interstitial archiannelid *Protodrilus symbioticus*. Samples of sterile sand were innoculated with the bacteria shown and exposed to *Protodrilus*. The numbers represent the percentage of animals in each experiment which chose the sand in question. The transformed means were calculated to obtain homogenity of variance Differences between these means of less than 4·2 units are not significant at a probability level of 0·05. (after Gray 1966)

Species of bacteria	Experiment									Mean	Mean transformed
added to sand	A	B	C	D	E	F	G	H	I	%	value in angular units
Corynebacterium erythrogenes	9·1	5·3	5·9	3·7	4·5	3·8	5·3	5·1	4·2	5·2	13·1
Pseudomonas sp.	23·2	14·2	13·8	17·7	7·5	16·5	12·0	5·1	13·0	13·7	21·3
Flavobacterium sp.	13·1	8·0	5·9	4·7	5·3	5·1	5·3	5·1	1·4	6·0	13·7
Serratia marinoruba	10·1	8·0	5·9	3·7	2·3	3·8	5·3	10·1	1·4	5·6	13·2
Soil bacteria	10·1	22·2	31·4	13·1	21·8	11·4	20·0	17·0	17·0	21·0	25·0
Sand bacteria	12·1	25·7	15·7	19·6	37·6	19·0	21·0	20·1	35·3	22·9	28·3
CONTROL A (no bacteria)	4·0	1·8	2·0	1·9	2·3	1·3	2·7	3·4	0	2·2	7·8
CONTROL B (untreated sand from beach)	18·2	15·1	19·6	35·5	18·8	39·2	28·0	34·5	29·6	26·5	30·7

is better than none and the good results with soil bacteria suggest that a mixture of bacterial species is more effective than monocultures.

For effective colonisation by larvae Wilson (1955) proposed that the number of micro-organisms should be neither too great nor too small and that certain species are more attractive than others. Dead organisms and non-living organic matter are not attractive and may even be repellant.

The position is, however, not as simple as this for mere physical removal of the bacteria does not interfere with the attractiveness of sand. This is shown by testing the sand after the sample has been vigorously shaken in sea water, a process which removes most of the bacteria from the sand grains. A similar result is achieved by soaking sand in glycerol, sucrose or sodium sulphate solutions. It has been suggested that it is the surface film left by the bacteria which is the attractive factor and not the bacteria themselves although Meadows (1964a) found that sand from which the bacteria had been removed by shaking lost its attractiveness but possibly the bacterial film had also been destroyed by the process.

The precise method, if any, by which the larva tests the substratum or what property of the environment is selected remains to be elucidated but presumably it is chemical. What little is known of the role of surface chemicals in the settlement of marine larvae has been reviewed by Crisp (1965). The presence of a metabolite which is released into the water by metamorphosed larvae and which attracts other larvae is a likely probability. This mechanism has been demonstrated in organisms from other marine environments, e.g. in barnacles and in tubiculous polychaetes. The gregariousness shown in these groups and in so many other sedentary organisms, suggests that metabolites from already settled individuals, including adults, attract the free-swimming larvae. Such a reaction ensures that larvae settle in ecologically suitable areas but although the adaptive significance is clear with filter feeders, it is less obvious with deposit feeders because the young forms would appear to be at a disadvantage in dense populations as they are in danger of elimination through competition with the adults for food.

The size and shape of the sand particles themselves have to be of a suitable nature before a sand becomes attractive but these factors are probably subordinate to the biotic ones mentioned above. The physical environment must obviously be suitable before a population can survive but particle size will be considered later as belonging to our fourth environmental category—a place in which to live. The particle size is perhaps of greater significance to the interstitial fauna than it is to the larger forms although the distribution of the latter is nevertheless controlled to a large extent by the grade of the deposit.

Desiccation

A habitat which is covered with water for half the time and exposed to the air for the other half might be expected to present problems of water balance to the animals living there. The duration of emersion depends, of course, on the level of the shore on which the animal lives. Those low down at the extreme low water springs level may be uncovered for only a few hours a month while those at the extreme high water mark will be exposed to the air for most of the monthly cycle. Animals at this level are virtually terrestrial but we are concerned here with those truly marine forms which need to remain damp for respiratory or other reasons. The problem of desiccation is more acute on sandy than on muddy shores owing to the difference in drainage. Mud flats rarely dry out, even on the surface, but sandy beaches can become quite dry when the tide is out. Desiccation is unlikely to prove a lethal factor to an established beach fauna since the risk of drying up follows a regular pattern under which the animals have evolved. However, intertidal animals have had to develop certain physiological adaptations in order to withstand the risk of desiccation. One corollary is the need for an efficient osmoregulatory system, for one of the first effects of moisture loss is an increase in the concentration of the remaining body fluids. Often, although not always, desiccation is accompanied by an increase in temperature which necessitates a tolerance of high temperature in the intertidal animal. Against this, one should point out that the latent heat of evaporation helps to cool an animal which is losing water, often to a significant degree as was shown earlier. Respiratory complications also arise since the gills of many sea animals are dendritic or feathery with a shape maintained by the supporting action of the water. When removed from the water, such gills collapse and present a smaller area for gaseous exchange.

Such considerations as those above apply to animals which have to endure desiccation but, as we have seen, the fauna of sand or mud is much more likely to avoid the hazard by burrowing. Many rocky shore animals do in fact have to endure and a number of adaptations have been evolved to resist desiccation, for example the development of a thick shell or the secretion of mucus, but some avoid it by sheltering beneath vegetation or under rocks. Those animals which remain on the surface of a depositing shore are usually immigrants from the rocky shore such as winkles or crabs; the truly adapted forms burrow into the substratum where, at least in mud, desiccation presents no problems. Even those forms high up on sandy shores which are rarely wetted take avoiding action. Thus, the ubiquitous sand hopper (*Talitrus*) remains under stranded seaweed which, although it may be baked hard on the upper surface, is usually moist underneath. The comparable inhabitants of tropical shores, the ghost crabs, are much less secretive and, perhaps by virtue of their higher organisation, have become amphibious, being equally at home on dry sand and under water. Even so, these crabs tend to be more active at night when the air is cooler and more humid.

Interaction of weather factors

So far we have considered the various limiting factors only in isolation but in nature, factors never act in isolation and accordingly the modifying effect of one upon the other must be considered. Most environmental factors in the sea are interdependent, for example if the temperature increases, the specific gravity—and hence the salinity—decreases as does the oxygen concentration. As far as marine animals are concerned, the lethal limits of an environmental factor can vary with the values of the other factors. In the sea, the two most

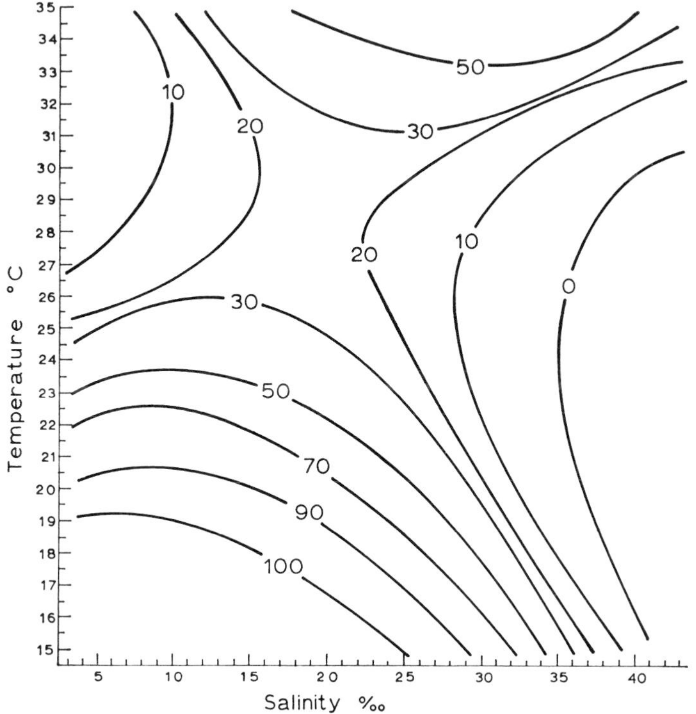

Fig. 30 Example of a two dimensional survival model. Each line shows the percentage mortality under given conditions of temperature and salinity. The model is derived from experiments with the megalops stage of *Sesarma cinereum* at 12 different combinations of temperature and salinity (from Kinne 1964 after Costlow, Bookhout & Monroe 1960)

important interacting factors are temperature and salinity and this is doubly true of the beach where these are the most widely fluctuating factors in the environment.

The concept of a tolerance zone in which species can exist is a familiar one in ecology. In many cases, data are sufficiently precise for a two or three dimensional model to be postulated showing the environmental conditions under which a species can survive. A two dimensional model showing temperature salinity interactions is shown in Fig. 30 while a three dimensional model incorporating a third factor, oxygen, is given in Fig. 31. Diagrams of these types can be constructed from the results of survival experiments in which the effect of a fixed value of one factor is assessed while another is being varied. The experiments are then repeated with a different value for the first factor or it may be that the second factor is kept constant while the first is varied. After various combinations of factors have been studied, it is possible to draw up graphs similar to those shown in the Figs. 30 and 31. The accuracy

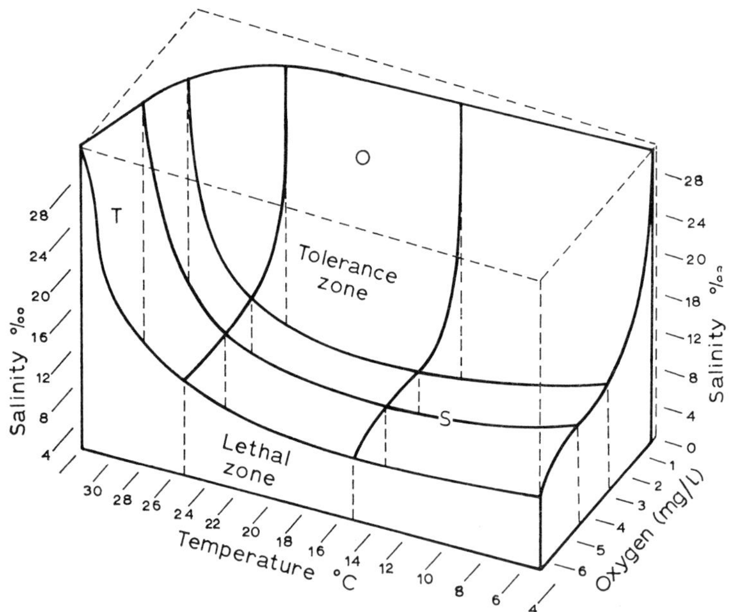

Fig. 31 Example of three dimensional survival model based on the reactions of lobsters. The diagram shows the interaction of temperature, salinity and oxygen. The tolerance zone occurs when all three factors are optimal. T, S and O show the regions in which respectively either temperature, salinity or oxygen acts alone as the lethal factor. (from Kinne 1964 after McLeese 1956)

of these figures depends, of course, on the number of combinations which have been assayed and the smoothly drawn models may exaggerate the accuracy of the data on which they are based. A further source of error is that the experimental data are acquired in the artificial conditions of the laboratory where at the most, only three factors are involved. In the field, there are many other factors which could invalidate the conclusions of the experiment, for example an optimal combination of factors in the laboratory might also stimulate a predator to more effective hunting or favour the spread of a parasite or disease organism. Another possible source of error is the use of constant conditions in most laboratory experiments. It is not unlikely that many animals need a gently fluctuating temperature or salinity for their well-being and their reaction to an unwavering factor may not be typical. It is, however, pedantic to press these points for it is obviously impossible to conduct experiments under completely natural conditions. The validity of experimental results can, of course, be tested by comparing the environmental conditions obtaining in areas where the species is found with the experimental model. In most cases that have been studied, the agreement between field observations and experimental results is reasonably close. It would be possible, theoretically, to devise a survival diagram from field observations alone but the disadvantage of this approach is that for most of the time, the animals are living somewhere near the centre of their tolerance zone and it would be difficult to measure the extreme values unless the ecologist was lucky, or unlucky, enough to be in the field during the most inclement of bad weather.

Unfortunately there are few comprehensive studies of temperature/salinity interactions with animals from the depositing shore but there is no reason to believe that they are different from organisms from other ecosystems in their pattern of environmental tolerances. The effects of temperature/salinity combinations have been reviewed by Kinne (1964) for marine organisms some of which are at least temporary members of the shore fauna. It is worth summarising some of Kinne's conclusions with the proviso that these details may not be mirrored in every respect by dwellers in mud and sand. Not surprisingly, one of the first principles to emerge is that acclimation to salinity change leads to greater tolerance than a sudden transference. A second principle is that resistance to extreme temperatures is decreased at salinities lower than the normal range but increased at salinities higher than normal within reasonable limits. Very extreme salinities at either end of the scale are themselves stress factors which will depress the whole metabolism including resistance to heat or cold. A third principle, which is complementary to the last, is that abnormally low salinities are tolerated best at low temperatures and abnormally high salinities at high temperatures. This generalisation does not apply to all species, however, and several species of shrimp, including some which are found on beaches, may survive low salinities better at high temperatures and high salinities better at low temperatures. Kinne found that the estuarine amphipod *Gammarus duebeni* showed a reciprocal relationship between salinity and temperature in that the tolerable salinity range is widest at optimal temperatures, while the tolerable temperature range is widest at optimal salinities. This strikes one as an eminently reasonable conclusion which may be of general application.

The significance of the oxygen concentration of water as an interacting factor is that at reduced salinities, the oxygen consumption of many euryhaline species increases. It has been suggested that this is due to the extra work done in osmoregulatory activities although there is some doubt whether the energy required for osmoregulation is sufficiently high to account for the relatively large increase in oxygen consumption. Croghan (1961) suggests that the two phenomena may be unrelated for if ion transport during osmoregulation is via an energy-rich bond, the energy required will depend upon the rate at which solution is transported and not upon the difference between the concentrations of the body fluids and the external medium. If this is so, no increase in osmoregulatory work would occur at reduced salinities. Whatever the reason, however, it seems that many species respire at a higher rate at low salinities rather than at high salinities and presumably need more oxygen. It is worth recalling that oxygen is more soluble in cold than in warm water and this may not be unrelated to the fact that some animals can best tolerate low salinities at the colder end of the tolerable temperature range. On the other hand, it seems that organisms which are confined to estuarine regions in the temperate zone became fresh-water forms in warm areas. Lockwood (1962) quoted the example of the temperate brackish water isopod *Jaera nordmanni* and the decapod *Palaemonetes varians* which have probably been the ancestral forms of the Mediterranean fresh-water species *J. balearica* and *P. antennarius* respectively. Similarly, the isopod *Cyathura*, which occurs on the eastern shores of the United States, becomes progressively more a fresh-water form the more southerly its position (Burbank 1959). Lockwood suggests that the stimulating effect of temperature upon the rate of ion uptake from the surrounding water is responsible for the greater tolerance of low salinities at high temperatures, which these observations reveal.

This discussion of temperature/salinity reactions has shown that much more practical work needs to be done before the tentative conclusions drawn above can be put forward as generalised principles. In particular, research on the mud and sand faunas is disappointingly meagre and we cannot assume that they react in a similar way to that of the forms which have been studied. It is, however, legitimate to assume that the reactions of all marine animals to temperature or salinity is modified by the other factor and that we cannot think of a single limiting salinity and limiting temperature, whether high or low, but must adopt the concept of a limiting range.

We have now concluded our consideration of the various factors which make up the 'weather' in the marine environment. The validity of using the term in the context of the shore seems, to the author, to be justified by this discussion. At least there is no difference between the concept of the sum of the physical environmental factors which is called weather in the terrestrial ecosystem and a similar combination of factors in the sea whatever the term used to describe it. We will now turn to the other components which make up the environment according to Andrewartha and Birch's classification and discuss these in relation to their status as limiting factors.

Food

The subject of the food of shore animals has already been considered in some detail in Chapter 3 and, having summarised the conclusions, the need is to relate them to the concept of limiting factors. The fauna of mud and sand can be broadly divided on a nutritive basis into two categories, namely the filter feeders and the deposit feeders. This division is, however, not absolute since some animals can feed by both methods and frequently alternate between the two. Nevertheless, these are the two broad categories and it is as well to appreciate their essential differences. Filter feeding makes use of nutritive particles in the water mass itself while deposit feeding utilises material in or on the surface of the substratum. An important corollary is that the food available to the deposit feeder is limited to that present in its immediate surroundings. Some implications were discussed in Chapter 4 in relation to the amount of time which could be spent in feeding because of the tidal cycles and some evidence was found to suggest that filter feeders grow best low down on the shore where they are covered by water for much of the tidal cycle and hence can feed for longer periods than would be possible higher up the shore. However, our concern here is with food as a limiting factor and the question to be asked is: Does the supply of food ever reach such a low level that it limits the population in circumstances where the supply of other necessary resources are perfectly adequate? To answer this question, it is necessary to separate the filter and deposit feeders. Theoretically, the supply of food to filter feeders should always be adequate if it is present at all, with the important proviso that there may be a seasonal variation in the food. On the whole, filter feeders take members of the nannoplankton, i.e. small, drifting organisms. In temperate zones, the density of the plankton fluctuates regularly throughout the year according to the seasonal variations of the primary producers, the diatoms and algae which make up the phytoplankton. Essentially, the cycle is controlled by light. In spring the increasing day length triggers off a spectacular reproductive increase in the phytoplankton which has been described as the spring 'flowering'. This is followed by a burst in the zooplankton population. There is a gradual decline throughout the summer due to a progressive exhaustion of the nutrients above the summer thermocline, but when this breaks down with the autumnal gales, the nutrients are replenished and since there is still sufficient light, a second, autumn 'flowering' of the diatoms occurs. Thus, throughout the spring, summer and autumn, there is ample food for the filter feeder but in winter, there is very little and at this time of the year, food could possibly become limiting. There is, however, little evidence that starvation during the winter controls populations of filter feeders. The period of shortage is relatively short and can be endured without much harm. The general metabolic rate is, in any case, lowest during the cold weather and energy requirements are reduced. Other adaptations exist such as the storage of fat during the season of plenty or the resort to other methods of feeding.

The position of the deposit feeder is rather different. Its food is usually the film of bacteria which covers the organic and inorganic particles of its substratum. Most bacteria are saprophytic organisms which derive their energy from the breakdown of complex organic materials. It follows, therefore, that a supply of organic detritus or debris is essential for the continued existence of deposit feeders. On the shore, this organic debris can come either from the sea in the form of dead plankton and other organisms or from the land as organic particles carried in rivers or blown by the wind. In the case of detritus from the sea, the supply is likely to be greatest during the spring and summer when the plankton is at its peak but the amount of detritus from the land will be greatest during the winter when the rivers are in flood and gales most frequent. An important subsidiary supply of detritus results from the decay of seaweeds which are broken off from the rocky shores and deposited on the beach. This may occur throughout the year but is more common during periods of rough weather. Thus, at all times of the year, an adequate supply of detritus is available to the deposit feeders and food should not be a limiting factor for the population through seasonal shortage. There remains the possibility of the population over-eating its food supply. This is more likely with sedentary forms than with those which can move through the substratum. The former, usually bivalve molluscs, feed by clearing the surface of the mud with some form of a siphon which sucks in the particles. The feeding area, therefore, is confined to a circle whose radius is equal to the length of the siphon. If several animals were feeding in the same area there would be competition for food and normally, deposit feeders are sufficiently spaced out so that their feeding areas do not overlap. Presumably, the length of the siphon has evolved to ensure an optimum feeding area, sufficiently large to support an individual but no more. Establishment of other organisms in the feeding area is prevented by the occupant sucking them up as soon as they settle as larvae. No such spacing out is necessary in the case of filter feeders since food is normally present in excess and hence they tend to occur much closer together.

Our conclusions, therefore, are that food rarely becomes a limiting factor for the shore fauna. Indeed, one of the reasons for the vast populations of animals in mud is that food is so very plentiful and any species which can overcome the exacting conditions of life in the intertidal zone finds an abundant supply. A distinction must be drawn here between muddy and sandy shores. The organic content of the former is high and favours the establishment of deposit feeders, but the sandy shores are poor in organic matter and are not suitable for such forms. For this reason, most sedentary animals of the sandy shore are filter feeders. To this extent, therefore, the food supply can limit populations on the shore but only in an evolutionary sense.

3 Other Animals and Pathogens

The other animals which make up the component of an animal's environment include members of the same species as well as other kinds. They may react with the animal under consideration either as prey, in which case they would be considered in the last section, as predators or as competitors for some essential resource. It is obvious that either predation or competition could theoretically limit a population and we will consider how far this is true with the shore fauna.

Predation

The dense populations of animals which make up the fauna of depositing shores present a standing invitation to predators to exploit them and, in fact, predation is very heavy. The most important predators are not the other permanent members of the shore but the temporary members particularly those of category 2b ii of our classification on page 18—the obligate, terrestrial, temporary members—or, in other words, the birds which frequent the sea shore. This is not to say there are not many predators amongst the permanent members, but these are of small size and hence, do not account for the removal of the same biomass of prey as do the vertebrate predators. Most of the invertebrate predators are either gastropod molluscs or polychaete worms. The former tend to feed on other molluscs, largely lamellibranchs, which they penetrate by boring small holes through the shell with their radulae. The polychaetes, on the other hand, are more opportunistic and will eat any animal they are able to overpower. These tend to be small soft-bodied forms such as other worms or crustaceans, but not the molluscs whose hard shells protect them from attack. As far as I know, there has been no quantitive study of the amount of animal matter taken by the invertebrate predators and it is not, therefore, possible to say whether such predation can limit the prey population. In some areas the density of predator populations is very high, for example Stephen (1929) records 36 specimens of *Nepthys caeca* per square metre at Kames Bay, Millport, and such numbers might well control the population of their prey species but more data are needed. The role of the predator in a prey population of other ecosystems is by no means clear. In the simplest model, the prey population is reduced to such a low level that it can no longer support the predator, which consequently suffers widespread mortality from starvation. This allows the prey population to build up and the cycle is repeated. Such predator/prey oscillations seem to occur in simple environments such as the arctic tundra with its lemming and vole cycles coupled with that of foxes and birds of prey; but this situation is perhaps not widespread. In some cases, as in biological control of pests, a predator can eliminate or drastically reduce its prey but such conditions are usually unnatural in that neither predator nor prey has evolved in the ecosystem concerned. More probably, it is the prey species which controls the predator by simple density dependent mechanisms. On theoretical grounds, one would not expect the predator to reduce the prey population for this would be to kill the goose which lays the golden egg. Most predators are good conservationists in that they take only the surplus increment, but in certain cases they can prevent the prey population from expanding. Whether or not this occurs on the shore remains to be discovered as far as the invertebrate predators are concerned.

More information is available about the predation by birds on the shore fauna. A detailed comparative survey of the food of British maritime ducks has been made by Olney (for example 1965a, b.). Fig. 32 shows the extent to which ducks feed in salt water and it will be seen that all the species investigated spend a large proportion of their time feeding in the intertidal zone. Fig. 33 shows the food items taken in this region. Only the wigeon (*Anas penelope*) is entirely vegetarian while animal matter forms the bulk of the food of the

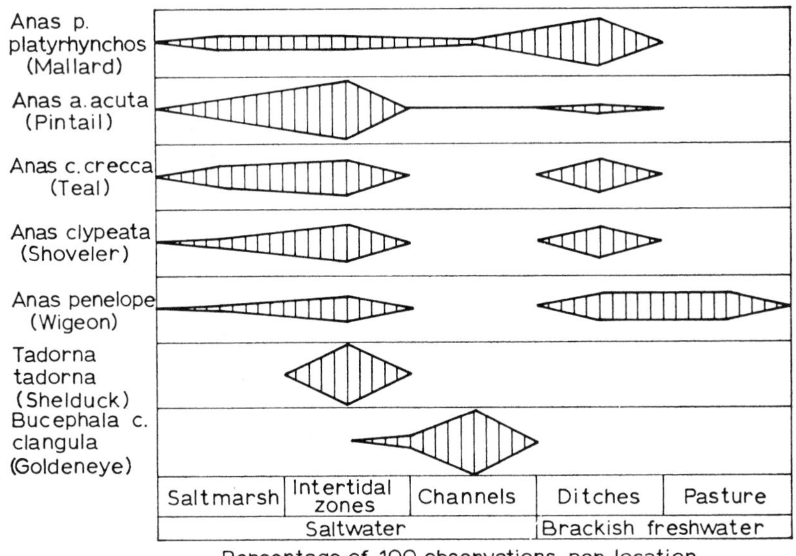

Fig. 32 Extent to which various species of ducks in the River Medway feed in salt water (after Olney 1965b)

other species with the gastropod *Hydrobia ulvae* bearing the brunt of the predator's activities. This mollusc was found in every shelduck and pintail examined. The feeding activity of the birds is related to the behavioural pattern of the prey which, as we have seen (p. 76), undertakes a movement up and down the shore each tidal cycle. The molluscs are at their maximum density (which has been measured at over 12 000 to the square metre) after they have settled on the ebb tide and begun to crawl upon the surface of the mud. The predators have evolved the habit of feeding on the ebb tide and most intensively at the mid-tidal level where most of the molluscs are concentrated. Several of the duck species have been studied in greater detail by Olney. He found (1965a) that the shelduck *Tadorna tadorna* feeds overwhelmingly on

Hydrobia ulvae and on little else, so much so that it suffers high mortality when the weather conditions prevent it from feeding on this mollusc species, as for example, when prolonged onshore gales keep the feeding areas covered with water. The shelduck was especially hit by the hard winter of 1962/3 when the intertidal zone froze and many birds starved to death. The bird feeds by walking on the mud and swinging its bill from side to side over the surface. Olney found that one bird could have upwards of 3000 specimens in its stomach but as the average number of shelducks in any one place is rarely

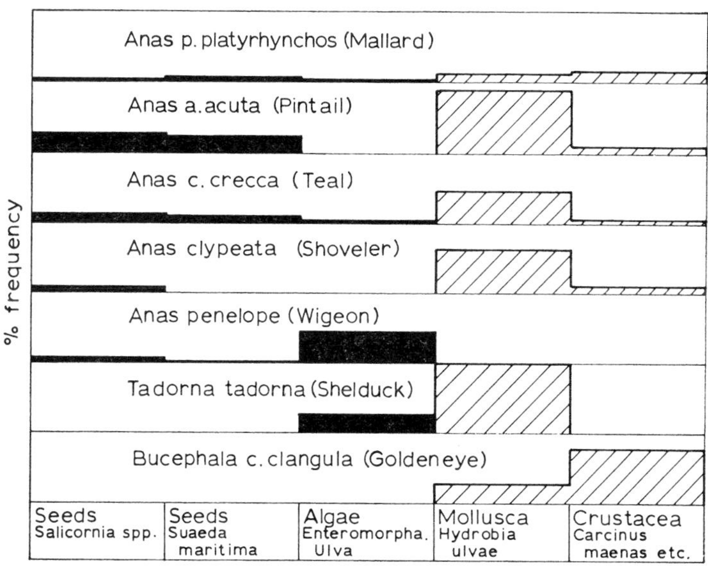

Fig. 33 Food items taken by ducks in the intertidal zone, River Medway (after Olney 1965b)

more than four or five thousand, it is likely that a total of no more than 10–15 000 000 *Hydrobia* is taken by the birds during each feeding period. At a population density of perhaps 10 000 snails to the square metre, this total could be supplied by an infinitesimal fraction of the available feeding area. Hence, it is most unlikely that the number of *Hydrobia* taken by shelduck, prodigious though it is, can in any significant way limit the population of the mollusc.

Olney (1964) also studied the stomach contents of 177 mallards which had been shot over coastal and estuarine areas of Britain. The mallard (*Anas platyrhynchos*) is an omnivorous duck feeding largely on the seeds of salt-marsh plants but some were found to have taken bivalves and crustaceans such as *Littorina littorea* (1·7% frequency) and *Corophium volutator* (also

1·7% frequency). Although the frequency of a particular species in the gut contents might be low, individual birds may sometimes eat large numbers, e.g. one mallard was found to have over 70 specimens of the common shrimp (*Crangon vulgaris*) in its stomach. Another group of mallard collected from mudflats in the Severn Estuary was found to be feeding on *Scrobicularia plana*, *Macoma balthica* and *Corophium volutator* and Olney suggests that these species form the bulk of the diet of mallard feeding on muddy shores. However, the mallard is unlikely to be a controlling factor of mud fauna because of its predominantly herbivorous diet.

Wildfowl are not the only avian predators of the shore for most waders take large numbers of intertidal animals. Goss-Custard (1967) studied the winter predation of *Corophium volutator* by the redshank (*Tringa totanus*). He found that each bird took up to 40 000 items of food each day, mostly specimens of *Corophium volutator* above 4 mm in length. This is a heavy predation pressure and the estimated depletion of the number of large *Corophium* present at the beginning of the winter was put at 38%. This is balanced, to some extent, by recruitment from the smaller size groups but even so, it was estimated that 16% of all *Corophium* present in the previous October were taken during the winter. It was concluded that even this heavy predation did not deplete the prey population to an extent whereby the predator's feeding efficiency was reduced. One may assume, therefore, that predation by redshank does not reduce the population of *Corophium* unduly but it must surely be an important controlling factor.

Incidental information emerging from this study, which was carried out on the Ythan Estuary in Scotland, is that the species of prey taken and the feeding rate depend upon the temperature of the mud. *Corophium* was the most important item in the diet at temperatures above 6°C. The pecking rate of the birds was found not to be dependent upon the density of the prey.

Davidson (1967, 1968) investigated the possibility that the oyster-catcher (*Haematopus ostralegus*) is important in depleting the stocks of shell fish. The preferred food is the commercial cockle, *Cardium edule*, with *Macoma balthica* as the second choice. In the Burry Inlet, South Wales, Davidson found that the food consists almost entirely of cockles about 18–22 mm in length which is just below the legal minimum size for marketing. Larger cockles are not taken even though they may be present in great abundance. The average feeding rate was found to be one *Cardium* per minute for 55% of each low water period. With an average daylight exposure of 510 minutes each 24 hours, this would amount to approximately 280 cockles per bird per day. The number of oyster-catchers varied from 1000 to 16 000 with an average winter population of 8–9000. In other words, the population of oyster-catchers would remove 2 to 3 million cockles each day and it was estimated that the depletion rate of second winter cockles could be as high as 75%. Drinnan (1957) found that the number of oyster-catchers on the oyster-beds in Morecambe Bay could reach up to 30 000 during the winter. He estimated the daily feeding rate to be between 214 and 315 cockles per bird, that is the daily off-take might be somewhere between 6 and 10 million cockles amounting over the winter to a 22% reduction in the total cockle population. Clearly, predation at these levels must exert a controlling factor on the cockles but presumably not to an extent beyond the recuperative powers of the prey population. That the bird may compete with the human predator is another matter but such considerations need not concern us here.

The oyster-catcher is also an important predator of *Mytilus*, the edible mussel, but as these molluscs are not strictly members of the fauna of mud and sand, we will not consider them further.

The beach holds many other waders. Thus, in the Scolopacidae, we have the curlews, godwits, sandpipers, greenshanks, knot, dunlin, sanderling, stints as well as the redshanks. Recurvirostridae has the avocet while the Charadriidae includes many plovers, dotterels and turnstones. Less common coastal birds in other orders are the phalaropes (Phalaropidae) and stone curlews (Burhinidae). These waders do not necessarily spend the whole time on the shore. In fact, during the summer breeding season, many species are found in inland areas and take terrestrial food but during the winter, vast aggregations occur on the shore and as most are carnivorous, the sum total of the invertebrate biomass consumed must be prodigious. With such a diverse community of predators, it is difficult to see how ecological separation is effected but it is most likely to be associated with feeding habits, and it is significant that the length of bill varies widely so that certain food species buried deeply in the mud are physically inacessible to some birds but available to others. However, more detailed information is required of the food and feeding habits of waders before the question can be properly answered.

Finally, we must consider the mammalian members of the temporary fauna. These comprise a surprising number of species including rats, mice, voles, otters and even badgers but their significance is probably negligible compared with that of the birds. Such mammals visit the shore only because they happen to live nearby and the amount of food they take there is unlikely to comprise the bulk of their daily intake. It is difficult to be more precise as not much work has been carried out on the feeding habits of such species. The rocky shore with its surface fauna is more typically visited by mammals than is the beach since few of them are adapted to digging into the substrata. Herbivorous rodents are common in saltmarshes but rare on the beach because of the absence of vegetation.

Competition

Competition with other animals for some necessary resource such as food or living space could limit the spread of a population as effectively as predation. The concept of competition is, however, questionable and its reality has been critically discussed by Andrewartha and Birch (1954). Competition implies a struggle between two contestants in which one is victorious and the other vanquished. In other words, competition is a transitory condition which alters the situation. This conflicts with the concept of an ecosystem which is a balanced system in dynamic equilibrium. Competition can certainly be seen in novel situations such as occurs when a fresh habitat is created by a natural catastrophe or by human interference, for example the raw earth of a railway cutting, but once the climax community is reached, it is doubtful that competition occurs in any real sense. It is true that the presence of one species may prevent the establishment of another but the two are hardly competing. It is simply that species A renders the environment unsuitable for species B in the same way that too low a salinity might also be unsuitable. One does not look upon species as competing with the salinity or with any other inanimate part of the environment and it seems an unnecessary complication to use a separate term for the animate part particularly as it obscures the essential similarity in the way in which the two classes of environmental factors interact with the organism concerned. Competition for space may seem obvious in, for example, sessile organisms crowding onto a limited area of rock face but invariably, either the first species to settle will predominate or another will always succeed in dominating the others. In the latter case, it is elimination, not competition, which has occurred. Competition between animals of the same species is another matter. This is often overt, as in fights between male animals, and would be expected on theoretical grounds from the dogma of natural selection. This form of competition is simply the survival of those individuals which can best adapt to the environment in which they find themselves.

Nevertheless, by whatever name it is described, the presence of animals and plants of other species can profoundly influence a population. We have already considered the case in which the other species is a predator or provides food and we will shortly consider the special cases of parasites and pathogens. The circumstances involved in what we call competition is the effect of two or more species sharing the same essential resource. Apart from food, these include living space, water, oxygen, possibly light and essential elements. Competition in the classical sense will ensue if any of these resources come to be in short supply or changes quantitively. In marginal environments, in which there is a chronic shortage of resources, something approaching competition will be seen. A species may succeed in establishing itself in a 'good' year when the balance is slightly tipped in its favour but it will be eliminated by another species in a 'bad' year when there may be a grave shortage of the resource. However, the presence of the other species is no different in its effect from the absence of the necessary resource as far as the species attempting to establish itself is concerned. On the rocky shore, living space can be limited for organisms which need a hard substratum for attachment, but it is doubtful whether this is true of the animals living in mud and sand. The density of such species is so high that one might be tempted to think that there would be room for no more, but most populations are strongly aggregated and not spread evenly throughout the beach. This means that although the beach may seem full to overflowing in one area, it is quite empty not so far away. The beach itself is also always changing. New loads of silt and sand are carried by rivers or in sea currents and deposited on top of the old to create virgin habitats. Even within the densely populated beaches, single species populations are the rule, at least among the macrofauna, so that there is little opportunity for inter-specific competition. It is true that settlement of other species is often prevented by the established molluscs because the larval forms are swept into the inhalent siphons of the residents if they attempt to settle, but this is simply a case of failure to survive because of the selection of an unsuitable substratum.

Of the other essential resources, water, salts and oxygen are likely to be available to one species as much as to another. Oxygen as well as any other factor may become limiting but there is no reason to think that, above the sulphide layer, the presence of one species reduces the chances of any other of obtaining an adequate supply. If there are shortages, any competition is as likely to be intra-specific as inter-specific. Such considerations reinforce our earlier conclusions that shore animals do not compete with other species although this is not to say that the presence of one species does not affect the well-being of others, in addition to the special cases of predator/prey or host/parasite relationships. Because molluscs are not immortal, a large number of empty shells is a by-product of a molluscan population and these can be present in sufficiently large numbers so as to alter radically the character of the beach. Some beaches are, in fact, largely composed of dead shells or their fragments. The mollusc population is, therefore, influencing the other members of the beach fauna by controlling the abiotic part of their environment. The influence is not limited to the immediate locality of the beach for shell particles may be carried by currents and deposited many miles away from the sire of their formation. The burrowing activity of many members of the beach fauna is another example of the way in which species may interact. *Arenicola*, the lugworm, is the marine counterpart of the earthworm and through its burrowing and mud-eating habits, ensures a uniform mixing of the upper layers of the sediment. This has obviously important implications for organisms living within the surface mud, for example the numbers of bacteria do not begin to decline with depth until the limits of the burrowing animals are passed (Hayes 1964). The metabolic products of the macrofauna can cause significant changes in the environment of the meio- and microfauna. The same is true of the products of decay when the large animal dies. In these and many other ways, the activities of one animal profoundly influence the environment of other species but such changes have always occurred and always will. Consequently, the animals living in the habitat have adapted to such fluctuations and the population suffers no long-term ill effects. If they were unable to cope, the species would not have become established there in the first place.

To sum up this discussion of competition on the shore, one can conclude that other species do not constitute a limiting factor to any animal. Populations are controlled by other species in the sense that the latter may make existence impossible but this is no different from the situation in which a species is unable to exist because of a permanently unfavourable physical factor. As far as animals of the same kind are concerned, there is as yet insufficient evidence to conclude whether or not there is any self-regulating mechanism in populations of marine invertebrates. This possibility for animal populations in general has been extensively discussed by Wynne-Edwards (1962) who believes that animals become aware of their numbers through behaviour specifically evolved to make themselves conspicuous. Such displays, termed epideictic, include the mass flights of starlings or the apparently meaningless calls in which so many species indulge. There are plenty of behaviour patterns in marine invertebrates which could be epideictic in nature but as yet, there is no evidence that they are. The tendency to hold territories amongst aggressive decapod crustaceans in particular (Reese 1964) may quite probably lead to some measure of control over the population size.

Pathogens

The other factor in the particular aspect of the environment which we are considering in this section is the pathological one. Parasites occur in most groups of shore animals and may cause debilitating diseases. In some cases it is not easy to differentiate between parasites and epizoites or animals which are simply attached to the host and not seriously inconveniencing it. Many sedentary animals of the sea will settle on a living mollusc or any other hard organism as readily as on a piece of rock and there is a continuous gradation between these and the genuine ectoparasite. However, it is not intended that this section should comprise a treatise on marine parasitology and our interest here is simply to see whether pathogens or parasites seriously limit the populations of shore animals.

The most common parasites of mud or sand inhabiting invertebrates belong to the Trematoda, a class of the phylum Platyhelminthes, and to the Copepoda, a group of crustaceans whose free living members are either planktonic or mud-dwelling.

Trematodes have the common name of fluke and usually have three consecutive hosts although from two to four are known. Normally, the primary host, defined as that in which the parasite is sexually mature, is a vertebrate while the secondary hosts are invertebrates. In the marine environment, the primary host is usually a fish or a bird and the first secondary host a gastropod mollusc or, more rarely, a bivalve mollusc. The second intermediate host may belong to a variety of taxonomic groups including the coelenterates, turbellarians, annelids, crustaceans, molluscs, insects or even fish. However, many of the flukes which are found on the shore have a copepod as the intermediate host. The trematodes have evolved a complex system of several larval forms each capable of reproducing asexually. This is a typical parasitic insurance against extinction because of the slim chance of being transferred from host to host. The life cycle begins when the egg hatches into a free-swimming larva, the miracidium, which actively penetrates the first host, normally an aquatic snail. Multiplication of the parasite occurs within the host through sporocyst and redia stages, until finally, free swimming ceraria larvae emerge and actively seek and penetrate the intermediate host in which they encyst. Further development of the parasite can occur only if the intermediate host is eaten by the final host and when this occurs, the adult fluke is released in the gut where it may remain or perhaps travel to other parts of the host's body. Variations in the way of transferring from host to host occur but the above account is fairly typical of the life cycle of flukes infesting shore animals.

An exception to the general rule that adult trematodes occur in vertebrates was found by Freeman & Llewellyn (1958) who described the adult form of *Proctoeces subtenuis* in the kidney of *Scrobicularia plana* at Chalkwell in the Thames Estuary, where 100% of all bivalves examined were infected. This is probably an example of an abbreviated life cycle caused by the absence of a suitable vertebrate to serve as a final host.

Hydrobia ulvae acts as the first host for a number of trematodes infecting coastal fish and birds. This is not surprising considering its ubiquity and abundance. An example of such a fluke is *Cryptocotyle jejuna* which occurs as the adult in the intestine of the redshank. The eggs of *Cryptocotyle* pass out with the faeces but do not, as most trematode eggs do, hatch on the shore. Instead, they are eaten by the snail, possibly accidentally, and hatch into miracidia which penetrate the gut wall. On emerging from the snail, the cercaria of *C. jejuna* penetrates the second intermediate host which is one of the gobies.

Another species which infects *Hydrobia* is *Catatropis verrucosum*, whose final host is a goose or duck. In this case, there is no intermediate host and the cercaria, on escaping from the snail, encysts on its shell or on vegetation, usually eel grass, which is eaten by the final host along with the *Hydrobia*. Another fluke which has eliminated the intermediate host is *Maritrema ocoysta* whose cercariae never leave the first host, again *Hydrobia*, but encyst within its body. The life cycle is completed when the snail is eaten by the final host, the redshank. Flukes of the family to which this species belongs frequently use the shore crab *Carcinus maenas* and other intertidal crustaceans as their intermediate hosts.

The copepods are most familiar as active members of the zooplankton both in the sea and in fresh water. We have also seen how one large sub-order, the Harpacticoidea, has become adapted for life in the interstices of mud and sand, but there are a number of other sub-orders which have adopted the parasitic way of life. Such parasitic forms are greatly modified from the typical structure of a copepod, with a worm-like body and greatly exaggerated ovisacs. They are frequently found on the gills of fish but they occur also in shore invertebrates. One of the best known of these, because of its economic significance, is *Mytilicola intestinalis* which lives in the gut of *Mytilus edulis*. As far as it is known, there is no secondary host, which is usually but not invariably found in the life cycle of a parasitic copepod. Examples of other copepod parasites on the shore include *Parenthessius rostratus* which lives in the mantle and testis of *Cardium edule*, *Sarsilenium crassirostris*, an ectoparasite of the polychaete *Harmothöe impar* and *Sphaeronella danica* in *Corophium volututor*.

Not much is known about bacterial or viral infections in the shore fauna. Bacteria abound in the sea and bacterial growth can often be found on moribund animals, but whether these are pathogenic or simply saphrophytic is not always clear. Fungal infections have been reported. Thus, Atkins (1954) recorded the presence of *Leptolegnia marina* on two bivalves, the pholad *Barnea candida* and *Cardium echinatum*. Although pholads are not usually denizens of mud or sand, *Barnea* does bore into firm mud and *C. echinatum*, the prickly cockle is found on the shore although less commonly than the edible variety. The fungal hyphae were present in the gills of *Barnea* and, as white pustules, in the foot and mantle edge of *Cardium*. The same fungus was also found on *Pinnotheres pisum*, the pea crab, which is itself a parasite, or at least a symbiont, of the common mussel *Mytilus edulis*. The fungus attacks the body tissues of the host, especially the gills, as well as the eggs and developing embryos. Fungal growths on the surface of marine animals are not unusual but as with bacteria, it is difficult to separate the cause and effect.

The above examples are sufficient to demonstrate that the beach fauna is infected by pathogenic organisms. We have not considered the temporary members of the shore, the birds and fishes, except in so far as they are involved in the life cycles of shore parasites, although they are, of course, heavily laden with parasites of their own and vulnerable to many diseases. However, the point of interest in this chapter is whether or not pathogens and parasites control the population of their hosts. Unfortunately, there are no data on the incidence of disease in shore populations nor have there been any pathological studies on the shore, but one can draw conclusions from similar investigations in other ecosystems. It is certainly true that shore animals can be killed by disease. *Cardium* infected by *Leptolegnia* mentioned above died, as did the infected *Pinnotheres* and the deaths were almost certainly the result of the fungal attack. In the case of *Pinnotheres*, the direct cause of death is probably interference with the respiratory functions since the gills are widely invaded by fungal mycelia. Fungal and bacterial attack of damaged organs probably accelerates necrosis and hasten the death of the animal.

The preoccupation with the effect of parasites on the final host, often man or his domestic animals, has led to the neglect of their effect on the first host. In many cases, it appears that the snail infected by the larval stages of trematodes is little effected. There is the famous example of a specimen of *Littorina littorea* which lived for seven years while giving off cercariae of *Cryptocotyle lingua* at the rate of over 3000 a day (Meyerhof & Rothschild 1940) without, it seems, coming to much harm. This example is also of interest in demonstrating the amazing powers of asexual reproduction in larval flukes for although the snail was probably infected by more than one sporocyst, there was no chance of reinfection. Once inside the primary host, the miracidium may encyst in a variety of organs but most often in the digestive glands or liver. The sporocyst and the following stage, the redia, move about in the host tissue absorbing nourishment and their activity can cause widespread damage but whether this results in the death of the host or seriously interferes with its metabolism is not clear. In other groups of animals, disease can act as a population regulator particularly at high densities, for then the chance of transmitting the parasite from host to host is enhanced. However, on the shore it is not possible, at present, to state whether or not disease regulates the populations, and this is an obvious field for future research. It is, perhaps, pertinent to mention the phenomenon of parasitic castration which is as effective as death in removing the individual from the reproductive population. A familiar example is the infection of the shore crab, *Carcinus maenas*, by the degenerate cirripede parasite *Sacculina*. The crab becomes sterile in the case of males, and alters its appearance to that of the female. Heavy infestations by a parasite causing such loss of reproductive powers could act as a powerful controlling factor of a population. An analogous phenomenon is the elimination of a pest by releasing sterile males into the population.

This discussion of disease as a population regulator on the shore has unfortunately but necessarily been vague. This is primarily due to lack of data on the subject but by analogy with other ecosystems, the possibility of population control by parasites or disease cannot be discounted.

A Place in Which to Live

A place in which to live is the last of the four environmental categories of Andrewartha and Birch. As was mentioned at the beginning of this chapter, it refers to the presence of some entity in the environment without which the organism would be unable to survive. In a sense, it is a home, although this description is not completely accurate. As a population regulator, this factor can only act in a negative sense, that is it becomes significant by its absence.

A place in which to live can be of two kinds; it may simply be some object which the animal has taken over or it can be manufactured by the animal itself. On the shore, most of the examples come into the first category but the burrows and tubes made by so many of the shore fauna are of the second class. There are many examples of the utilisation of an *objet trouvé*. The case of rocky shore animals being able to exist on the depositing beach because of the presence of stones was mentioned earlier. In developed countries, where much of the shore line is artificially stabilised, there are many surfaces available for attachment varying from concrete seawalls to piers, piling, groynes and slipways. Even the empty shells of the dead molluscs are utilised by barnacles or hydrozoa for such purposes. As a result, many of the typically rocky shore animals and plants are present in large numbers on the depositing shore but without such places to live, their existence on sandy or muddy beaches would be impossible. The full significance of the introduction of such aliens on the ecology of the depositing shore has yet to be worked out but there must be some effects. It is possible that they take food which would otherwise be destined for the filter feeders in the mud, although they themselves probably provide food in the form of planktonic larvae or carrion for the indigenous species. On the other hand, the presence of rocky shore animals may not seriously impinge upon the life of the mud or sand fauna because they live in such a different environment and they may do no more than increase the overall energy level of the ecosystem. The hard surfaces provide attachment for marine algae as well as animals. In an ecosystem so short of primary producers, these seaweeds must increase the level and diversity of the energy flow.

A group of organisms which one would not expect to find on a mud flat is the rock boring pholads yet several species of these bivalve molluscs (e.g. *Pholas*, *Zirfaea* and *Barnea* spp) may bore into hard compacted clay. Indeed, this habitat may illustrate the route through which the rock boring behaviour was evolved. In the context of the present discussion, the compacted clay is their place in which to live for with a slightly more dilatant soil, their presence would be impossible.

Other borers, in this case of wood, may be found on mud flats or sandy beaches if a suitable place in which to live in the shape of a piece of drift wood is present. On the British coast, this is very likely but even if drift wood is absent, wooden boats or breakwaters, which are usually inadequately protected, provide homes for the borers. Under more natural conditions there is rarely a shortage of stranded tree trunks or branches which have been carried down to the sea in rivers. Consequently, the wood borers must be included in the list of species from sand and mud. In Britain, these include both crustacean and molluscan borers, the former the well-known gribble, *Limnoria*, and the latter the even better known shipworm, *Teredo*, as well as *Xylophaga*.

The numerous boulders and stones on a depositing shore provide homes for more than sessile organisms. If one is turned over, it is quite likely that a shore crab will be revealed or a sea slater will crawl hurriedly away. Such stones are instrumental in protecting larger animals from desiccation when the tide is out. Very often the stone has settled into the mud forming a small depression which retains some water as the tide recedes. Such miniature pools may harbour shrimps or even small fish which could certainly not remain on the beach in their absence. The surface of the intertidal zone, including mud flats, is rarely level and often quite extensive pools are left behind by the tide. These form ideal places in which crabs, shrimps, isopods, amphipods, molluscs and fish live which would otherwise have to burrow or retreat with the tide. Thus in semi-permanent pools, a whole community of animals and plants which could not survive on an exposed beach may become established.

The strand line at high tide level provides a further example of a specialised habitat with a fauna which is not typical of the shore. The strand line is more pronounced on sandy beaches than on mud flats and consists of dead or decaying seaweed broken off in storms as well as dead fish, drift wood and, unhappily, empty bottles, tin cans and other rubbish. Usually there may be a number of strand lines for each high tide following the monthly maximum reaches a slightly different level and deposits its contribution a little lower down the beach. This is an ideal place for many animals to live because of its rich organic content while the water retaining capacity of the strand line contrasts markedly with the relatively sterile and fast draining sand. That the habitat is fully exploited is evident to anyone who turns over a heap with his foot. The most noticeable animal is likely to be the sand hopper, *Talitrus saltator*, which jump into the air in myriads. This is a scavenger which spends the day in burrows under the weed and comes out in the evening to feed on any organic matter it can find. *Orchestia gammarella*, a closely related species of the strand line, also hops out of the way when uncovered. These are amphipods and the hopping is achieved by a sudden extension of the abdomen which, as in most other animals of this order, is normally held in a flexed position under the thorax. As they are so high up on the shore, these amphipods are almost terrestrial in their habits and in fact, their home is shared with animals which are usually considered as truly terrestrial, for example beetles, mites and spiders. There is usually a gradation between the lowest of the strand lines whose fauna is mostly marine and the highest in which the animals are terrestrial (Table 13), and this zonation is a reflection of the varying amount of time for which each strand line is wetted by the tide.

Many of the terrestrial species in the strand are dipterous flies, often collectively known as wrack flies. Two species *Fucellia maritima* and *F. fucorum*, occur in Britain and the biology of the former has been studied by Egglishaw (1960). The eggs are deposited in the upper, drier wrack beds and three larval instars are passed there before pupation. The larvae are not adapted to living in wet wrack beds for they lack hairs on their posterior spiracles and the large spines on the ventral surface which are present in other wrack flies. The adults emerge fairly suddenly between the middle and end of March and remain abundant until September after which there is an abrupt decline in numbers. Unlike the larvae, the adults are attracted to the wettest wrack beds and are even found on living *Fucus* growing on concrete structures. *Fucellia maritima* is the only species of wrack-breeding fly known to be attracted to decaying organic matter such as decomposing fish and other unmentionable material. Flies of the family Coelopidae are also very common on the strand line. During the autumn and winter, *Coelopa frigida* and *C. pilipes* are present in particularly large numbers (Egglishaw 1960). *Thoracochaeta zosterae*, a species of sphaerocerid fly, is one of the most abundant species present during the summer months. As all these flies breed in the wrack beds, none of them would be present on the shore but for this 'place in which to live'.

TABLE 13

The relative numbers of terrestrial and marine species in 5 strand lines on a sandy beach in the Menai straits, North Wales. Note that the number of marine species remains roughly constant but that the number of terrestrial species declines markedly from the highest strand line (1) towards the lowest (5). The biomass was highest in the central strand line and fell off to either side. This suggests that the strand line fauna is specifically adapted to its mode of life and declines if conditions become either too terrestrial or too marine. (Data collected by students on a field course, April 1967)

Taxonomic group	Strand line 1		Strand line 2		Strand line 3		Strand line 4		Strand line 5	
	Terrestrial species	Marine species	Terrestrial species	Marine species	Terrestrial species	Marine species	Terrestrial species	Marine species	Terrestrial species	Marine species
Oligochaeta (*Enchytraeus*)		1		1		1		1		1
Amphipoda (*Talitrus*)		1		1		1		1		1
Harpacticoida				1						
Isopoda	1*									
Collembola	2		1							
Coleoptera										
Staphylinidae	3		2		2		1		1	
Other beetles	3		4		1					
Diptera										
adults			1		4		3			
larvae	1		1				1			
Acarina	8		4		4		3		2	
Pseudoscorpionidae	1									
Mollusca (Pulmonata)			1							
Totals	19	2	14	3	11	2	8	2	3	2
Subjective estimate of biomass	low		high		very high		high		very low	

*woodlouse (*Oniscus asellus*)

Other terrestrial forms which occur in the wrack beds are beetles, particularly staphylinid beetles. Some of these such as *Bleduis spectabilis*, are subsocial, whose behaviour in saltmarshes has been described by Bro Larsen (1952). These beetles live in tunnels with side chambers for the developing larvae. The parents keep the young supplied with food and ventilate the burrows. Such parental care is thought, by Bro Larsen, to be related to the need for an adequate oxygen supply for the larvae in the deoxygenated sand in which the tunnels are made. Should the female be killed, the sand lining the tunnel becomes reduced to the sulphide state and the eggs or developing larvae die.

Mites form another group of organisms normally considered terrestrial which owe their presence on the shore to the strand line. Many of these are predators but the oribatid or beetle mites, which are detritus feeders, are present in large numbers. The predators feed on animals even smaller than themselves such as *Anurida maritima*, a member of the Collembola which is an order of primitive, wingless insects. Other arachnids of the strand line include a variety of spiders many of which are strictly terrestrial species able to tolerate the high salt content of their environmen' It is doubtful that they are ever submerged in sea water because they retreat up the shore or ascend plant stalks or other objects during the very few hours each month when the spring tides reach their level of the shore.

Centipedes may also be found in the wrack beds. Some centipede species are truly maritime and have become adapted to life in the littoral zone as has *Hydroschendyla submarina*. The female of this species lays impermeable eggs which are not, therefore, affected osmotically by immersion in sea water. On the other hand, *Strigamia maritima*, another intertidal centipede which lays permeable eggs, moves out of the littoral zone to breed (Lewis 1962). These species are more typical of rocky shores although they seem to prefer crevices in which there are deposits of silt. Other centipedes are more terrestrial and probably avoid contact with sea water. This is true of those in the wrack beds but *Necrophloeophagus longicornis*, a typical arable and grassland species, has been reported from mud flats and shingle beaches and probably experiences regular immersion. A feature of shore centipedes is that they are present in denser populations than those inland, a difference attributed by Lewis to the fewer predators on the shore and to a reduction in parasites because of the difficulty which parasitic mites have in completing their life cycle in the sea. An increased food supply and a more favourable microclimate may well be contributing factors.

Sufficient has been said to demonstrate that a place in which to live is a meaningful concept in the context of the depositing shore. The existence of such places make it possible for certain species to survive in areas where they would otherwise perish. To this extent, this aspect of the environment effects a controlling influence over the population of the shore.

Conclusion

Before this chapter is concluded we can re-examine in terms of the marine environment, the classification of the environmental elements which, it will be remembered, was devised for terrestrial populations. In the light of the examples on the previous pages, it seems clear that in each category, the classification applies to the sea shore. The only objection is a semantic one; 'weather' is not a very appropriate term for the physical features of the sea but there seems to be no extant English word which describes them better. In terms of controlling factors, each category of the environment theoretically can, and in many cases quite definitely does, control the population size of the inhabitants of the shore.

CHAPTER SEVEN
THE SPECIAL CASE OF ESTUARIES

An estuarine shore is a special case in one respect for although it may be structurally similar to other shores with much the same appearance, it differs in that the salinity of the water is subject to fluctuation. It is not necessarily true to say that an estuary is a low salinity area; when the tide is flowing, the fresh-water entry may be dammed back so that the salinity remains that of the open sea. Under certain conditions, much as a hot Sun on an isolated pool at low tide, the salinity may even rise above normal. Nevertheless, it is true that estuarine regions are characterised by periods of low salinity with, in general, a lower overall salinity than that of the neighbouring sea. One must distinguish clearly between estuarine and brackish waters. A typical brackish sea is the Baltic where the salinity may reach a level much lower than that in many estuaries. The basic distinction, however, is that the salinity of brackish water tends to be permanently low whilst that of estuarine water is very variable over short periods. Accordingly the osmotic problems confronting organisms living in the two environments are obviously very different.

Estuaries are formed in flat areas where rivers meet the sea. Typically, they have a funnel-like topography so that it is difficult to say where the river ends and the sea begins; the Thames and Severn illustrate this point very well. But even if the river debouches directly into the sea over a waterfall, it is still correct to describe the conditions where the fresh-water meets the salt as estuarine. The degree to which the sea becomes diluted by the incoming fresh water depends of course, on the size of the river. The two examples quoted above, the Thames and Severn, each deliver many millions of litres of water a day and cause considerable reduction in the salinity far out to sea. On the other hand, there are estuaries which have such a small freshwater intake that they are virtually inland arms of the sea. A good example is the Kingsbridge Estuary in South Devon where salinity conditions are fully marine. Such estuaries are often drowned river valleys or are the consequence of the drying up of a previously large river.

The classical funnel shaped estuary is usually found in areas with a pronounced tidal regime. The tide, racing up the funnel twice a day, prevents the settlement of mud in the centre of the estuary. Instead, it is deposited at the sides to form the extensive mudflats which are so characteristic of estuarine regions. One of the features of this type of estuary is the tidal 'bore' which sweeps up the river mouth at high tide every day and becomes particularly spectacular at the time of the equinoctial high tides. A well known example is the Severn 'bore' which may be 2·5 metres high as it sweeps past Stonebench, a favoured view point, at about 25 km/h. Its effects extend as far inland as Gloucester, well beyond the region where any salt water influence is detectable. The necessary prerequisite for 'bore' formation includes a funnel-shaped mouth with progressively shallower water as the estuary is ascended.

In near tideless areas with little wave action, the sediment is deposited all over the river mouth and overflows onto the bed of the sea. This tends to block the mouth of the river which must, therefore, break through as a number of channels to form a delta. There are several examples in the Mediterranean and its neighbouring inland seas as in the Nile, Rhône, Danube and Volga while extensive deltas typify the mouths of the Amazon, Mississippi, Indus, Ganges and Irrawaddy rivers.

Although salinity is the feature which most clearly distinguishes estuarine from other types of shore, it is not the only one. Other factors in which there are qualitative differences include temperature, water currents and wave action, turbidity and to some extent oxygen. A model of these interactions in an estuary is given by Arthur (1969). In order to understand why these factors should differ it is necessary to deal in a little more detail with the physical characteristics of an estuary.

In the case of salinity, it must not be supposed that there is a gradual and regular change from fresh water in the river to fully marine conditions at sea. To do so is to overlook the dynamic nature of estuaries. Not only is fresh water moving down the river, but salt water is moving up with the tide. At low water, a point far up the estuary of a big river will experience almost fresh-water conditions but during a high spring tide the sea will advance up the estuary and holding back the river water, bathe the same point with scarcely diluted sea water. Even so, the situation is not so simple. If the fresh-water is at all massive, the incoming tide will be incapable of damming it back but instead of mixing with the sea water, the river water, being lighter, will simply flow over the top of it. There will thus be a gradation from fresh water at the surface to sea water on the bottom. Alternatively, the river may push the sea water over to one side (right in the northern hemisphere) so that conditions are marine on one bank and fresh water on the other. Other variations include a three-dimensional tongue of salt water proceeding up the estuary. Some of these possibilities are illustrated in Fig. 34 but there is an infinite number of variations on these general themes including a mixing of fresh-and salt-water if the river flow is fast and the estuary small. This is the nearest one gets in practice to the idealised estuary grading from fresh to salt water.

It should be borne in mind that the estuarine models in Fig. 34 are intended to show the salinity conditions at the time of high tide. During low tide the water high up the estuary will have only a small salt content. A glance at the diagrams will make it clear that an organism in these regions may experience wide variations in salinity within periods as short as a few hours. The methods by which marine invertebrates cope with such a situation has been dealt with

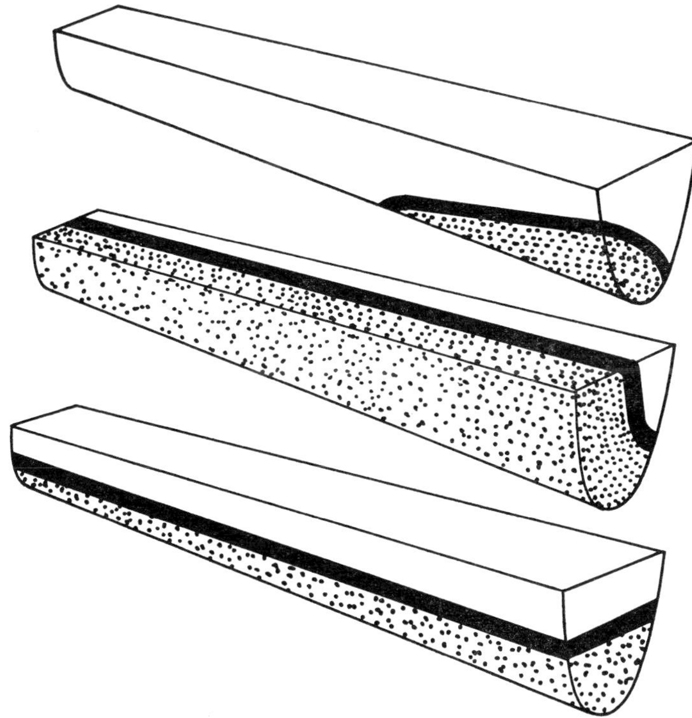

Fig. 34 Structures of estuaries. The diagrams show some typical estuarine conditions within the body of the river, i.e. as if the water had been frozen and lifted from the river bed. White—fresh water, black—mixed or brackish water, stippled—salt water. River flow is from left to right

in the previous chapter but it is interesting to examine here the degree to which regulating mechanisms can be categorised according to the classification of estuarine waters.

An attempt to do this was made by Beadle (1959). He was interested in inland saline waters as well as marine so that many of his conclusions are not relevant to the present discussion. However, he found that the waters of both environments could be classified into three categories with, in the case of brackish water, two extremes and one extensive intermediate zone. The first zone is marine in character with poikilosmotic (polystenohaline) animals unable to tolerate more than a slight reduction in the salinity of the water. The second is the most extensive zone reaching down to salinities of about 5‰. The inhabitants of these waters include both conformers and regulators with varying degrees of control over the blood concentration. The third zone is that with salinities below 5‰. In contrast to the fauna of the previous zones, the third contains animals which are basically fresh water in origin. These, therefore, already have highly efficient osmoregulatory mechanisms, which involve the uptake of ions as well as the production of a hypotonic urine. Beadle makes the interesting point that studies of the fauna of inland saline waters, which have also been colonised by fresh-water animals, show that the upper limit of 5‰ for the penetration of brackish waters by fresh-water animals would be considerably higher if salinity alone was acting as the limiting factor. It is most probable that the presence of other animals is the real limiting factor in estuaries. If this is true, it suggests that animals which originated from the sea have adapted better to estuarine conditions than those that have entered from fresh water or else have got there first.

Beadle also considers a classification of osmoregulatory mechanisms in relation to these three zones. As was mentioned above, the first (marine) zone is populated by animals with little or no osmoregulatory powers and the second by animals with a variety of ways of meeting the problem of a low salinity environment. Some are simply conformers whose tissues can tolerate body fluids of low ionic content. An example is *Mytilus edulis* which can exist in diluted sea water corresponding in salinity to the blood of *Anodonta cygnea*, the swan mussel, a freshwater animal with the lowest known blood concentration. Most animals from this second zone, however, have developed osmoregulatory powers. These may depend on ionic exchange or on the production of hypotonic urine as outlined in the last chapter. The third zone is occupied by animals which are primitively freshwater and whose tissues have become irreversibly adapted to blood of low osmotic pressure. Such organisms seem incapable of maintaining their blood at an osmotic level above that of the surrounding water and consequently, they are rarely found in areas where the salinity of the water rises above 10‰ for any prolonged period while 5‰ is the limit for continual exposure.

Beadle summarised his ideas in a consolidated diagram which is reproduced here. (Fig. 35). It shows the possible relationships between the osmotic concentrations of the body fluids of brackish and fresh-water animals and the external medium. It will be noted that none shows hyposmotic regulation since no plots fall to the right of the isosmotic line. The first group A contains the brackish water animals with wide variations in osmoregulatory ability from that of *Holoecius cordiformis*, the most efficient homoiosmotic species, to the common jelly fish, *Aurelia aurita*, which has little osmoregulatory control. The irregular left-hand edge of area A reflects the failure of these brackish water species to colonise fresh water. B and C represent categories of fresh-water animals. B includes but a few species which are capable of living in fresh-water without the evolution of a renal osmoregulatory mechanism. The group includes crabs of the genus *Potamon* and the Chinese mitten crab *Eriocheir sinensis*. Regulation in these species is by ionic uptake and the loss of salts in the water is low relative to that from across the body surfaces. Thus in *E. sinensis*, only 14% of the total salt loss is via the urine and in *Potamon* this figure is 1%. By contrast, the loss of salts through the urine in *Gammarus*

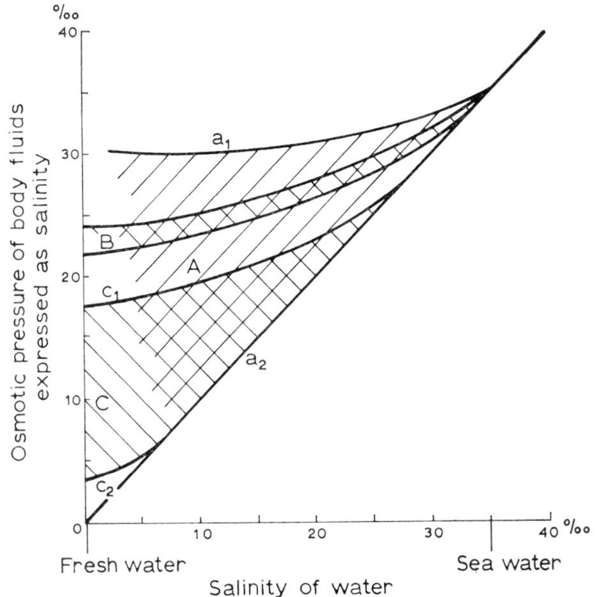

Fig. 35 Diagram representing the relation between the osmotic pressure (expressed as salinity) of the blood and external water in brackish (A) and fresh-water (B, C) animals (after Beadle 1959)

duebeni when producing isosmotic urine is 80% (Lockwood 1964). The significance of these figures is that by avoiding an isosmotic urine, the total salt loss is cut down. The third category C represents the majority of fresh-water animals which effect osmoregulation both by the active uptake of ions and by a renal mechanism coupled with a general lowering of the blood concentration, the lowest being found in *Anodonta* and the highest in the crayfish *Astacus astacus*.

The particular relevance of Beadle's conclusions to our present discussion is that they justify the consideration of estuarine animals as a separate entity distinct both from the faunas of the sea and of fresh water. It is his extensive intermediate zone and its fauna which is the truly estuarine division and the one to which most attention will be paid. It should not, however, be forgotten that estuaries contain other categories of animals which could be classified according to the convention of permanent and temporary members drawn up in Chapter 2. Complications arise in that the terrestrial element is of greater significance on estuarine shores than on other types. This is due largely to the increased vegetative cover found there. Saltmarshes are a feature of the ecosystem and naturally the increased primary productivity has attracted an extensive and diverse fauna. Salt-marshes are not confined to estuaries but they are usually associated with peculiar features such as spits, offshore barrier islands or sheltered bays with shallow water. They are periodically inundated by the sea and could be considered intertidal but their phanerogamic vegetation associate them inevitably with the land and their fauna is terrestrial rather than marine. Nevertheless, the terrestrial forms, principally arthropods, tend to stray onto the beach where with their general tolerance of saline conditions, they are likely to flourish. The probable reason why they are so obvious on estuarine shores is that the organic debris which offers them food and shelter is more persistent because of the absence of major water movements.

As we are concerned only with the shore, we will not consider the estuarine fish but these too, constitute a distinct biological grouping.

Morphologically, the estuarine shore offers the same variety of habitats as the sea coast including rock, although rocky shores are rare since a deposit of mud has generally clothed rocky outcrops. Mud and sand compose the usual estuarine shore away from saltmarshes with mud by far the more common type.

The features, therefore, which distinguish the estuarine from the coastal shores are not structural but are concerned with 'weather' factors of which salinity is the most significant. It follows that the differences between estuarine and marine animals are physiological, not morphological. There is no question of a *Lebensformtyp* as was found for example, with interstitial fauna and it is not possible to identify an estuarine species simply from its general appearance. There may be minor structural modifications associated with the estuarine habit and these will be mentioned below.

In Chapter 3, an analysis of the taxonomic composition of the fauna from muddy, sandy and rocky shores proved to be a useful method for assessing and comparing the animals from each environment. It should be equally illuminating to make a similar analysis of the estuarine fauna. Green (1968) gives lists of animals taken from various habitats in British estuaries and a breakdown of this list according to the taxonomic classification is given in Table 14. Green includes two lists for estuarine sand and mud, the first dealing with the macrofauna and the second with the microfauna although, on the classification adopted in this book, most of the latter would be considered as meiofauna.

In the case of the macrofauna, we find a situation similar to that with the beach fauna, namely a high proportion of annelids and molluscs (14 and 6 respectively out of a total of 32 species). The presence of two oligochaete species amongst the annelids should be noted as this class is not represented on the coast. The molluscs species, however, are proportionally fewer than is the case with coastal communities. On the other hand, the arthropods with 10 species (31 %) are better represented due principally to the insects, which are

TABLE 14

The taxonomic affinities of species living in estuarine sands and muds in Britain (compiled from data in Green 1968)

Macrofauna	
Nemertini	2
Annelida	
Polychaeta	12
Oligochaeta	2
Mollusca	
Lamellibranchiata	5
Gastropoda	1
Crustacea	
Amphipoda	4
Isopoda	2
Insecta	
Coleoptera (Staphylinidae	3
Diptera	1
Meiofauna	
Coelenterata	1
Platyhelminthes (Turbellaria)	5
Nematoda	37
Annelida	
Archiannelida	3
Polychaeta	1
Oligochaeta	5
Crustacea	
Ostracoda	11
Copepoda (Harpacticoida)	21

rare on marine beaches. Decapod crustaceans are not a feature of estuarine shores in Britain and none is recorded by Green although it would be by no means surprising to find *Carcinius maenas*, the shore crab, on an estuarine mud flat.

A total of 67 protozoan species, most of which are ciliates, is included as microfauna. It is unlikely that the list is exhaustive. I have not been able to find a similar list of marine protozoans from mud and sand but very probably, ciliates would again prove to be the dominant group.

The meiofauna shows a preponderance of nematodes and harpacticoid copepods. This almost certainly repeats the species composition of the coastal shores. Again, a full species list of coastal meiofauna is not easy to find but Boaden (1966) lists a total of 115 interstitial animals from various sites, not all intertidal, around Strangford Lough, a land-locked arm of the sea in Northern Ireland. Of the 115 species, 2 were cnidarians, 52 turbellarians, 28 gastrotrichs, 20 annelids (including 14 species of archiannelid) while 13 belonged to various other groups. The absence of nematodes and copepods from this list can only mean that these groups were not recorded for they must certainly have been present. McIntyre (1964) discusses the meiobenthos of sub-littoral muds and his results are, most likely, not untypical of the marine intertidal meiofauna. He found that nematodes were the dominant group followed by copepods of which all but 3 of the 81 species were harpacticoid. The other groups found were Kinorhyncha (5+ species) and Polychaeta (at least 8 species)

The principal conclusions from these comparisons of the meiofauna is that estuarine communities resemble those from fully marine conditions in being dominated by nematodes and harpacticoid copepods, but differ in that Kinorhyncha are not represented in estuaries. The Kinorhyncha is a group of minute animals less than 1 mm in length with spiny segmented bodies. It is often given phyletic rank but Hyman (1951) includes it as a class together with nematodes and four other groups, in the phylum Aschelminthes. It is said to be an exclusively marine group and it would appear that Kinorhyncha are unable to tolerate brackish conditions.

The absence of ostracods from the marine examples given above is of no significance for they are certainly present in large numbers in marine soils. The group contains both fresh-water and marine forms but the estuarine ostracods tend to belong to two families, one primarily marine and the other fresh water. This suggests that the estuarine forms have invaded the ecosystem from both directions. The basically marine family is the Cytheridae while the Cypridae, to which the textbook species *Cypris* belongs, is the fresh-water group. Ostracods may be found as free-swimming animals in pools or they may be part of the meiofauna within the mud.

An interesting study of recent ostracods was made by Barker (1963), a palaeoecologist who had studied fossil ostracods. He had noticed that there were size differences within two species of *Fabanella* found at several sites near Aylesbury. Suspecting that the size difference reflected variations in salinity of the fossil environments, Barker studied recent species of ostracods in the Tamar Estuary near Plymouth to ascertain if a similar situation occurred. His results (Fig. 36) shows that individuals from the more saline reaches of the estuary are larger than those higher up. Although other factors related to salinity could be operative in this size difference, it seems likely that it is salinity itself which is responsible. The precise reasons for the relationship are not clear but they could possibly be concerned with the increased salt content which might permit the growth of a larger carapace.

The harpacticoid copepods have already been mentioned in the discussion of the meiofauna in Chapter 3 but they are as typical of estuarine muds as of the open coast. The number of estuarine species is very high with most of them

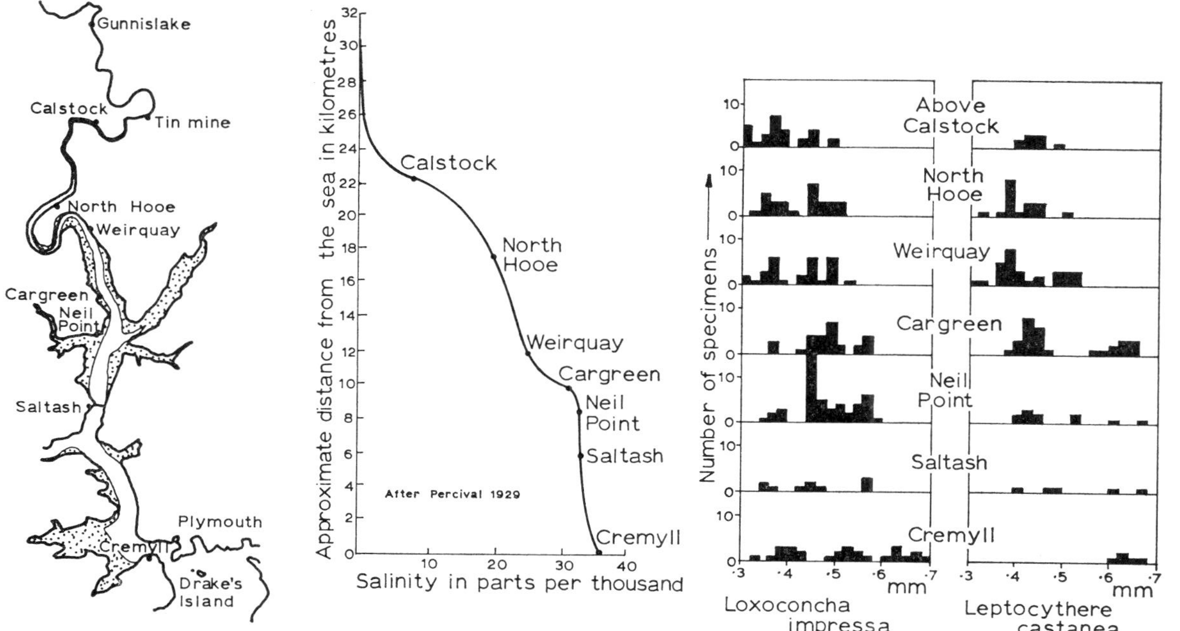

Fig. 36 Size distribution of ostracods in the Tamar Estuary, South Devon. On the left is a map of the estuary; in the centre is a graph showing the salinity of the water at various points in the estuary while on the right are size frequency histograms for two species of ostracods at these points. Note that the size decreases as the estuary is ascended (after Barker 1963)

physiologically adapted to life in brackish water. It is possible to rank the species according to their tolerance to dilution of the media from the near marine *Longipedia* to *Canthocamptus* which is a fresh-water form able to tolerate some degree of salinity.

The annelids amongst the meiofauna are dominated by the archiannelids although these are by no means exclusively estuarine forms. Those species which have become dwellers in brackish water tend to show a remarkable tolerance to wide salinity fluctuations implying a highly developed osmoregulatory mechanism. This and other physiological sophistication, argues against the ancestral or primitive nature often assigned to the group.

Of the nematodes, one can say little except that they are extremely numerous both in species and individuals but this is true of every habitat which the class inhabits. They are typically interstitial forms in estuaries as on the coast. Their success as a group can probably be attributed to their extreme adaptability in feeding behaviour for the various species exploit all possible sources of food from bacteria to metazoan prey. Weiser (1953) gives a classification of feeding methods in free-living nematodes.

The macrofauna is worth considering in some detail to see which species are diagnostic of estuarine conditions. Amongst the polychaetes, *Nereis diversicolor* is as typical of estuaries as a kangaroo is of Australia. It is an extremely euryhaline species which can survive both in fresh water and in fully marine conditions although in nature it avoids both extremes. Most probably, it is its inability to breed in fresh water which has prevented it from colonising that habitat while the presence of the many other annelids of coastal beaches is the probable deterrent to a seward expansion of its range. Much of its success as a species must be due to its versatile feeding activities. It is a scavenger as well as a predator and it will also behave as a detritus feeder. Most likely, the detritus is taken from the substratum but the worm has been observed to secrete a mucous net at the mouth of its burrow to filter out particles from the water (Harley 1950). The adaptations to the estuarine conditions by *N. diversicolor* are not confined to its osmoregulatory powers. Although the larvae can swim, they do not do so but remain in the surface layers of the mud. This is most likely a precaution against being carried away from the estuary and out to sea on the ebb tide or in the river's flow. The price which has to be paid, of course, is reduction in the opportunities for dispersion.

The appearance of oligochaetes in the annelid list is a novelty. Oligochaetes are basically fresh-water forms with some terrestrial members. The Tubificidae is the brackish water family *par excellence* and under suitable conditions they can occur in astronomical numbers. Mention was made earlier of the extreme abundance of tubificids in polluted waters where the absence of other organisms has enabled them to exploit the rich food supply to the full. Some species are tolerant of high salinities; *Tubifex costatus* for example, is found in full strength sea water as well as in lower salinities. The high biomass of the brackish water oligochaetes makes them very important members of the estuarine fauna.

Brinkhurst (1964) described the biology of *Tubifex costatus* from a man-made moat at Hale, Lancashire, which is periodically flooded with brackish water from the River Mersey. It is a common oligochaete in many estuaries but at the site chosen for study, it is present almost in pure culture at a density of about 40 000 per square metre. Most of the worms were found in the top 5 cm of the mud although they could penetrate to deeper regions. The population breeds each year in early summer after a maturation period of two years which is the life span of the individual. *Nereis diversicolor* occurs there at a density of about 1000 per square metre and is the only likely predator.

The estuarine molluscan fauna is essentially lamellibranch and the only gastropod of any importance is *Hydrobia ulvae* whose biology has been extensively covered in earlier chapters. Its upstream distribution in estuaries has been studied by Newell (1964) in the estuary of the River Crouch in Essex. Here, there is a wide variation in salinity which at Battlesbridge ranges from 2·6‰ at low tide to about 15·4‰ at a tidal level eight feet above the river bed. It was found that *Hydrobia* avoids salinities less than about 2·8‰. Since the overall salinity decreases and the higher salinities occur further up the shore as the estuary is ascended, it would be expected that the intertidal range of *Hydrobia* would be progressively restricted to the upper shore in the upper reaches of the estuary. Fig. 37 shows that this does occur and demonstrates that it is the 2·9‰ isohaline which appears to be the limiting factor in the

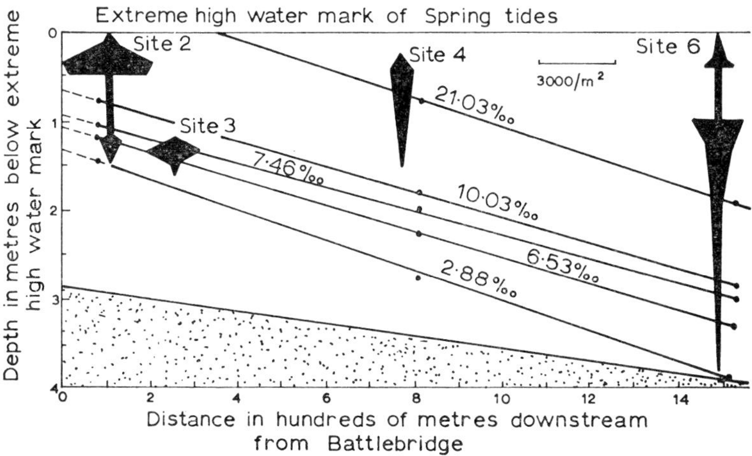

Fig. 37 Intertidal range of *Hydrobia ulvae* in relation to the position occupied in the Crouch estuary. Note that the molluscs are confined to the upper shore in the higher reaches of the estuary. The most likely controlling factor is the 2·88‰ isohaline which is also found higher up the shore as the estuary is ascended (from Newell 1964)

downshore spread of the species. It may also be the factor which ultimately limits the upstream distribution but the limitation does not act in a lethal manner, that is the snail is not prevented from moving into lower salinities because they are lethal. Rather, the restriction is a behavioural one. The method by which *Hydrobia* maintains its position on the shore by floating in and out with the tide was described in Chapter 5. However, there is a danger high up in estuaries that a floating *Hydrobia* would be carried into water of lethal low salinity, but the risk is avoided by the organism losing its floating response at salinities below about 2·1‰. At this value and below, the animal simply closes its operculum if it is swimming and sinks. There is in any case a progressive decline in general activity as the salinity is reduced until eventually the snails on the shore do not respond by floating when the tide covers them.

Apart from the danger of a floating snail being carried into water of lethal salinity, there are other reasons why floating should be avoided under these conditions. *Hydrobia* feeds during the floating period by trapping suspended matter in a mucous net and normally much of this suspended matter consists of diatoms and the bodies of dead zooplankton. But such items are absent from low salinity areas and it would therefore be wasteful for the animal to attempt to feed by this method. It would be far better for it to concentrate on deposit feeding durings its crawling period. For similar reasons, it is likely that the proportion of deposit feeding species in the shore community increases as the estuary is ascended (Newell 1964).

Popham (1966) came to similar conclusions concerning the location on the shore of the fauna in the upper reaches of estuaries. He found that the euryhaline species tend to be more and more restricted to the high tide level towards their upstream limits. This is more true of the deposit feeders than of the filter feeders, which perhaps occur further down the shore because they require longer periods of immersion even though this involves exposure to more dilute conditions. Popham found that salinity and the increasing silt content suspended in the water were the most significant factors in limiting the upstream distribution of the fauna. Of 33 species in the Ribble estuary, 18 do not occur east of Fairhaven where the maximum salinity is under 75% sea water and where the silt content of the water exceeds 15% as measured by volume in a measuring cylinder. Only four of these species occur east of Freckleton where the salinity does not often exceed that of half-strength sea water. The silt content of the water depends, to a large extent, on the degree of wind and wave action which also tends to stir up the bottom fauna itself. As a result, the animals are inclined to be concentrated in depressions on the uneven shore where there is some protection from water movements.

Howells (1964) found a similar tendency for species to occur further up the shore in the upper regions of estuaries. Thus the greatest densities of the isopod *Bathyporeia pilosa* were recorded at 1·21 metres above Ordnance Datum near the mouth of the Towy Estuary and at just over 2·43 metres above datum at a point about 8 kilometres upriver.

Howells' data (Figure 38) show how abruptly the various species drop out as the estuary is ascended. He found also that by far the majority of species had disappeared before a third of the way up the Towy Estuary. The number of individuals, on the other hand, showed a marked increase towards the

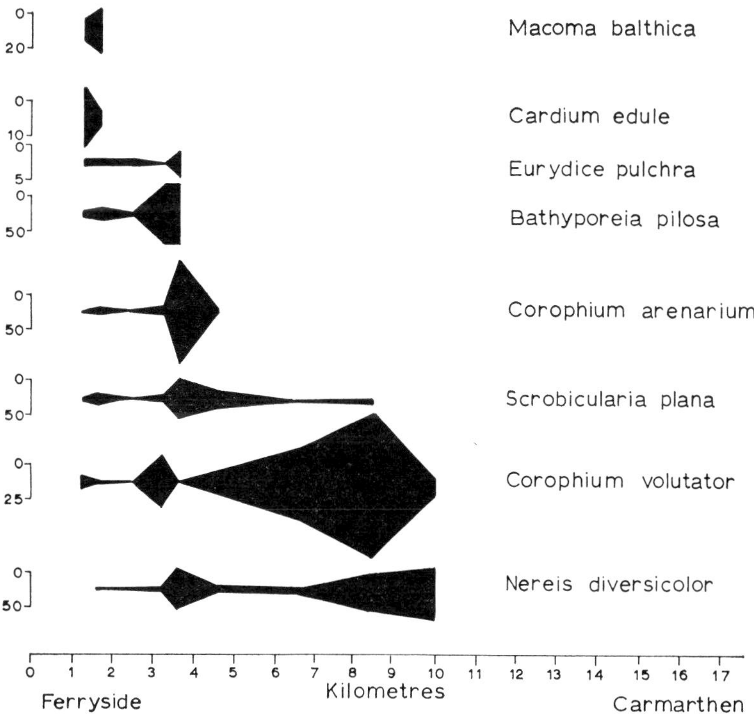

Fig. 38 The degree of up-river penetration of some species in the Towy estuary, South Wales. The sharp limit to the upstream distributions is believed to be due to a salinity factor. Note how each species tends to have its greatest population density at one of three reaches of the estuary viz. lower, middle or upper. These are the areas of optimal salinity for the species (after Howells 1964)

head of the estuary. This tendency is similar to other situations which have often been pointed out elsewhere in this book of few species but many individuals. The common factor throughout is an exacting environment.

Two of the most typical, but not excusively, estuarine lamellibranch species are *Scrobicularia plana* and *Macoma balthica* both of which are inhabitants of mud, both are deposit feeders, or principally so, in the same mid-tidal level and both may have very high population densities—resemblances which may lead to the conclusion that competition occurs between them. However, the principal method by which one species excludes the other is through the ingestion of the larvae as they settle from the plankton. Such ingestion is not selective and cannibalism is as likely as predation. Obviously an established high density population of one species will exclude the other but it will also exclude any other settling organism. This is no doubt the explanation of the frequently observed phenomenon that *Macoma* may be dominant in one area and *Scrobicularia* in another under apparently similar conditions. It is simply that the species with the first massive spat fall in a virgin area is the one which

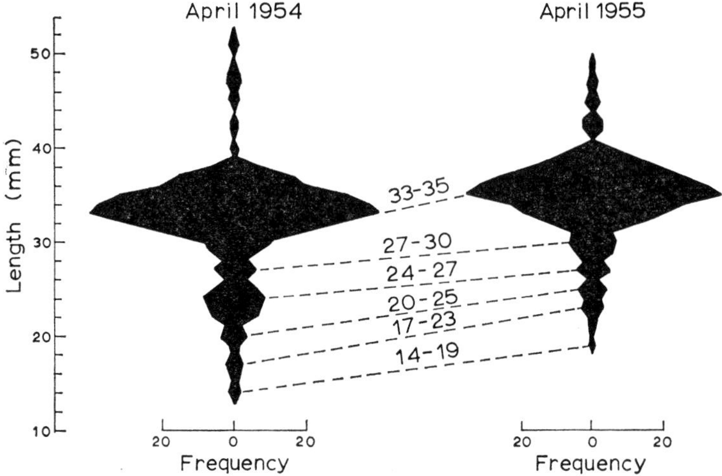

Fig. 39 Length frequency distributions of *Scrobicularia plana* in the Gwendraeth estuary, South Wales. Note how the peaks shift up after one year showing that each represents the spat fall of a particular year. The increase in size for each year group is equal to the growth in one year as measured from the growth rings on the shell. The large 'wings' in the kite diagram show that most of the population examined was hatched in the same year and that, therefore, years of successful development are rare (after Green 1957)

will become dominant. Green (1957) found that there was very little successful settlement of spat of *Scrobicularia plana* in an area of the Gwendraeth Estuary, South Wales, where the density of adults was of the order of 500 per square metre. Less than 2% of the shells were under a year old. Fig. 39 shows quite clearly that most of the animals are of about the same age and that good spat fall years are few and far between. Incidentally, Green deduced that some *Scrobicularia* lived for 18 years.

An obvious ecological difference between the two species is that *Scrobicularia* is a deep burrowing form while *Macoma* is a shallow burrower, but as both species feed on the surface and respire surface water, this difference is not as significant as it may seem. It does, however, have a bearing on their relative vulnerability to predation since *Scrobicularia* is out of reach of all but the longest billed birds, while *Macoma* can be taken by waders with quite short beaks.

It is likely that *Macoma* and *Scrobicularia* are not natural neighbours, because the former is a boreal, cold water form while the latter is a southern species, but in Britain, their ranges overlap. *Macoma* is the more hardy under intensely cold conditions and its specific name of *balthica* refers to the Baltic Sea which regularly freezes over. During the extremely cold British winter of 1962–3, *Macoma* was hardly affected but *Scrobicularia* was wiped out in many areas and decimated in others (Crisp 1964).

Most estuarine crustaceans belong to the isopod or amphipod classes. Only a few are genuine inhabitants of mud or sand for the majority occur under stones and algae or as free-swimming animals. On the coast, isopods are more typical of rocky areas than of beaches but there are some sand dwelling species. *Eurydice pulchra* is one. This animal spends a good deal of the time swimming actively in the water but it also burrows into sand. *Cyathura carinata* is a mud-dweller which occurs in high densities in some areas as in the Tamar Estuary where Spooner and Moore (1940) recorded as many as 378 per square metre. Most isopods are scavengers eating dead material from both animals and plants but some of them are predators capable of overpowering worms or other smaller crustaceans.

Most estuarine amphipods belong to the family Gammaridae of which the 'freshwater shrimp' *Gammarus pulex* is a well known example. The gammarids show an elegant gradation of species from the fresh-water *G. pulex* to fully marine species. *G. duebeni* is a remarkably euryhaline animal which can tolerate both fresh and full strength sea water. It is not found with *pulex* in fresh water since its reproduction rate is lower than that of the latter species (Hynes 1955), but where *pulex* is absent it is the fresh-water representative of the group. At the other end of its range, *G. duebeni* overlaps with *G. zaddachi* and *G. salinus*, two brackish species well adapted to saline conditions. The existence of *G. duebeni* in the head waters of estuaries probably reflects its ability to withstand rapid and massive salinity changes better than other gammarid species. Its reactions to salinity changes have been closely studied as it has become the guinea-pig for students of osmoregulation. It has been shown experimentally to be able to tolerate concentrations twice that of sea water even though it is unlikely that it would meet such conditions in nature except perhaps in a pool exposed to prolonged evaporation. Nevertheless, such a wide tolerance of salinity changes is the hallmark of an estuarine animal which *G. duebeni* indubitably is. A curious feature of its biology, which is not fully understood, is that the sex of the young produced varies with the environmental temperature; below 5°C the young are all males and above 6°C they are females both sexes are produced at intermediate temperatures.

Gammarus zaddachi and *G. salinus* used to be considered as two races of *G. zaddachi* but they are now recognised as separate species. *G. zaddachi* proper occurs further up the estuary than does *G. salinus* but there is a considerable degree of overlap. *G. salinus* also overlaps the distribution of *G. locusta* at the seaward end of its range. *G. chevreuxi* is found with *G. zaddachi* and has similar salinity requirements. Although these closely related species show such an extreme overlapping of their distributions, no interbreeding occurs.

G. locusta is basically a marine animal but it regularly penetrates for some distance up the mouths of estuaries. The most marine of the gammarids used to be known as *G. marinus* but this name confuses five common shore species which are now put into the separate genus *Marinogammarus*. The most tolerant of dilution of the medium is *M. marinus* which may penetrate half way up an estuary, but the other species are more definitely marine although a few, such as *M. stoerensis* and *M. pirloti*, are found only on shores where there is some form of fresh-water flow or seepage onto the beach.

Melita is another amphipod genus found in estuaries, although a very common shore representative, *M. hergensis* is not found in brackish water. *M. pellucida* on the other hand, is estuarine and is typically associated with *Gammarus zaddachi* and *G. chevreuxi*, where the salinity is low. *M. palmata* is a shore species which prefers areas of fresh-water seepage. It also penetrates estuaries and is particularly common in stony areas but it cannot tolerate much dilution of its marine surroundings.

All these amphipods can be found on muddy or sandy beaches but they usually live under stones rather than in the substratum. A true burrowing amphipod from estuaries is *Corophium volutator* and although genuinely estuarine, it can also be found on the seashore. It is well able to tolerate low salinities and can even survive in fresh water for several days. *C. volutator* is an inhabitant of mud while a closely related species, *C. arenarium*, prefers sand.

Some recent studies on estuarine *Corophium* have been published by McLusky (1967, 1968). Laboratory experiments had shown that a salinity of 2‰ was the minimum which could be tolerated by *Corophium* for long periods (at least 500 hours) but that 5‰ was necessary for moulting and growth. In the higher salinity range, it was found that *Corophium* can survive at a salinity of 50‰, almost $1\frac{1}{2}$ times the value of full strength sea water, but 30‰ was the maximum for normal metabolic functioning. These results show a close correlation with field observations made by McLusky (1968) in the Ythan Estuary in Aberdeenshire. He showed that an interstitial salinity of 2‰ was critical in limiting the distribution of *Corophium* up the estuary although few animals were found below 5‰. Breeding took place only at salinities above 7·5‰. Salinity, therefore, was considered to be the 'ecological master factor' controlling the distribution and abundance of the species but in areas where salinity conditions were favourable, the quality of the substratum became important.

An interesting morphological adaptation to low salinity is shown by estuarine amphipods. Species found higher up the estuary are distinctly more bristly than those occurring lower down in less dilute surroundings (Fig. 40). This difference is particularly obvious in the series of *Gammarus* species. The

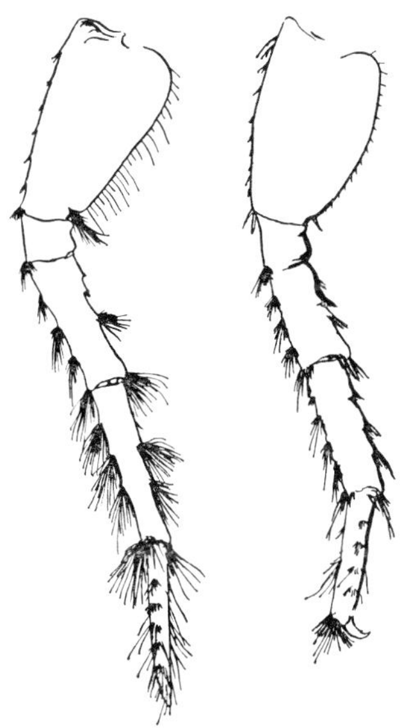

Fig. 40 The seventh peraeopod of the amphipod *Gammarus zaddachi*. The one on the left is from a fresh-water sub-species while the other is from a race which lives near the mouth of the estuary. The increase in hairiness with dilution of the medium may reflect the need for a greater surface area for propulsion in the less dense medium (after Spooner 1947)

most likely explanation of this phenomenon is that in less dense brackish water, more and longer bristles on the limbs are necessary to exert adequate purchase on the water during swimming movements.

CHAPTER EIGHT
THE PLANTS OF THE SHORE

The vegetation of depositing shores between the tide marks is meagre compared with that of rocky shores and its description would not occupy many pages unless one included saltmarshes and other semi-terrestrial habitats along the coast. A proper consideration of the maritime flora is outside the scope of this book and beyond the competence of its author, but a brief treatment is necessary in order that the contribution of phanerogams to the productivity of the shore may be adequately assessed in the next chapter. A full account of coastal vegetation has been published in book form by Chapman (1964).

One of the most noticeable differences between the depositing and erosion shores from a vegetation standpoint is that the flora of rock is climatic and stationary whereas that of the depositing shore is often in a state of flux. This is particularly true of the upper reaches of muddy shores which are giving way to saltmarsh. In this sense, the vegetation of the rocky shore is often termed *static* while that of the depositing shore is called *dynamic*. A further difference is that the flora, as well as the fauna, shows a well marked zonation on rocky shores whereas zonation is a doubtful or non-existent phenomenon on depositing shores. Instead, there is a succession of vegetation types beginning with the first algal colonisers of bare mud and leading to the full climax community of a mature saltmarsh. Consequently, a description of the vegetation must be sequential and in phase with the physical changes which take place when a saltmarsh is formed. The flora of each of the three basic types of depositing shore is distinct and will be considered in turn.

Shingle beaches

The flora of shingle beaches has the property of rarity (in common with the fauna). The reason is the same as with the fauna. On a constantly shifting surface composed of large pebbles, no rooted plant can hope to exist for long. Even regions well above normal high tide mark can be affected during periods of storm and shingle beaches frequently experience rough seas. During bad weather, the shingle bank is often breached by the tide and the pebbles spread out inland as a shingle fan which destroys any evanescent vegetation growing there. 'Destroy' is not strictly correct, for although the plant body is completely covered, the prostrated branches put out new roots and vertical shoots soon push through to the surface. Consequently, a sparse vegetation develops on the pebbly soil behind a shingle bank. These plants are usually brought to the shore as seeds in the rubbish deposited on the strand line. This rubbish will be considered in more detail below, but it may be mentioned here that its humus is certainly of great importance in enabling germinating seeds to develop. The variety of species concerned is not great but the most important is the shrubby seablite, *Suaeda fruticosa.* Other plants, which may be more typical of sand dunes but which nevertheless appear on shingle, include sand couch-grass (*Agropyron junceiforme*) and sea couch-grass (*A. pungens*) as well as creeping fescue (*Festuca rubra*), sea campion (*Silene maritima*), sea sandwort (*Honkenya peploides*) and such typically dry-land plants as groundsel (*Senecio vulgaris*) and sowthistle (*Sonchus oleraceus*). Salinity is not a problem to the shingle beach flora for the plants are rooted in fresh, not salt water. The reason for this is that they are growing well above sea level because of the steep slope of the beach. Their water supply comes from three major sources: rain, normal dew and internal dew which forms in the cool interstitial spaces within the beach even in warm weather. Whatever its source, the water soon percolates down between the pebbles and comes to rest on *top* of the salt water table with little mixing. Undoubtedly, the major environmental hazard is destruction by wave action and new communities are constantly reforming as the old ones are destroyed. Not all plants are able to regenerate from the covered shoots as *Suaeda* can, but those which are killed are important in that they decay and provide humus to a substratum notably lacking in this commodity. There is, however, no question of the pioneering plants binding the shingle together and stabilising the beach as happens with marram grass and sand dunes. Although *Suaeda* originates in the strand line it can grow only well up the shore because the roots of the seedlings are very susceptible to water-logging.

It should be clear even from this brief account that the flora of shingle beaches are not of any great significance to the animals or plants of the intertidal sector of the shore. Not only does the rooted plant community occur above the high tide mark, but it is essentially terrestrial in composition and behaviour with few connections with the marine environment.

Sandy shores

A pure sandy shore or one with very little mud or organic matter does not support vegetation. Apart from other considerations, the unstable nature of the surface makes it very difficult for seeds to germinate or for delicate seedlings to become established. Even on muddy sands, seed germination is unknown between the tide marks and all reproduction is vegetative. On some sandy shores with sufficient organic content, diatoms can be found on the surface but as these are more characteristic of muddy shores, their consideration will be left until later. This section would now be closed if it were not for the strand line which occurs even on the most barren of sandy coasts. Although the strand line is found on all types of shore it is, perhaps, of greatest significance on sand because of the absence of any other organic material.

The fauna of the strand line was considered in Chapter 6, in which the presence of several strand lines, resulting from successive high tides at different levels, was pointed out. The highest line is left by the equinoctial spring tides which occur only twice a year, so that conditions within the debris become quite terrestrial. This is true only on more sheltered coasts, for storms backing up normal high tides will often raise the sea level well above its expected height and affect the equinoctial strand line. However, as a general rule there may be anything up to half a dozen strand lines all close together and even touching, but with a distinct zonation from marine conditions at the seaward side to almost fully terrestrial at the top of the shore. This zonation is reflected, too, in the distribution of the animals with amphipods and other marine creatures lower down and fully terrestrial forms such as spiders, woodlice and beetles in the upper strand line.

Because of the danger of storm tides, existence is precarious for any rooted plants growing in the strand line. It is not simply that they may suffer burial or physical damage, for in addition most are unable to tolerate immersion in salt water for any length of time. However, their environment must necessarily be brackish to some extent, even above high tide mark, because of spray or fine droplets of sea water carried on an inshore wind and recognisable morphological adaptations to these conditions may be found. Curiously enough, these are similar to the modifications shown by some desert plants although, on reflection, it will be appreciated that the desert and shore environments are physiologically not so very different. In both habitats there is a shortage of water, in the first case because there is very little rain and in the second because any water in the plant cells is in danger of being lost by osmosis either through the roots or through the leaf surfaces which are often encrusted with salt. Consequently, both desert and shore plants characteristically have thick waterproof cuticles with fleshy leaves adapted for the storage of water.

The organic debris carried in by the tides is of prime importance to the establishment of rooted plants in the strand line. The seeds themselves are brought in on the tide, but they would be unable to germinate were it not for the humus resulting from the decay of the organic matter from previous strandings. Much of this organic matter is detached seaweed rich in nitrogenous material. Successful germination, therefore, requires that the seed should be deposited in a mature, well rotted strand line and that there should be a period of calm during the germination to reduce the risk of mechanical damage to the delicate seedling. Once the plant is established, it can tolerate a certain degree of burial in the sand and can find its way through to the surface. Many of the mature plants have their roots not in the strand line through which they are growing but in the humus of an earlier strand line some distance below the surface.

Very little work has been done on the plant succession of the drift line and this remains a fertile field for investigation. However, most of the species are annuals and belong to three families, the Cruciferae, Chenopodiaceae and the Polygoniaceae. Typical colonisers of the strand line include saltwort (*Salsola kali*), sea rocket (*Cakile maritima*), sea sandwort (*Honkenya peploides*), sand couch-grass (*Agropyron junceiforme*) and two species of orache, *Atriplex littoralis* and *A. hastata.* The couch-grass is a particularly aggressive species since it is not incapacitated by occasional immersion in sea water nor by being buried in the sand. In fact, its growth is stimulated under such conditions and consequently its rapidly growing, ramifying rhizomes tend to stabilise the drifting sands. This is the first step towards the formation of sand dunes.

A consideration of the formation of dunes and their vegetation is out of place here since sand dunes are never inundated by the tide and are outside our definition of the shore. However, sand dunes originate on the shore and it is as well to be aware of the wide tract of plant life which backs an otherwise vegetationally sterile environment. Dune formation begins after the sand which piles up around the drift line plants becomes sufficiently stabilised by sea couch-grass for other colonisers to come in. One of the first of these is marram grass (*Ammophila arenaria*) which enters either as seeds or as regenerating rhizomes as soon as the risk of tidal flooding recedes. Marram grass is perhaps the most important plant to complete the stabilising process begun by sea couch-grass. As long as sand continues to accumulate around the vertical shoots of the marram grass, the plant grows vigorously to maintain itself on the surface. The tussocky growths so produced trap further sand and the level is progressively built up. The rhizomes which are now well below the ground degenerate and are replaced by adventitious roots at the surface. Eventually, dense growths of marram grass are formed behind the shore, but nevertheless there are still extensive expanses of open sand. At this point, the succession has progressed to the yellow dune stage with conditions passing from marine to terrestrial and away from the subject of this book.

Muddy Shores

In this section, all types of shore with an appreciable mud content will be considered. These are the only shores with a phanerogam flora between the tide marks. The dominant plant low down on the shore is *Zostera*, the eel grass, with two species, *Z. marina* and *Z. nana*, which are completely submerged for long periods during high tide. Some are permanently covered since the *Zostera* community can extend into the sublittoral regions. In more brackish waters a second grass, *Ruppia maritima*, may be almost as abundant as *Zostera*. The *Zostera* beds used to be much more extensive than they are now but in the early 1930s a wasting disease spread rapidly through the populations. The decline of the *Zostera* beds has had serious consequences on those birds which feed extensively on them, particularly brent geese and widgeon. The latter adapted rather better than the geese to new foods, but now both species have recovered satisfactorily and the *Zostera* beds themselves are becoming re-established.

The phanerogamic vegetation of the sea shore has derived from terrestrial plants which normally cannot survive prolonged submersion in water. Death usually results from gross interference in the respiratory metabolism even if it is only the roots that become water-logged. There are plenty of fresh-water plants which have overcome these difficulties, but the intertidal plant has the added problem of the high salt content of the water to contend with as well as the extremes of temperature and other environmental factors which were discussed earlier in relation to the animals. The plants have adjusted rather differently from the animals to these conditions. Usually the intertidal animal is so well adapted to intertidal conditions that it cannot do without them, thus an estuarine animal cannot survive if salinity conditions go beyond certain limits; even fully marine conditions may be lethal. In other words, it needs the conditions under which it exists. This is not necessarily true of plants. Many, perhaps most, do not need to be doused regularly in sea water nor do they require saline conditions; they may as easily grow in the corner of one's garden as on the sea shore. All that has happened is that they have learnt to tolerate the saline conditions and are, therefore, at an advantage over those plants which have not. It is a further example of a harsh environment which has rich rewards for any organism able to adapt to it. Plants have usually colonised the shore from the land while animals have come from the sea, but whereas animals have adapted physiologically to the new conditions, plants have merely evolved a toleration of them. As with all generalisations, exceptions can be found to this rule but on the whole it remains true.

The most extensive area of phanerogam vegetation on muddy shores are the saltmarshes. These are produced by the colonisation of bare mud flats by pioneering grasses which trap mud particles between their stems and leaves, in a way similar to that by which marram grass builds up the sand dune. As a result of this deposition of mud, the level rises so that the expanses become flooded less and less frequently. Under these conditions, plants less tolerant of salt water than the early colonisers are able to come in and through the production of vegetable detritus and the trapping of more mud, the level is raised still further until only the very highest of equinoctial spring tides flood the marsh and the climax vegetation is reached. There are, therefore, two zonations concerned with saltmarshes. The first is spatial and is used in the normal sense of a progression from bare mud low down on the shore to mature marsh a long way inland. The second is temporal and is known as a succession in the sense used earlier in this Chapter. The fact that a succession occurs with bare mud giving place to saltmarsh means that the process is a constructive one with land taking over from the sea. It is doubtful whether this is of any great significance in the global balance between land and sea under natural conditions since the area of saltmarsh is relatively small, but economically it is an important process. Many of the saltmarshes have been reclaimed through human engineering and cultivated, although a considerable time must lapse before the soil is sufficiently leached of salt for crops to grow. A more common economic use to which saltmarsh development is put is the rough grazing of sheep and other livestock, particularly on the salt pasture which develops on sandy marshes.

Because even the apparently terrestrial salt pasture is occasionally inundated by flood tides, there is a greater justification for us to consider the development of saltmarshes than was the case with sand dunes. There is certainly a transfer of nutrients from the shore to the marsh when a high tide deposits its load of silt and much dead plant material is carried away from the marsh on the ebb.

The necessary precursor for the development of saltmarsh is quiet water. For this reason, saltmarshes are confined to estuaries and protected bays or to any area sheltered by sand spits or similar formations. Most of these areas are muddy for reasons which were outlined in Chapter 1. Sand or shingle spits or barrier islands are unstable structures and their mobility has a complex influence on the development of saltmarsh associated with them. Offshore islands often grow parallel with the shore and throw off a series of landward hooks behind which saltmarshes in various stages of development may be found. Thus saltmarsh formation is not always from shore to land but may be found running along the shore. The size of saltmarshes can vary exceedingly depending on the local topography, from pockets of a few hectares in extent to areas of several square kilometres. The steeper the slope of the shore, the smaller the area of marsh; thus really large marshes are confined mainly to estuarine regions and particularly to those rivers with wide mouths.

To trace the development of a saltmarsh, it is expedient to begin at the bare mud stage. On muddy sand the first coloniser may be *Zostera*, although this occurs so far down the shore that it has little influence on the development of marsh. The more typical colonisers are species of *Spartina*, the cord grass, which is less tolerant than *Zostera* of prolonged immersion in salt water The significance of *Spartina* in saltmarsh formation cannot be overstated, for it is the most important agent in the consolidation of bare mud. Although it is less marine than *Zostera*, *Spartina* is, nevertheless, extremely tolerant of salt and is a genuine intertidal plant. Before describing the subsequent succession, it is worth looking at the cord grass community, or Spartinetum as it is called, a little more closely. A point which should be stressed, with the following Chapter in mind, is that the *Spartina* forms a complete carpet over the mud and is not discontinuous as is the cover of marram grass on the young sand dunes. Area for area, therefore, the young saltmarsh is more productive than the young dune. The ability of *Spartina* to stabilise mud flats has been exploited commercially in land reclamation schemes. The species used is *S. townsendii* whose curious history has been reviewed by Lambert (1964). This species is a natural hybrid between the native *S. maritima* and *S. alterniflora* which was probably introduced accidentally from North America to the Southampton area early in the 19th century. The hybrid was described as a new species from material collected in 1878 at Hythe in Southampton Water and it has proved to be a particularly aggressive form, illustrating to a high degree the phenomenon of hybrid vigour. It is unusual for a hybrid in being fertile, or at least some strains are fertile. These proved on examination to be diploid. The number of chromosomes of the parent species are 60 and 62 respectively, while counts of 62 and about 120 have been returned for the hybrid. Those plants with 62 chromosomes are infertile while the fertile plants have counts of 120, 122 (commonest) and 124. *S. townsendii* spread rapidly along the south coast of England assisted by artificial transplanting, and it has also been introduced into Holland and elswhere for land reclamation purposes.

The rapid spread of *Spartina* has been arrested over the past forty years by a disease known as 'die-back'. The symptoms include death of the underground buds and rotting of the rhizome apices. Two types of the disease have been recognised: the first is called 'channel die-back' as it occurs along the edges of large creeks, and the second is termed 'pan die-back' because it develops as a pan in low-lying areas within the centre of the marsh. There is no evidence that 'die-back' is caused by pollution or by pathogenic organisms and it is most probably due to unfavourable habitat changes associated with poor drainage and water-logging of the soil. Although *Spartina* is much more tolerant of wet soils than most other saltmarsh plants, it is not completely unaffected.

One consequence of the spread of *S. townsendii* is that it has replaced other plants as the primary coloniser of bare mud. The native *Spartina* is not an important coloniser, although locally it can form dense stands. Normally the first colonisers belong to a community which is dominated by species of the glasswort (*Salicornia*). *Suaeda maritima*, the seablite, may also be very abundant and can even become dominant. In other areas the sea poa, *Puccinellia maritima*, becomes co-dominant with *Salicornia*.

The Salicornietum can be the primary community of sandy marshes as well as of mud, but sandy pioneers of the west coast of Britain are grasses which form two main communities. The first is dominated by *Puccinellia* and the second by *Festuca rubra*, the creeping fescue.

There is no spectacular change from the primary communities to later stages, but there is a progressive invasion of other species, particularly as the level of the marsh rises and flooding becomes less frequent. In some areas, the mature saltmarsh vegetation has been recognised as a General Saltmarsh Community with a lower sector which later develops into the upper near the mean high water mark. The lower community is more frequently covered by the tide and contains species with a tolerance to sea water greater than that shown by species of the upper community. The General Saltmarsh Community is displaced by *Spartina townsendii* in those areas where the species has become established while the development of the Community on sandy marshes in the west is often retarded through grazing by livestock and, in the days before myxomatosis, by rabbits.

The General Saltmarsh Community is not a climax community and gives way to others, as conditions become more terrestrial. It must be remembered that each tidal flooding deposits a layer of silt over the marsh and raises its level until a stage is reached when the sea can no longer reach it. Even so, the soil remains impregnated with salt for a period which varies with the original salinity of the sea water and with the local climatic and other conditions affecting the leaching rate. A saltmarsh situated far up an estuary is obviously much less saline than a coastal one and its succession may be quite different. Very often, the General Saltmarsh Community is omitted altogether and the Spartinetum develops directly into a fresh water swamp dominated by *Phragmites communis* or, if conditions remain brackish, to a swamp composed of *Scirpus maritimus*, the sea club-rush, as the dominant with species of the bullrush (*Schoenoplectus*). The development of a fresh-water swamp depends on the influx of fresh-water streams into the saltmarsh. In the absence of fresh water, typical high level communities succeeding the General Saltmarsh Community include a Juncetum dominated by either *Juncus geradi* or *J. maritima*, the sea rush. Alternatively, both species may be present. As the Juncetum is one of the final communities in the development of a saltmarsh, it is found in the region where the strand line is located. The consequent rich organic content of the soil is responsible for the presence of many nitrogen-loving plants such as *Atriplex hastata* and *Artemisia maritima* (sea wormwood). The presence of these plants also depends on the seeds being carried to the area by the tide.

Some of the possible successions which may be found on the south coast of England are shown in Fig. 41. It is, of course, impossible that any one salt marsh will show each type of community, and not all communities will be clear cut. The rate at which the communities develop depends on a number of environmental factors related to the tidal regime, climate and topography of the area, while human interference can terminate the succession at any point or even completely change its course. Usually, however, the development of the saltmarsh from bare mud to the general community stage occurs within a matter of years rather than decades and is easily studied.

From the standpoint of the shore fauna, the later stages of saltmarsh development are not of great significance. The terrestrial and marine influences in the saltmarsh fauna bear a complicated relationship with each other. It is a subject which cannot be followed here, for to do it justice at least another book would be required. However, the fauna of the upper marsh is almost fully terrestrial or fresh water with little interchange with the marine fauna of the lower marsh. It is probably true that there is little exchange of organic matter apart from the rare tidal inundations. These are more likely to bring organic matter from the sea to land rather than the reverse, for the ebb is so gentle that any particulate matter picked up by the tide is soon stranded as the water recedes.

An interesting faunistic habitat found throughout saltmarshes is provided by the salt pans which in many ways are comparable to rock pools. They are small areas which remain flooded at low tide for one reason or another and

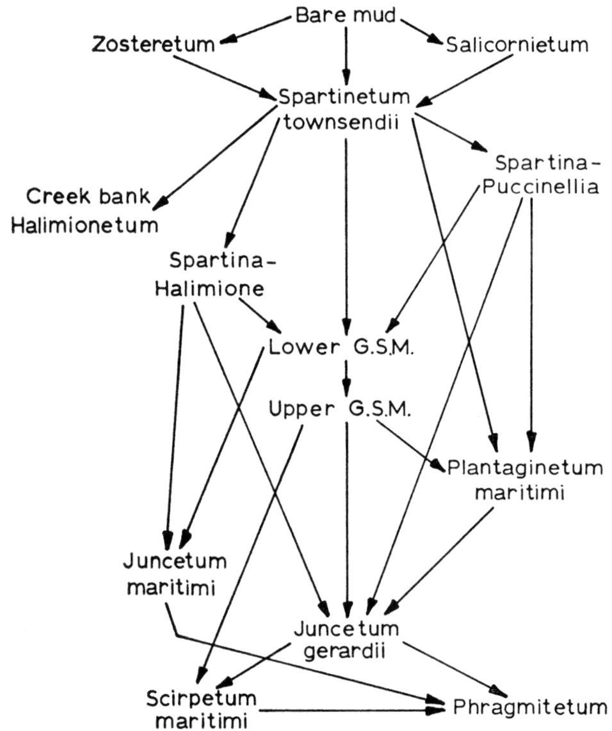

Fig. 41 Generalised succession of a salt marsh in the South of England. In regions of reduced salinity e.g. Poole Harbour or Lytchett Bay, there may be a direct transition from the Spartinetum to either a Phragmitetum or Scirpetum maritimi. G.S.M. = general saltmarsh community (after Chapman 1964)

thus are devoid of vegetation. Pans have been classified into various categories depending on their mode of formation. Many pans high up on the marsh experience periods of prolonged evaporation when salinity values may reach a very high level. Alternatively, but less frequently, the salinity may fall during times of heavy rain. Conditions in the pans are so different, both from the surrounding marsh and from the sea, that the fauna is equally distinct. Very little work has been done on this subject, but it does not seem likely that the pans are of any great relevance to the shore fauna.

The lower or early stages of the saltmarsh are, however, of great significance to shore life. The phanerogamic flora of these regions has been described above, but there are also a number of algal species which are very typical of the lower marsh. Many of these are present in the *Zostera* beds including, on the more sandy shores, green algal species belonging to *Enteromorpha*, *Ulva*, and *Cladophora*. There are also free-living or unattached species of brown algae which are typical of the lower saltmarsh communities. The fucoids include varieties of such well known rocky shore seaweeds as *Fucus vesiculosus*, *Pelvetia canaliculata* and *Ascophyllum nodosum*. There is an obvious advantage, as far as the productivity of the shore is concerned, in having both algae and phanerogam vegetation. The lower portions of the salt marsh are more frequently flooded by the tide than any other. During the period of immersion, photosynthesis by the phanerogams is considerably reduced, not only because of the decreased light intensity caused by silt suspended in the water, but also because the plant is unable to obtain supplies of carbon dioxide with which to build up carbohydrates. The seaweeds, however, are affected differently since they have evolved in the intertidal area and have adapted their photosynthetic processes accordingly, for example the yellow chloroplasts make use of those light rays which penetrate best through turbid water and they are physiologically adapted to taking up carbon dioxide from the sea. Consequently, the algae are able to undergo photosynthesis when covered by the tide and, in their case, it is during periods of low water that photosynthesis is reduced. Thus, between them, the algae and phanerogams are able to maintain an uninterrupted sequence of photosynthetic activity which must make the saltmarsh one of the most productive of all intertidal areas.

There remains one other great group of plants which occurs in the intertidal zone. This is composed of the microscopic unicellular algae most of which are diatoms (Bacillariophyceae) although dinoflagellates (Dinophyceae) can be equally numerous. Not all dinoflagellates are producers, however, for some are colourless and presumably saprophytic in their nutrition. Together, the diatoms and dinoflagellates, with other unicellular algae such as desmids, make up the phytoplankton which comprises the primary producers of the open sea. The phytoplankton is a temporary segment of the shore fauna and is found there only when the tide is in. It is doubtful that it is a significant part of the shore ecosystem since the production is rarely utilised by the permanent members of the shore fauna. Diatoms are, nevertheless, important producers of the shore because they occur also on the surface of the mud. No photosynthetic organism can survive other than in lighted areas and diatoms must not become buried under silt. This is likely to happen in the turbulent inshore waters, but diatoms have the ability to move and, if not buried too deeply, they can find their way back to the surface. Indeed, diatoms are under the surface for much of the time and they bury themselves regularly.

The diatom cell is a unique construction as it consists of two halves or valves which fit into one another like a box and its lid. In size, diatoms range from tiny cells too small to be retained by the finest plankton nets to quite large forms visible to the naked eye, but normally they fall within the range of 15–500μ. The valves, which form the cell wall of the plant, are heavily impregnated with silica and are often intricately sculptured. A very typical feature of the planktonic forms is the attenuation of the cell walls into the long spindle-shaped and branched structures which probably act as anti-sink devices by increasing the surface to volume ratio. It is not known for certain how diatoms move, but the cytoplasm protrudes from small perforations in the cell wall and may be responsible for locomotion. Reproduction is normally by simple cell division with the two valves separating and each forming a new cell by the development of a further valve *inside* the old one. It follows that the original smaller valve will produce a smaller daughter cell and this in turn will produce a yet smaller cell so that there is a progressive reduction in body size as reproduction continues. Before the process has advanced too far, however, the *status quo* is restored by the formation of an auxospore which is basically a diatom that has shed its valves. The auxospore grows until the original size is regained when it develops new valves to enclose itself.

One of the reasons for the success of diatoms on the shore is their ability to form resistant spores when conditions become limiting. The cytoplasm becomes less vacuolar and secretes a thick covering around itself. In this form the diatom is able to tolerate freezing temperatures and exposure to air for long periods.

In common with many other algae, the chromatophores of diatoms are brown and not green. This may be related to the fact that they, like seaweeds, spend so much of their photosynthetic life under turbid water through which there is a differential absorption of light waves. Consequently, diatoms are brown in the mass and are readily apparent as patches or streaks on the mud surface. A certain amount of light filters through the top layers of even the thickest mud, and diatoms do not necessarily have to be on the surface in order to photosynthesise. An account of the biology of littoral diatoms has been given by Aleem (1950) while a recent study of the ecology of mud-flat diatoms was carried out by Hopkins (1963, 1964, 1966) in the estuary of the River Ouse in Sussex. He found that more than 60% of the diatoms live in the top 2 mm of mud. He also found that the number present in coarse mud was less than that in fine mud, the actual figures being $0{\cdot}78 \times 10^5$ and $1{\cdot}08 \times 10^5$ cells per cubic mm respectively. The reason for this is given as the lack of protection afforded by the coarse mud relative to the fine mud with a consequent loss of diatoms from the larger interstitial spaces through the movements of the tide. On the other hand, light penetration is greater through coarse mud and diatoms are able to exist at deeper levels than they can in fine mud. Thus, the light value at 1 mm below the surface was found to be 21% of the surface measurement in coarse mud but only 13% in fine mud. Hopkins considers that this difference is reflected in the distribution of the diatoms of which the proportions found in the top millimetre were 68% and 58% in the case of fine and coarse muds respectively. When the tide is in, the diatoms are normally under the surface of the mud and, presumably, not producing but when the tide recedes, the diatoms come to the surface. Measurements of the speed with which diatoms can move have shown that even diatoms from below 2 mm can come to the surface in less than two minutes at summer (July) temperatures, and even during the winter (February) they take only eight minutes. These are negligible compared with the total intertidal period. Normally, the diatoms leave the surface some 5–18 minutes before the tide reaches their position. The adaptive significance of this reaction is not altogether clear. Obvious ly, it is necessary for them to be at the surface while the tide is out in order to carry out photosynthesis, but there is no real necessity for them to burrow when the tide is in. Admittedly, there is a danger that the cells would be washed away by tidal currents if they were on the surface, and this may well be the reason, but it would seem to be to their advantage to come to the surface of the mud during the period when the tide was in, i.e. to burrow only as the tide was passing. A possible explanation is that photosynthesis under these conditions would be impossible or uneconomic, but there is evidence that diatoms will rise to the surface if conditions of illumination become suitable. Hopkins found that diatoms will surface when still covered by 25 cm of water on bright days during calm weather. Presumably light penetration under these conditions is adequate for photosynthetic activity. On the other hand, diatoms can be inhibited by too strong an illumination and will not surface when the tide is out on very bright summer days. Obviously light is a factor which regulates the movement of diatoms.

Hopkins (1965, 1966) analysed the factors controlling the movement of diatoms by means of a series of laboratory experiments with *Surirella gemma*. He found that there is a regular up-and-down movement—a so-called vertical migration—which persists even in the absence of the stimuli inducing the response. The basic rhythm is diurnal and is controlled by day length, but there is also a tidal rhythm which is superimposed upon the diurnal movements. Of the more obvious factors associated with the tidal ebb and flow, hydrostatic pressure was eliminated as a conditioning influence since *Surirella* was found to be insensitive to pressure changes. There was, however, a reaction to mechanical disturbance and to wetting of the mud, stimuli which would be associated in nature with the passage of the tide. It seems likely, therefore, that the tide-induced movements begins with the downward movement of the diatom in response to the incoming tide. The subsequent rise to the surface is endogenous and based on the familiar biological clock mechanism. In the absence of any further stimulation, the next downward movement is also automatic or endogenous. Under natural conditions, of course, endogenous mechanisms rarely come into play since the diatom normally reacts to the tidal stimuli and it is only when the plant is brought into the laboratory under constant conditions that the endogenous rhythm becomes apparent. The interaction between illumination and tidal factors in determining the onset of migration is complex. Thus Hopkins (1966) showed that a persistent tidal rhythm occurs only if the following conditions are satisfied: first, there must have been a period of at least three hours of suitable illumination prior to the tidal wetting, and second, that the wetting must occur during daylight or within 20 to 40 minutes of adequate illumination. Under these conditions, a persistent rhythm with a period of about 25 hours is maintained. Should, however, the wetting occur during darkness, the resulting downward movement of the diatom is transient and is not repeated on the next day when the rhythm reverts to one synchronised with the night/day cycle.

Translated to field conditions, one can see how beautifully these reactions are linked to the ecological needs of the diatom. As a photosynthetic organism, there is no purpose in its being on the surface other than in daylight when the tide is out. If a diatom on the surface experiences an inadequate amount of illumination on one day, it is likely to experience even less on the following day, bearing in mind that the tidal cycle has a period of rather more than twelve hours, so that the time of high tides gets progressively later each day. Consequently, it would be to its advantage to 'forget' about the tidal cycle and to start afresh with the diurnal cycle. It would then be able to take full advantage of the whole of the intertidal period of that day. As we have seen, the reactions of *Surirella* enable this programme to be realised. The possibility of the diatoms coming to the surface during darkness appears to be prevented by a reaction which inhibits migration to the surface on two successive tides. Even if the two tidal exposures occur within daylight, the diatoms will not move up during the second period. There is probably evolutionary wisdom here since the second period is likely to be overtaken by the failing light of evening and not worthy of the expenditure of the energy required to come to the surface. In any event, it is much more likely that the second low tide will occur during the hours of darkness and the complexity of a reaction which will inhibit upward migration on alternate tides during the night but not during the day has apparently outweighed its advantages.

The control of the zonation of diatoms on the shore is complex. A photosynthetic organism will function best if it is in an illuminated medium for the maximum amount of time. In the case of shore diatoms, such conditions are found near the high water mark bearing in mind that photosynthesis does not take place when the tide is in. However, such a location is not ideal because an intertidal organism exposed for long periods during the day is in danger of desiccation. A position further down the shore might solve this problem, but introduces another. Not only is an adequate amount of illumination essential for photosynthesis, but an absence of light or a low level of illumination may be positively inimical to a photosynthetic organism. Thus, diatoms have the conflicting problems of low light intensity and desiccation to contend with and these two factors, with a third, seem to control their distribution on the shore. The third factor is the depth of the black reducing layer beneath the surface. It is thought that the sulphurous layer releases substances of an acid nature which can reach the surface if the layer is shallow enough. There is certainly a positive correlation between the distribution of *Pleurosigma balticum*, a resistant species, and the shallowness of the reducing layer, but *P. aestuarii* is not resistant and avoids areas where the sulphide layer is near the surface. Table 15 gives a list of diatom species and their relative tolerance of these

TABLE 15

The toleration by diatoms of desiccation, low illumination and sulphureous materials (after Hopkins 1964)

Species	Reactions to *Desiccation*	*Low illumination*	*Sulphureous material*
Navicular ammophila var. *flanatica*	Not resistant in July	Tolerant	Tolerant
N. cyprinus	,, ,, ,, ,,	,,	Not tolerant
N. cancellata	Resistant	,,	,, ,,
Stauroneis salina	Not resistant in July	,,	,, ,,
Nitzschia closterium	Not resistant	,,	?
Tropidneis vitrae	Resistant	Poor tolerance	Not tolerant
Pleurosigma angulatum	,,	Intolerant	?
P. aestuarii	,,	,,	Not tolerant
P. balticum	Poor resistance	Intolerant February	Tolerant
Triceratium favus	Not resistant	Tolerant	?
Campylodiscus spp.	Resistant	Intolerant	?

three environmental factors. This table is reproduced from Hopkins (1964), who tested their reactions during the winter (February) and in summer (July). In general, those species which are resistant to desiccation are found higher up the shore than those which are not, and those which are tolerant of low levels of illumination are concentrated around the mean low water level. The reaction to the depth of the reducing layer is not determined by tidal factors and is responsible for the absence of some species from places where they might be expected to occur.

The tolerance of low illumination and of the harmful products from the reducing layer is presumably physiological, although those species low down on the shore have developed behavioural adaptations; for example, they move faster than diatoms from other areas and continue to move at low temperatures when other diatoms would be immobile. These reactions ensure that they derive maximum benefit from what little illumination is available since little time is wasted in coming to the surface.

The resistance to desiccation, on the other hand, is mechanical as well as physiological. Certain species, such as *Nitzschia closterium* and *Pleurosigma angulatum*, are able to secrete mucilage which was found to cover the colonies completely within two hours at 15°C (Hopkins 1964). The mucilage would obviously be of value in preventing desiccation while the tide is out but it may not always have this function because one of the species in which it is found, *Nitzschia closterium,* occurs low down the shore and is not resistant to desiccation. There is some evidence that the diatoms remove the mucilage by extra-cellular digestion when the need for it has passed. Not all diatoms from the upper shore secrete mucilage and in such species as *Tropidoneis vitrae* and *Navicula cancellata*, resistance to desiccation must be physiological.

Data from other parts of the country suggest that Hopkins' observations on the migration of diatoms in the River Ouse are not of universal applicability. Thus Perkins (1960) found that *Pleurosigma aestuarii* and *Surirella gemma*, species which occur in the Ouse, did not move down when covered by the tide in the Eden Estuary, Fife, where he studied them. The difference may, perhaps, be attributed to the cleaner water of the Eden Estuary which permits sufficient light to reach the mud surface for photosynthesis to continue for some time after the tide has risen.

In estuarine regions, salinity is important in controlling the distribution of diatoms and a distinctive brackish water community has developed. These are truly estuarine species and do less well under fully marine conditions. They display a similar pattern of vertical movement as that shown by species from more saline habitats (Round & Palmer 1966) but the extreme turbidity of most estuarine water ensures that photosynthetic activity is strictly confined to intertidal periods.

In the upper reaches of estuaries where prolonged low salinity conditions prevail, the primary producers of the shore are supplemented by members of a typically fresh-water group. These are species of the unicellular green alga *Euglena*. *E. obtusa* was studied in the River Avon at Bristol by Palmer & Round (1965), who found a diurnal migration based on light but with an underlying endogenous rhythm similar to that of the diatoms.

CHAPTER NINE

THE SHORE AS AN ECOSYSTEM

The term 'ecosystem' was first coined by Tansley in 1935 with particular reference to terrestrial vegetation. Various definitions have been offered and all agree that an ecosystem is a self-contained ecological unit, but there is in fact only one truly self-contained ecosystem—the global. In all other so-called ecosystems there is always some form of interchange of energy or materials. An ecosystem has often been likened to a machine which requires energy (fuel) to drive it and materials (metal, oil) to enable it to function. The global ecosystem is powered by solar energy and functions on the familiar carbon-nitrogen-water biochemical system. The ecosystem is a machine because it performs work. It builds up from simple chemicals the complex organic molecules which form the bodies of living creatures. The energy necessary for this is provided by solar radiation which is utilised by autotrophic organisms, the green plants, to produce organic chemicals through photosynthesis. Green plants, therefore, are described as the producers of the ecosystem or as forming the first trophic level. Without plants, animal life would be impossible. The animals which feed directly on the plants are called primary consumers or as belonging to the second trophic level. Not all animals, of course, feed on plants and many are predators consuming other animals. These are known as secondary consumers or as members of the third (tertiary) trophic level. It is possible to have tertiary consumers—carnivores which feed on carnivores—and members of even higher trophic levels, for example parasites of a polar bear which has killed a seal which has fed on fish which have consumed phytoplankton belong to the fifth trophic level. This example illustrates the familiar food-chain, but usually the situation is not so simple. Very often an animal belongs to more than one trophic level, thus it may be omnivorous—partly feeding on plants and partly on animals, or it may prey on animals from different trophic levels. This is recognised by the general rejection of the food chain concept and the substitution of the term 'food web', but in very simple environments examples of the classical food chain can be traced.

This series of trophic levels rises in a pyramid from the basic primary producers, and its structure is inescapable since there must always be a loss of energy between one trophic level and the next higher one. Each trophic level contains a finite quantity of potential energy, but all of this cannot be transmitted from one level to the next since much is used up in performing work such as locomotion or in growth and reproduction. Very often, this basic fact is revealed by simply counting the organisms in an ecosystem—there are millions of grass plants for each antelope and hundreds of antelopes for each lion—but such a pyramid of numbers breaks down when one considers the number of aphids on a rose plant. In such cases, a consideration of the biomass (the weight of living matter) restores the pyramid structure. An even more accurate picture is given by the pyramid of energy in which the utilised and potential energy of each trophic level are represented. We are not dealing here in abstract terms since it is perfectly possible to measure the energy at each trophic level. To be more precise, the energy equivalent of a sample from each trophic level is measured and the results extrapolated according to the estimated biomass of the animals and plants present. The potential energy can be measured by igniting the sample in a bomb calorimeter which is an instrument which registers the amount of heat given off by a substance when it is burnt. The amount of energy utilised by the organism during its normal biological activities can be estimated by measuring its respiration. All energy required by an animal, whether to produce heat or to enable it to move or to grow, comes eventually from the oxidation of blood sugars, and this oxidation is directly reflected by the amount of oxygen consumed. All these forms of energy are ultimately lost from the ecosystem by radiation of heat into space. It is important to realise this fact since it means that energy is constantly passing through the system and being used up. Hence, the not too fanciful analogy of solar energy as the fuel of the ecosystem machine. The passage of energy through the ecosystem is known as energy flow.

The ecosystem, as we have described it, seems to contradict the second law of thermodynamics which states that heat cannot be transferred by any continuous, self-sustaining process from a colder to a hotter body. In more general terms, it infers that a complex system cannot be built up from a simple one although this is what happens within an ecosystem. However, the operative word in the definition is 'continuous'; the components of the ecosystem are not immortal and eventually the animals of the highest trophic levels die and are broken down to their component simple elements or compounds. Death is essential to life, for without death all the nutrients of the world would become locked up in animals and none would become available for plants which would, therefore, die. The death of all animals would occur soon afterwards and the ecosystem would run down. That it does not is due to the recycling of the chemicals essential to life, the nutrients, which are returned to the ecosystem after death. The breakdown of dead animals in the return of nutrients to the soil is achieved by a variety of small animals and heterotrophic plants such as fungi and bacteria. So important are they that a distinct de-

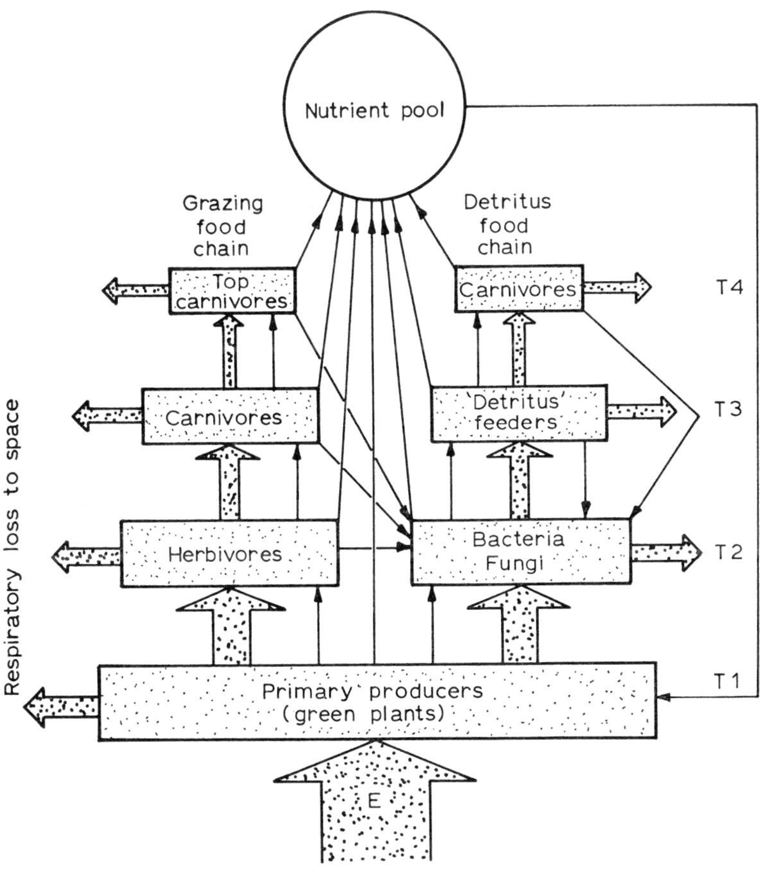

Fig. 42 Generalised diagram of a simple ecosystem. The lightly shaded blocks represent the biomass of each type of organism. The stippled arrows show the direction and magnitude of energy flow while the single line arrows indicate the transference of nutrients. *E* is the amount of solar energy taken up by the primary producers. Note that this energy is eventually lost from the ecosystem (as radiant heat to outer space). On the other hand, the nutrients are recycled within the system either by death and bacterial decay or by elimination of material during defaecation, urination etc. In practice, much of the material is deposited outside the ecosystem but an equal amount is introduced under balanced conditions. Hence, it is more realistic to show the nutrients passing through a nutrient pool which other ecosystems use and to which they contribute. Ecosystems are usually divided into a grazing food chain of large animals (left in the diagram) and a detritus or decomposer food chain of microorganisms. If the corresponding trophic levels (T2, T3, and T4) of each chain were joined, the classical ecological pyramid would result

composer cycle may be recognised in ecosystems. Not only animal matter is decomposed, for the majority of living plants are not consumed by herbivores but pass into the soil as dead material and are broken down by the decomposing organisms.

The ecosystem, therefore, consists of a number of entities. These are portrayed in Fig. 42, which is a diagrammatic model of an ecosystem. First, there is a community of green plants and their dependent animals. Then there are the decomposing organisms which may be considered either as part of the general biotic community or as a separate unit. Throughout the community or communities there is a flow of energy which passes from the Sun into the green plants and through the heterotrophic organisms and is lost eventually as heat radiated into space. Finally, we have the cycling of the abiotic nutrients. Within the ecosystem there are but finite amounts of nutrients which are taken up by the plants, passed through the animals and decomposers and back to the plants. Hence there is a nitrogen cycle, a carbon cycle, a water cycle and many others. All these cycles involve the transfer of nutrients to and from the organisms of the ecosystem and the inanimate environment. On land, for example, the nutrients are returned to the soil after being taken up by the plants and passed through the animals and organisms of decomposition. Because of this involvement with abiotic environment, these are known as biogeochemical cycles—'chemical' because the process is chemical, 'bio' and 'geo' because living and non-living elements are involved.

The purpose of most ecological research these days is the construction of a model of energy flow through the ecosystem. This is not merely an academic exercise, for it can bring to light some revealing facts about the way in which energy is being utilised; e.g. Golley (1960) has shown that in a grass field which is used to raise beef cattle, only about 0·5% of the total productivity is utilised by man. 75% of the productivity goes to the soil. These sort of figures might surprise the average farmer. Another advantage of energy flow diagrams is that they enable different ecosystems to be directly compared, for by reducing the data to mathematical concepts, differences are readily apparent. Thus, it might be difficult to say in what general way a forest ecosystem differs from a shallow marine bay. One could give a list of species or describe the nature of the soil but these are matters of detail, not principle. However, the energy flow diagram reveals that whereas the plant biomass is rather smaller than the animal biomass (about 50% of the latter) in the sea, the plant biomass is huge in the forest relative to that of animals (Odum 1963). Further, the diagrams show that two-thirds of the energy flow passes through the 'grazing' food chain in the sea compared with only about 12% in the forest (Fig. 43). These are fundamental differences which might not be apparent from a classical ecological survey of the two areas. The figures given here are approximate since the data from which they are compiled are partly hypothetical. No ecosystem has yet been studied so thoroughly that a complete and accurate energy flow diagram can be drawn up for it. Incidentally, it might be wondered why

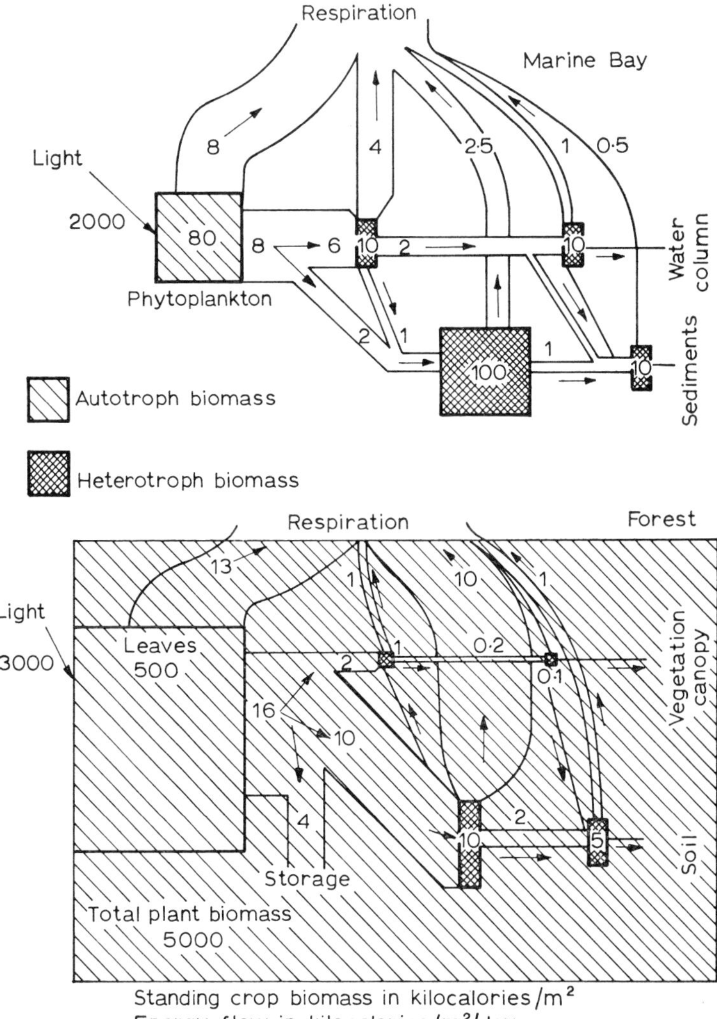

Fig. 43 Comparison of two distinct ecosystems. Above, a marine bay and below a forest. The shaded boxes represent the biomass in terms of kilocalories per square metre per year and the energy flow is shown as kilocalories per square metre per day. Note the relatively vast proportion of the incident energy which passes into the plants (autotrophic biomass) of the forest (after Odum 1963)

the animal biomass exceeds that of the plants in the shallow sea. This does not contradict the theory of pyramids since much of the animal biomass is found in the sediment where the organisms are breaking down dead plant material laid down over a long period. It can even happen that the 'grazing' biomass of the sea can exceed that of the plants for short periods in the spring when the reproductive rate of zooplankton is so great that the phytoplankton is consumed almost as soon as it is produced. The observed plant biomass represents no more than the standing crop which is not a true measure of primary production.

The marine diagram is worth examining a little more closely since it depicts an ecosystem closely related to the shore. The autotrophic organisms in the shallow sea are members of the phytoplankton which differ from most terrestrial producers in being unicellular. The advantage of this in the marine environment is that a greater surface area per unit volume is available for the uptake of nutrients which are much scarcer in the sea than in soil. It will be seen that only two-thirds of the amount of light energy captured by the terrestrial plants is absorbed by the phytoplankton. This is a measure not of the inefficiency of algal chloroplasts but of the absorption of light by the water. However, the energy flow through the marine ecosystem is at a lower level than is the case with the forest. Only 16 kilocalories per square metre per day flows through the primary producers in the sea compared with 29 in the forest, and only 8 pass on to the next trophic level. This is just half the energy which the primary consumers of the forest receive. Note how all the energy passed to the primary consumers is eventually lost as respiration except in the forest where a substantial proportion (25%) remains locked up in the form of undecayed wood.

Both ecosystems are alike in having two major pathways for energy flow. First there is the route through the heterotrophs (animals) which feed on living plants. This is the *grazing food chain*. Secondly there is the *detritus food chain* which represents the flow through those heterotrophs which feed on dead plant material. This division seems to be a fundamental one which results in a Y shape to all energy flow models. The distinction is important because of the difference in the period between the production of organic matter by the plants and its consumption by the heterotrophs. In the grazing food chain, the energy flow is much more rapid as the animals eat living plants but the heterotrophs in the detritus cycle feed on dead plant material which may have been produced weeks or even months before. A further difference is that the detritus pathway is much more complex with elaborate interactions between organisms forming intricate food webs. Although food webs occur in the grazing pathway, they are simpler with a tendency towards the food chain type of relationship. In terrestrial ecosystems, the two types of energy pathways are associated with animals that are above and below the ground respectively. The above-ground animals are generally large vertebrates, although insects are often equally important. The detritus pathway is associated with the soil fauna which is largely composed of micro-arthropods together with saprophytic fungi and bacteria. A similar situation is found in the marine environment in the sense of an above- and below-ground fauna, but many of the larger animals are below the ground while the organisms belonging to the grazing pathway are small, being members of the zooplankton. While it is not true, therefore, that the grazing food chain is always to be found above ground in the marine environment, it is probably correct to say that the majority of the decaying organisms are present in the soil as in terrestrial ecosystems.

An interesting development in the study of ecological energetics has been the energy relationship between the various trophic levels. There is some evidence to suggest that the energy flow between the different levels follows a constant pattern in many ecosystems. This has become apparent through an analysis of the 'efficiency' of the ecosystem in the transfer of energy from one trophic level to another. The terminology has not become completely stabilised and a number of definitions of 'efficiency' have been put forward, but it may help in the understanding of the following pages if a brief evaluation of these is made.

The efficiency of any system is a measure of its success in deriving the greatest benefit from its motivating forces. To the average motorist, an efficient car engine is one which enables the vehicle to travel farthest on a gallon of petrol, but to the racing driver it is one which drives the car fastest irrespective of economy. Thus, efficiency is a subjective term and should not, perhaps, be applied to a natural ecosystem which, to one of orthodox Darwinian faith, must presumably be operating at the optimum level relative to the conditions obtaining. Nevertheless, one would regard a predator which utilised only x% of the prey population as being less efficient than one using 2x%, always provided, of course, that the prey population was not permanently harmed in the process. The study of ecosystem efficiency is simply to discover whether such differences exist.

Much of the pioneer work on the efficiency in ecosystems was carried out by Slobodkin who studied laboratory populations. This was inevitable since natural ecosystems are so complex that it is extremely unlikely that clear-cut data can ever emerge from field observations alone. The conclusions from laboratory studies can always be tested in the field and, indeed, this is the sequence through which most major advances in ecology have been made.

Two principal concepts in ecosystem efficiency have arisen; the *food chain efficiency* (F.C.E.) and the *gross ecological efficiency* (G.E.E.). Food chain efficiency was studied by Slobodkin (1959) on a predator/prey relationship, specifically a *Chlamydomonas*/*Daphnia*/Man food chain with *Chlamydomonas* as the primary producer, *Daphnia* as the herbivore and with the experimenter acting as the predator and removing the prey at different rates. The food chain efficiency is defined as the percentage, in calorific terms, of the food available to the prey which is passed on to the predator. This is a measure not only of the feeding efficiency of the predator but also of the prey because it is the amount of food available to the prey, but not necessarily consumed, which is considered. The percentage of food which passes to the predator in terms of the amount *consumed* by the prey is the gross ecological efficiency, i.e.

Food chain efficiency (F.C.E.)

$$= \frac{\text{calorific content of food consumed by predator}}{\text{calorific content of food available to prey}} \times 100$$

Gross ecological efficiency (G.E.E.)

$$= \frac{\text{calorific content of food consumed by predator}}{\text{calorific content of food consumed by prey}} \times 100$$

Slobodkin's experiments with *Daphnia* showed the importance of both factors concerned in the food chain efficiency (Fig. 44). At a low level of predation, there was little difference between the values of the F.C.E. whatever the population of *Chlamydomonas*, the primary source of food. However, as the predation rate increased, the amount of food available to the prey became significant. When predation rose above 50%, the F.C.E. fell although there was still an abundant supply of the primary food (*Chlamydomonas*). At a low food level, the F.C.E. did not decline until the predation rate had passed 75%. These results are due to the fact that under an increased predation rate, insufficient *Daphnia* are left to consume the *Chlamydomonas* when the latter are numerous but when the algal population is small, a greater proportion is eaten by the *Daphnia.* On reflection, these conclusions appear to be common sense although it should be remembered that the size of a *Daphnia* population varies with its food supply.

When all the food available to the prey is consumed, the denominators in each of the above expressions become identical and the F.C.E. equals the G.E.E. In nature, it is extremely unlikely that this ever occurs except under the most transient conditions. However, it was possible to achieve this in the laboratory with *Daphnia* and it was found that the G.E.E. is then independent of the amount of algal food available (Fig. 45). This figure shows that the maximum G.E.E. obtained was about 13% and such data as are available suggest that this value is of the order of the G.E.E. in most ecosystems, a round figure of 10% being suggested by Slobodkin. This is the hypothesis which needs to be tested under field conditions, but except for the most simple ecosystems, it would be a formidable task to collect accurate data. The G.E.E. concept should be confined to animals since it is very difficult to relate the amount of food available to plants in any meaningful way to that available to higher trophic levels. Difficulties in the estimation of the G.E.E. arise from the omnivorous habits of so many animals, but in the grazing food chain, the producer/primary consumer/prey relationship can usually be followed. In natural ecosystems, many species will be involved at each trophic level. Indeed, the G.E.E. of a single species would not be very useful and almost certainly not consistent, for example one can think of large herbivores such as elephants or gorillas on which predation is extremely low, and consequently whose G.E.E. would be correspondingly low. It may be that the present concepts of ecosystem efficiency are too restricted because they have been

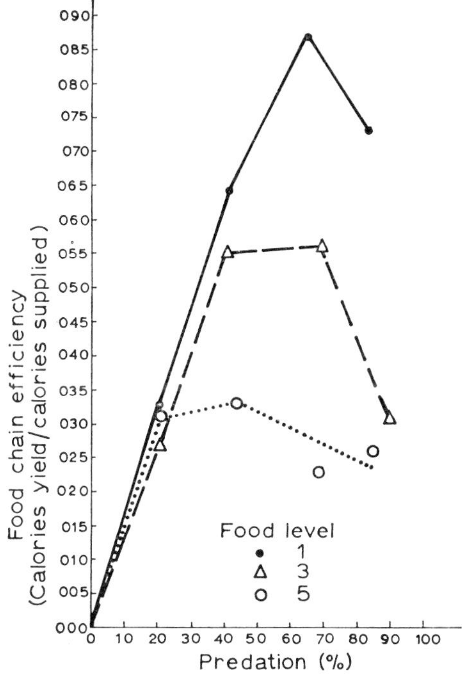

Fig. 44 Effect of different levels of predation on the food chain efficiency of a *Chlamydomonas*/*Daphnia*/man experimental 'ecosystem'. The food levels refer to the quantity of *Chlamydomonas* available to *Daphnia* (1 being low and 5 high) and the predation to the percentage of *Daphnia* removed from the system by the experimenter. Note that there is little difference in the food chain efficiencies below 25% predation but above this value, the *Daphnia* become less efficient at high food levels than at low in exploiting the *Chlamydomonas*. The reason is that there are too few *Daphnia* to take sufficient algal cells to maintain the food chain efficiency (after Slobodkin 1959)

derived from simple laboratory or field experiments. After all, one is interested in the transfer of energy from one trophic level to another, and it matters little whether it is a predator or a scavenger which consumes the prey. Thus most elephants finish up inside the stomachs of lions, hyaenas or vultures although they are not killed by them. Therefore it would be better to define the gross ecological efficiency as

$$\frac{\text{amount of energy passing to trophic level x}}{\text{amount of energy passing to trophic level x–1}} \times 100$$

The difference between the values of the denominator and numerator represents the amount of energy which is lost as respiratory heat plus that which passes to the detritus food chain.

If it can be shown that the G.E.E. of all ecosystems approximates to 10%, it may be assumed that all are operating at the same level of efficiency and our earlier faith in evolutionary competence would be justified. It cannot be

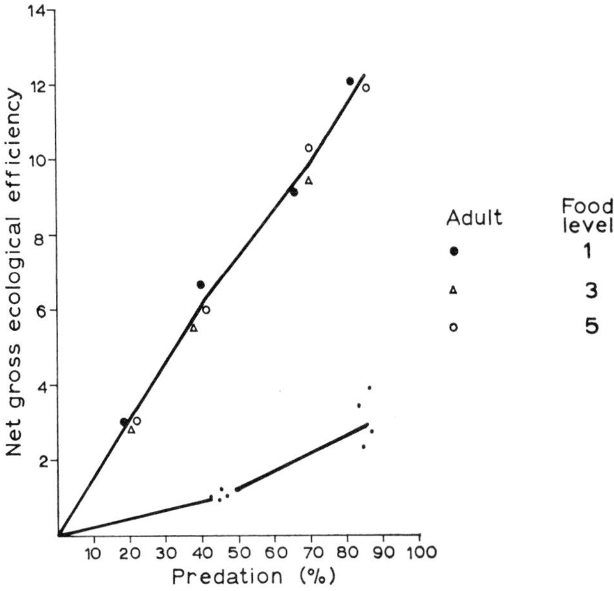

Fig. 45 Gross ecological efficiency in relation to predation in a *Chlamydomonas*/*Daphnia*/man 'ecosystem'. Note that at all food levels, the gross ecological efficiency of the *Chlamydomonas*/*Daphnia* relationship rises with increased predation of the *Daphnia* but that it is independent of the amount of *Chlamydomonas* available. The upper graph refers to experiments in which only adult *Daphnia* were removed and the lower to experiments in which the predation was restricted to young *Daphnia*. The first relationship is linear while the second is distinctly curved. c.f. Fig. 44 (after Slobodkin 1959)

expected that accurate data will emerge from field studies of complex ecosystems but there is room to hope that in time sufficient information will accrue to reveal the degree of probability that the G.E.E. of all ecosystems is the same. A further point, which may be easier to answer, is whether or not the G.E.E. is similar at all trophic levels in the same ecosystem.

It was stated above that there is only one true ecosystem which is on a global scale, but although this is true, it is usual to consider the natural environment as being composed of a number of distinct ecosystems. Some examples will illustrate the point. Thus a lake, a forest or mountain are each considered to be an ecosystem. To qualify for the title, an ecosystem must have all the components mentioned above, i.e. a biotic community within which there is an energy flow and a recycling of nutrients. Obviously an ecosystem must be big—a small pond is not an ecosystem since a high proportion of organic matter enters and leaves it by such channels as falling leaves, metamorphosed tadpoles or fish-eating birds. There can never be a completely isolated ecosystem because an exchange of organic matter is bound to take place at the borders while the migration of large mammals and birds from one ecosystem to another is a widespread phenomenon. However, it suffices for practical purposes to accept as an ecosystem any ecological unit most of whose nutrients recycle within it. Thus the trees of a forest shed most of their leaves on to the floor where the breakdown products are taken into the soil and from thence back to the trees. However, some leaves may blow into a neighbouring lake while browsing animals may pass into and out of the forest, alternately contributing and removing organic matter. It is perhaps best to think of a global nutrient pool drawn upon by all ecosystems which, however, contribute as much as they take.

The ecosystem concept as outlined here may seem elementary but unfortunately it seems not to have been understood by many of the educated and intelligent men who govern our lives. The essence of conservation is the preservation of the nutrient cycles and the maintenance of the flow of energy through the ecosystems at maximum levels. Much of the energy flow can be directed through channels useful to man such as agriculture, but to ignore ecological principles altogether leads to habitat degradation and environmental problems which, as air and water pollution, are already with us. It is salutary to realise that many desert areas of the world are man-made as a result of failure to maintain the nutrient cycles. Fortunately, the need for environmental studies is beginning to be appreciated and it is upon these rather than the more glamorous investigations into the nature of life at the molecular level that the welfare of mankind is likely to depend.

However, this is a digression. This discussion of ecological principles has been given as a background to an examination of the depositing shore to see how far it possesses the properties of an ecosystem. The question is whether or not the shore is sufficiently distinct from the neighbouring land and sea to be considered as a separate ecosystem. That there is a transfer of energy between the shore and these environments is obvious from the previous chapters, but we need to establish only that the majority of the energy turnover takes place within the shore habitat for it to qualify as an ecosystem. The shore is certainly a habitat, sufficiently different from all others to carry a distinct community of animals, but this in itself is not diagnostic of an ecosystem. In the following pages, the various components of an ecosystem will be considered in the context of the depositing shore and at the end an attempt will be made to assess the evidence for or against the concept of a shore ecosystem.

The Primary Producers

Compared with those of the rocky shore, the primary producers of a depositing shore are not at all obvious; at best, there may be an occasional clump of *Zostera* visible on muddy sand. The major primary producers of the depositing shore are the inconspicuous diatoms, but this is true only if one confines the definition of the shore to that area which is frequently covered by the tide. The problem of saltmarshes has been discussed in the previous chapter. These are intertidal in the sense that they are periodically inundated by the tide but the long intervals between flooding and the short duration of each immersion renders the habitat radically different from that of the shore proper. In the early stages of saltmarsh formation, the tidal regime is more frequent and the plants can with justification be considered as contributors to the primary productivity of the shore. However, in relation to the total area of the intertidal zone, excluding rocky shores, the area of saltmarshes is quite small as they are confined to sheltered regions particularly of bays and estuaries. In most cases, saltmarshes are found on fine grade deposits; sandy saltmarshes do occur but they are rare.

It is necessary to distinguish clearly between sandy and muddy shores in respect of the primary productivity for remarks made about one type are unlikely to apply to the other. The primary production of sandy shores is very low indeed if the sand is at all clean. Even *Zostera* requires some mud in which to grow and few diatoms are found on the shifting surface of a clean sandy beach. Steele & Baird (1968) give a figure of 5 gm of carbon per square metre per year for the productivity of a moderately exposed sandy beach. This is much lower than in most terrestrial deserts. One bizarre source of primary productivity on sand is the algal cells found as symbionts in the subdermis of *Convoluta roscoffensis*, a flatworm, from the lower shore. The precise contribution which these algae make to the primary productivity has never been evaluated but *Convoluta* can occur in such numbers as to form a prominent green band on the shore, and locally their influence must be considerable. The species of alga found in *C. roscoffensis* is *Platymonas convolutae* which also occurs in a free-living form amongst other monads in the vicinity of the worms (Parke & Manton, 1967). *Convoluta* is colourless when hatched and presumably is infected by algae from the environment. This probably occurs early in life since *Platymonas* is chemotactically attracted to the empty egg cases of the worm. Its tendency to stick to surfaces by means of its flagella must be of value in facilitating penetration of the host. The behaviour of *Convoluta* is adapted to make the maximum use of its symbiont's photosynthetic capabilities. It is attracted toward light and positions itself in the sunniest part of its favoured habitat which is the water trapped between the ripple marks of the sand at low tide. It constantly readjusts its position relative to the incident light to derive optimum benefit from the changing illumination. The alga itself is adapted for symbiotic existence. It can withstand without harm long periods of darkness, such as it must experience when the worm buries itself between tides, and it shares with its host a broad tolerance of temperature extremes.

In most situations, however, diatoms are the primary producers of the depositing shore. These are confined to the finer grained deposits or to soils which have a high proportion of organic matter. In those places where salt-marshes or *Zostera* beds are found or where the mud flats are covered with dense beds of Cordgrass (*Spartina*) the spermatophyte vegetation assumes the significant role in primary productivity.

The phytoplankton of the sea becomes temporarily part of the shore ecosystem when the tide is in. In those areas where the high tide tends to occur around the middle of the day, this source of primary production will obviously assume a greater importance than would be the case when high tides occur at night. However, on muddy shores where the tide ebbs quietly, some of the phytoplankton is stranded on the mud and may contribute to the productivity while the tide is out. These plants of the plankton are predominantly diatoms.

The role of the sulphur bacteria should not be ignored in a discussion of primary productivity. These organisms occur in the black sulphide or reducing layer which has already been described. They are chemosynthetic and not photosynthetic because they utilise the energy released from the chemical reaction when ferric oxide in the sand is reduced. The energy is used to build up complex organic matter from simple elements in the same fashion as the conventional green plant uses light energy from the Sun. The organic matter produced by the sulphur bacteria is made available to the ecosystem through the agency of nematodes which live in the reducing layer and feed on the bacteria. Their numbers are very large—Perkins (1958) records a density of $5{\cdot}56 \times 10^4$ nematodes per sq. m. at a depth of between 4 and 6 cm at Whitstable—and the nematodes are an important link in a food chain from which man benefits, viz. sulphur bacteria—nematodes—*Nereis diversicolor*—fish—man. A muddy shore is unique amongst environments in having these two productive layers, one on the surface and another a few centimetres below the ground.

Enough has been said to establish the fact that primary producers do occur on depositing shores, but this in itself is not sufficient to qualify the habitat for the status of an ecosystem. The situation is complicated by the periodic appearance on the shore of primary producers from other ecosystems. These are invariably algae which have either become detached from nearby rocky shores or are unattached species, which were described in the last Chapter. In either case, they are washed up on the beach, especially during periods of storm. Such vast quantities of brown seaweeds are deposited on sandy shores that they become a nuisance in holiday resorts and it is often economically profitable to harvest them for use as manure. Although these seaweeds eventually die, they can survive for considerable periods since they are able to absorb all their nutrient requirements through the surface and they are not dependent upon roots, as the terrestrial plants are. Conditions may not be ideal but nevertheless these stranded algae continue to produce. Should they then be considered as primary producers of the depositing shore? Eventually they decompose on the beach under natural conditions and contribute to the energy budget of the shore ecosystem, but most of the decomposing material has been produced elsewhere. Brown seaweeds are not often found on mud flats since these are usually some way from the rocky shores where seaweeds grow. However, members of the Chlorophycae, the green algae, are frequently present. The species concerned belong primarily to two genera, *Ulva* and *Enteromorpha. Ulva*, the sea lettuce, is usually fixed to a hard substratum but it is easily detached and can survive while floating in the water. Sometimes such large quantities are deposited on the shore that the mud surface is completely covered. The algae may appear overnight and can as quickly be carried away on another tide, although normally they remain on the mud flat for long periods, sometimes for several months. Thus, although much of the algal material brought to the shore is broken down there and contributes to the supply of organic matter, not all of it is; indeed, some organic matter produced while the algae is on the shore may even be lost to the ecosystem, but the loss must be negligible compared with the total turnover.

More commonly, the green algae on the mud belong permanently to the environment. Fronds of *Enteromorpha* can often be found lightly buried under the mud during the winter months and these seem to act as centres for the spread of the alga during the spring and summer. At these times, a thick felt of *Enteromorpha*, which is loosely attached to the mud, may spread all over the flats and its contribution to the productivity of the shore must be very great, perhaps more so than that of the diatoms. Data on the spatial and temporal distribution of *Enteromorpha* on muddy shores are badly needed and it may prove to be so widespread as to discount our earlier statement that diatoms are the major primary producers of the shore. A curious feature of its biology is that the *Enteromorpha* seems to be hardly utilised by the consumers. To be sure, it is eaten by brent geese but these are not present during the summer when the algal growth is at its height. It is also grazed by *Littorina* but not to an extent that makes any noticeable difference to the mass of plant material present. It is unusual to find a potential food resource which is not exploited by the animal community.

A further source of primary production on the depositing shore is the typical rocky shore vegetation which finds a footing on the hard man-made structures which are so abundant around our coasts. Not all of these structures are covered by seaweeds, however, for some are in polluted waters or are deliberately kept clear of weed, but in the mass these 'rocky shore' plants are probably important contributors to the productivity of the beach. Even on 'natural' mudflats unaffected by human artefacts, there are often rocky outcrops and boulders on which seaweed will become established.

The question which is raised by this discussion of productivity in the shore ecosystem is what proportion of the organic matter available to the consumers of the first trophic level has been derived from the primary producers of the shore itself, and how much has come from outside. The problem may be clarified by Fig. 46, which shows the various sources of the organic matter found in muddy and sandy shores. When set out in this form, the local

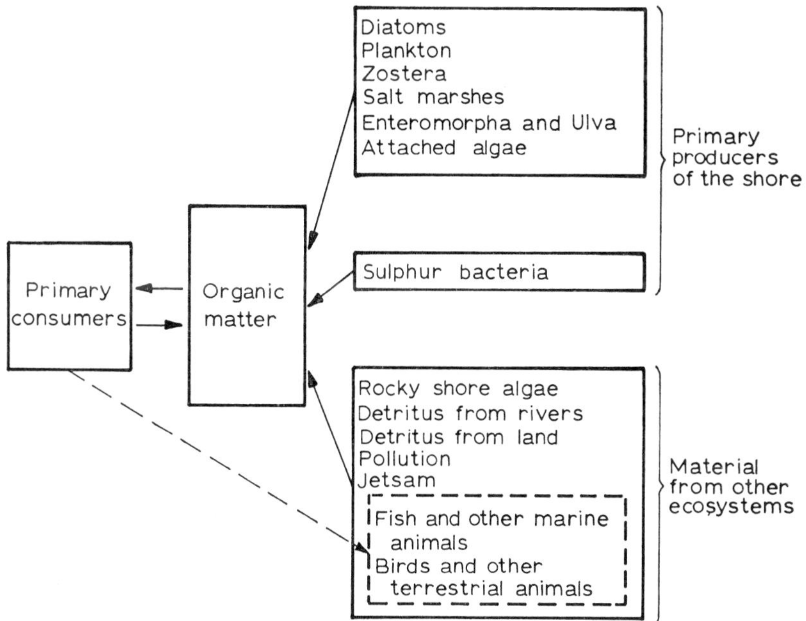

Fig. 46 Sources of organic matter on the depositing shore. Note that most of the high energy material on the shore originates in other ecosystems. Note also that much of the secondary production is removed from the shore by predation (dotted lines) or by emigration

sources of organic matter seem rather meagre beside the sources from outside the shore ecosystem. Most of the primary producers from the shore itself are of minor or of strictly local importance, for example, plankton, *Zostera*, salt-marsh plants, and only diatoms and sulphur bacteria are of universal significance. By contrast, the amount of organic matter originating from outside may be very large indeed. The contribution of the rocky shore algae has already been mentioned. Many thousands of tons of material are carried to the sea each year from the rivers and although much of this is deposited on the sea bed, much more settles on the shore, particularly in estuaries where almost all the material can be alluvial in origin. The suspended matter in rivers derives ultimately from the land but the 'detritus from land' in Fig. 46 refers to material which is blown onto the shore. This consists of dust or soil particles but also of leaves and other pieces of vegetation. This sort of debris can be very abundant, particularly high up on the shore where sticks and leaves may make it difficult to obtain a core sample. As vegetable matter does not remain undecomposed for long, the amount of organic matter which must pass into the soil from this source is prodigious.

The importance of pollution cannot be exaggerated particularly in a country as industrialised as Britain. Domestic sewage is often discharged untreated into the sea and, unhappily, it all too frequently finds its way back to the shore. Even treated sewage has a high organic content and much of this is transported to the shore in the river systems. Industrialised wastes may be of greater significance than domestic sewage; for example, the mud flats of the Swale in North Kent are grossly contaminated with a flocculent pollutant from a nearby paper mill. Oil, it need hardly be mentioned, is a common pollutant which is also organic and is eventually broken down by the action of bacteria. The variety of sources of pollution in the Thames estuary has been reviewed by Arthur (1969).

Fish and other marine animals are of some importance in presenting faeces and dead remains to the shore ecosystem, but they are not as significant as the birds. The vast population of sea birds which frequent the intertidal regions have been described in an earlier chapter. The enrichment of the mud by the faeces of these birds is very great and is obvious from casual inspection, but as much of the food of the birds originates in the intertidal zone, this is basically a cycling of material within the ecosystem—a fact which is recognised by the dotted line in Fig. 46.

Jetsam is used here to describe organic material which is washed up on the shore. The strand line is jetsam and, as we have seen, a very important factor in the shore ecosystem particularly on sandy beaches where it may be the only significant source of organic matter. A large number of tree trunks are stranded on the shore and usually are decomposed there, although it may take several years before the wood is broken down and the contained energy released to the shore organisms. Occasionally vast quantities of organic matter are deposited as jetsam as, for example, when a school of whales is stranded.

The Consumers

We will leave the primary producers for the time being and turn to the consumers to see how far they constitute a community within an ecosystem. Divisions must be made amongst the primary consumers from a nutritive standpoint and the categories follow those adopted earlier in the discussion of their natural history, i.e. macro-, meio- and microfauna. The macrofauna needs to be further divided into deposit and filter feeders.

The deposit feeders have been fully covered in Chapter 3, in which it was shown that they feed on bacteria in the film surrounding organic and inorganic particles and not upon the particles themselves. It is a peculiar feature of the shore ecosystem that such a large body of primary consumers should feed not upon the primary producers directly, but upon the products of their decomposition. The deposit feeders, therefore, feed on material—bacteria—which originate within the ecosystem, but the material on which the bacteria feed is quite likely to have come from outside the shore. Filter feeders, on the other hand, take much of their food from outside the ecosystem. The filter feeders are usually non-selective in that all particles in the overlying water are drawn into the body although, as in the case of bivalve molluscs, sorting and selection of the particles may take place before they pass into the gut. The major sources of food comprise plankton and organic particles on the surface which are disturbed and taken into suspension by water movements. As plankton filter feeders tend to be found in coarse grain deposits, the quantity of suspended organic matter consumed is small and plankton forms the bulk of the food in most cases. Much of this is phytoplankton which, for the most part, represents the primary producers of another ecosystem.

The feeding behaviour of the meio- and microfauna has not received much attention, but some studies have been made by Perkins (1958) at Whitstable, in Kent. Here, deposits vary from mud to sand. His techniques consisted of (i) the examination of gut contents of metazoans and of the cell inclusions of protozoans and (ii) a study of the feeding behaviour of beach organisms kept in cultures based on wheat grain infusions. The results of the culture experiments were inconclusive but they suggest that bacteria and protozoa are important in the diet of nematodes and harpacticoid copepods. Other interesting conclusions which emerged are not strictly relevant here, although they are probably applicable under field conditions. Thus, the nematodes were the most viable organisms in culture closely followed by ostracods, with copepods the least viable. There was, however, a slow but general decline in numbers which, apart from the artificial conditions of the experiment, was probably due to the absence of some important nutritive factor. The results of the gut and food vacuole analyses were more clearcut although it was not always possible to identify with certainty the material concerned.

The protozoans were found to be feeding extensively on diatoms, although bacteria were also present in large numbers in their vacuoles. One large ciliate, *Condylostoma*, whose size of up to 500 μ brings it into the range of meiofauna, was sometimes found to have the eggs of copepods within its vacuoles. The microfauna of Whitstable, therefore, are true primary consumers in that they feed for the most part upon the primary producers.

The nematodes were amongst the most numerous members of the meiofauna. The feeding habits of marine nematodes had been studied earlier by Weiser (1953) who divided them into four groups according to the structure of the oral cavity which was assumed to reflect the feeding habits. These groupings were: selective and non-selective deposit feeders, epigrowth feeders and carnivores. In general, the nematodes at Whitstable conform to this classification, although the carnivores were found to be also feeding extensively on diatoms. The predators are dominant in muds and form up to 48% of the nematodes in sand with a poor organic content. The other nematodes in this habitat comprise nearly 40% deposit feeders and 18% epigrowth feeders. The third major habitat at Whitstable is sand with a rich organic deposit. Here the composition of the nematode community is about 33% epigrowth feeders, 53% non-selective deposit feeders and about 14% carnivores. The flourishing epigrowths, largely bacteria, are responsible for the high proportion of epigrowth feeders. The community of nematode species living in the reducing layer of marine soils and their importance in feeding on the sulphur bacteria has already been mentioned.

It is obvious that the nematodes, as a group, are of the greatest significance in the ecology of the shore. Not only are they present in great abundance but they feed extensively on material which originates within the ecosystem. Most of them feed in one way or another on the bacteria which are breaking down the organic matter in the soil but some are primary consumers. A large minority are carnivores and hence form an important segment of the third trophic level.

Data for the other members of the meiofauna were less clear-cut. No significant results were obtained from the turbellarians whose guts are not the easiest to examine. Ostracods were found to contain diatoms when the gut contents were identifiable and these algae probably form the bulk of their food. Many of the harpacticoid copepods contained diatoms within their guts plus much vegetable matter of uncertain origin as well as sand and detritus. The latter suggests that the copepods are deposit feeders and digest the bacterial film around the soil particles. They are probably less selective than the nematodes in their feeding habits. Although these results are incomplete, they strengthen the conclusion that meiofauna feeds primarily on two basic resources, the primary producers, diatoms, and the bacterial film surrounding organic and inorganic particles. Perkins found no evidence of predation amongst the meiofauna, apart from the nematodes, but several of the ostracods and harpacticoid copepod species are believed to be carnivorous and the same may well be true of the turbellarians. The nematodes tend to be restricted to a particular feeding pattern by the structure of the pharynx, but most of the meiofaunal species are probably omnivorous although with a predeliction towards one or other type of food.

Perkins was not concerned with the macrofauna, although some of the species mentioned, for example *Corophium* and *Hydrobia*, belong to that category. Diatoms were found in both these species, although in the laboratory the former fed freely on bacteria. However, we saw earlier that both these species are deposit feeders with *Corophium* also acting as a filter feeder at times. It is, no doubt, during filter feeding that diatoms are taken in by

Corophium while *Hydrobia* catches the diatoms during the period it spends floating in the water, as described in Chapter 4.

So far, only the permanent fauna of the shore has been considered. This contains animals belonging to all trophic levels but the temporary members, whether terrestrial or marine, are almost exclusively predators. The only exceptions are the brent geese, which feed on *Zostera* or *Ulva* in the intertidal zone, and the primary consumers in the plankton which join the shore fauna when the tide is in. A few vegetarian fish may also be included. It is not necessary to detail the food of the temporary members here since this should have been covered adequately in Chapter 6.

The Bacteria

The bacteria of the depositing shore deserve separate consideration since they play such an important role in the ecosystem. Although they are usually classed as plants, they are technically consumers since their method of nutrition is heterotrophic. In terrestrial ecosystems, they belong to the detritus food chain and are responsible for the conversion of dead organic matter into nutrients which eventually are taken up by green plants, the primary producers. On the shore, their role is similar in that they break down dead organic matter but differs in that the principal benefactors from their activities are the primary consumers, not the producers. The implications of this conclusion will be taken further later in this Chapter, but first it is necessary to examine more closely the manner in which the bacteria influence the shore ecosystem. From what has already been written, it is obvious that their most significant function is to convert the organic matter in the soil into food for the permanent members of the shore. All the animals of the shore, whether detritus feeders or carnivores, are ultimately dependent upon the activities of the bacteria, with the exception of those which feed directly on diatoms. As we have seen, some of the bacteria are chemosynthetic and belong, consequently, to the first trophic level but most are heterotrophic and ecologically behave as animals. Concentrations of up to 5 million bacteria per ml have been reported from sand (Pearse, Humm & Warton, 1942) and the concentration in mud is almost certainly much greater. Zobell & Feltham (1942) quote, with a cheerful impartiality towards the metric and duodecimal systems, that 1 cubic foot of mud produces 0·4 grammes dry weight of bacteria per day. As most of the bacteria are in the upper layers, the production at the surface is very much greater. It is this vast bacterial community which contains the virtual herbivores of the shore and supports the bulk of the macrofauna. Most of the primary consumers on the shore are, therefore, bacteria.

Organic Material

An attempt to determine the amount of organic material available to a non-selective deposit feeder was made by George (1964b). The animal concerned was the polychaete *Cirriformia tentaculata.* This worm feeds on particles under 0·4 mm in diameter and George compared the carbon content of particles below this size before and after digestion with enzymes known to be present in *Cirriformia,* namely protease, amylase and lipase. The difference between the carbon content before and after digestion was taken to be the amount of organic material available to the worm. A second experiment compared the carbon content of the faeces of the worm with that of the mud from which they were taken. The results showed that 14% of the organic matter that was available for ingestion was digested by the enzymes but only 8% was actually assimilated by the worms. If the particles used in the experiment were inorganic, these results represent the amounts of bacteria that were digested, but as carbon and not nitrogen content was measured, it is possible that some of the organic matter was detritus not directly available to the worm (see page 22). However, it is probable that the generally low values are typical since the results approximate to figures of 10–20% obtained by other workers from analyses of the benthic communities of the open sea. The energetics of the bacteria/polychaete relationship were not considered, but with digestion rates of the order quoted, a gross ecological efficiency close to the ideal 10% would not be unexpected.

Food Webs on the Shore

Enough is known of the inter-relationships of the muddy shore community to justify an attempt at the construction of a food web diagram. This was provided by Perkins (1958) and a modified version of his diagram is given in Fig. 47. Changes have been made to accommodate the recent findings that detritus is important chiefly as a substratum for bacteria and not as a food source in itself. The central position of bacteria in the food web is clear from this figure; the primary producers, the diatoms, assume quite a peripheфral importance as a food source. The other most striking feature of the diagram is the large number of energy channels which converge on *N. diversicolor.* This worm is extremely abundant at Whitstable, as it is anywhere where conditions tend towards the estuarine. Its ecological importance in these semi-estuarine areas stems from its omnivorous feeding habits and its ability to dominate the community. Its catholic tastes were revealed by Perkins, whose examination of the gut contents showed the presence of diatoms, turbellarians, nematodes, polychaete chaeta, copepods, ostracods, appendages of other arthropods, copepod eggs, vegetable matter and detritus. In other words, almost all the plants and animals, living and dead, as well as the inorganic detritus in the soil are taken. The dominant position of *Nereis* is of economic significance since it is one of the major food items of fish, many of which are commercially exploited. Although the fish do not feed exclusively in the intertidal zone, they take a large proportion of their food from the shore ecosystem. *Hydrobia* and *Corophium* are also preyed upon by fish. These three organisms, *Nereis,*

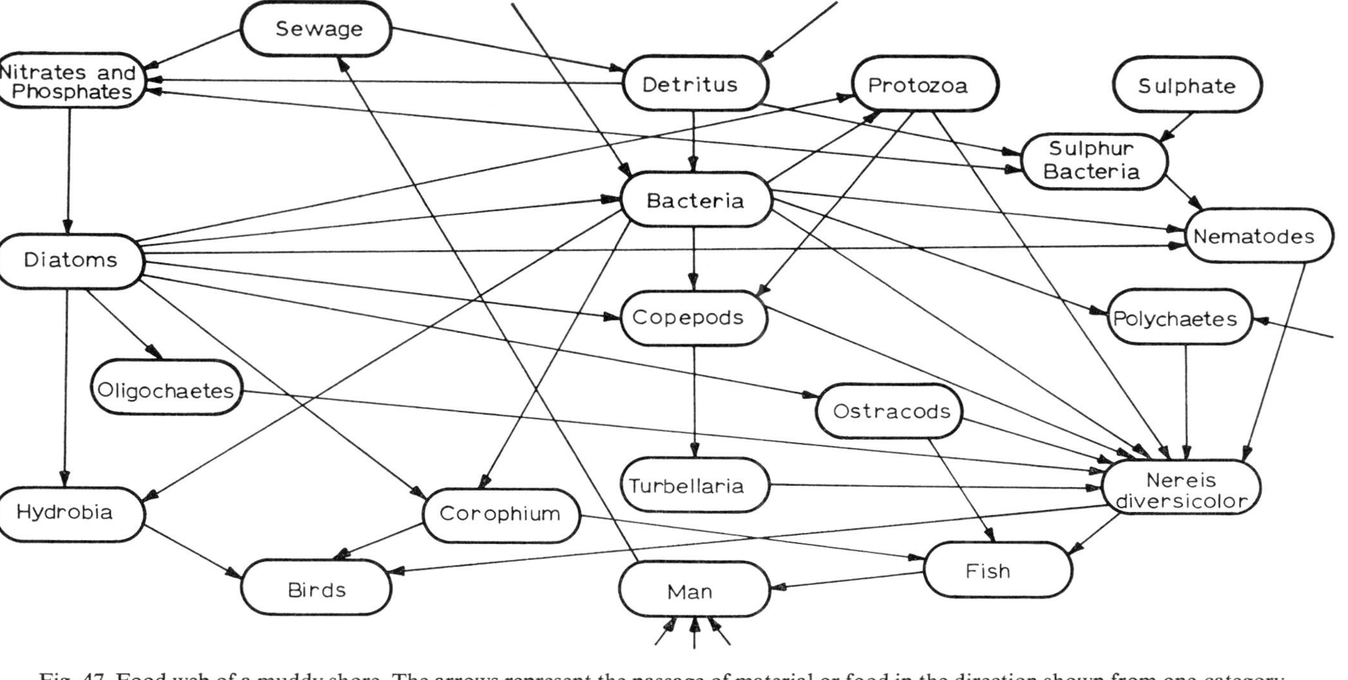

Fig. 47 Food web of a muddy shore. The arrows represent the passage of material or food in the direction shown from one category to another. Note the central position of bacteria as a food source and the importance of *Nereis diversicolor* as a consumer. High energy material is brought into the ecosystem from outside mainly through man and as detritus (adapted from Perkins 1958)

Hydrobia and *Corophium*, are, because of their abundance, the most important members of the macrofauna in the energy budget of the shore. All three are taken also by birds to an extent which was described in Chapter 6.

The Depositing Shore as an Ecosystem

The point has now been reached where it is necessary to sum up the conclusions from the previous pages and to decide whether the depositing shore conforms sufficiently closely to the concept of an ecosystem for it to be accorded the status. The term has already been applied to the shore many times in this discussion, but its use has been indiscriminate for purposes of convenience and was not intended to be significant. Superficially, the depositing shore has all the features of an ecosystem. It has in the diatoms a primary producer and a well defined, recognisable community of consumers at all trophic levels. There are large animals and the micro-organisms corresponding to the herbivores and soil fauna of terrestrial ecosystems. However, it is when one attempts to trace the grazing food chain and the detritus food chain of the classical ecosystem that difficulties appear. Can one in fact separate the two? What is the intertidal equivalent of a cow or, by a closer analogy, of the limpets and other herbivores of the rocky shore? Those which are often taken to be the primary consumers—the large molluscs—are as often as not feeding on bacteria. Even those animals which feed directly on the producers derive only a proportion of their energy resources from them, with the possible exception of the specialised conditions in the sulphide layer. Furthermore, the primary producers themselves contribute only a small fraction towards the total organic matter of the shore.

These anomalies make it very difficult to relate the depositing shore to the conventional ecosystem pattern. Thus there is really no grazing food chain and an energy flow diagram of the shore would not be the typical Y-shaped one but would have a complex webbed pattern. Either the depositing shore is a radically different kind of ecosystem from all others studied or it is no ecosystem at all.

The problem may perhaps be resolved by a simple effort of the imagination. Suppose the animals of the permanent macrofauna were very small, to be measured in microns rather than in centimetres, but otherwise with the same ecological habits, what conclusions would be drawn? Surely they would be recognised as part of the soil fauna and the depositing shore would be considered as the below-ground portion of a wider ecosystem. It is the opinion of the author that this is the case. The absence of an obvious primary consumer from the shore and the fact that bacteria provide the principal source of food render the system so unusual that it can be accepted as an ecosystem only by greatly extending the concept. It has already been pointed out more than once that much of the high energy material on the shore originates elsewhere and that much of the secondary production leaves the habitat. A glance at Fig. 46 demonstrates this clearly. The sizes of the blocks are meant to reflect the relative importance of the major sources of organic matter. The proportions are hypothetical but it is probably correct to show the greater contribution as coming from outside the system. The amounts entering as detritus are certainly large. Besides, much of the energy accumulated by the consumers is lost to other ecosystems, particularly through the activities of birds and fish. Admittedly, much of this energy is returned to the shore through defaecation or by death, but much of it is not.

If the depositing shore is not an ecosystem by itself, it must be part of a larger one. If this is accepted, or let it for the sake of argument be accepted, what is the larger ecosystem? The rocky shore is usually considered to be an ecosystem. Its first trophic level comprises a wide range of algae which entirely supports a rich and varied community of large herbivores with all the carnivores, parasites and other types of organisms typical of ecosystems. Yet it lacks one important feature, a detritus food chain. A below-ground, detritus-based community is obviously not possible on rock but there is no equivalent of a community feeding upon dead plant material. The reason for this is that most plant remains are removed from the area by tidal action soon after death and transported to mudflats or sandy beaches where they are eventually broken down. Thus if the rocky and depositing shores are put together, one has all the components of an ecosystem with a grazing food chain based on herbivores feeding on living plants and a detritus food chain based on herbivores feeding on dead plant material. The first community is found on the rocky shore and the second in the sand or mud of the depositing shores. The only major anomaly is that the herbivores of the detritus food chain are bacteria rather than animals. A further distinction of lesser significance is that the two types of food chain communities are separated horizontally instead of vertically as is more usual in other ecosystems.

It is concluded, therefore, that the intertidal zone as a whole constitutes a legitimate ecosystem although reference to Fig. 46 (p. 193) shows that much of the energy flow into the system originates from outside and many of the nutrient cycles are only partly within the shore environment. However, no ecosystem is self-contained although here the degree of overlap with other ecosystems is rather greater than usual. The principal interaction is between the river ecosystem and the mudflats. This is probably greater now than it was in the days before civilised man deposited large amounts of organic waste into the rivers, but many more data on the relative composition of muddy, sandy and rocky shores are required before firm conclusions can be drawn on the significance of the contribution made by the rivers.

The shore ecosystem can be made more self-contained if it is extended to include both the sub-littoral continental shelf and the coastal strip of vegetation. This could be called the coastal ecosystem. Such a solution removes many problems of definition. It was shown in the first chapter that the shore cannot be defined accurately and many of the shore organisms extend into the sub-littoral area on the one hand and on to land on the other. It would still be difficult to draw the line dividing the coastal from the terrestrial ecosystems, but most shores are marked by cliffs or some equally sharp discontinuity. The principal advantage of the concept of a coastal ecosystem is that salt-marshes, salt pastures and sand dunes may be included. These can also be considered as independent ecosystems, but there are considerable interchanges of material between them and the shore. The seaward border of the coastal ecosystem is more satisfactory. The continental shelf fauna is a distinct one and the continental slope forms an abrupt barrier to the ocean depths. The free-swimming animals and plankton are largely confined to the continental shelf and there is relatively little migration to and from the area.

The coastal ecosystem as defined here is a super-ecosystem containing distinct components which many ecologists would regard as ecosystems in their own right. The main justification for adopting the concept is that the energy flow through various channels and the circulation of nutrients take place almost entirely within the system. An attempt to demonstrate this diagrammatically is given in Fig. 48. The most significant feature of this diagram is that the import channel is wider than the export. This is a hypothetical diagram, but if it is a true reflection of natural conditions, it means that the coastal ecosystem is gaining energy and nutrients from outside. The major source of the imported material is shown to be the detritus carried down in rivers or blown directly from the land. This implies that there is a continuous energy drain from the land to the sea and it is quite likely that such an energy sink exists. The major return pathways before the advent of man were via the fish-eating birds, but as many of these are maritime in habit, they are not really removing material from the coastal ecosystem. Nowadays, the major return path is through the commercial fisheries but to offset this, the amount of organic matter passing from land to sea as a result of pollution has increased tremendously through the growth of human populations. The long-term consequences of these simple facts are enormous and should be exercising the minds of all persons concerned with human welfare.

Having defined the ecosystem and produced a crude energy flow diagram for it, how near are we to assigning values to the various quantities of energy which pass through it? Unfortunately, it is almost impossible to make even the wildest guess. The subject of ecological energetics is comparatively new and, with few exceptions, has been confined to terrestrial or freshwater ecosystems. It is too much to hope that even a whole laboratory of marine biologists will be able to estimate an overall energy budget for the coastal ecosystem in the near future. This may seem a depressing conclusion, but some of the most significant of the many facets are quite feasible for small-scale studies. The ecology of sandy shores and mud flats is at a threshold, over which lie unlimited opportunities for research, which should yield results of the greatest ecological significance.

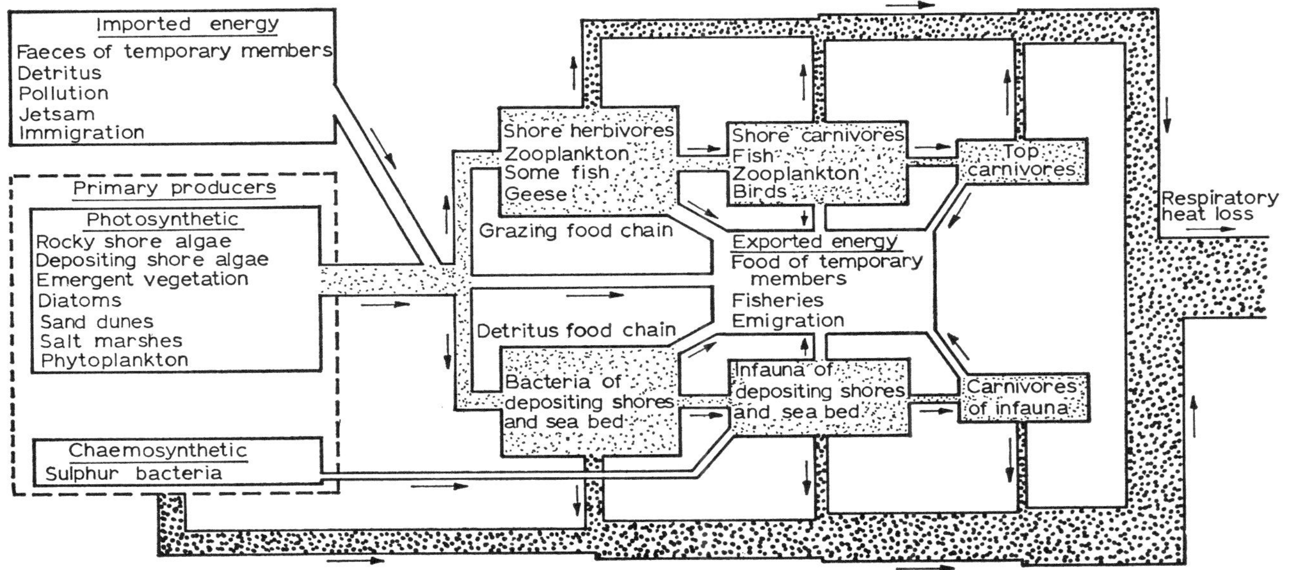

Fig. 48 The coastal ecosystem. An energy flow diagram of a super-ecosystem consisting of the shore, the fringing terrestrial regions and the sub-littoral zone. Such an ecosystem has the typical Y-shaped structure (shown in fine stipple) with the grazing food chain consisting of rocky shore herbivores and zooplankton and a detritus food chain of the bacteria of depositing shores and sublittoral muds with their dependent detritus feeders and carnivores. The width of the energy pathways are hypothetical but probably reflect the relative proportions accurately enough. Note that the import of energy exceeds the export and that consequently, the ecosystem is continuously gaining energy. The coarse stippled columns show respiratory heat loss

REFERENCES

ALEEM, A. A. (1950) Distribution and ecology of British marine littoral diatoms. *J. Ecol.*, **38**, 75–106.

ANDREWARTHA, M. G. and BIRCH, L. C. (1954) *The Distribution and Abundance of Animals*. University of Chicago Press, Chicago, xvi, 782.

ANSELL, A. D. (1961) Reproduction, growth and mortality of *Venus striatula* (da Costa) in Kames Bay, Millport. *J. mar. biol. Ass.*, **41**, 191–215.

ANSELL, A. D. and TRUEMAN, E. R. (1967) Burrowing in *Mercenaria mercenaria* (L) (Bivalvia, Veneridae). *J. exp. Biol.*, **46**, 105–15.

ARTHUR, D. R. (1969). *Survival; Man and his Environment*. English Universities Press, London.

ATKINS, D. (1954) Further notes on a marine member of the Saprolegniaceae *Leptolegnia marina* infecting certain invertebrates. *J. mar. biol. Ass. U.K.*, **33**, 613–25.

AVENS, A. C. (1965) Osmotic balance in gastropod molluscs. II the brackish water gastropod *Hydrobia ulvae* Pennant. *Comp. Biochem. Physiol.*, **16**, 143–53.

AVENS, A. C. and SLEIGH, M. A. (1965) Osmotic balance in gastropod molluscs. 1. Some marine and littoral gastropods. *Comp. Biochem. Physiol.*, **16**, 121–41.

BARKER, D. (1963) Size in relation to salinity in fossil and recent euryhaline ostracods. *J. mar. biol. Ass. U.K.*, **43**, 785–95.

BARNETT, P. R. O. (1958) The zonation of harpacticoids on a mudflat. *Ann. rep. Challenger Soc.*, 10.

BARNETT, P. R. O. (1968) Distribution and ecology of harpacticoid copepods of an intertidal mud flat. *Int. Revue ges. Hydrobiol.*, **53**, 177–209.

BARRAS, R. (1963) The burrows of *Ocypode ceratophthalmus* (Pallas) (Crustacea, Ocypodidae) on a tidal wave beach at Inhaca Island, Mocambique. *J. Anim. Ecol.*, **32**, 73–85.

BATTLE, H. I. (1932) Rhythmic sexual maturity and spawning of certain bivalve molluscs. *Contr. Canad. Biol.* (*N.S.*), **71**, 255–76.

BEADLE, L. C. (1959) Osmotic and ionic regulation in relation to the classification of brackish and inland saline waters. *Archivio di oceanografia e Limnologia*, XI, 143–51.

BEANLAND, F. L. (1940) Sand and mud communities in the Dovey estuary. *J. mar. biol. Ass. U.K.*, **24**, 589–611.

BOADEN, P. J. S. (1963a) The interstitial fauna of some North Wales beaches. *J. mar. biol. Ass. U.K.*, **43**, 79–96.

BOADEN (1936b). Behaviour and distribution of the archiannelid *Trilobodrilus heideri*. *J. mar. biol. Ass. U.K.* **43**, 239–50

BOADEN, P. J. S. (1964) Grazing in the interstitial habitat: a review. Crisp, D· J. (ed) Grazing in terrestrial and marine environments. *Symp. British Ecol. Soc.*, **4**, 299–303.

BOADEN, P. J. S. (1966) Interstitial fauna from Northern Ireland. *Veroff. Inst. Meeresforsch. Bremerhaven*, **2**, 125–30.

BRADY, F. (1943) The distribution of the fauna of some intertidal sands and muds on the Northumberland coast. *J. Anim. Ecol.*, **12**, 27–41.

BRAFIELD, A. E. (1964) The oxygen content of interstitial water in sandy shores. *J. Anim. Ecol.*, **33**, 97–116.

BRAFIELD, A. E. (1965) Quelques facteurs affectant la teneur en oxygène des eaux interstitielles littorales. *Vie et Mileu*, **16**, 889–97.

BRAFIELD, A. E. and CHAPMAN, G. (1967) Gametogenesis and breeding in a natural population of *Nereis virens*. *J. mar. biol. Ass. U.K.*, **47**, 619–27.

BRAFIELD, A. E. and NEWELL, G. E. (1961) The behaviour of *Macoma balthica*. *J. mar. biol. Ass. U.K.*, **41**, 81–7.

BRINKHURST, R. O. (1964) Observation on the biology of the marine oligochaete *Tubifex costatus*. *J. mar. biol. Ass. U.K.*, **44**, 11–16.

BRO LARSEN, E. (1952) On sub-social beetles from the salt-marsh, their care of progeny and adaptation to salt and tide. *Trans. 9th Int. Congr. Ent.*, **1**, 502–6.

BURBANK, W. D. (1959) The distribution of the estuarine isopod *Cyathura sp.* along the coast of the United States. *Ecology*, **40**, 507–11.

CHAPMAN, G. (1949) The thixotrophy and dilatancy of a marine soil. *J. mar. biol. Ass. U.K.*, **28**, 123–40.

CHAPMAN, V. J. (1964) *Coastal Vegetation*. Pergamon Press, Oxford. pp. viii, 245.

CLUTTER, R. I. (1967) Zonation of nearshore mysids. *Ecology*, **48**, 200–8.

COSTLOW, J. D., BOOKHOUT, C. G. and MUNROE, R. (1960). The effect of salinity on the larval development of *Sesarma linereum* (Bosc.) reared in the laboratory. *Biol. Bull.* Woods Hole, **118**, 183–202.

CRAIG, G. Y. and JONES, N. S. (1966) Marine benthos, substrate and palaeoecology. *Palaeontology*, **9**, 30–8.

CRISP, D. J. (1956) Surface chemistry, a factor in the settlement of marine invertebrate larvae. *Botanica gothoburgensia*, **3**, 51–65.

CRISP, D. J. (1964) (Ed.) The effects of the severe winter of 1962/63 on marine life in Britain. *J. Anim. Ecology*, **33**, 165–210

CROGHAN, P. C. (1961) Competition and mechanism of osmotic adaptation. *Symp. Soc. exp. Biol.*, **15**, 156–67.

DAHL, E. (1952) Some aspects of the ecology and zonation of the fauna on sandy beaches. *Oikos*, **4**, 1–27.

DALES, R. P. (1958) Survival of anaerobic periods by two intertidal polychaetes, *Arenicola marina* (L) and *Owenia fusiformis* Delle chiaje. *J. mar. biol. Ass.*, **37**, 521–9.

DAVIDSON, P. E. (1967) The oystercatcher as a predator of commercial shellfisheries. Ibis, **109**, 473–4.

DAVIDSON, P. E. (1968) The Oyster Catcher, a Pest of Shell-fisheries. Murton, R. K. and Wright, E. N. (Eds). *The problem of birds as pests*. Institute of Biology and Academic Press, pp. xiv, 254.

DELAMERE-DEBOUTTEVILLE, C. (1960) *Biologie des Eaux Souterraine Littorales et Continentales*. Hermann, Paris, p. 740.

DIGBY, P. S. B. (1967) Pressure sensitivity and its mechanism in the shallow marine environment. Marshall, N. B. *Aspects of Marine Zoology*. (Symposia of the Zoological Society of London No. 19). Academic Press. pp. 157–88.

DRINNAN, R. E. (1957) The winter feeding of the oyster catcher (*Haematopus ostralegus*) on the edible cockle. (*Cardium edule*.) *J. Anim. Ecol.*, **26**, 441–69.

DUNCAN, A. (1960) The spawning of *Arenicola marina* (L) in the British Isles. *Proc. Zool. Soc. London*, **134**, 137–56.

EDWARDS, R. L. (1958) Movements of individual members in a population of the shore crab *Carcinus maenas* L. in the littoral zone. *J. Anim. Ecol.*, **27**, 37–45.

EFFORD, I. E. (1965) Aggregation in the sand crab. *Emerita analoga* (Stimpson) *J. Anim. Ecol.*, **34**, 63–75.

EGGLISHAW, H. J. (1960) The life history of *Fucellia maritima* (Haliday) (Diptera, Muscidae) *Entomologist, London*, **93**, 225–31.

EVANS, R. G. (1948) The lethal temperatures of some common British littoral molluscs. *J. Anim. Ecol.*, **17**, 165–73.

FREEMAN, R. F. H. and LLEWELLYN, J. (1958) An adult digenetic trematode from an invertebrate host: *Proctoeces subtenuis* (Linton) from the lamellibranch *Scrobicularia plana* (da Costa) *J. mar. biol. Ass. U.K.*, **37**, 435–57.

FREUNDLICH, H. and RODER, H. L. (1938) Dilatancy and its relation to thixotropy. *Trans. Faraday Soc.*, Vol. XXXIV, 308–16.

GEORGE, J. D. (1963) Equipment for facilitating movement on an intertidal mudflat *Nature*, **200**, 921–22.

GEORGE, J. D. (1964a) The life history of the cirratulid worm *Cirriformia tentaculata*, on an intertidal mudflat. *J. mar. biol. Ass. U.K.*, **44**, 47–65.

GEORGE, J. D. (1964b) Organic matter available to the polychaete *Cirriformia* (Montagu) living in an intertidal mudflat. *Limnol. Oceanogr.*, **9**, 453–5.

GEORGE, J. D. (1967) Cryptic polymorphism in the cirratulid polychaete, *Cirriformia tentaculata*. *J. mar. biol. Ass. U.K.*, **47**, 75–9.

GOLLEY, F. B. (1960) Energy dynamics of a food chain of an old-field community. *Ecol. Monogr.*, **301**, 187–206.

GOODRICH, E. S. (1912) *Nerilla*, an archiannelid. *Quart. J. micr. Sci.*, **57**, 397–425.

GOSS-CUSTARD, J. D. (1967) The winter predation of *Corophium volutator* by redshank. Ibis., **109**, 475.

GRAY, J. S. (1966) The attractive factor of intertidal sands to *Protodrilus symbioticus*. *J. mar. biol. Assoc. U.K.*, **46**, 627–45.

GREEN, J. (1957) The growth of *Scrobicularia plana* (da Costa) in the Gwendraeth Estuary. *J. mar. biol. Ass. U.K.*, **36**, 41–7.

GREEN, J. (1968) *The Biology of Estuarine Animals*. Sidgwick and Jackson, London, pp. x, 401.

HARLEY, M. B. (1950) Occurrence of a filter feeding mechanism in the polychaete *Nereis diversicolor*. *Nature*, **165**, 734–5.

HAYES, F. R. (1964) The mud-water interface. *Oceanogr. Mar. Biol. Ann. Rev.*, **21**, 121–45.

HOLME, N. A. (1949) The fauna of sand and mud banks near the mouth of the Exe Estuary. *J. mar. biol. Ass. U.K.*, **28**, 189–237.

HOPKINS, J. T. (1963) A study of the diatoms of the Ouse Estuary Sussex I. The movement of the mudflat diatoms in response to some chemical and physical changes. *J. mar. biol. Ass. U.K.*, **43**, 653–63.

Hopkins, J. T. (1964) A study of the diatoms of the Ouse Estuary Sussex II. The ecology of the mudflat diatom flora. *J. mar. biol. Ass. U.K.*, **44**, 333–41.

Hopkins, J. T. (1965) Some light-induced changes in behaviour and cytology of an estuarine mud-flat diatom. *Light as an Ecological Factor*, pp. 205–29. Ed. R. Bainbridge. Oxford: Blackwell Scientific Publications.

Hopkins, J. T. (1966) The role of water in the behaviour of an estuarine mud-flat diatom. *J. mar. biol. Ass. U.K.*, **46**, 617–26.

Howells, W. R. (1964) The macrofauna of the intertidal soils of the Towy Estuary, Carmarthenshire. *Ann. Mag. nat. Hist.*, **7**, 577–607.

Howie, D. I. D. (1959) The spawning of *Arenicola marina* (L) I. The breeding season. *J. mar. biol. Ass. U.K.*, **38**, 395–406.

Hyman, L. H. (1951) *The Invertebrates: Vol.* 3 *Acanthocephala, Aschelminthes, and Entoprocta. The Pseudocoelomate bilataria.* McGraw Hill, New York, pp. viii, 572.

Hynes, H. B. N. (1955) The reproductive cycle of some British freshwater Gammaridae, *J. Anim. Ecol.*, **24**, 352–87.

Ingle, R. W. (1966) An account of the burrowing behaviour of the amphipod *Corophium arenarium* Crawford (Amphipoda: Corophiidae). *Ann. Mag. nat. Hist. Ser.*, 13, **9**, 309–17.

Kinne, O. (1963) The effects of temperature and salinity on marine and brackish water animals. 1. Temperature. *Oceanogr. Mar. Biol. Ann. Rev.*, **1**, 301–40.

Kinne, O. (1964) The effects of temperature and salinity on marine and brackish water animals. II Salinity. *Oceanogr. Mar. Biol. Ann. Rev.*, **2**, 251–339.

Kriss, A. E. (1963) *Marine Microbiology* (*Deep Sea*), Oliver and Boyd, Edinburgh, pp. xviii, 536.

Lambert, J. M. (1964) The *Spartina* story. *Nature*, **204**, 1136–8.

Lewis, J. G. E. (1962) The ecology, distribution and taxonomy of the centipedes found on the shore in the Plymouth area. *J. mar. biol. Ass. U.K.*, **42**, 655–64.

Lewis, J. B. (1963) Environmental and tissue temperatures of some tropical intertidal marine animals. *Biol. Bull.*, **124**, 277–84.

Lewis, J. R. (1964) *The Ecology of Rocky Shores.* English Universities Press, London, pp. xii, 323.

Lockwood, A. P. M. (1962) The osmoregulation of Crustacea. *Biol. Rev.*, **37**, 257–305.

Lockwood, A. P. M. (1964) Activation of the sodium uptake system at high blood concentration in the amphipod, *Gammarus duebeni. J. exp. Biol.*, **41**, 447–58.

McIntyre, A. D. (1964) Meiobenthos of sub-littoral muds. *J. mar. biol. Ass. U.K.*, **44**, 665–74.

McIntyre, A. D. (1969) Ecology of marine meiobenthos. *Biol. Rev.*, **44**, 245–90.

McLeese, D. W. (1956) Effects of temperature, salinity and oxygen on the survival of the American lobster. *J. Fish. Res. Bd.* Canada, **13**, 247–72.

McLusky, D. S. (1967) Some effects of salinity on the survival, moulting and growth of *Corophium volutator* (Amphipoda). *J. mar. biol. Ass. U.K.*, **47**, 607–17.

McLusky, D. S. (1968) Some effects of salinity on the distribution and abundance of *Corophium volutator* in the Ythan Estuary. *J. mar. biol. Ass. U.K.*, **48**, 443–54.

MARE, M. F. (1942) A study of a marine benthic community with special reference to the micro-organisms. *J. mar. biol. Ass. U.K.*, **25**, 517–54.

MARINE BIOLOGICAL ASSOCIATION (1957) *The Plymouth Marine Fauna.* Marine Biological Association of the United Kingdom, Plymouth. pp. xliv, 547.

MEADOWS, P. S. (1964a) Experiments on substrate selection by *Corophium* species: films and bacteria on sand particles. *J. exp. Biol.*, **41**, 499–511.

MEADOWS, P. S. (1964b) Experiments on substrate selection by *Corophium volutator* (Pallas): depth selection and population density. *J. exp. Biol.*, **41**, 677–687.

MEADOWS, P. S. (1967) Discrimination, previous experience and substrate selection by the amphipod *Corophium. J. exp. Biol.*, **47**, 553–9.

MEADOWS, P. S. and REID, A. (1966) The behaviour of *Corophium volutator* (Crustacea: Amphipoda) *J. zool., Lond.*, **150**, 387–99.

MEYERHOF, E. and ROTHSCHILD, M. A. (1940) A prolific trematode. *Nature*, **146**, 367–68.

MORGAN, E. (1965) The activity of the amphipod *Corophium volutator* (Pallas) and its possible relationship to changes in hydrostatic pressure associated with the tides. *J. Anim. Ecol.*, **34**, 731–46.

MORGAN, E., NELSON-SMITH, A. and KNIGHT-JONES, E. W. (1964) Responses of *Nymphon gracile* (Pycnogonida) to pressure cycles of tidal frequency. *J. exp. Biol.*, **41**, 825–36.

NEWELL, G. E. (1948) A contribution to our knowledge of the life history of *Arenicola marina* L. *J. mar. biol. Ass. U.K.*, **27**, 554–80.

NEWELL, G. E. (1958a) The behaviour of *Littorina littorea* (L). under natural conditions and its relation to position on the shore. *J. mar. biol. Ass. U.K.*, **37**, 229–39.

NEWELL, G. E. (1958b) An experimental analysis of the behaviour of *Littorina littorea* (L) under natural conditions and in the laboratory. *J. mar. biol. Ass. U.K.*, **37**, 241–66.

NEWELL, R. C. (1962) Behavioural aspects of the ecology of *Peringia* (=*Hydrobia*) *ulvae* (Pennant) (Gastropoda, Prosobranchia). *Proc. zool. Soc. Lond.*, **138**, 49–75.

NEWELL, R. C. (1964) Some factors controlling the upstream distribution of *Hydrobia ulvae* (Pennant) (Gastropoda, Prosobranchia). *Proc. zool. Soc. Lond.*, **142**, 85–106.

NEWELL, R. C. (1965) The role of detritus in the nutrition of two marine deposit feeders, the prosobranch *Hydrobia ulvae* and the bivalve *Macoma balthica. Proc. zool. Soc. Lond.*, **144**, 25–45.

ODUM, E. P. (1963) *Ecology*. Holt, Rinehart & Winston, New York. pp. viii, 152.

OLNEY, P. J. S. (1964) The food of mallard *Anas platyrhynchos* collected from coastal and estuarine areas. *Proc. zool. Soc. Lond.*, **141**, 399–418.

OLNEY, P. J. S. (1965a) The food and feeding habits of shelduck. *Tadorna tadorna.* Ibis., **107**, 527–32.

OLNEY, P. J. S. (1965b) The autumn and winter feeding biology of certain sympatric ducks. *Trans. VI Congr. Int. Union Game Biol.*, The Nature Conservancy, Lond., 309–20.

OLNEY, P. J. S. and MILLS, D. H. (1963) The food and feeding habits of goldeneye *Bucephala clangula* in Great Britain. Ibis. **105**, 293–300.

PALMER, J. D. and ROUND, F. E. (1965) Persistent vertical migration rhythms in benthic microflora I. the effect of light and temperature on the rhythmic behaviour of *Euglena obtusa*. *J. mar. biol. Ass. U.K.*, **45**, 567–82.

PARKE, M. and MANTON, I. (1967) The specific identity of the algal symbiont in *Convoluta roscoffensis*. *J. mar. biol. Ass. U.K.*, **47**, 445–64.

PEARSE, A. S., HUMM, A. J. and WHARTON, G. W. (1942) Ecology of sand beaches at Beaufort N.C. *Ecol. Monogr.*, **12**, 135–90.

PENNAK, R. W. (1951) Comparative ecology of the interstitial fauna of freshwater and marine beaches. *Colloques int. Cent. natn. Rech. Scient. Ecol. Paris*, 449–80.

PERKINS, E. J. (1957) The blackened sulphide-containing layer of marine soils, with special reference to that found at Whitstable, Kent. *Ann. Mag. nat. Hist.*, **10**, 25–35.

PERKINS, E. J. (1958) The food relationships of the microbenthos, with particular reference to that found at Whitstable, Kent. *Ann. Mag. nat. Hist.* Ser. 13, vol. 1, 64–77.

PERKINS, E. J. (1960) The diurnal rhythm of the littoral diatoms of the River Eden Estuary, Fife. *J. Ecol.*, **48**, 725–8.

PETERSEN, C. G. J. (1918) The sea bottom and its production of fish foods; a survey of the work done in connexion with valuation of Danish waters from 1883–1917. *Rep. Dan. Biol. Sta.*, **25**, 1–62.

POPHAM, E. J. (1966) The littoral fauna of the Ribble Estuary, Lancashire, England. *Oikos*, **17**, 19–32.

PROSSER, C. L. and BROWN, F. A. (1961) *Comparative Animal Physiology*. 2nd Edition. Saunders. Philadelphia and London, pp. ix, 688.

REES, W. J. and LUMBY, J. R. (1954) The abundance of *Octopus* in the English Channel. *J. mar. biol. Ass. U.K.*, **33**, 515–36.

REESE, E. S. (1964) Ethology and marine zoology. *Oceanogr. mar. biol. Ann. Rev.*, **2**, 455–88.

REMANE, A. (1940) Einführung in die zoologische Ökologie der Nord- und Ostsee. *Tierwelt d. N.- u. Ostsee, Leipzig*. pp. 238.

RICE, A. L. (1961) The responses of certain mysids to changes of hydrostatic pressure. *J. exp. Biol.*, **38**, 391–401.

ROUND, F. E. and PALMER, J. D. (1966) Persistent vertical-migration rhythms in benthic microflora. II Field and laboratory studies on diatoms from the banks of the River Avon. *J. mar. biol. Ass. U.K.*, **46**, 191–214.

RUSSELL, F. S. and YONGE, C. M. (1963) *The Seas*. 3rd Edition. Warne, London, pp. xiv, 376.

SLOBODKIN, L. B. (1959) Energetics in *Daphnia pulex* populations. *Ecology*, **40**, 232–43.

SMITH, R. I. (1955) Salinity variations in interstitial water of sand at Kames Bay, Millport, with reference to the distribution of *Nereis diversicolor*. *J. mar. biol. Ass. U.K.*, **34**, 33–46.

SOUTHWARD, A. J. (1958) Notes on the temperature tolerances of some intertidal animals in relation to environmental temperatures and geographical distribution. *J. mar. biol. Ass. U.K.*, **37**, 49–66.

SPOONER, G. M. (1947) The distribution of *Gammarus* species in estuaries. Part I *J. mar. biol. Ass. U.K.*, **27**, 1–52.

SPOONER, G. M. and MOORE, H. B. (1940) The ecology of the Tamar Estuary. VI

An account of the microfauna of intertidal muds. *J. mar. biol. Ass. U.K.*, **24,** 283–330.

STEELE, J. H. and BAIRD, I. E. (1968) Production ecology of a sandy beach. *Limnol. Oceanogr.*, **13**, 14–25.

STEPHEN, A. C. (1928) Notes on the biology of *Tellina tenuis* da Costa. *J. mar. biol. Ass. U.K.*, **15**, 683–702.

STEPHEN, A. C. (1929) Studies of the Scottish marine fauna: The fauna of the sandy and muddy areas of the tidal zone. *Trans. Roy. Soc. Edinb.*, **56**, 291–306.

STEPHEN, A. C. (1930) Studies on the Scottish marine fauna: Additional observations on the fauna of the sandy and muddy areas of the tidal zone. *Trans. roy. Soc. Edinb.*, **56**, 521–35.

STEPHEN, A. C. (1932) Notes on the biology of some lamellibranchs in the Clyde area. *J. mar. biol. Ass.*, U.K., **18**, 51–68.

STEPHEN, A. C. (1953) *Life on Some Sandy Shores.* Essays in Marine Biology. Oliver and Boyd, Edinburgh. pp. 50–72.

STEVEN, G. A. (1930) Bottom fauna and the food of fishes. *J. mar. biol. Ass. U.K.*, **16,** 677–705.

SWEDMARK, B. (1964) The interstitial fauna of marine sand. *Biol. Rev.*, **39**, 1–42.

THORSON, G. (1946) Reproduction and larval development of Danish marine bottom invertebrates with special reference to the planktonic larvae in the Sound (Oresund). *Medd. Komm. Hav. Plankt.*, **4**, 1–523.

TODD, M. E. (1964) Osmotic balance in *Hydrobia ulvae* and *Potamopyrgus jenkinsi* (Gastropoda: Hydrobiidae). *J. exp. Biol.*, **41**, 665–77.

TRUEMAN, E. R. (1968) Burrowing habit and the early evolution of body cavities. *Nature*, **218**, 96–98.

TRUEMAN, E. R., BRAND, A. R. and DAVIS, P. (1966) The dynamics of burrowing of some common littoral bivalves. *J. exp. Biol.*, **44**, 469–492.

WADE, B. (1964) Notes on the ecology of *Donax denticulatus* (Linne) *Proc. Gulf Caribb. Fish Inst.*, **17**, 36–41.

WATKIN, E. E. (1942) The macrofauna of the intertidal sand of Kames Bay, Millport, Buteshire. *Trans. Roy. Soc. Edinb.*, **60**, 543–61.

WELLS, G. P. (1945) The mode of life of *Arenicola marina. J. mar. biol. Ass. U.K.*, **26**, 170–207.

WIESER, W. (1953) Die Beziehung zwischen Mundhöhlengestalt, Ernährungsweise und Vorkommen bei freilebenden marinen Nematoden. *Ark. Zool.*, **4**, 439–84.

WILSON, D. P. (1953) The settlement of *Ophelia bicornis* Savigny larvae. *J. mar. biol. Ass. U.K.*, **31**, 413–38.

WILSON, D. P. (1954) The settlement of *Ophelia bicornis* Savigny larvae. The 1952 experiments. *J. mar. biol. Ass. U.K.*, **32**, 209–33.

WILSON, D. P. (1955) The attractive factor in the settlement of *Ophelia bicornis* Savigny. *J. mar. biol. Ass. U.K.*, **33**, 361–80.

WYNNE-EDWARDS, V. C. (1962) *Animal dispersion in relation to social behaviour.* Oliver and Boyd, Edinburgh. pp. xi, 635.

YONGE, C. M. (1949) *The Sea Shore.* Collins, London. pp. xvi, 311.

YONGE, C. M. (1953) *Aspects of Life on Muddy Shores.* Essays in Marine Biology. Oliver and Boyd, Edinburgh, 29–49.

ZOBELL, L. E. and FELTHAM, C. B. (1942) The bacterial flora of a marine mud-flat as an ecological factor. *Ecology*, **23**, 69–77.

INDEX

(Figures in heavy type refer to pages on which illustrations occur.)

Acronida 37
Actinic rays 118
Activity range 101
Adhesive organs 49
Agropyron 165, 167
ALEEM, A. A. 174
Algae 52, 173 ff
Along-shore distribution 64
Ammodytes 45
Ammophila 167
Ampharete 37
Amphipoda 35, 65, 66, 161 ff
Anaerobic conditions 112
ANDREWARTHA, M. G. 97, 135
Annelida 35, 43 ff, 68 ff, 154, 156
ANSELL, A. D. 40, 103, 120
Anti-crushing devices 49
Anurida 60, 145
Arachnida 145
Archiannelida 10, 48, 54, 156
Arenicola
blood concentration 106
burrowing habits 43 ff, 44, 137
grade of deposit 31
mortality 99
reproduction 88–89
respiration 112, 116
zonation 68–71, 73, 93
Artemia 105
Artemisia 171
ARTHUR, D. R. 184, 194
Ascophyllum 173
ATKINS, D. 140
Atriplex 167, 171
Audouinia 68, 85
Aurelia 150
AVENS, A. C. 23

Bacteria 52, 121–123, 137, 140, 191, 197
BAIRD, I. E., 190
Bar, **2**, 8
BARKER, D. 154
Barnacle goose 61
Barnea 142
BARNETT, P. R. O. 73, 75
BARRAS, R. 79
Bathyporeia 66, 67, 158
BATTLE, H. I. 90
BEADLE, L. C. 149
BEANLAND, F. L. 17
Beetles 145
Bicarbonates 117
Biogeochemical cycles 182
Biomass 180
BIRCH, L. C. 97, 135
Bleduis 145
Blenny 45
Blepharipoda 66
Blood concentration 106
Blood, oxygen capacity of 11, 111
Blood pigments 111
Borers 142
BRADY, F. 70, 92
BRAFIELD, A. E. 31, 81, 90, 113
Brent goose 61, 197
BRINKHURST, R. O. 157
BRO. LARSEN, E. 145
BROWN, F. A. 106
Bucephala 59
BRUBANK, W. D. 128
Burrowing methods 38 ff, **14**, **39**

INDEX

Caecum 48
Cakile 167
Callianassa 36, 104
Calliostoma 35
Canthocamptus 156
Canuella 73
Carbon content of soil 20 ff, **21**
Carbonates 117
Carbon dioxide 116
Carbonic acid 117
Carcinus 36, 62, 80, 105
Cardium 38, 40, 42, 62, 68, 73, 74, 134, 199
Catatropis 139
Centipedes 145
Cercaria 139
Cereus 9, 36
Chaetopterus 31, 43
CHAPMAN, G. 12, 90
CHAPMAN, V. J. 164
CHEMOSYNTHESIS 11, 191
Ciliary gliding 52
Cirolina belt 66
Cirrifornia 85 ff, 90, 198
Cladophora 173
Clay 12
CLUTTER, R. I. 74
Cnidaria 35 ff
Coastal ecosystem 201, **203**
Coelogynopora **51**
Coelopa 143
Competition 135
Condylostoma 195
Conformers 103
Consumers 179, 194 ff
Convergent evolution 52, 66
Convoluta 190
Copepoda 73, 154 (harpacticoid) 139 (parasitic)
Cordylophora 62
Corophium
  as prey 134
  behaviour 25 ff
  burrowing mechanism 28, **29**
  downshore movements 77
  estuarine 162
  reproduction 26, 32
Corystes 36, 75
Crangon, 17, 62, 105
CRAIG, G. Y. 46
CRISP, D. J. 99, 123, 161
CROGHAN, P. C. 128
Crustacea 35 ff, 65 ff
Cryptocotyle 139
Cumopsis 67
Currents **7**
  as a limiting factor 119
  longshore 6
  rip 7
  undertow 4
Cyathura 128, 161
Cylindropsyllis **50**
Cypris 154

Dab 56
DAHL, E. 65
DALES, R. P. 112
DAVIDSON, P. E. 134
Day length 88
Decapoda 35, 79
Decomposers 182
DELAMERE-DEBOUTTEVILLE, C. 48
Delta 148
Density 20, 74
Deposit feeding 30, 47, 53, 129
Desiccation 123
Detritus 10, 20, 130
Detritus eaters 52
Detritus food chain 181, 184
Diatoms 129, 173 ff, 191
Die-back 170
DIGBY, P. S. B. 77, 118
Dilatancy 12
Dinoflagellates 173
Diptera 37, 143
Dispersal 16, 24, 26, 34, 52, 84, 87, 120
Dissociation curve 111
Distribution on fauna 64 ff
Diurodrilus 48
Donax 40, 42, 68, 119
Downshore distribution 64 ff
  seasonal changes in 92 ff
Drainage 124
Dreissensia 62
DRINNAN, R. E. 134
Ducks, 59, 132 ff
DUNCAN, A. 88
Dune 167

Echinocardium 36
Echinodermata 35 ff
Ecological efficiency 185
Ecological pyramids 180
Ecosystem 179 ff, **181**, **183**, 189
EDWARDS, R. L. 79
EFFORD, E. 79

EGGLISHAW, H. J. 143
Eggs 24 ff
Eider duck 61
Emerita 66, 79
Enchytraeus 92–94
Endopsammon 16
Energy flow 180, 184
Ensis 40
Enteromorpha 14, 173, 192
Environment 95, 97 ff
Epideictic display 138
Epifauna 15, 19 ff
Epipsammon 16
Eriocheir 151
Erosion 2
Estuaries 63, 147 ff
Eteone 69, 93, 94
Euglena 178
Eunice 90
Eurydice 77, 66–67, 92–93, 161
Euryhaline 103
Eurythermal 87
EVANS, R. G. 101

Fabenella 154
Facultative shore fauna 59 ff
Fauna
  of shore 15 ff
  estuarine 152 ff
  origin 61
Feeding methods
  benthos 47
  birds 59
  flatfish 56
  Hydrobia 76
  roundfish 57
  shelduck 77
FELTHAM, C. B. 197
Festuca 165, 170
Field mouse 59
Filter feeding 30, 47, 129
Fish 45 ff, 55 ff
Flatfish 55 ff
Flounder 56
Food
  as a limiting factor 128 ff
  of *Cirriformia* 198
  of meiofauna 52, 195
  of microfauna 195
Food chain 179
Food chain efficiency 185
Food web 79, 198, **199**
FREEMAN, R. F. H. 139
Freshwater see page 107
FREUNDLICH, H. 12
Fucellia 143
Fucus 173
Fungi 140

Gadidae 56
Gammarus 127, 161
Gastropoda 35, 37
Gastrotricha 18, 48
Geese 60 ff
General Saltmarsh Community 170
GEORGE, J. D. 13, 85–87, 197
Ghost crab 65, 124
Gibbula 35
Glycera 43
Glycolysis 112
Goby 45
Goldeneye 59
GOLLEY, F. B. 182
GOODRICH, E. S. 119
GOSS-CUSTARD, J. D. 134
GRAY, J. S. 121
Grazing food chain 181, 184
GREEN, J. 152, 160
Gribble 84, 142
Gross ecological efficiency 185
Growth rates 74
Gulls 59
Gwynia 53

Habitat 95
Haddock 57
Hake 57
Halammonhydra **51**, 52
Halibut 55, 56
Halichoerus 58
HARLEY, M. B. 156
Harpacticoida, 73, 154
Harpacticus 73
Haustoriidae 66
Haustorius 66, 67
HAYES, F. R. 137
Hedyopsis 49
Helicoprorodon 48
Hemigrapsus 104
Hippidea 66
Holeuryhaline 104
HOLME, N. A. 37
Holeocius 150
Honkenya 165, 167
HOPKINS, J. T. 174
Horizontal distribution 64, 74

HOWELLS, W. R. 158
HOWIE, D. I. D. 88
HUMM, A. J. 53, 197
Hyas 104
Hydrobia
  estuarine distribution 157
  movements on shore 76 ff
  nutrition 19 ff, 196
  osmoregulation 22 ff, 31 ff
  parasites 139
  prey 132
  reproduction 24 ff
Hydrogen sulphide 11, 117
Hydroschendyla 145
Hydrostatic pressure 77, 117
HYMAN, L. H. 154
HYNES, H. B. N. 161
Hypersomotic 103
Hyposmotic 103

Infauna 15, 34 ff
Interaction of factors 124 ff
Interstitial fauna 16, 48, 73, 153
INGLE, R. W. 28
Ions in sea water 106
Irritability range 100
Isopoda 35 ff, 66 ff
Isosmotic 103

Jaera 128
JONES, N. S. 46
Juncetum 171
Juncus 171

KINNE, O. 99, 100, 127
Kinorhyncha 154
KNIGHT-JONES E. W. 78
KRISS A. E. 55

LAMBERT J. M. 170
Lamellibranchiata 38, 73
Lanice 43
Leptolegnia 140
Leptosynapta 36, 48
LEWIS, J. B. 102
LEWIS, J. G. E. 145
LEWIS, J. R. 2
Liebig's law of minimum 96
Life cycles 84
Light
  and diatoms 175 ff
  as limiting factors 118 ff
Limiting factors 95 ff
desiccation 123 ff
  food 128 ff
  gases 116 ff
  interactions of 125
  light 118 ff
  other animals 135 ff
  oxygen 111 ff
  pathogens 138 ff
  predators 131 ff
  salinity 103 ff
  temperatures 99 ff
  water currents 119 ff
  weather 98 ff
Limnoria 142
Littoral zone 1
Littorina
  as prey 233
  mortality 99, 102
  movements on shore 77
  osmoregulation 23
  parasites 140
  reproduction 24–25
LLEWELLYN, J. 139
Lobster 57
LOCKWOOD, A. P. M. 104, 128, 151
Locomotion (see also movements down shore)
  of *Macoma* 80 ff
  of Meiofauna 52
Longipedia 156
Longshore currents 6
Lumbricillus 104
LUMBY, J. R. 58
Lumpsucker 46
Lunar rhythms 90

Macoma
  as prey 134, 161
  burrowing habits 40, 42
  in estuaries 159 ff
  mortality 99
  movements in shore 80 ff
  nutrition 22, 31
  reproduction 90
  zonation 70, 71, 73
Macrofauna 16, 34 ff, 152 ff
Macrostomum 104
Mallard 59, 133
Mammals 135
MANTON, I. 190
MARE, M. F. 16
Marine birds 60 ff, 132
Marine borers, 142

Marinogammarus 162
Maritrema 139
Meadows, P. S. 26–28, 121
Meiofauna 16, 48 ff, 73, 153
Melita 162
Mercenaria 40
Mesopsammon 16
Meyerhof, E. 140
Microfauna 16, 54 ff, 153
Microhedyle 54
Microjaera **50**
Microstomus 56
Mid-littoral zone 65
Miracidium 139
Mites 145
Mollusca 35, 68, 70, 153
Monobryozoan 53
Moore, H. B. 161
Morgan, E. 77–78
Movements down shore 76 ff
Mud 12 ff, 35 ff, 68 ff
Mullet 56
Mya
  burrowing habits 42, 75
  mortality 99
  position on shore 70, 73
  reproduction 90
Mysidacea 37, 74
Mytilicola 140
Mytilus 36, 62, 99, 140, 150
McIntyre, A. D. 54, 154
McLusky, D. S. 162

Nannoplankton 129
Nassarius 35
Necrophloeophagus 145
Nelson-Smith, A. 78
Nematoda 18, 156, 191, 195
Nemertini 35 ff
Neoteny 49
Nepthys
  feeding habits 45, 131
  zonation 68–70
Nereis
  as prey 198
  estuarine 156
  feeding habits 45, 157, 198
  reproduction 90–92
  zonation 68–70, 73, 107
Nerilla 119
Nerine 69, 93, 94
Newell, G. E. 31, 79, 81, 88
Newell, R. C. 13, 20, 77, 158

Nitrogen 20 ff, **21**, 116
Nitzschia 178
Nutrient cycles 182
Nutrients 120 ff
Nutrition (see food)
Nymphon 78

Obligate (terrestrial) fauna 60 ff
Octopus 57
Odum, E. P. 182
Oedicerotidae 66
Oligochaeta 156
Oligostenohaline 103
Oligothermy 87
Olney, P. J. S. 59, 132, 133
Onchidoris 35
Oocyte (seasonal changes) 86, 91
Ophelia 69
Opisthobranchiata 35
Orchestia 143
Orientation on the shore 79 ff
Ostracoda 62, 154
Osmoregulation 22 ff, 31 ff, 103 ff, 105 ff
Otter 59
Owenia 112
Oxygen
  and meiofauna 75
  as limiting factor 96, 111 ff
  in interstitial water 113
Oxypode 65, 79
Oyster-catcher 60, 134

Palaemonetes 128
Palmer, J. D. 178
Palolo worm 90
Parenthessius 140
Parke, M. 190
Particle size 20
Patella 102
Pathogens 138 ff
Peachia 36
Pearse, A. S. 53, 197
Pectinaria 43
Pelvetia 173
Penaeus 104
Penetration anchor 38, **39**
Pennak, R. W. 119
Pereiopods 26
Peringia see *Hydrobia*
Perkins, E. J. 11, 178, 191, 195, 199
Permanent fauna 17, 19 ff
Petersen, C. G. J. 17
Petrobius 60

INDEX

Phoca 58
Pholas 12, 142
Phragmites 171
Phyllodoce 68–72, 93
Phytoplankton 191
Pinnipedia 58
Pinnotheres 140
Pipe fish 48
Place in which to live 141 ff
Plaice 56
Plankton 24, 191
Platychelipus 73
Playtmonas 190
Pleopods 26
Pleurosigma 177–178
Pollack 57
Polychaeta 38, 43 ff, 68 ff, 154
Polyplacophora 35
Polystenohaline 103
Polythermy 87
Polyzoa 35
Pontocrates 66, 67
POPHAM, E. J. 158
Porifera 35
Potamon 151
Praunus 78
Predation 313 ff
Proctoeces 139
Producers 179, 190 ff
PROSSER, C. L. 106
Protein 106
Protochordata 35
Protodrilus 121
Psammodriloides 48, **50**
Psammodrilus 48, **50**
Psammohydra 48
Pseudocuma 67
Pseudovermis **51**
Puccinellia 170
Pycnogonida 35
Pygospio 37
Pyramids, ecological 180

Quicksands 12

Rat 59
Raven 59
Redia 139
Redshank 134, 139
Reducing layer 11, 177, 191
REES, W. J. 58
REESE, E. S. 138
Regressive evolution 49
Regulatory factors 95
REID, A. 26, 28
REMANE, A. 48
Remanella 49
RENAUD-DEBYSER 76
Reproduction
  co-ordination of 90
  Corophium 26
  duration of 90
  gadoids 57
  Hydrobia 24
  Littorina 24
  Macoma 90
  meiofauna 53
  Mya 90
  Nereis 90
  seasonal effects 84
  shore fauna 32
  shore fish 46
Reproductive stenothermy 87
Respiration 127
Rhodope 49
RICE, A. L. 78
Ringed plover 60
Rip currents 7
Ripheads **7**
Rock-borers 142
Rockling 57
Rock pippit 61
RODER, H. L. 12
Rook 59
ROTHSCHILD, M. A. 140
ROUND, F. E. 178
Ruppia 167
RUSSELL, F. S. 3

Sabellariidae 43
Sabellidae 43
Sacculina 14
Salicornia 170
Salinity 22 ff, 103 ff
Salsola 167
Saltmarsh 168 ff
Saltpan 172
Salt pasture 169
Sand 10, 11, 35 ff, 65 ff
Sand eel 45
Sand goby 45
Sand hopper 65
Sand lickers 53
Sarsilenium 140
Schistomysis 78
Schoenoplectus 171

Scirpus 171
Scololepis 69, 94
Scoloplos 68, 99, 114
Scrobicularia
  as prey 134
  mortality 99
  nutrition 31
  parasites 139
  zonation 70, 73, 159 ff
Sea scorpion 45
Sea spider 77
Seals 58
Seasonal changes 83 ff
Selective feeding 30
Self-regulatory mechanisms 138
Senecio 165
Settlement of larvae 121 ff
Shelduck 60, 77, 132
Shipworm 142
Shore
  definition 1
  formation of 3 ff, **5**
  types of 3 ff
Shrew 59
Silene 165
Silt 12
Sipunculoidea 35
Siriella 78
Size in relation to zonation 73
SLEIGH, M. A. 23
SLOBODKIN, L. B. 185
SMITH, R. I. 107
Sole 55
Sonchus 165
Sorting of particles 5
SOUTHWARD, A. J. 100
Spartina 169 ff
Spat fall 85
Species 34, 71
Sphaeronella 140
Spider crab 57
Spio 70
SPOONER, G. M. 161
Sporocyst 139
Standing crop 184
Statocyst 52
STEELE, J. H. 190
Stenhelia 73
Stenohaline 103
Stenothermy 87
STEPHEN, A. C. 3, 68, 73, 74, 85, 131
Sting fish 45
Strand line 143 ff, 166–167
Strigamia 145
Sub-littoral zone 65
Sub-terrestrial zone 65
Succession 168 ff, **172**
Sueda 165, 170
Sulphide layer 11, 177, 191
Sulphur bacteria 191
Supercooling 100
Survival range 100
Suspension feeding 52
Swans 59
SWEDMARK, B. 48, 52, 53, 75

Talitrus 65, 124, 143
TANSLEY, A. G. 179
Tardigrades 53
Taxes 76
Tellina
  locomotion 41 ff
  size 73–74
  spatfall 85
  zonation 68–71
Temperature
  adaptations to 100 ff
  and reproduction 87 ff
  as limiting factor 99 ff
Temporary fauna 17, 55 ff
Terebellidae 43
Teredo 142
Terminal anchor 38, **39**
Terrestrial shore fauna 58 ff
Thermodynamics, second law of 180
Thiobacillus 11
Thiospira 11
Thixotropy 12
Thoracochaeta 143
THORSON, G. 89
Tidal bore 148
Tidal rhythms 175
TODD, M. E. 22
Tolerance zone 125
Trachinus 45
Trematoda 138
Trilobodrilus 118, 120
Trophic level 179
Tropidoneis 178
TRUEMAN, E. R. 38, 40, 41, 42
Tubifex 97, 156
Turbellaria 35
Turbot 55, 56

Ulva 14, 173, 192
Undertow 4

INDEX

Upogebia 36, 104
Urodasys **51**
Urothoë 66, 67

Vegetation 164 ff
Veliger larva 24
Venus 90, 103, 120
Vermiformity 49
Vertical distribution 75 ff
Vertical migration of diatoms 175
Viviparity 25, 54
Vole 59

WADE, B. 40
Waders 60, 134
Water conservation 66
WATKIN, E. E. 66
Wave-built terrace **5**, 6
Waves
  constructive **9**
  destructive (plunging) **9**
  wave length 3
  wave of translation 4, **7**
  wave structure **4**
Weather
  aquatic 110 ff
  atmospheric 98 ff
Weever, lesser 45
WELLS, G. P. 43, 112
WHARTON, G. W. 53, 197
Whiting 56, 57
WIESER, W. 156, 195
Wigeon 132
WILSON, D. P. 121, 123
Wood-borers 142
Wrack flies 60, 143
WYNNE-EDWARDS, V. C. 138

Xylophaga 142

YONGE, C. M. 3, 62

Zebra mussel 62
Zirfaea 142
ZOBELL, L. E. 197
Zonation
  effect of salinity 105
  effect of temperature 102
  in estuaries **157**
  maintenance of 82
  of diatoms 176
  of meiofauna 73, 120
  on muddy shores 73
  on sandy shores 65
Zostera 36, 61, 167, 190

Daily Guideposts, 1986

GUIDEPOSTS

Carmel, New York 10512

All Scripture references herein are from the King James Version of the Bible unless otherwise noted.

ACKNOWLEDGEMENTS

Scripture quotation within text marked (NIV) for June 29th is from the Holy Bible, New International Version. Copyright © 1973, 1978, 1984 International Bible Society. Used by permission of Zondervan Bible Publishers.

The quotation for September 30th is reprinted by permission of Dodd, Mead & Company, Inc. from MY UTMOST FOR HIS HIGHEST by Oswald Chambers.

The quotation for October 13 is a selection from THOUGHTS IN SOLITUDE by Thomas Merton. Copyright © 1956, 1958 by the Abbey of Our Lady of Gethsemane. Reprinted by permission of Farrar, Straus & Giroux, Inc.

Photograph of Phyllis Hobe by Gladys Mitchell.
Photograph of Elizabeth Sherrill by Perry Alan Warner.

Designed by Elizabeth Woll
Mid-month Illustrations by Gail Rodney
Indexed by Mary F. Tomaselli

Printed in the United States of America

Table of Contents

PREFACE 5
INTRODUCTION 7
JANUARY 9
- *His Healing Word* 9
- *Practicing His Presence* 10
- *Wilderness Journey* 21
- *Praise Diary* 33

FEBRUARY 35
- *His Healing Word* 35
- *Practicing His Presence* 36
- *Seven Pillars of Marriage* 42
- *Wilderness Journey* 49
- *Praise Diary* 59

MARCH 61
- *His Healing Word* 61
- *Practicing His Presence* 62
- *Wilderness Journey* 72
- *A 2,000-Year Odyssey in Faith* 79
- *Praise Diary* 90

APRIL 93
- *His Healing Word* 93
- *Practicing His Presence* 94
- *Wilderness Journey* 105
- *Praise Diary* 118

MAY 121
- *His Healing Word* 121
- *Practicing His Presence* 122
- *Wilderness Journey* 133
- *Praise Diary* 146

JUNE 149
- *His Healing Word* 149
- *Practicing His Presence* 150
- *Wilderness Journey* 161
- *Praise Diary* 174

JULY — *177*
- *His Healing Word* — 177
- *Practicing His Presence* — 178
- *Wilderness Journey* — 189
- *Praise Diary* — 202

AUGUST — *205*
- *His Healing Word* — 205
- *Practicing His Presence* — 206
- *Wilderness Journey* — 216
- *My Savannah: Behold Its Goodness* — 219
- *Praise Diary* — 232

SEPTEMBER — *235*
- *His Healing Word* — 235
- *Practicing His Presence* — 236
- *Wilderness Journey* — 247
- *Praise Diary* — 260

OCTOBER — *263*
- *His Healing Word* — 263
- *Practicing His Presence* — 264
- *Waiting Through an Illness* — 268
- *Wilderness Journey* — 275
- *Praise Diary* — 288

NOVEMBER — *291*
- *His Healing Word* — 291
- *Practicing His Presence* — 292
- *Wilderness Journey* — 302
- *What is Christmas All About?* — 313
- *Praise Diary* — 315

DECEMBER — *317*
- *His Healing Word* — 317
- *Practicing His Presence* — 318
- *Wilderness Journey* — 329
- *Praise Diary* — 342

THE MEETING PLACE — *345*

THE READER'S GUIDE (Indexes) — *359*

Preface

Celebrating Ten Years of Friendship

FOR the past ten years, my wife Ruth and I have not only enjoyed writing for *Daily Guideposts*, but we've been reading it faithfully, too. We use it after breakfast, taking turns reading the inspiring message of that day.

Hundreds of thousands of these books have been used across the decade touching the lives of several million people. And we are thrilled by the many who tell us how *Daily Guideposts* helped them to meet life's problems. As one man said, "It is strange how it seems to be just what I need so many times." And so it is that a special group of gifted writers have the skill to turn a little story out of everyday life into spiritual truth and guidance.

By our reading of *Daily Guideposts,* all of the writers have become cherished personal friends. So, too, have all of you who have joined us through the years. And especially this year, more than ever, we are thankful for the warm, loving friendship of spending a new year with you. What a powerful fellowship when together we form a friendship with Jesus through these pages three hundred and sixty-five days of the year.

My wish for you then, friend, is for peace, joy, courage, faith and that you fill your life to overflowing with love of God and people.

—NORMAN VINCENT PEALE

Introduction

FELLOWSHIP. It was part of God's plan from the very beginning—that each of us should gain strength and support not only from Him, but from one another.

As a sign of this, God sent a rainbow to His friend Noah. Moses and the Israelites saw the hand of God part the Red Sea and lead them into the Promised Land. A devoted Ruth, who so loved Naomi, pledged to be with her "whithersoever thou goest." And a brave Jonathan remained a trusted friend to David, even unto death.

Even Jesus needed the fellowship of twelve disciples. And after His death and resurrection, He sent the Holy Spirit, with the promise:

I will not leave you comfortless.

We still can receive that comfort and friendship today—from God, from others, from the book you now hold in your hands. In it, twenty-eight writers have reached across its pages in friendship. Eagerly they share their faith, their hopes and dreams, their victories and successes. Their hurts and disappointments, too. They reveal themselves openly. And with friends like that, we can most easily be ourselves.

These friends want to meet you. In the days to come, travel with them into the peaks and valleys of the new year and heed their call to unbroken fellowship:

Come, we will not leave you comfortless.

CHARTING THE COURSE AHEAD

An adventure awaits you as *Daily Guideposts, 1986* celebrates ten years of growing friendship.

His Healing Word, monthly prayer-poems by Fred Bauer, open each chapter. Each poem begins as a meditation, and ends with reassuring reminders of God's love. Your personal *Praise Diary* at month's end will help you track your spiritual journey in 1986. Use the space to write a word, a phrase, a short prayer of praise or thanksgiving. From time to time return to the diaries and see the miracle of God's friendship at work.

On the first of each month, in *Practicing the Presence of God,* Sue Monk Kidd helps you draw closer to God every day in new ways. At mid-month, Marilyn Morgan Helleberg, in *Wilderness Journey,* relives the Exodus story to find ways of trusting God in all circumstances. *A 2,000-Year Odyssey of Faith* with Jeff Japinga takes you through Holy Week to meet ordinary people who were changed by Jesus. Then during Advent, learn *What Christmas Is All About* as Patricia Houck Sprinkle discovers the God Who never stops giving.

Early in the year, Dr. and Mrs. Norman Vincent Peale share secrets for a happy marriage; while Arthur Gordon invites you south for a summer week; and in the fall, Sue Monk Kidd helps you cope with illness.

Then, if you tire and need a waystop, turn to *The Meeting Place,* the authors' section, for fellowship and long visits. And most important is a new feature just for you. *The Reader's Guide,* a comprehensive index to all the selections in *Daily Guideposts, 1986,* to help you meet that special need.

Didn't we say a spiritual adventure awaited you? With anticipation and fellowship, let's begin. A year of unknown possibilities lies ahead. No more delays. Only a blessing: *Godspeed!*

—*THE EDITORS*

January

S	M	T	W	T	F	S
			1	2	3	4
5	6	7	8	9	10	11
12	13	14	15	16	17	18
19	20	21	22	23	24	25
26	27	28	29	30	31	

The Tree of Friendship

Friendship is the oak of life
it grows through rain and drought,
It stands against the storms of night
deflecting fear and doubt,
It comforts us when illness comes
—reassures when conflicts rage,
And spreads its shading summer arms
in deference to age,
But like all earthly things God made
His creations aren't forever,
Our friends are pruned much too soon
as if we friended never,
Oh, lonely is the one who's lost
a love from friendship's tree,
But lonelier by far is he
who missed the opportunity.

I came so you would have life
and have it abundantly,
I wrapped each day in newness:
live it unredundantly.
Take time to plant a tree of hope
to tend a life that's bare,
And I will bless you richly
with friends beyond compare. —*FRED BAUER*

1 *Practicing His Presence in January*

ACCEPTING THE CHALLENGE

Thou wilt keep him in perfect peace, whose mind is stayed on thee. —*ISAIAH 26:3*

On January 1, 1937, a man named Frank Laubach opened his diary and wrote an amazing resolution for that year. "God, I want to give You every minute of this year. I shall try to keep You in mind every moment of my waking hours...." I read those lines from his journal utterly intrigued. *Keep God in mind every moment?* What an extraordinary challenge for the year! Laubach even wrote percentages at the top of his journal pages—*40 percent...70 percent*—to note how much of the day he was aware of God.

As I thought about this experiment, I remembered a seventeenth-century kitchen monk named Brother Lawrence, who wrote of "practicing the Presence." He, too, worked at being aware of God throughout the day as he went about his ordinary tasks. This desire for constant communion belongs to all of us, but for many, like Laubach and Brother Lawrence, it became a practical reality.

And what about this year of our Lord nineteen hundred and eighty-six? What if we embark on an adventure like that? How would we go about it, especially when our lives are so busy and fast-paced? Could we, together, spend the year practicing God's Presence...filling as much of our waking moments as possible with a new and wider awareness of Him?

I know it can't happen overnight. Such a bold resolve unfolds slowly as we work at it. It takes a conscious plan. So on the first of each month this year, let's work at it. The idea is not to stop our regular activities in order to be aware of God, but rather to keep our thoughts toward God *in the midst of them.*

Perhaps our efforts will be imperfect. But that's okay. We will grow from the smallest effort. And at the end of the year, who knows? We may find our everyday world brimming with newness and aliveness. For God's Presence is here, just waiting to be "practiced."

Lord, I want to make this a year of practicing your Presence every day. Help me spend this month committing myself to this challenge.

—SUE MONK KIDD

2 THURSDAY

...They lifted up their voice with the trumpets and cymbals and instruments of music... —*II CHRONICLES 5:13*

In my hometown of Pasadena, California, January 1 means the Rose Parade. Ever since I can remember, I've watched the parade—from a parent's lap, a car, or the grandstand, along with thousands of others who brave the early morning to line the streets every year. We cheer at the bands, wave at the horseback riders and clap for the exquisitely flowered floats. I've missed viewing the parade only once—and that was the time we were in it.

January 1, 1983 was "Dad's parade." That year he was president of the volunteer group that organizes the Rose Parade. And so we, his family, led the parade, filling up two cars that rolled down Colorado Boulevard. We waved, yelled, cheered, and clapped. It was a brief moment of glory that we'll always remember.

Since then, we've returned to our noncelebrity status. As in years past, we sit on a car parked along the parade route applauding uproariously. And I find I actually have more fun watching the parade than I had being in it. Of course, it's nice to be the center of attention once in a while, but I find that the times that I can really feel His Presence closest to me is when I applaud the best in others.

Whatever Your call may be, Lord, I shall follow... enthusiastically!

—RICK HAMLIN

3 FRIDAY

Behold the day, behold, it is come... —*EZEKIEL 7:10*

The other day I heard about an inventor who has taken out a patent on a new kind of alarm clock that wakes you up by releasing a cloud of perfume. If it happens that on a particular day the sleeper's nose is clogged by a cold, there's a backup light that will flash. Apparently clock manufacturers know that exciting any of the five senses will do the trick on a slumbering human.

As for me, well, maybe the aroma of coffee and frying bacon wouldn't be so bad as an eye-opener, but I'd hate to be "flashed" awake, certainly not shaken awake, and I can't imagine how I'd ever be "tasted" awake. Actually, if I can't wake up on my own, I think I'd choose the "sound" approach. I don't want the loud shrill of an old Big Ben (not by chance is it called an *alarm* clock), but some soft music, perhaps. Or better still, I'd like a clock that talks to me sensibly and gives me a good reason for getting up. Yes, that's it, a clock that speaks to my superior intelligence, sort of like this:

"Good morning, Van. Did you know that if you lived the Biblical three score and ten years, you'd have 25,550 days in which to accomplish something on this earth? Come on, forget about all those days you've wasted. Here's a brand-new one. It's a day the Lord hath made and He's made it for *you*...."

Now Lord, sound out another day and let me make it a day for You.

—VAN VARNER

SATURDAY

Hast thou entered into the treasures of the snow?

—JOB 38:22

The freshly fallen January snow covering the lawn today glimmers in the sunlight, like thick meringue sprinkled with thousands of tiny crystals. It reminds me of the game of "Fox and Geese" we played during the winters when I was a child.

What we did first was to dig the heels of our boots into the spotless snow to make a huge circle. Then, from the center we'd tromp out radial lines like a huge wheel, with the center hub the "Safety Zone." One child would be the "Fox" who tried to tag the rest of us, the "Geese," who tried to run safely to the center hub unscathed. The whole game was carried out *only* on those tromped lines, and you were out if you stepped into the untouched snow. But if you were caught, then *you* had to be the fox. It was great fun and sometimes so many children would join us that the circles could be fifty feet across.

Today, as I remember this childhood game on this fresh, new January day, it brings back to mind an old poem that's a befitting reminder for this time of year:

The New Year lies before you
  Like a spotless track of snow.
Be careful how you tread it,
  For every mark will show.

Lord, this year I want to follow Your tracks instead of man's.

—ISABEL CHAMP

5 SUNDAY

Fear not...I am thy shield... —*GENESIS 15:1*

Ralph Waldo Emerson once said that "the wise man in the storm prays to God not for safety from danger, but for deliverance from fear. It is the storm within which endangers him, not the storm without."

I have a neighbor in Pawling, New York, who would agree wholeheartedly with that statement. My friend's name is Smith Johnson and he's the retired head of The Pawling Rubber Company. Smith told me once that when he was a youngster he was deeply afraid of thunder and lightning. He looked upon this fear as a personal weakness and he was determined to do something about it.

One night a violent storm swept in over the fields where he lived. Smith got up out of bed, put on his raincoat and heavy rubber boots, and went outside. While the thunder rolled and the lightning flashed he walked across a meadow and through a dark wood. When he had trudged half a mile he sat on a fence and turned his face to the wind and rain until the storm quieted. Then he went home, climbed back into bed and slept more peacefully than he had ever slept before.

Thunder and lightning never frightened Smith again—and he is now well past his ninetieth birthday. Smith had laid his fear to rest by following still another piece of Emersonian advice—do the thing you fear, and the death of fear is certain.

Let me be wise enough and strong enough, Lord, to look fear straight in the face.

—NORMAN VINCENT PEALE

MONDAY

...And peace, and love, be multiplied. —*JUDE 2*

Right after New Year's my wife Judy begins to eye our Christmas tree disapprovingly and starts preparing young Dave and me for the inevitable. She'll drop hints like, "Mmm, the tree's losing a lot of needles, and the branches are beginning to droop." Dave and I know that by Twelfth Night she'll get straight to the point: "Okay, you guys, it's me or that ratty old tree. Out with it!"

And every year it's been tears for Dave and gnashing of teeth for me. Both of us hate to see it go. The year before last, as David wailed, I dismantled our tree in record time, nearly strangling myself on our string of lights. We got it down all right, but as always it's hard for me to let go of Christmas.

Last year Judy was strangely silent on the subject of our tree and Dave and I were very suspicious. But on the afternoon of Twelfth Night, she surprised us by calling us into the parlor for cookies, egg nog and the sounds of Christmas carols.

"This year we're going to do it right!" she announced. "We'll each take a turn taking down an ornament, just as we do putting them up." As we did, we played a game of trying to remember who had given us different ones. We were surprised at how many memories they evoked. Some brought us laughter and joy, others brought a quick prayer to our lips: better health for one, comfort for another now widowed, a new job for one out of work....

Then, when everything was put away, we joined in a final prayer of thankfulness for *all* we'd received during the holidays, for family and friends...and especially to carry on the Christmas spirit into the New Year.

Blessed Jesus, with Christmas behind us, help me to extend into the year ahead Your peace, joy and love to others. —JAMES McDERMOTT

TUESDAY

Peace, be still... and there was a great calm. —*MARK 4:39*

I woke up this morning with laryngitis. Absolutely no sound came out of my mouth when I shaped the words.

"What's a mother to do without a voice?" I moaned to myself as I pulled on my slippers and padded down the hall to start breakfast for my husband and our three children.

Within minutes, familiar sounds and actions filled the kitchen. "I slept on my bangs wrong," wailed ten-year-old Lindsay. "Where is the homework that I left right here?" demanded twelve-year-old Derek, frantically pounding the counter. "My knees won't work and I can't move," complained seven-year-old Kendall, sitting cross-legged and half-dressed in the middle of the floor. The phone rang, the dog barked, somewhere a radio blared.

Usually, I direct this typical scene by shouting orders above it. But not today. I had no voice. I responded only in weak whispers. And slowly, a funny thing began to happen. Instead of escalating into the usual bedlam, the whole place began to calm down. And believe it or not, everybody else started whispering too.

So, right there in the middle of buttering an English muffin, the truth hit me. Good or bad, the tone of my voice can control the mood of the morning. Somehow I thought of how Jesus touched and changed the lives of those around Him. Not with a loud voice or demanding orders, but with gentle strength. I hope I remember that even when my laryngitis goes away.

Dear Lord, let me reflect Your gentle loving Spirit in my voice and actions today. —CAROL KUYKENDALL

WEDNESDAY

... They trusted, and thou didst deliver them. —*PSALM 22:4*

"Come look at this," my husband called from the dining room. "We have a visitor!"

I ran to join him and we gazed at it in wonder, this tiny spider dangling from the ceiling, with nothing visible to support it. It was simply *there*, alive but absolutely still, so motionless we thought at first it might be dead.

George reached out an experimental finger just above it, and sure enough the wee body swayed back and forth on its invisible thread. And yet it did not stir. "He *must* be dead," I decided. "Or he would run away."

The telephone rang; we went about our other activities and forgot it. But later, we discovered our little visitor had disappeared.

"Did you sweep it down?" my husband asked.

"No, I thought perhaps you did."

We both realized then: The spider had indeed been living! Simply waiting, trusting that we would not harm him, and then when we turned away, scampered back up to the ceiling, drawing his lifeline with him. What faith he had in something we couldn't see. Invisible to us, yet real and vital to him.

If only our own faith could be like that! Never mind that others can't see it, or that the storms of life will sway us. If our faith is tough and strong, like that little spider's, we too only need to be patient and hang on.

Lord, You are my lifeline. Give me the faith of that tiny spider—to wait, to hang on, to trust You. —MARJORIE HOLMES

THURSDAY

...I will speak of excellent things; and the opening of my lips shall be right things... —*PROVERBS 8:6*

At one point, the stack of correspondence, forms and other papers on my desk was nearly two inches high. The number of hours I'd spent writing, figuring, checking and rechecking the information added up to more time than it would have taken me to fly around the world twice. And that didn't include the number of hours I'd spent worrying.

Getting a mortgage on our new house was the cause of all this agitation. More than once, I found myself wondering out loud whether the house was worth all these headaches, whether, in fact, we could even afford it.

Then a friend heard my mutterings. "I think it's great," he said. "How would you ever know the benefits of owning your own home if you didn't take the risk?"

Good words, at the right time. They became the shovel with which I battled that blizzard of paper. And, now that we are in our new place, I know how right my friend was.

Good words. You and I believe in a God and share a faith and Bible that's filled with good news, good advice, good living. Maybe there's someone you know who's in need of a good word. And maybe you're the one to give it to him (or her).

Father, I thank You for Your good words, and, even more, for the joy of being able to share them with others. —JEFF JAPINGA

10 FRIDAY

Make a joyful noise unto the Lord... —*PSALM 100:1*

Some people have such nice smiles that I don't think it's fair that I have so much trouble turning up the corners of my mouth. Maybe some of us are predestined to be "Square Mouths."

My youngest son always has a ready smile, but my eldest almost never smiles. My wife Ginny says he is just like me. Of course, if somebody tells a joke, I'm as likely to laugh as the next. But then I'm back to "Prune Face," as Ginny sometimes calls me.

Sometimes it drives me to the bathroom mirror. I close my eyes and assume what I believe to be my normal expression. Then I open my eyes. Ugh! "Prune Face" is a good description. I try for a little improvement and smile at the mirror. Horrors!

Finally, in desperation one day, I turned my smile—or lack of it—over to the Lord. Now, I do it every morning. I'm not all the way home, but I managed to get this reaction from my grandson: "Hey, what's got you so happy?"

I had an answer: "Jesus loves me—and you, too!"

Lord, let Your radiance be reflected through my smile. —SAM JUSTICE

SATURDAY

I heard the voice of the Lord, saying, Whom shall I send, and who will go for us? Then said I, Here am I: send me.
—*ISAIAH 6:8*

Last June, just after those disastrous tornadoes had swirled through Ohio, Pennsylvania and Ontario, a woman on television said something that affected me deeply. She was one of the survivors in a little hard-hit Pennsylvania town called Atlantic. The camera showed her surrounded by men in large, black, flat-rimmed hats clearing debris, hauling lumber, rebuilding wrecked homes. These men were members of the Amish sect. They had come *en masse*, by buggy and by bus, to help the people of Atlantic—strangers to them—in distress. Their presence was not surprising, for with the Amish, helping others is a tradition, a way of life.

"When the Amish came," the woman on television said, "it was as though God had reached down with His own helping Hand."

His own helping Hand, I mused to myself, and for a moment as I watched one of the Amish taking a strong grip on a hammer, it seemed to me that I was seeing God's Hand at work. Then I looked at my own hand, and I saw it in a way I had never seen it before. Have you ever thought, just as I did then, of your hand as God's helping Hand?

Lord, give me the strength to work, and the wisdom to work for You.

—RUTH STAFFORD PEALE

12 SUNDAY

When he was yet a great way off, his father saw him, and... ran and kissed him. —*LUKE 15:20*

My husband and I have a system for exploring new hiking trails. One of us will start at the beginning while the other takes the car to the opposite end and hikes back toward our rendezvous point somewhere deep in the hills or the forest.

We've seen some of America's loveliest wilderness this way, and added something special to our relationship. All along the path I'll find myself saving up small adventures to share when we meet. Questions, too, about unfamiliar sights and sounds.

As our steps draw nearer, my anticipation mounts. Around any curve now he may appear...perhaps beyond that hemlock grove, or over that rise. When the trail's been long and the landscape strange, my expectation turns to longing. I break into a run....

I've never walked this day's path, Father. Keep me looking for You around each bend!

—ELIZABETH SHERRILL

13 MONDAY

A friend loveth at all times... —*PROVERBS 17:17*

Dear friends were moving halfway across the country and we had a farewell party for them. As we recalled the many joys that bound our friendship, they expressed apprehension over finding new friends. "Oh, don't worry, you'll soon make new friends," we reassured them.

"But how?" someone asked. "I've never moved, but I may have to some day."

Well, in the conversation that ensued, we all gained some good suggestions for not only how to make a friend, but how to be one. Let me share some of them with you:

How to *make* a friend:

1) Invite someone to have coffee with you...or invite her child to play with your child. Just *invite* her (or him).
2) Take a small gift—maybe a jar of homemade jam, or even a flower.
3) Write thank-you notes for small pleasures received.

How to *be* a friend:

1) Listen, really listen, to your friends. People need to know their feelings matter.
2) When a friend is down, write, call, or extend an invitation for tea or just a quiet stroll together.
3) Be tolerant of a friend's oddities—we all have them.

So what is friendship really all about? Someone at the party quoted an Arabian proverb: "A friend is one to whom one may pour out all the contents of one's heart, chaff and grains together, knowing that the gentlest of hands will take and sift it, keeping what is worth keeping and with the breath of kindness blow the rest away."

Today, Father, I'm going to make *a new friend...and* be *a friend to an old one.* —ZONA B. DAVIS

14

TUESDAY

And the Lord said unto Satan, Behold, he is in thine hand; but save his life. —JOB 2:6

Rereading the story of Job recently, I discovered something I had missed before. Job is not just a story about the man Job; it's a story in which Job is the believer on whom God stakes His own character. Satan said to God, "If Job were to suffer, he'd turn away from you." God said, "Try him and see" (Job 2:3-6).

Whenever I suffer, I think about how my body aches or how my mind is troubled. I have a hard time believing that all this may have some part in a plan of God's far-larger-than-my-own-small world. I may even have a hard time finding God through my pain. I certainly don't want somebody spouting platitudes about how this is "all God's will." What I want is relief!

Job did, too. He cried out for relief. He wondered why he suffered. But, even in his agony, he did not deny God. Through it all, he believed that God was God and must have had *some* reason for the suffering. Because Job accepted these two things, God won a victory over Satan.

When I hurt, I usually can't do much for God. But I *can* acknowledge that He *is* God and has a reason for everything He does. Perhaps that is all He expects or needs of me then.

Help me hold fast to faith in You, God, in days of pain.

—PATRICIA HOUCK SPRINKLE

15

WEDNESDAY

The Lord God is my strength, and he will make my feet like hinds' feet, and he will make me to walk upon mine high places. —HABAKKUK 3:19

Having just read the biography of civil rights leader Martin Luther King, Jr., I am in awe. What a brave man, what a man of vision! But he was human, too, with moments of weakness. There were times when inwardly he doubted his ability to carry out his dream, when it seemed simpler—and safer—just to quit. At other times he encountered worldly temptations, which took all the strength he could muster to resist. Nevertheless, through constant prayer, he rose above his frailties to march for justice under God's banner, and in so doing he changed the course of American history.

That's an inspiration to me, to know that a strong brave man can sometimes falter, yet still be used so powerfully by the Lord for good. Instead of worrying, as I have sometimes in the past, about whether or not I'm worthy to do God's work, I think I'll acknowledge my faults, turn them over to Him and then just do the best I can. I may not accomplish history-changing deeds, as did Dr. King, but I *can* extend God's love to those around me, even if I reach only my own neighbors.

Strengthen me this day, O Lord, that I may better serve You.

—MADGE HARRAH

16 *WILDERNESS JOURNEY* (*Exodus 11–13*)

Following the Pillar-Shaped Cloud

I will bring you up out of the affliction of Egypt unto the land of the Canaanites... —*EXODUS 3:17*

I've just finished reading about that momentous morning long ago when the Israelites, after 430 years as slaves in Egypt, gathered together their flocks and herds and a few possessions and began the long and treacherous journey away from bondage. Tribe by tribe, family by family, some three million men, women and children followed Moses, their God-chosen leader, into the desert.

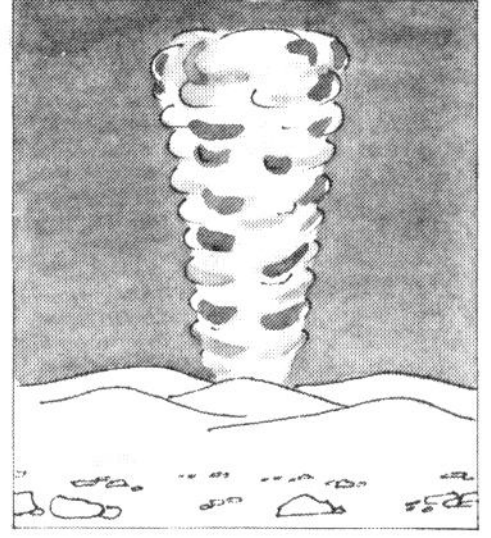

I've known that story since childhood, but it's never seemed to have much to do with my twentieth-century life. I wonder. If I could somehow project myself into the scene, would it come alive? Would it be possible for me to take my everyday concerns with me into the Exodus setting and let the events of that story act upon them? Would I discover, in the pillars of cloud and fire by which God led His people, that He offers *me* help *now*, with my most agonizing decisions? Could I let the Red Sea swallow up some stubborn fear of mine as it destroys the

Egyptian army? Would I find water springing from rocks, strange food on the ground in the morning and flocks of quail plopping down at my feet at sunset? And would these signs and wonders help me to release to God some nagging need of this day? If I could stand with Moses on Mt. Pisgah and see the Promised Land stretched out before me, could I envision the answer to a long-standing prayer, and so help it come to pass?

When Joan of Arc's accusers said that the voices that led her were only her imagination, she quickly replied, "Yes. That is the way God speaks to me." Perhaps you and I could allow God to speak to us through our imaginations, too. I invite you to travel back with me over three thousand years, to join that ragtag band of pilgrims, now somewhere between Egypt and the Red Sea. The sun is just beginning to spill over the eastern rim of the desert, even though we've been walking for many hours. There's not much talking—only the sound of trudging feet, a child's cry, and the bleating of sheep.

A mysterious cloud hovers over the sea of heads in front of us. Can you see it? It's pillar-shaped and tinged with a bright, golden light. It is the Lord's Presence sent to guide us. Shall we follow it, and each month experience a different event of the Exodus? Perhaps we'll find that there's a wilderness—*and a promised land*—within each of us. Maybe the Lord will use the Exodus story to lead us to new freedom in 1986!

Show me thy ways, O Lord; teach me thy paths. (Psalm 25:4)

—MARILYN MORGAN HELLEBERG

FRIDAY

And he is before all things, and by him all things consist. And he is the head of the body, the church...

—COLOSSIANS 1:17, 18

This morning I got a letter from my nephew Dennis. He is in the oil business in Texas. If you think that means he is wallowing in wealth, think again. He wrote: "We are running out of capital, the price of crude oil has slumped because there's so much of it, and I've had to dismiss most of my staff."

But then he added, "I've decided that there is nothing more I can do, so I am turning the company over to the Lord Jesus. He is now the chairman of my board of directors, and I am the only other member. I hope He knows how to do the job."

I'm sure He will. Relinquishment to the Lord is not a surrender. It is an *acknowledgment*—one that always results in triumphs.

Lord, You gave me Your life. Today, I give You mine. —GLENN KITTLER

18 SATURDAY

For whosoever shall do the will of God, the same is my brother, and my sister... —*MARK 3:35*

As a young man at the turn of the century, my father William Sass worked for the artist Louis Comfort Tiffany, who was famous for his unusual glass creations. Tiffany enjoyed showing the eager-to-learn William how to fire and shape the iridescent material into lampshades, bowls, vases and other decorative objects, including beautiful, glowing church windows. Often he'd urge the lad to take small items home to his mother. My grandmother cherished these pieces. Eventually, they were handed down to me.

When I wanted a special gift for my friend Marguerite, I thought about the Tiffany glass. Marguerite and I have known each other many years, and she knew my father well. So this gift seemed appropriate.

The day I presented her with a small scalloped-edge candy dish, I said, "My father made it."

Marguerite nodded. "Yes. That's why I will treasure it always."

Since then, Marguerite and I have grown closer still. Through a new relationship in Christ, we have discovered the joy of praying together. The other day I gave her another gift—a Bible. As I handed it to her I smiled: "My Father made it."

For an instant, a puzzled look crossed her face. Then she brightened in understanding. "Yes, and thanks to Him, we have truly become sisters."

Help me to love others in Your name, dear Father, so that we may be one family in You. —ELEANOR SASS

19 SUNDAY

There is nothing better for a man than that he should... make his soul enjoy good in his labor.

—ECCLESIASTES 2:24

Last year, as part of a program to help sixth graders feel easy with their "Elder Elders," I was invited to be their Senior Guest at the local public school.

I answered a lot of odd questions: What kind of candy bars did we have in my youth? (Only three, plus licorice whips and jawbreakers.) How much did we pay for chewing gum? (5¢ a pack.) For movies? (25¢.) Did we really believe the moon was made of green cheese? More seriously, what was life like before TV? Before airplanes? Were the changes good or bad?

As recess approached I was about to award myself a laurel or two when a precocious young man demanded one last question. "So, you've told us all about your past. Give us a tip for our future." It had to be a secular answer and I was stumped. I wafted a prayer and said hesitantly, "Well, I can tell you a story I heard in the sixth grade that helped me.

"It happened during the reign of the Merry Monarch, Charles II of England, when Sir Christopher Wren was London's favorite architect. A country gentleman visiting the city paused to watch a group of men working on a large cleared area. He asked one of the men what he was doing. 'I'm laying bricks, mate,' the man said sourly, 'for a couple of farthings. I oughter be gettin' more.' The country gentleman strolled on and asked another man. This one raised a cheery face and said, 'I'm helping Sir Christopher Wren build a cathedral.'"

"Gotcha!" sang out my precocious young man in the audience. "The price of success doesn't change. Work hard and think big."

Father, grant that our young men and women may see visions and dream dreams, and help us give them a peaceful world in which to fulfill them.

—ELAINE ST. JOHNS

20 MONDAY

Except a man be born again, he cannot see the kingdom of God.

—JOHN 3:3

Some people say they have trouble seeing God's hand in the world;

others say that they see evidence of Him everywhere they look. Lately, I seem to be falling into the latter camp.

You see, I'm fascinated with antique cars and never miss an opportunity to talk with someone who owns one. Not long ago, a man who has restored several old automobiles was telling me about the great change that took place in travel when manufacturers turned from hard rubber tires to pneumatic or air-filled tires. He continued to talk about old cars, but my mind made a quick U-turn to a place in the Bible where the Greek word *pneuma* appears. It is in John 3 where Jesus is trying to explain to Nicodemus the mystery of being born again, comparing God's Spirit to the wind—you can hear the sound, but don't know from whence it came or where it goes.

Some people are God-blind and can't find Him anywhere; others see Him in the intricate design of a wildflower's petals, in the configuration of the stars, even in such mundane things as car tires, inflated by that strange phenomenon Jesus was talking about: the wind...which, like God's Spirit, is everywhere.

Help us to not only see Your hand, Lord, but to take it and let it lead us. —FRED BAUER

21 TUESDAY

And he took butter, and milk, and the calf which he had dressed, and set it before them... —*GENESIS 18:8*

It had been seven years since I had seen my relatives in the Northwest, where Dad had grown up. I had changed a lot and I wondered how we would get along now—whether, in fact, they would even remember me.

Knowing I would be in their area on business, I called to ask my uncle if I could drop by for a visit. I guess I knew everything would be okay when I drove up to Uncle Chester's house. He's a big man with blue eyes and sandy gray hair, and he was out on the sidewalk waiting for me. The first thing he said when I got out of the car was, "Let's have a hug!"

The rest of my relatives were just as warm and hospitable. We caught up on years of news over the kitchen table; they toured me around and even asked me to move out west. I left feeling like a real part of this family.

That's the great thing about family (whether real or extended).

There's a closeness that defies distance and time. And in this New Year I've promised to tighten that bond with family, to write, to visit, to call—to let them know in every way I can that I love them.

Dear Heavenly Father, help us to nurture and care for our loved ones, to keep the family bond that You've given us strong.

—SAMANTHA McGARRITY

22 WEDNESDAY

Go to the ant, thou sluggard; consider her ways, and be wise. —*PROVERBS 6:6*

"Guess what," our friend Barbara cried, "Charlie has joined our prayer group!"

I was surprised, myself. Charlie wasn't a joiner. So the next time I saw Charlie I asked him about it.

"Well," said Charlie in that deliberate way of his, "I had problems and I was praying about them, but I didn't seem to be getting anywhere. The problems got bigger than my prayers, if you know what I mean. Then one day I picked up a magazine that had an article about bees. The article said that when it gets too hot in a hive, a group of worker bees all face in one direction, anchor themselves to the floor and begin to fan their wings rapidly. One bee alone wouldn't make much difference but a lot of bees can produce an air current strong enough to draw fresh cool air into the hive and blow the stale air out.

"So I said to myself, if a group of bees working together can activate a healing current that changes everything for the better, maybe a group of people can do the same thing."

Lord, help us to remember that we can't always rely solely on our own efforts, and that sometimes we must seek "where two or three are gathered...."

—ARTHUR GORDON

23 THURSDAY

But ye are a chosen generation, a royal priesthood, a holy nation... —*I PETER 2:9*

When I was a child I was constantly told how I resembled my paternal grandmother, who died before my birth. I grew very tired of hearing it, and one day told my mother, "I don't want to be like Grandma Hall. I want to be *me*!"

Mother put her arm around me and said, "You *are* you. It doesn't matter what you look like. What's important is what you make of your life. God made you for a particular purpose. What He planned for your life, no one else can do."

I've never forgotten her words and they have made me search for His plan for my life. Sometimes it seems so obscure that I get discouraged and frightened. I dream of doing great things, most of which elude me, and I wonder if I'll ever accomplish anything of value.

That's when I have to remind myself that it was He Who designed my life and He Who knows what I must do with it. I take my dreams to Him, along with my anxieties and failures, and we talk about them. And always He sends me the peace and reassurance that only He can give.

Do you need a lift today? Try my springboard. Remember that He created you and made you somebody special with a niche all your own in His great universe.

Thank You, God, that I didn't just happen, *that You created me as a part of Your plan.* —DRUE DUKE

24 FRIDAY

Give ye ear, and hear my voice; hearken, and hear my speech. —*ISAIAH 28:23*

I was excited, flattered, and only twenty-three when my Aunt Doris invited me to become a junior partner in her new beauty shop. For many years she had been a hairdresser in motion pictures, working with such stars as Sonja Henie, Janet Blair and Hedy Lamarr. And now she planned to branch out on her own.

She named the shop "Doris of Hollywood," and her clientele included those same lovely actresses. I soon discovered that they followed her to gain more than physical beauty. You see, there was something special about my aunt. She was not only a skilled beautician, but a loving and patient listener. Along with facials, pedicures and permanents, she dispensed "big ear" therapy. While clients poured out their defeats and victories, she listened with interest, understanding and love.

My aunt has since died, but her theory lives on. She taught me a lesson worth remembering. We all have friends who desperately need someone to talk to. So let's give them the gift of the "big ear," and watch friendship grow—and be cherished.

Thank You, Father, for listening so patiently as I pray. Let me do the same for others. —DORIS HAASE

25 SATURDAY

They helped every one his neighbor; and every one said to his brother, Be of good courage. —*ISAIAH 41:6*

For many, many years my husband and I have enjoyed our daily walks together. Sometimes we talk quietly, other times we're just together in silence.

But these walks lost their glow when, after an illness, I *had* to walk as therapy. Where before we had gone along casually wherever our inclination took us, now I suffered with each step. It changed our gait, our buoyancy. Moreover, it was changing my attitude toward getting well. *What's the use?* I thought to myself. *It's doing no good....*

Then one day, a little note, centered between two lovely doves of peace, arrived in the mail. It said:

> Keep it up, you two lovebirds! You make my day as you pass my window every afternoon, looking so fit and happy. I am a shut-in and don't know what I would do without the inspiration you give. "Things will be better," your very presence seems to say....

Don't you see? It just shows that each and every one of us is an instrument of God, serving one another in our way, sometimes inspiring each other without knowing it. Today would be a good day to let someone know what they secretly give to you. Why not tell them?

Let me be of good courage, Father, for the sake of others.

—JUNE MASTERS BACHER

26 SUNDAY

And if he trespass against thee seven times in a day, and seven times in a day turn again to thee, saying, I repent; thou shalt forgive him. —*LUKE 17:4*

International Forgiveness Week starts today and runs through this coming Saturday. On each of the days one is supposed to forgive in a specific area. For instance, tomorrow one is to forgive one's family; on Tuesday, one's friends and associates; and so on through the week to Saturday, when one forgives other nations. But the first step, to be taken today, is: *Forgive yourself.*

How important that is. And how difficult. I think of times in the past that I've groveled in self-condemnation over an error or an argument, unable to reconcile with a loved one because I felt embarrassed and ashamed. At those times I just didn't believe I was worthy of forgiveness.

But God must consider forgiveness very important, for there are one hundred eight references to it in the Bible, as listed in my concordance. In 1 John 1:9 we read: *"If we confess our sins, he is faithful and just to forgive us our sins, and to cleanse us from all unrighteousness."*

I've found that's what I have to do—confess my error to the Lord and ask His forgiveness first, before I can forgive myself. Once I've taken that first step, I can move on to the next step of forgiving others and asking their forgiveness.

International Forgiveness Week. A wonderful idea. But why stop at a week? Wouldn't it be great to make 1986 a Forgiveness Year?

Forgive me, Father, and help me to forgive. —MADGE HARRAH

27 MONDAY

He giveth power to the faint; and to them that have no might he increaseth strength. —*ISAIAH 40:29*

One wintry afternoon I watched ice skaters at a local rink. The coach worked them with patience but strict supervision. One teen-age girl was trying to perfect an extremely difficult jump. Bam! Over and over she came down on the ice. Each time the coach instantly motioned with his hand: *Get up. Try again!*

Once she fell so hard she couldn't get up. The coach skated over and helped her up, at the same time nodding for her to try again. When I saw the expression on her face, I knew just how she felt—discouraged. I had been attempting something difficult in my life too. Each time I fell, God seemed to say to me: *Get up. Try again!*

The young skater fell again, but this time sprang up and skated around and around the rink, gaining the speed she needed for the jump. Up in the air she leaped, spinning, twirling. I couldn't help but pray for her...and for myself as well. This time she landed on her feet in perfect position, and stood there like a graceful statue for a few marvelous seconds. The coach broke into applause, shouting "Bravo!" I too was clapping for the courageous skater and for the inspiration she had given me. Tears of joy stung my eyes.

Yes, I would try again...and again....

Father, give me the courage not to give up after my failures.

—MARION BOND WEST

28 TUESDAY

Be ye all of one mind, having compassion one of another...

—I PETER 3:8

Like many people, I worry about the future of the family. Will it survive all the changes and pressures in our world? Yet it's amazing how hope and reassurance can come to us unexpectedly.

For instance, Claire and Ben are a young couple I know. She has a good job, which is important because Ben is in medical school and they need her salary. But just the other day they told me they were expecting a child. They were very excited and so was I, but I wondered how parenthood would affect Ben's plans to become a doctor.

"I know what you're thinking," Claire said to me. "How are we going to manage, right?" I nodded, and she grinned. "I won't have to give up my job. I'll take a leave of absence and then go back to work for two years until Ben gets his M.D."

"But who will take care of your baby?" I couldn't help asking.

"My mother," Claire said. "She's as excited as we are. She can't wait to be a baby-sitter."

I remembered that Claire's mother had planned to go back to school and get a degree.

"She's putting it off—for two years," Claire told me. "Then, when I come home, Mom will go to school."

I should have known. God never meant families to become obsolete. He meant them to live with change by finding new ways to love each other.

Sometimes, Lord, when our love becomes a bit rigid, remind us that working together makes our world go round. —PHYLLIS HOBE

29 **WEDNESDAY**
But it is good for me to draw near to God... —*PSALM 73:28*

Living with our two small boys is sometimes like living in a *Star Wars* outpost. Our couch is a spaceship, the hall is a launching pad, our poor old broom doubles as a laser sword. But I don't mind. It stimulates their imagination and adds to their vocabulary. Last week "space talk" even enriched my prayer life!

I was sitting still, trying to gather my thoughts for prayer, but my mind roved restlessly. Suddenly I thought of tractor beams—powerful magnetic forces that in space movies are sent out by a large ship to draw in a smaller one.

"That's what I need," I muttered. "God's tractor beams."

As I sat in stillness, I pictured them—long rays of love beaming from God into my room and my life, drawing me irresistibly to Him whenever I bring myself into His path.

Do you know what? I found them!

Draw my soul, O Christ, closer to You on Your "tractor beams" of love. Amen. —PATRICIA HOUCK SPRINKLE

30 **THURSDAY**
Trust in the Lord with all thine heart; and lean not unto thine own understanding. —*PROVERBS 3:5*

Retire? How could I? With the uncertain economy, high taxes and dwindling Social Security, retirement would slice our income to half of what we were already struggling to live on. Besides, I had a mortgage and had fallen behind in some income-tax payments. Company rules allowed me to work until age seventy, but at sixty-three, I was exhausted. When I suggested retirement to my wife Ruby, she was apprehensive. "Oh, Oscar," she said, "how will we manage?"

In desperation I sought our pastor. He listened, then said: "Oscar, if you're really thinking of retiring, let me tell you this: *You* are the only one who can make that decision."

And that was just the problem: making a decision. Retiring meant letting go, trusting the future and accepting whatever might come. I was afraid.

Finally, after several more weeks of worry and prayer, I made the decision. I retired. And—to my utter amazement—the unexpected started to happen: Gifts began to flow from the Lord. Retirement gave me the time to serve, a tax-free income to clear up our mortgage and tax debt, and, best of all, I discovered hitherto unknown precious hours to spend with my family, enjoy my home...and move closer to God.

Are you troubled by indecision? Why not let go...and He will prepare the way for you.

Gracious Father, strengthen me when I face the unknown, and guide me into the path You know is right for me. —OSCAR GREENE

31 FRIDAY

There shall be showers of blessing. —*EZEKIEL 34:26*

It is snowing this morning around our house in Boulder, Colorado: big, fluffy, lazy flakes floating gently down from the sky. I pull on my coat, gloves, boots and hat to walk down our long driveway to get the newspaper. As I walk, I am fascinated by the falling flakes. Left alone in their downward journey, they swirl around and softly land on the ground where they become lost in the mounds of snow already there. But if I reach out my hand to catch some of them, at least for a fleeting moment, I can appreciate the unique, delicate beauty of each individual flake.

God's blessings in my life are the same way. He generously rains them down upon me, but unless I pause and reach out my hand to catch them, they fall unnoticed, and too often unappreciated, on a blanket of other blessings down around my feet.

Maybe today's a good day to remember. Shall we stretch out our hands, open our eyes and watch closely for our nearest blessing?

Help me, Lord, to pause often today to watch for Your coming.

—CAROL KUYKENDALL

Praise Diary for January

1

2

3

4

5

6

7

8

9

10

11

12

13

14

15

16

17

18

19

20

21

22

23

24

25

26

27

28

29

30

31

February

S	M	T	W	T	F	S
						1
2	3	4	5	6	7	8
9	10	11	12	13	14	15
16	17	18	19	20	21	22
23	24	25	26	27	28	

Winter of the Soul

'Tis a month of reddened nose,
of benumbed hands and tingling toes,
'Tis also time when spirits sink,
renounced by God, we sometimes think,
Do you not wonder in darkness alone,
weary of soul and chilled to the bone,
Where the Lord of healing's gone,
why His touch has been withdrawn?
Speak to us, Father, if You're there
tell us that You hear our prayer.

Trust in Me, My questioning one,
I'm constant as the rising sun,
A faithful shepherd, I care for My sheep,
I watch o'er you always, awake or asleep,
I note each prayer and I answer each, too,
but the hearing, beloved, depends upon you,
Open your heart and receive My sight,
lean on My wisdom, walk in My Light.

—*FRED BAUER*

1 Practicing His Presence in February

HOLDING OTHERS IN THE LIGHT

Pray one for another... —*JAMES 5:16*

One afternoon while doing research in the library, I came upon an odd sight. Sitting on a little stool among the deep rows of books, a woman sat sound asleep. She sagged against the shelf, lines of exhaustion and worry drawn in her face. And suddenly a thought surfaced abruptly in my mind: *Pray for her.*

Having no idea of her needs, I simply envisioned her bathed in the light of Christ. I imagined it surrounding her on all sides, giving her new energy, life, joy, hope, insight—whatever she needed. Then I walked on.

Around the corner I saw a man studying a page in a book. *Pray for him.* As I walked on, I lifted him into the light of Christ, too. Next there was a young mother...the librarian...a man at the card catalog. The adventure grew until I found myself picturing each person I came across bathed in the light of Christ, having a personal, unknown need met.

And I noticed something. Not only was I interceding for those around me, I was communing with God at the same time. In fact, it was one of the most God-conscious afternoons I'd ever spent. And it occurred to me that one of the best ways to keep my mind anchored upon God was by aiming prayers of light at those around me.

As I left the library and stepped into the stream of the street, I beamed a prayer at an approaching stranger, finding God's Presence fresh as the winter air.

Jesus, as I go about my normal activities during the days of February, help me lift the people with whom I come in contact into Your light. —SUE MONK KIDD

2

SUNDAY

For they that are after the flesh do mind the things of the flesh; but they that are after the Spirit the things of the Spirit. —*ROMANS 8:5*

A few thousand years ago, farmers used to watch what happened when a small furry animal emerged from hibernating underground. If the animal saw its shadow, probably because the sun was bright, it was said to run back to its shelter. Supposedly that meant winter was not yet over. If the animal remained above ground, most observers believed that spring would come early. The reason for the farmers' anxiety was that they had to plant their seeds before the end of winter, and they knew that prolonged cold weather could destroy young plants. Their only hope for survival was an early spring.

Groundhog Day was named for that old superstition. And some of us still think a groundhog can tell us when spring will arrive. No matter that a hibernating animal can't see a thing for at least eight hours after awakening. No matter that in some areas it is called forth prematurely with loud noises. We want to know when winter will be over!

The answer is that winter was over a long time ago. The cold and darkness of death, the soul's winter, can no longer touch us—because, with Christ, we live in eternal spring. When we know Him, we know that nothing can destroy the seed of His love in the soil of our hearts.

I trust You, my Savior, to meet the light of each new day and circumstance with me. —PHYLLIS HOBE

3 **MONDAY**

...Thus saith the Lord God; Behold, I will kindle a fire in thee... —*EZEKIEL 20:47*

Most people like a roaring fire as long as it is contained in a fireplace. On cold wintry nights, there is something about blazing logs that kindles the imagination.

Recently I became entranced by logs that had burnt down to embers. I was getting ready to retire and had turned out the light, when the brightly burning embers caught my eye. Almost hypnotized, I sat down in the darkened room to watch as the last log disintegrated into scores of brightly glowing embers. An hour later, I was still there, watching in fascination and thinking how the embers relate to our lives.

As young people embarking on careers, we are fired up and burn with ambition and energy. In mid-life, we tend to settle down to a lower, but steady flame in our careers and our walk with the Lord. Later on, we are often reduced to embers, but they can glow brightly as long as we keep contact with our Source. For many, the glow of old age can burn as beautifully and as impressively as the fires of youth.

Lord, may Your fire burn brightly within me and may it warm and comfort others through the passing years. —SAM JUSTICE

TUESDAY

Thus saith the Lord God, Behold, I lay in Zion for a foundation a stone, a tried stone, a precious corner stone, a sure foundation: he that believeth shall not make haste.

—*ISAIAH 28:16*

One winter I lived in Scotland near the royal family's Balmoral Castle. My landlady had hired one of the Queen's employees to paper her three-story hallway in the evenings after his regular work. The first night he spent all his time measuring and putting up only the first strip. *How slowly he works*, I thought. Then, night after night, I watched in fascination as he matched other strips to the first with incredible precision. Never before had I seen such meticulous workmanship. Fit for a queen, the glowing new hall was impeccable.

God works that way in my life, too. Having laid the cornerstone, Jesus Christ, He slowly gives me, piece by piece, what I need to build who I am going to become. But how slow God's work seems sometimes. I don't want to learn patience bit by bit. I don't want to work on my temper today and jealousy tomorrow. I want perfection *now*!

I need to learn from that Scottish workman to trust God's painstaking work in my life. In Greek, the end of the verse above reads, "the believer shall not be ashamed." When there's a Master at work, we can anticipate the finished product will be perfect!

Dear God, help me to trust Your expert Craftsmanship in my life. I know You are building someone we both can be proud of. Amen.

—PATRICIA HOUCK SPRINKLE

5 WEDNESDAY

Sing praises to God, sing praises: sing praises unto our King... —*PSALM 47:6*

You know how people are always saying, "Write it down so you don't forget"—like grocery lists, birthdays, or telephone messages. Well, for the past several years I've applied this to my devotional life, using the Praise Diary at the end of each month. There I write down daily thoughts of prayer and praise.

I never realized there were so many things worth remembering—and being thankful for—until I recently reread my past entries. My heart was stirred and I found myself smiling.

Here's the entry for the day I mailed four stories to a writing contest:

> *I praise You, Father, for typewriters, photocopy machines and the U.S. Post Office.*

Or for the weekend that Meghan came home from college for a visit:

> *Praises, Father, for our beautiful daughter who sits across the breakfast table from me in a long blue nightgown.*

Another entry that is self-explanatory:

> *I praise You, Lord, for blessing us with the income to pay all these bills.*

I notice, as I read these entries, how mundane most of the things may seem, but they represent the nitty-gritty details of living, the nuts and bolts that hold together the structure of my days. Nothing big or significant—and yet these gifts from God add up to a richly blessed life.

Why not turn to your Praise Diary today—and every day—and watch your blessings increase daily?

I praise You, Father, Lord of my life. —MADGE HARRAH

THURSDAY

I have planted, Apollos watered; but God gave the increase. —*I CORINTHIANS 3:6*

From an elevated position in our tour bus while in China, we looked out over sidewalks teeming with people. Their movements reminded me of an anthill that had been disturbed. It seemed that we were viewing all of Shanghai's twelve-million population at once. And when the traffic light changed, I was sure that most of its three million bicycles were moving, too.

Christianity has been slow to take root in this vast country, where the centuries-old Eastern religions and philosophies prevailed. I am afraid that my aunt and uncle, who were missionaries to China, did not live to see the fruit of their labors. So I felt especially privileged when an elderly Chinese man, now a Christian pastor and leader, told me, "We have many members scattered over many places because of the faithfulness of workers, like your relatives, who told us of God's love."

Now God sometimes allows His children to see the results of their "planting" and "watering." But not always. And should we ever feel that our labors in His name have been fruitless, that is the very time we must remember that, after all, it is God Who produces the harvest...in His Own way, in His Own time.

Father, let me sow with diligence, leaving the harvest to You.

—ISABEL CHAMP

7 FRIDAY

Behold, how good and how pleasant it is for brethren to dwell together in unity! —*PSALM 133:1*

Here in New England we have harsh winters. Early on Tuesday, February 7, 1978, the weather bureau warned us that the snowstorm of the century was roaring up the coast our way. From five o'clock in the evening until Wednesday midday, the snow, churned by furious winds, bombarded everything in sight. Cars vanished beneath five-foot drifts, and on the ground, snow piled three feet high.

We couldn't even open our storm doors—the snow barricaded them. In the street, a city snowplow lay overturned and half-hidden. I turned to Ruby and said, "What do you know? We can't get out—we're really snowbound!"

For two days we shoveled until we had a path to the street. Other people were busy too, and soon footpaths appeared throughout the city. Ruby and I, like giddy schoolchildren, wended our way down through the drifts. As we walked along, something curious happened. Strangers paused to chat with us. People smiled, called out to one another and waved.

For one whole week we all were a friendly and cheerful people. We cared. We were part of a family—God's family. That was the Blizzard of '78—wherein we walked in faith and trusted one another.

Don't you, as I, wish that all our days could be peace filled?

Lord, I don't want to wait for a snowstorm. Help me go forth into this day, welcoming others, responding with love and trust.

—OSCAR GREENE

SATURDAY

If we live in the Spirit, let us also walk in the Spirit. Let us not be desirous of vain glory... —*GALATIANS 5:25, 26*

Last February my church celebrated the 125th anniversary of the Sunday (February 26, 1860) that Abraham Lincoln worshipped in our Brooklyn, New York, sanctuary. Lincoln's visit was his way of honoring the ardent crusade of our founding pastor, Henry Ward Beecher, for the abolition of slavery.

One would think, given their prominence and common goals, that Lincoln and Beecher would have admired one another. But this was not the case: Lincoln felt that Beecher was too flamboyant and sensational; Beecher thought Lincoln indecisive and unsophisticated. Nonetheless, Lincoln made an extra effort to pay a respectful visit, and Beecher treated Lincoln as an honored guest. They were able to put aside their personality differences in order to pursue what both perceived to be the common good.

And now, 125 years later, what do we remember about these men? Few recall their private opinions about one another. Most remember that they dedicated themselves to ideals of personal freedom that shaped our nation. I've often thought that one of the keys to their success as leaders was the fact that each could set aside petty personal differences (or ego) in favor of a greater end.

And that's something we can all try to remember when differences of opinion with others threaten mutually held goals. Then we can go one step further and ask ourselves: *Is there some small sacrifice I can make that could ensure victory for all?*

Lord, I need reminding that, if I truly seek common ground with another, I will always find it—and that it will most surely lead to a common good. —JAMES McDERMOTT

9 **SUNDAY**

No man, having put his hand to the plough, and looking back, is fit for the kingdom of God. —*LUKE 9:62*

When I was twelve, I wandered into Golden's Feed Store and found myself before a display of flower seed packages. As I looked at the bright pictures, I felt something pulsing inside me. I wanted to grow something. I wanted to plant a flower garden. So right then and there I committed myself to the task. I bought five packages of seed.

When I got home I raced out to the garden, eager to begin. But there I found an untended clump of ground spilling over with weeds and hard, lumpy Georgia clay. I stood beside the miserable little plot as reality set in. It would take a lot of work—pulling weeds, hoeing, tilling. I went inside and laid the seeds on my window sill. In August they were still there...unplanted, reminders of a garden that never grew.

Ash Wednesday is the first day of Lent. It's a day when Christians traditionally look at their lives and seek to renew faith and make changes and commitments. It marks a season of growing. And I've discovered that starting on a Lenten commitment is not that different from starting on a garden. Often the hardest part is simply beginning.

So when I make my Lenten promise this season, I know there will be work in getting started. I will have to pull up my resistance to change, muster discipline, break up old patterns and initiate brave new ones. And sure, it may not be easy, but I've learned since I was twelve that it's a small price to pay for the joy of growing.

Lord, do not let my spiritual life become a garden that never grew.

—SUE MONK KIDD

ADVENTURES IN LOVING:

Seven Pillars of Marriage

In anticipation of Valentine's Day, we asked Ruth and Norman Vincent Peale to share some of their secrets of staying in love. In the week ahead, the Peales tell us how their marriage has grown steadily stronger and more fulfilling since they embarked on this "adventure in loving" some fifty-six years ago. —*The Editors*

10 Day One—***Love is Trusting***

The heart of her husband doth safely trust in her.

—PROVERBS 31:11

Some years ago T. E. Lawrence—the famous "Lawrence of Arabia"—wrote a book called *The Seven Pillars of Wisdom*, based on his wartime experiences in the Arabian desert. When the editors of *Daily Guideposts* asked my wife Ruth and me to write about our views of love and marriage, the title of Lawrence's book came to mind and I wondered if there might be "Seven Pillars of Loving" or even "Seven Pillars of Marriage" that we might discuss in the seven days ahead.

There are many pillars that support a good marriage, but after some thought and discussion, Ruth and I came up with seven. The first one was *Trust*. No marriage can be called a good one unless there is total trust between the partners. You have to trust your partner to be faithful, to be honest, to be generous and understanding and kind. And the partner has to be *trustworthy* in all these and many other areas of living.

When that trust is applied to your children also, it becomes the glue that holds the whole family together. I remember once being asked in a television interview whether I thought I could trust one of our children who was at Mt. Holyoke College to behave as a Christian young lady should. I replied without hesitation that I trusted her completely: morally, ethically, intellectually and every other way. I said I had not the slightest doubt about her conduct at all. None whatsoever.

It just so happened that up at Mt. Holyoke our daughter and some of her friends were listening to that broadcast. They said that the glow of happiness on our child's face was marvelous to behold. And I'm sure it was.

Trust. The first of the "Seven Pillars." And one of the strongest—when it comes to bearing heavy loads.

Lord, let our trust in You be reflected in the way we trust each other.

—NORMAN VINCENT PEALE

11 Day Two—*Love is Sacrifice*

...By love serve one another. —*GALATIANS 5:13*

One of the pillars that support a good marriage, I'm sure, is the willingness to make sacrifices. Not just major, dramatic sacrifices—although sometimes those are necessary—but the little almost unseen adjustments where you change your behavior or your thinking because it will help or benefit your partner. Love is give-and-take, they say, but when you're truly in love you're willing to give more than you take.

I remember as a young bride I soon discovered that Norman liked neatness in his home. Anything disorderly disturbed him a great deal. I had been brought up more casually perhaps, and it didn't bother me if a hat was tossed on a chair or a newspaper wasn't folded just so. And at first I thought, "Why should I change my ways to suit his?"

But then I found that Norman's creativity was closely linked to his desire for orderliness. If the house was in disarray, his flow of ideas simply disappeared. So I had to say to myself, "Which is more important: your pattern of easygoingness or the creativity in Norman that helps literally millions of people?" The answer to that was very plain. So I made myself become a neat housekeeper.

Has it become second nature to me now? No, it hasn't. It takes a determined effort of will on my part today, just as it did over fifty years ago. But out of that little willingness to make a quiet sacrifice, much happiness has come. For me—and for countless unseen people around the world.

Lord, You gave up everything for us. Help us to give up our own self-centeredness for those we love. —RUTH STAFFORD PEALE

12 Day Three—*Love is Forgiving*

...I say not unto thee, Until seven times: but until seventy times seven. —*MATTHEW 18:22*

All the stories in Guideposts Magazine are memorable, I think. Some are unforgettable. In the unforgettable category I would place a story of a Pennsylvania farmer, Jay Meck, and his family. He and his wife had a little boy whom they adored. The school bus always let him out near their home. One day when the school bus was stopped with warning lights flashing, a driver from New York City failed to heed the lights. The Mecks' little boy was killed.

Losing a child that way may well be the profoundest sorrow human beings can feel. The Mecks were devastated. Angry friends urged them to sue the careless driver. But, grieved though they were, the Mecks knew that the driver was suffering too. They prayed about their situation, and when they prayed they knew what they had to do. They invited the driver and his wife to dinner. In faltering tones, they offered forgiveness. And, insofar as was possible, a terrible wound was healed.

Forgiveness. Successful marriages need it and use it all the time. Not in great tragedies, perhaps, but in little failings and shortcomings. And what is forgiveness, really? It's offering the person who has offended a chance to do better. And then another chance. And another. Up to seventy times seven.

Lord, forgive us our debts as we forgive our debtors.

—NORMAN VINCENT PEALE

13 Day Four—*Love is Reinforcement*

Bear ye one another's burdens... —*GALATIANS 6:2*

What does a pillar do? It supports something, doesn't it? That's what marriage partners should do, too. They should hold each other up. They should be supportive.

Most men need a lot of emotional support, I think, and I'm no exception. I need a wife who keeps reassuring me that I can achieve the goals I've set out to achieve, that my mistakes aren't fatal, that I have what it takes to keep trying to do better. Countless times after I've given a talk somewhere and we've hurried to a car to rush to the airport, Ruth will lean over to me and whisper "Terrific!" She means that the talk I just gave was terrific. Well, sometimes I know it wasn't, but when I hear her say that all my qualms and fears and feelings of inadequacy subside.

Sometimes during those talks I look down at Ruth in the audience, and there she is laughing at my jokes! Now she has heard those jokes at least two thousand times, but there she is laughing anyway, enthusiastically, joyously, as if she had never heard them before. That's what I call being supportive! When too many people or problems come crowding in on me and I begin to get tense, she quietly diverts some of the pressure away. When I get into a gloom, she waits patiently until I come out of it, or if I stay too long she gives me a nudge that pushes me out of it.

She's my balance wheel, my shock absorber, my compass, my North Star. She's the pillar that holds up my life.

Thank You, Lord, for creating someone like Ruth—and letting me find her.

—NORMAN VINCENT PEALE

14 Day Five—*Love is Closeness*

Ye are my friends... —*JOHN 15:14*

There's a four-letter word always associated with happiness in marriage, and certainly with Valentine's Day. That word is *love*. But there's also a ten-letter word that in my opinion is equally important. That word is *friendship*.

What is a friend? Someone having lunch with Henry Ford asked him that question one day. Ford took a pencil and wrote on the tablecloth: "He who brings out the best in you is your friend." Emerson expressed the same idea. "Our chief want in life," he wrote, "is someone who will make us do what we can." And he added that this is the function of a friend.

In a truly successful marriage the partners are not just good friends, but best friends. That means they're totally comfortable with each other. They're never bored because they share interests, viewpoints, values, goals. They like being together; they feel incomplete when they're separated. This doesn't mean that they never disagree. But they don't fight about differences; they compromise and work them out.

Ruth and I are like that. We love each other deeply and truly, but we're also friends. Best friends. When you're married to your best friend, that's just about the surest guarantee of happiness that this uncertain world has to offer.

Lord, on this Valentine's Day we thank You for the joy of being best friends—with each other. —NORMAN VINCENT PEALE

15 Day Six—*Love is Laughter*

...He that is of a merry heart hath a continual feast.

—PROVERBS 15:15

One of the strongest unifying elements in our family life has always been a capacity for shared fun. Norman has the great gift of laughing at himself, and I think to some extent this has rubbed off on all of us.

Certainly our three children always displayed a lively sense of humor—sometimes a bit *too* lively, perhaps. I remember the time the doorman of our building came to see me with a solemn face. "Mrs. Peale," he began, "there's a very angry lady downstairs...." It seemed that John and Margaret, determined not to be meek and mild goody-goodies, had been dropping small paper bags filled with water on the heads—or at least close to the heels—of passersby. I had to be stern and punitive, of course, but I remember hoping that I wouldn't destroy their irrepressible high spirits.

Family meals were always occasions for the exchange of lively ideas and humor. I remember once we were discussing whether some property we owned should be turned into a subdivision. What could such a venture be called? Up spoke Elizabeth without hesitation, "Why, we'd call it Peale's Positive Plots!"

At mealtime, when the children were small, Norman would tell them stories about imaginary characters who had all sorts of dramatic adventures. There was a favorite trio named Larry, Harry and Parry who had an inflatable airplane. They would blow it up and fly away to the moon or other romantic places. Our children still have affectionate recollections of Larry, Harry and Parry.

Shared fun. Shared laughter. These may not be the biggest and strongest pillars that hold up the edifice of marriage and family closeness, but they certainly carry their share of the load.

Dear Lord, give us merry hearts that can lift another's spirits through good times and bad. —RUTH STAFFORD PEALE

16 Day Seven—***Love is God's Presence***

Except the Lord build the house, they labor in vain that build it.... —*PSALM 127:1*

"Mother," said our teenage daughter, "I don't want to ride the bus to school anymore! I'm too old for that. I want to take the subway to school, the way my friends do!"

In those days New York City subways were a lot safer than they are now, but still I had qualms about them, and I said so. But Elizabeth begged and pleaded. "All right," I said finally, "you can go with a group of friends. But just remember this. My prayers are going with you, every day, every step of the way."

And I did pray, placing Elizabeth in God's hands every morning and thanking Him for bringing her home safely every afternoon. Elizabeth knew this; she felt protected and loved. Once she did tell me that two men got into a noisy altercation not far from her group. "But we just got off at the next stop," she said, "and waited for another train!" I was sure that the Lord, watching over her, put this simple and sensible idea into her head.

My point is a simple one: Of all the pillars that support and sustain a marriage and family life, this is the strongest—this willingness to bring God into every situation, every problem, every aspect of living. He loves us; He cares about us; He wants us to be happy.

All we have to do is ask!

Be with us, dear Lord, in everything we do—this day and every day. —RUTH STAFFORD PEALE

17 WILDERNESS JOURNEY (Exodus 13)

Led by a Pillar of Fire

He took not away the pillar of the cloud by day, nor the pillar of fire by night... —EXODUS 13:22

A dream is beginning to come true! Slaves for hundreds of years, the Israelites are finally on their way to the land God promised to their great-great-grandfather's great-great-grandfather. What excitement there must be in that throng of ragged, whip-scarred ex-slaves! Yet what courage it must take to leave all that is familiar and set out on foot across a barren desert, bound for a land that may be only a dream.

It's so hard to face the unknown! I know. After thirty-four years as his own boss, my architect husband has sold his business and taken a position with the state. Will we be able to adjust? Our daughter was married in October and will soon be moving away from Kearney. What will it be like without Karen? A prolonged illness has interfered with my son John's schooling, and I'm uneasy about his future.

Carrying all of my uncertainties with me, I close my eyes now and join that band of desert wanderers. Maybe you have uncertainties, too. Bring them along. It's late afternoon. Our legs ache. There are blisters under our sandal straps and needling qualms in our hearts. Will there be enough food? Will we die of thirst? What if a plague strikes? Each time we mumble a doubt, the old man next to us points wordlessly ahead, to the pillar of cloud that leads us. *Ah...God-with-us.* What blessed assurance!

Now darkness begins to settle over the land as we prepare to camp for the night. Are you as uneasy as I am? But look! Up ahead where the cloud has been! It's a pillar of fire now, rising up to the heavens! Come. Let's move closer. Can you feel the flame's warmth on your skin, see its golden glow, sense its power? We stand on holy ground! As I reach down to remove my sandals, I again think about all those uncertainties in my present life. Are doubts about the future keeping me in bondage to the past? Perhaps I could feed them, one by one, into the pillar-shaped flame of God's Presence and let them go—right now. Perhaps you could do that with your anxieties, too. Oh, to be free!

My journey is in Your hands, Lord. Trusting You to lead me, I relinquish every uncertainty. —MARILYN MORGAN HELLEBERG

18 TUESDAY

And they that know thy name will put their trust in thee...
—*PSALM 9:10*

He stood before the fireplace, tall, gaunt in a seedy frock coat, his hair long and unkempt, his face thin and drawn.

His large melancholy eyes had a solemn, faraway, troubled look. Gazing at the handwritten document before him, he recalled how courage to act had failed him three times before. He faltered, "I have been driven many times to my knees by the overwhelming conviction that I had nowhere to go; my own wisdom and that of all around me seemed insufficient for the day."

This was President Abraham Lincoln, and the document before him was the Emancipation Proclamation that, together with the Thirteenth Amendment, freed 3,895,172 slaves.

My wife Ruby and I think President Lincoln would have rejoiced to have foreseen one particular slave-descendant named Shawn Greene—our grandson—born February 12, 1973, *one hundred and ten years* following enactment of the Proclamation.

President Lincoln would have been amazed to know that Shawn has more books in his bedroom than Lincoln had in his law office. And one book Shawn and Lincoln could especially share is our beloved Bible. Lincoln said, "I believe the Bible is the best gift God has ever given to man."

This month we can all join Shawn in honor of Lincoln as we build renewed trust among ourselves, raise our voices in prayer, seek counsel from the Bible and walk humbly among our brothers and sisters.

Thank You, Father, for those especially courageous people who have braved adversity that Thy Will might be done. Give me the courage to follow their example. Amen. —OSCAR GREENE

19 WEDNESDAY

...A time to break down, and a time to build up.

—ECCLESIASTES 3:3

Because older women in my family have endured the stooped back and broken bones of osteoporosis, I decided recently to do some reading on prevention. In the library I found books with charts and chemical analyses, but one sentence stuck with me: *Like all living tissue, bone is constantly being broken down and re-formed.*

As I read on, I found the words seemed to apply to this season of Lent. Forty long days of repentance, repentance and still more repentance...until I'm ready for Easter and His triumph rather than my failure.

But it was the word *living* that leapt out at me: *living* tissue that is continually torn down and rebuilt. As long as my relationship to God is alive, this biological fact seemed to suggest, the tearing-down process is part of it. The break-up of outworn habits, the rejection of outgrown assumptions, go hand in hand with renewal. No growth without pruning, no rebirth without death.

The medical name for this continuous cycle of break-down and reformation, my reading informed me, is "bone-remodeling."

Soul-remodeling—isn't that what Lent is all about? Repentance, forgiveness, new life. An essential sequence, not just at this time of year, but for always.

This Lent, Father, transform my Spirit with new life. Amen.

—ELIZABETH SHERRILL

20 THURSDAY

With men this is impossible; but with God all things are possible.

—MATTHEW 19:26

All through January, despite my fierce determination, I just couldn't seem to keep my New Year's resolutions. I couldn't lose those ten pounds I had promised myself I would, or finish the writing goals I had set. Each morning I began a diet that lasted until my coffee break with the girls at work, and each evening I settled down with paper and pencil while words and ideas, like willful children, played hide-and-seek in my mind.

"Oh, well," I sighed again and again. "Maybe tomorrow...."

At last I stopped trying. "Lord," I said, completely discouraged, "I give up. I just can't do it."

Strangely enough, those turned out to be almost magical words, for as soon as I said them, the Lord seemed just to take over. The next day I stopped so desperately *wanting* the morning junk foods, and that evening orderly ideas began to march through my head. As the days went by, pounds dropped away and my notebook began to fill up with words—yet all I had done was to admit defeat.

Have you been failing to keep your New Year's resolutions? If so, why not let God keep them *for* you? Maybe you will discover, as I did, that sometimes we have to surrender first before we can win.

Lord God, help me where I am unable to do for myself.

—DORIS HAASE

21 FRIDAY

The word of the Lord came again unto me...

—EZEKIEL 37:15

Everyone loves a good storyteller, and Henry Holkenbrink of Sigel, Illinois, had a memory bag crammed with good yarns. I remember how much we enjoyed listening to his accounts of bygone days.

One of my favorites concerned the man with the woodpile. "All early settlers had woodpiles," Henry related. "A fine rick of wood was a man's pride. But one feller up near Sigel kept missing wood, so he decided to find out who the thief was. He whittled a hole in a stick of hickory, packed it with gunpowder, smeared mud over it and laid it on top of the rick. The next morning there was a booming explosion at a neighborhood house and stove lids flew right through the roof. After that you can bet no more wood was stolen."

Another story we especially liked was the one about the time Sigel got its first gaslights. "They put one in front of a little store where the men liked to loaf on the plank steps of an evening," Henry told us. "June bugs would fly through that gas flame, singe their wings and drop. There was an old toad that lived under the steps and he'd have a feast. The men would bet on how many bugs he'd eat. Maybe as many as sixty or seventy. That was what they did for excitement in Sigel in those days."

Although Henry has been gone for many years now, I remember him with special fondness because of his marvelous stories. To this day, they give me pleasure.

Lord, inspire my words that others might receive joy... or comfort ... or love... or some message that will linger in their hearts.

—ZONA B. DAVIS

22 SATURDAY

Thou therefore endure hardness, as a good soldier of Jesus Christ. —*II TIMOTHY 2:3*

What "hardens" us into becoming good soldiers for the Lord? George Washington became known as "the Father of His Country," even in his own lifetime. But he is revered more as a human temple than loved as a fellow human being.

Modern research has uncovered the real man behind the formidable legend. It has shown him to be a man of faults, but with the greatness to rise above them.

As a soldier, he made mistakes, but he won the war.

As a statesman, he spoke seldom, but when he did, spoke eloquently.

And so our first president was admired for his "soldierhood" of courage, energy, judgment, common sense, granite character and sense of honor.

Can I rise above my faults?

Can I accept my failures, get up, and continue the fight?

Can I hold my tongue until it has healing words to share?

Can I *endure*—crying quietly, if cry I must?

Then the Lord has need of me. And of you, too.

Here we are, Lord. Make good soldiers of us for Your Cause.

—JUNE MASTERS BACHER

23 **SUNDAY**

If a house be divided against itself, that house cannot stand. —*MARK 3:25*

Around our church in New Jersey, where my wife is the pastor, I'm known as "the minister's wife." And from time to time at least one person asks what kind of casserole I'll be bringing to the church pot-luck supper.

But being "the minister's wife" isn't always fun. It was especially tough in the beginning, when Lynn had meetings four nights a week and spent Saturdays polishing her sermon. I felt as if I were taking second place, and I didn't like it one bit.

One Sunday, an older gentleman in the congregation—to this day I wonder if he could have been a mind reader—told me about his friend's grandnephew, who also was a minister. When he graduated from college, a large corporation wanted him to work for them. Three times they offered him an increasingly higher salary. Three times he turned them down. "Isn't the salary big enough yet?" the exasperated company president finally asked.

"The salary is big enough," the young man replied, "but the job isn't."

"And you know," the old fellow finished, "that boy's family stood behind him all the way. Now he has a large church out west." Then he added, "Stand behind Lynn."

He seemed to know more about the secret of being a "minister's wife" than I did. And I soon discovered it's a secret anyone can use. It's all about goals and dreams and being there to help someone else reach his or her aspirations. A dream is a wonderful thing to have. But it's even more wonderful to share it with someone. Whose dream can you share today?

I thank You, Lord, for friends and loved ones who stand by me.

—JEFF JAPINGA

24 **MONDAY**

...For I will turn their mourning into joy... —*JEREMIAH 31:13*

At the southeast corner of Central Park, there's a promenade that leads to the children's playground. Walking there yesterday, I chanced to look up at the trees overhead. Since it was February, their limbs were skeletal and bare, or so I thought until something made me stop and examine them more carefully. *Why did they have all those little strings dangling from them?* I wondered. I looked around; there were similar threads on all the trees along the promenade. And then it hit me:

Balloons. Balloons escaping from children's hands, getting snared in the branches, irretrievable in the high trees. Oh, the anguish of a lost balloon—one of the tragedies of childhood.

Standing there in the wintry park, I thought of a summery afternoon many years ago when my goddaughter Elizabeth and I had stood in that very spot. I thought of how from that day on I'd known for certain what a fine young woman she would grow up to be (and she has). Elizabeth was very little then, and that may have been the first of our many outings together. We'd been to the famous soda shop, Rumpelmayer's, for hot fudge sundaes and were strolling toward the zoo when we acquired a yellow balloon. Elizabeth wasn't holding it tightly enough, for soon the balloon tugged and the string slipped free.

"Oh-h-h-h," she gasped, and I lunged, too late. The two of us watched helplessly as the yellow balloon soared away high into the heavens. I looked at Elizabeth, wondering about what godfatherly thing I could do to dry the tears that were sure to come.

But there were no tears, only resolution.

"Well," she said, smiling up at me and taking my hand, "I think God will have fun with our balloon, don't you?"

When disappointments come, Lord, show me how to look at them carefully, and in Your light. —VAN VARNER

25 **TUESDAY**

Come unto me, all ye that labor and are heavy laden, and I will give you rest. —*MATTHEW 11:28*

I went to a time management seminar a few weeks ago where the speaker advised us to clear the clutter out of our lives. "Power rake your homes," she encouraged. "Be unmerciful. Go through each room, closet and cupboard. Take out all the unnecessary things that bog you down. You'll feel so much better. I promise!"

The vivid image of power raking my life has stuck with me since that day, probably because I keep seeing areas that need to be power raked. Take my walk with God, for instance. I get so bogged down and burdened with unnecessary negative thoughts and fears and self-doubts that pile up and slow me down. "I can't...I don't know how... What if...If only...."

In God's eyes these thoughts and fears must look like a lot of unnecessary clutter. "Power rake them," I can hear the command. "You'll feel so much better. I promise!" And when I obey, I do feel so much better because I walk with Him more efficiently and effectively.

Lord, today I will clear out the clutter and let go of negative thoughts that slow me down in my walk with You.

—CAROL KUYKENDALL

26 WEDNESDAY

Our Father which art in heaven... —*MATTHEW 6:9*

In the book, *Handbook of Positive Prayer*, Hypatia Hasbrouck notes that according to some scholars, the earliest form of the Lord's Prayer was in more positive terms than in the translations we know. These scholars believe that the spoken language of Jesus' day used a verb tense that could be translated: "Thy kingdom *is coming*. Thy will *is being done....*"

Looking at the whole prayer that way reminds me that Jesus often petitioned His Father by affirming that the good He asked for *already was*. For example, He thanked God for hearing His prayer for Lazarus *before* He called his friend back from the dead: "Father, I thank thee that thou hast heard me" (John 11:41). And He advised His disciples that, if they could believe they *already had* whatever they prayed for, their prayers would be answered: "And all things, whatsoever ye shall ask in prayer, believing, ye shall receive" (Matthew 21:22).

I still like the old way of praying the Lord's Prayer because I'm used to it. But today, I'm going to personalize it and use the present tense that scholars say Jesus used. Why don't you pray it, now, too, carefully listening anew to each word, each phrase, each thought, believing that they are already being fulfilled for you.

My Father which **is** *in Heaven, Hallowed* **is** *Your name. Your king-*

dom **is coming**. *Your will* **is being done** *in earth, as it* **is** *in Heaven. You* **are giving** *me this day my daily bread, and You* **are forgiving** *me my debts, as I* **am forgiving** *my debtors. You* **are leading** *me not into temptation, but* **are delivering** *me from evil: For Yours* **is** *the kingdom, and the power, and the glory, for ever. Amen.* —MARILYN MORGAN HELLEBERG

27 THURSDAY

Now ye are the body of Christ, and members in particular.

—I CORINTHIANS 12:27

"Hi! It's me," I announce whenever I call home long distance. I know my parents will recognize the voice. Then we talk and get caught up on each other's news. At the end of the conversation I say, "We wish we were there," and Mom and Dad say, "We miss you." Finally, just before hanging up, they close by saying, "God bless."

Of all my siblings, I am the only one who lives thousands of miles from home. I love my adopted home, this city that affords me countless opportunities. But often I carry a quiet ache in my heart. I long to see my nieces and nephews, go sailing with my brother, eat one of Mom's casseroles. In short, I get homesick.

But lately I've discovered I have another family here. For instance, there's the housebound Irish lady whom I visit once a week. And the young friend (close in age to my sister) with whom I meet regularly to talk over his career. There's our pastor, our choir friends, the kids in Sunday School—and the body of Christians we call the Church. They are our extended family. And, like my family back in California, they're never farther than a prayer away.

Yes, Father, family is nice—especially those with whom I have spiritual ties. Bless them, everyone. Amen. —RICK HAMLIN

28 FRIDAY

Train up a child in the way he should go: and when he is old, he will not depart from it. *—PROVERBS 22:6*

Our first grandchild, Jessica, who is not yet two as of this writing, is developing quite a vocabulary. I realized this one day recently when

I was called upon to baby-sit for her, while her mother and grandmother went shopping. In addition to all her colors, numbers and many letters, Jessica can identify most animals in her books and knows the sounds they make. (The rabbit just wrinkles its nose and sniffs, in case you've forgotten.)

But the thing that impressed me most was her manners. When she was thirsty, she requested, "Juice, please," and when it was poured, she said, "Thank you." Obviously, Mama has been working overtime to teach this social grace. But then I suppose I shouldn't be surprised. My wife Shirley doggedly drilled Jessica's mother, Laraine, in politeness when she was a baby, and Laraine has only passed her legacy on to the next generation.

In the Old Testament many references are made to the sins of the fathers being visited upon the third and fourth generations (Exodus 34:7). Fortunately, virtues can be passed on, too. Our loving words can help offset inconsiderate rudeness, and our loving deeds can help balance the world's self-centeredness, regenerating from one person to another, like violets on a thousand hillsides, spring after spring after spring.

Thank You, Father, for all You've given us, both within and around us.

—FRED BAUER

Praise Diary for February

1

2

3

4

5

6

7

8

9

10

11

12

13

14

15

16

17

18

19

20

21

22

23

24

25

26

27

28

March

S	M	T	W	T	F	S
						1
2	3	4	5	6	7	8
9	10	11	12	13	14	15
16	17	18	19	20	21	22
23	24	25	26	27	28	29
30	31					

Temptation's Tender Trap

Temptation comes in many forms
in many shapes and guises,
But always it speaks softly
when asking compromises...
Each time we fall to tender trap
each time we take a bite
Of Eden's crimson apple
we vow in prayer contrite,
That no matter how appealing,
no matter what the prize,
Next time we'll be more careful,
next time we'll be more wise,
Why aren't we more suspicious, Lord,
of the dark one's devious ways?
Why aren't we more resistant
to transparent tricks he plays?

Only mindless puppets
on a ball of string
Could live a life that's perfect
free of wandering,
So choose to draw upon My strength
when you are sorely tested,
And I will steer you clear of harm
till evil has been bested. —FRED BAUER

1 *Practicing His Presence in March*

TYING A YELLOW RIBBON

Remember the Lord, which is great... —*NEHEMIAH 4:14*

As we were driving along I noticed a bright yellow ribbon tied around the trunk of a pine tree in someone's front yard. "Look at the bow around the tree," I said, pointing through the window.

"What's it for?" the children asked almost in unison.

I told them a yellow ribbon around a tree was usually put there as a reminder of something we wanted to keep in mind. "You mean like tying a string around your finger, so you won't forget," Bob said.

"Right!" I answered.

The next morning I opened my Bible to the reading for the day—Deuteronomy 8:11. "Beware that thou forget not the Lord thy God...." I blinked at the words, recalling the bright yellow ribbon the day before. Theologian Paul Jones once wrote, "I am discovering that the primary enemy of prayer is forgetfulness." That morning I realized one of my biggest obstacles to keeping God constantly in mind was that I simply *forgot.*

Tie a yellow ribbon. The thought pushed its way to the front of my mind. Was there something that could become a reminder of God's Presence in my life, something that would point me to Him whenever I saw it?

That same day I made the schoolhouse clock over the mantle my "yellow ribbon." Each time it chimes throughout the day I am reminded to pause and give God a loving glance, a moment of prayerful remembrance.

This month won't you tie a "yellow ribbon" in your day by finding some object that can be a reminder to you to keep God in your mind and heart throughout your day?

Lord, this month help me to find everyday reminders that turn my thoughts toward You. —SUE MONK KIDD

2

SUNDAY

This cup is the new testament in my blood, which is shed for you. —*LUKE 22:20*

Do you know why some Bibles have the words of Jesus printed in red?

If you don't, here's the reason.

Back in 1899, Louis Klopsch was caught by the words in the verse above. *Those were the words Jesus said at the Last Supper for His disciples,* he thought.

Dr. Klopsch reasoned that since blood was red, why not a Bible with the words of our Lord in red letters? He asked American and European Bible scholars to send him the passages they believed had been spoken by Christ. After collating their replies, he ran off the first red-letter edition of the Bible on his own presses. These Bibles were first advertised in November 1901, exactly 85 years ago. The sixty thousand first-edition copies sold so quickly the presses had to be run day and night to supply the demand for more. The King of Sweden sent Dr. Klopsch a congratulatory cablegram. And President Theodore Roosevelt invited him to dinner at the White House.

Somehow, when I read my red-letter edition, the words of Christ resonate with life and startling immediacy. Listen:

> *Nothing shall be impossible unto you.* (Matthew 17:20)
> *Peace, be still.* (Mark 4:39)
> *Thy sins are forgiven.* (Luke 7:48)
> *Follow me.* (John 1:43)

And there's more...turn to the Gospels in your Bible and soak up some Power.

Lord, empower my life with Your words. —ISABEL CHAMP

3 **MONDAY**

...That there be no complaining in our streets.

—PSALM 144:14

The other day a co-worker handed me a small white card. It read:

COMPLAINT BLANK

State nature of complaint
in this square

□

Write clearly
GIVE FULL DETAILS!

I laughed at the joke, and yet it started me thinking. *Was* it a joke? Or was it really a wise directive? How easy it is to complain about the many small things that go wrong every day. But just supposing the next time I'm tempted, I look at that card and try to figure out how to fit my complaint into that small space. I'll bet in no time I'll be laughing and the annoyance will fade.

How healing laughter can be! Sometimes it even leads to positive thinking. Instead of finding fault, I might even try to find something to praise. If I do, then that little white card would be God's way of transforming anger into joy in just a few short minutes.

The Psalmist said it in Psalm 126: "Then was our mouth filled with laughter, and our tongue with singing...The Lord hath done great things...."

Amen!

Turn my face from turmoil toward Your everlasting peace, Lord God.

—DORIS HAASE

TUESDAY

In all thy ways acknowledge him... *—PROVERBS 3:6*

I was at a basketball game, watching one of my sons play. I didn't know anybody there, so I sat alone. My other son Jeremy came up and sat with me.

I was surprised. None of the other kids was sitting with his mother;

it isn't "cool" to sit with your mother at a basketball game. But Jeremy acted as if it were perfectly natural for us to watch the game together. We talked, cheered, laughed, ate popcorn and drank sodas. After the game was over, he went off to talk with some of his friends, calling, "See you later," as though I were a friend, not his mother. Our fellowship was an unexpected joy.

I wonder whether God feels the same joy when we acknowledge Him before our friends without reservations. I know that when I'm at certain events and with certain people I seem to forget about Him. What's more, I'm afraid that, like Peter, I've been guilty of denying Him.

That day I decided that I must learn to acknowledge Him easily and naturally, at *all* times, no matter where I am or whom I'm with—the way Jeremy acknowledged me at the basketball game.

In everything I do and everywhere I go, Father, I must learn to say, "You are welcome here." —MARION BOND WEST

5 **WEDNESDAY**

To every thing there is a season... a time to plant, and a time to pluck up that which is planted. —*ECCLESIASTES 3:1, 2*

No question that anticipation is the greater part of realization for gardeners. My wife Shirley has gone through the winter seed catalog season and the waiting-for-seeds-to-arrive season, and is now into the planting-in-flats season. Yesterday, she placed vegetable and flower seeds in egg-carton-shaped containers filled with rich humus. Next she watered them and then turned on what she called a "grow" light, which will imitate a May sun, and help turn those seeds into seedlings that can be transplanted in the garden.

Nothing surprising about that scenario for people who grow things, yet there is a miracle about seeds that never ceases to amaze me. What potential they hold...just like people. But neither can reach full bloom without nurturing light.

For you and me, our spiritual grow light is Christ. Without Him we are dormant seeds, small and insignificant. But once we are exposed to His transforming light, we become new creatures...capable of a thousandfold increase. In the light of such a miraculous harvest, who in his right mind would choose darkness?

Praise God from whom all blessings flow,
The Light of Life by whom we grow. —FRED BAUER

THURSDAY

For all these have of their abundance cast in unto the offerings of God: but she of her penury hath cast in all the living that she had. —*LUKE 21:4*

Whenever I read the Bible story about the widow's mite, I recall the day I saw it in action.

It was during World War II, when meat was rationed. Every Saturday morning my mother would arise at five o'clock and head for Merkel's Pork Store, hoping to be one of the first in line when the store opened at eight. About seven o'clock, my father would send me up to the store with a thermos of coffee and the newspaper for Mother. Then I'd wait with her. By the time the store opened, the line reached around the block.

On this Saturday, Mother was fifth in line. When the doors opened, she could see that there were exactly five hams left. Soon it was our turn, and Mother told the butcher she'd take the last ham.

As we left, we spotted Mrs. Beck about twenty-eighth in line. Mrs. Beck lived across the street from us and had six children. Mr. Beck was having difficulty finding a job. "Oh, Mrs. Sass," she called out, "are there any hams left?" My mother shook her head. Then, after a moment's hesitation, she said, "I bought the last one. But we also have a pot roast and some chicken at home. So if you'd like to have my ham, Eleanor will bring it over to you when we get home."

I knew that we didn't have any pot roast or chicken at home, but something told me to keep silent until we had walked a few blocks. Then, clutching the parcel, I said, "But, Mommy, this ham is *all* we have for Sunday dinner."

My mother smiled. "Not really, Eleanor," she said. "We'll have the joy of knowing that the Becks are having it."

And she was right: joy and Spam were what we had for dinner.

Dear Father, Your lessons in giving are all around us. Help us to learn from them. —ELEANOR SASS

7 **FRIDAY**

Thou didst cause judgment to be heard from heaven; the earth feared, and was still... —*PSALM 76:8*

I have a tendency to race around thinking I can do anything, even in areas where my skills and knowledge are limited. It's called being impractical—but I have been quite stubborn about it.

Then recently, while renovating a house (one of those areas where I'm *really* limited), I fell off a ladder and broke some ribs and a wrist. For days I was completely undone by my immobility. So when a dear friend invited me to her home in the country for a weekend, I jumped at the chance to get my mind off my fractured bones.

One morning, I wakened to a quiet world covered with snow and ice. Nothing moved. Not a flake. Everything was beautiful and caught in an awesome, perfect stillness.

It seemed God was trying to say through the beautiful snowfall: "Be still and know that I am God." In other words, *surrender*. Although God may want me to work and be active at other times, *now*, while my body was healing, it was time for stillness. A kind of "time-out." To heal my spirit, to stop, to contemplate, to *know*—that He is God.

I returned to the busy city, with peace in my heart and healing in my bones.

Dear Father, I will take "time out" to be with You. Fill me now with Your peace. —SAMANTHA McGARRITY

SATURDAY

...Whosoever shall not receive the kingdom of God as a little child, he shall not enter therein. —*MARK 10:15*

Not long ago I had the following conversation with my three-year-old son, David, after he had said his prayers and gone through his considerable number of "God Bless's."

"Is God up in the sky, Dad?"

"Sometimes we think of Him that way, as in heaven that is somewhere up above us."

"You mean in the stars?"

"Well, yes, kind of."

"But is God *really* in the stars, sitting in a big house on a star?"

"Well, no, probably not. He's too big a Presence just to locate in a specific house...."

"If He's so big, then is He down here too in the world? Can we see Him?"

"Well, no, but sometimes we can see evidence of His work. We can, ahh...."

"Some things we just don't know about. Right, Dad?"

"Right, Dave."

"Good night, Dad."

"Good night, Dave."

And good night, Dear Lord, in all the places You are, in all the things You do, in all that great and glorious Mystery that is You.

—JAMES McDERMOTT

9 **SUNDAY**

...Be strengthened with might by his Spirit in the inner man. —*EPHESIANS 3:16*

I spent a lot of time looking through my closet this morning for the right outfit to wear. Nothing seemed to fit my need or mood. Reluctantly, I made a selection, but after I put it on, it didn't feel right. As I gazed into the mirror, I decided my face didn't look right either. I rearranged my hair, changed my makeup, but it made no difference. I felt just plain frumpy as I drove to my meeting.

I carried the same uncomfortable feelings into the roomful of women where I emptily greeted a few friends. If only I'd taken the time to change purses, I silently scolded, comparing myself to the women around me. Or chosen more stylish shoes. Or had a manicure. Moving stiffly through the room, I was practically paralyzed with self-concern. Once in my seat, I squirmed self-conciously and barely spoke to anyone else the whole morning.

Later when I got home, I sat down to spend my few quiet moments with the Lord, which had gotten squeezed out in my morning rush. "Do you not know that you are God's temple and that God's spirit dwells in you?" the verse told me. And then it hit me. The Lord loves me from the *inside out*, not the *outside in*.

I'll need to remember that—and with it gain the freedom and confidence to concentrate on others rather than myself.

Lord, I will remember to seek You first, so that my life will radiate from the inside out. —CAROL KUYKENDALL

10 **MONDAY**

And in the morning, rising up a great while before day, he went out... and there prayed. —*MARK 1:35*

I know a businessman in North Carolina who has a beautiful way of starting each day. On waking, he goes to the window, looks out at the day, and says, "Good morning, Lord!"

He has good reason for performing this daily ritual. A few years ago, George Shinn went broke. Everything he tried failed and he was deeply in debt. One day as he was driving near his home outside of Charlotte after another business fiasco, he burst into tears, and he said out loud, "Lord, if You can hear me, let me hear You."

Something happened to him. He could feel it. A calm. A peace. When he got home, he didn't tell anyone what had happened—he wasn't sure himself. He went to bed.

When he awoke in the morning he felt joy. He went to the window and cried out, for the first time in his life, "Good morning, Lord!"

He has been doing this every morning since. Today, George Shinn is a successful businessman. But that's not the point. He is a *happy* man. And he has given me the clue to making my life happier, too—starting first thing every day.

Good morning, Father. I'm going to begin my day with You.

—GLENN KITTLER

11 TUESDAY

For it is God which worketh in you both to will and to do of his good pleasure. —PHILIPPIANS 2:13

Following a bout of flu, I found myself in a depression I couldn't seem to snap out of. I had no energy, and just moped around feeling sorry for myself.

Then one morning I watched a kindergarten teacher distract a weeping child by saying, "Let me help you find something to *do*."

She led the boy to an easel and set him to work with brushes and jars of paint. As the bright colors flowed onto the paper, the child began to smile, his tears forgotten.

After watching that little episode, I decided I needed to try something like that on myself. I went to my own Teacher in prayer, saying, "Lord, help me find something to *do*."

Soon afterward a friend from church phoned to ask if I'd help an older woman who'd had surgery and who would be unable to drive for several weeks. I said yes; and sure enough, I chauffeured that woman around town (and in the process made a new friend), my spirits lifted and my energy returned.

Color me happy, Lord, when I do Your good pleasure.

—MADGE HARRAH

12 WEDNESDAY

Rejoicing in hope; patient in tribulation; continuing instant in prayer. —ROMANS 12:12

My robust, vivacious friend had reached retirement age.

"What are you going to do with all that time you'll have, now that you won't be working?" I asked him.

"I don't know," he replied. "God hasn't told me yet. I'll just continue starting every day the way I always have."

"And how is that?" I asked.

"As soon as I wake up, I say, 'All right, God, what do You and I need to accomplish today?' And then I wait until He shows me."

Isn't that a wonderful way to start off each day?

Father, what do You and I need to accomplish today? Amen.

—DRUE DUKE

13 **THURSDAY**

Every good gift and every perfect gift is from above...

—JAMES 1:17

It was late night when we arrived at the shore. "Nothing's going to be open out here, this time of year," my husband John warned.

Something comes over us after months of winter—a longing to hear ocean waves and sniff salt air. And so, for a little reprieve, we seek a room that's right near the water. But we'd driven through all the seaside towns without finding so much as a gas station open. Then, about to give up, we spotted it—a small sign in front of a turreted Victorian mansion. "Open All Year." Soon we were being shown up a carved oak stairway to a high-ceilinged room with a four-poster bed. Lulled by the sound of waves, we drifted to a peaceful sleep.

The next morning, we found the young owner vacuuming the room next to ours. "I'm sorry I didn't have this one ready last night," she said. "It's our biggest room."

It was big, all right. Twice the size of the one we'd had, with windows on three sides, and a choice collection of period furniture. And somehow as we strolled on the beach, the warmth had gone from the winter sun.

"It's because of that front room, isn't it?" John said. And of course as soon as he said it we saw the childishness of it—we who'd been given the sun and the sea, out of sorts because we couldn't have the moon too!

That little episode taught me something. I'd always thought the opposite of gratitude was ingratitude. But it's not—it's *greed*. Taking our eyes off God and His moment-by-moment provision for us, and fixing them on what looks bigger, better, more desirable.

Whenever we wish He would give us something different, we are not letting God be God...and ourselves be loved.

Thank You, Father, for just what You give today.

—ELIZABETH SHERRILL

14 FRIDAY

And be renewed in the spirit of your mind. —EPHESIANS 4:23

Today I woke up with a restless feeling pulsing inside. I want to dig in the dirt and make something grow. I want to clean out my closets and paint the bedrooms and chase butterflies. I am full of stirrings and longings born sometime during the night. Every year about this time, it happens to me. Somewhere between the last breath of winter and the first green leaf, I contract "spring fever."

Suddenly I am struck by a thought. Could spring fever actually be a spiritual phenomenon? Could my hunger to grow a flower, redecorate the house and follow the flight of a newly resurrected butterfly actually be a sign of some deeper need? Could it really be a hunger for my own growth and resurrection?

I begin to hear God speaking to me, showing me how insular, barren and static my spiritual life became through the cold winter. My prayers are frozen like the January ground and my commitment seems unfocused and neglected like a cluttered closet.

So I decide. Yes...I will go and plant seeds in the earth...follow a butterfly...and clean my closets. But I will also take time to cultivate the "spring" burgeoning within my spirit. Beginning now.

Father, there is something wonderful stirring deep within the cocoon in which I have wrapped myself—unfurl new life in me.

—SUE MONK KIDD

15 WILDERNESS JOURNEY (Exodus 14:1–15:22)

Crossing the Red Sea

And the Lord overthrew the Egyptians in the midst of the sea. —EXODUS 14:27

When I read or hear the story of the Israelites crossing the Red Sea, I recall the night Grandmother Banta told me that story as she tucked me under the feather comforter in the attic bedroom of her big old house in Oberlin, Kansas over forty years ago.

Grandmother was a wonderful storyteller. I could almost hear

God's great wind howling as it pushed back the waters of the Red Sea. I could almost see God's people hurrying across the ridge of dry land, with the terrible Egyptian army bearing down on them. Grandmother's eyes grew wide as she told me about the pursuing Egyptians with their chariots and their snorting horses and their screaming shouts. And when the sea closed in, drowning the enemy army and the Israelites were safe at last, I breathed a delicious sigh of relief.

But as Grandmother turned off the light and started out the door, I begged her to stay. I was at an age when there are monsters in the dark, so Grandmother held me close and told me to remember that God was with me and would not abandon me. "Any time you're afraid," she said, "walk across the Red Sea in your imagination. God will close it after you, so no evil can touch you." As I lay there listening to a cricket chirping, a parade of monsters and mice and spiders marched across my imagination, trying to scare me, but God let them all be swallowed up by the Red Sea.

I'm not afraid of monsters anymore, but I still have my fears. If you are sometimes afraid—of speaking in public, or making a difficult phone call, or being alone at night; of heights or closed spaces or dogs or anything else—why not walk across the Red Sea in your imagination and visualize God closing it after you, separating you from your fears. It has helped me to deal with frightening situations in my life for almost half a century.

Lord, You are my strength and my song. (Exodus 15:2)

—MARILYN MORGAN HELLEBERG

16 **SUNDAY**
Father, forgive them; for they know not what they do.
—*LUKE 23:34*

Recently our church was broken into *twice* in two weeks. The first time, the thieves took two silver chalices, two silver communion plates and tried to disassemble a large silver cross—fortunately without success. The second robbery came the day after our annual fair, and the thieves got away with two hundred dollars in cash.

As you can well imagine, the robberies incensed the parishioners. Our rector was so upset that she went on a two-day retreat to try to restore her spiritual perspective. The Sunday after she returned, she announced that she was going to preach a sermon about the robberies. Eagerly we settled back to await a thundering denunciation of the malefactors. But—and it took us a while to understand this—she was surprisingly calm.

"The Lord made me see," she said, "that with or without the stolen communion silver, the work of this church will still go forward."

She explained that although the things taken were valuable, they certainly were not necessary to carrying out the Lord's mission for the church. Gradually her words helped us understand that the essentials of Christianity can never be pilfered or damaged by outside forces. The citadel of our faith lies deep within our hearts.

After the sermon, many of us had lost enough of our anger to stay behind to pray for the thieves and ask the Lord to lead them to salvation.

O Father God, it is often difficult to ask pardon for those who have sinned against us, but we seek Your great wisdom and pray for the blessing of forgiveness to enter our hearts. —SAM JUSTICE

17 **MONDAY**
Thou preparest a table before me in the presence of mine enemies...
—*PSALM 23:5*

Whenever my life gets rough, two people come to mind: my friend Posy, who tells me, "Sam, there are no coincidences; there's a reason for everything," and St. Patrick, whose life demonstrated that. Born in Scotland about 389 A.D., he was raised in a Christian home where pagan beliefs were frowned upon.

What worse fate, then, could befall this sixteen-year-old boy than to be captured by Irish raiders and sold as a slave to a master who practiced human sacrifice? For six years, the boy watched over his owner's sheep while praying for deliverance. He did escape finally, and afterward he entered a monastery.

It was not until twenty-seven years after his capture that Patrick became a bishop and was sent by the Pope to Ireland to convert the Irish to Christianity. His former captivity there had given him an invaluable knowledge of the people and the land, and he traveled the island telling hostile people about the love of Christ.

And the same can be said for us. Sometimes when we look back, we can see how past troubles have prepared us for challenges in the present. They have been stepping stones, building blocks—part of God's plan. A plan without coincidences, but rather alive with meaning.

Lord, help me to remember that life is an education—a school of learning Your will and purpose for me. Amen.

—SAMANTHA McGARRITY

18 TUESDAY

Neglect not the gift that is in thee... —*I TIMOTHY 4:14*

The winter of '34 was severe in western Massachusetts. As temperatures plunged to −27 degrees, schools were closed because classrooms failed to reach even fifty degrees. But that was fine with me. Because I was failing, too. I just couldn't get motivated in school.

Father couldn't find work and our furnace stood idle. To survive, we huddled around a kerosene heater in the kitchen. The other rooms were frigid. Mother's eyes and Father's voice betrayed their anxiety and fear about the future. My heart went out to them and I wished I could do something. But what could I, a sixteen-year-old failing in school, do?

Well, I decided I would wake up at five A.M. to get the kerosene heater going, so that when the family awoke at seven, the kitchen would be toasty warm. One freezing morning, as I waited for the room to heat up, I picked up one of my schoolbooks and leafed through it. The words leaped from the pages. These were answers to examination questions! The next morning I tried it again. Soon, I was studying every morning from five to seven. It became my special time.

My grades rose from failing to honors and I began to feel better about myself. Especially since I was taking two worries off my parents' shoulders—the morning heat and *me*.

I discovered I had something to offer. And I had to take one small step to help out my parents, before I could find a way to help myself.

I'm learning, Lord, that even a small step toward You leads me to new discoveries in myself. —OSCAR GREENE

19 WEDNESDAY

...He that shall humble himself shall be exalted.

—MATTHEW 23:12

"It might be a good idea if, like the Chinese, we gave our years names," a friend suggested. "We should call 1986 'The Year of the *Towel*.' We're living in an era when people are continually asking, 'Who's got the authority? Who's got the power?' Why don't we ask, 'Where's the *towel*?'"

He went on to explain that he was referring to John's account of the Feast of Passover, when Jesus was having His last supper with His disciples. John tells us: "He riseth from supper, and laid aside his garments; and took a towel, and girded himself. After that he poureth water into a basin, and began to wash the disciples' feet, and to wipe them with the towel wherewith he was girded" (John 13:4, 5). Then he said unto them, "Ye call me Master and Lord: and ye say well; for so I am. If I then, your Lord and Master, have washed your feet; ye also ought to wash one another's feet" (John 13:13, 14).

After the Resurrection, a strange and glorious thing happened. The disciples no longer asked, "Where is my power?" Remembering that Jesus had said, "Whosoever of you will be the chiefest, shall be servant of all" (Mark 10:44), their concern became serving others.

Wouldn't the Year of the Towel be a good way to encourage us to think of how we can serve others, rather than who will serve us?

Lord, help me to see the needs of others and to serve with Christ-like love all through the year. —ZONA B. DAVIS

20 THURSDAY

If ye have faith as a grain of mustard seed... nothing shall be impossible unto you. —*MATTHEW 17:20*

Every few years, as they reach the right age, it is my delight to pass on to one of my grandchildren the story of how the wild mustard comes to cover our native California with a golden carpet each year. This spring it is Robin, about-to-be-five. We are sitting in a walnut orchard surrounded by the bright flowers.

"Once upon a time," I say, for thus all good stories begin, "only a few Indian tribes lived in all our state. Then the King of Spain sent a handful of soldiers and missionaries up from Mexico to explore. They got hungry and thirsty, the Indians were mischievous, but what really frightened them was that they might get lost and never find their way home.

"Then Father Junipero Serra had a wonderful idea. All along the way he planted wild mustard seeds. 'With the spring rains,' he said, 'this will become a glowing trail for us to follow.' He called his flower trail El Camino Real, 'the King's Highway.' The relieved soldiers thought he meant the King of Spain, but Father Serra meant the King of Kings, Lord Jesus."

Robin and I recall what Jesus promised us if we had a tiny mustard seed of faith, and Father Serra had a whole bagful. So he converted many Indians to Christ, cultivated the land and built missions up and down our coast. His Franciscan friars trod the mustard seed trail until *El Camino Real* became a real highway that we can drive on today.

"Let's tell Hilary," whispers Robin. She gathers an armload of delicate yellow flowers and runs to her about-to-be-four sister. "Here, Hil," says Robin. "This is a magic bunch. These are faith flowers. You can always follow them home."

Lord, give me grace to follow Your Highway until I am home with Thee. —ELAINE ST. JOHNS

21 **FRIDAY**
He first findeth his own brother Simon, and saith unto him, We have found the Messiah... —*JOHN 1:41*

Bob and I sat enthralled as a high school basketball team battled for the state championship. Three players seemed able to shoot the ball into the basket from anywhere on the court. Only gradually did I begin to notice Number Eleven.

Number Eleven did not make a basket the entire game. But when someone else made a basket, it was usually because Number Eleven had captured the ball and passed it to him. Number Eleven was not a recognized hero, but he was the best back-up player I've ever seen.

Andrew was a good back-up player, too. Andrew brought his brother Simon Peter to Jesus. Andrew found loaves and fishes for Jesus to multiply. Andrew was never a hero, but he gave heroes the back-up support they needed.

It's exciting to be a hero or heroine, to be the one everybody admires. But there are times when the best way I can help get God's work done is to be in the shadows, as a support to somebody else. When I am not in the spotlight, I'm going to think about Andrew and Number Eleven. And if just passing the ball is the quickest way to get to one of God's goals, then back-up I'll proudly be.

Dear Lord, help me to serve as well in the background as I do in the limelight. Amen. —PATRICIA HOUCK SPRINKLE

22 **SATURDAY**
...The kingdom of God is within you. —*LUKE 17:21*

Last Easter my husband and I hid eggs for the grandchildren, nesting them both in the house and in the yard. The youngsters ran all over the place, excitedly filling their baskets. But when it was over, and we assured them there were still a few left, they were baffled.

With renewed frenzy, they raced about, looking behind curtains, under bushes, high up on shelves. "You're hot! You're hot!" we kept encouraging, whenever they drew near us; or, when they got farther away, "You're cold! You're cold!" Because we had hidden those last few eggs in our pockets, they didn't know the prizes were right there on us, close enough to touch.

This experience reminded me of the words of Jesus: "The kingdom of heaven is within you." We can ask questions about the meaning of eternal life, but what a wonderful place God has chosen to hide the answer. He doesn't put it under that cup on the table, that plant on the windowsill. He didn't put it even in books or sermons or science. God has hidden it right here, in your own soul.

Look *within*—it's close enough to touch.

How wonderful, Lord, that through the power of Your indwelling Spirit, we can know all the treasures of Your kingdom.

—MARJORIE HOLMES

ASKING THE QUESTIONS OF HOLY WEEK: *A 2,000-Year Odyssey in Faith*

HOLY WEEK is an extraordinary time, one like no other in all of history. And yet the events that marked this week touched most deeply not the wealthy nor the powerful, but the lives of very ordinary people: a crippled widow, a young child, a merchant, a fisherman....

Theirs are the hesitant footsteps in which we will walk in the next week, as we ask their questions of the tumultuous happenings occurring in their midst, seek to share their response to the confusing welter of opinion over the One proclaimed by some to be the Messiah, and try to understand their perplexities as they witness the powerful forces of conflict abroad in their land.

And in the man they knew as Jesus of Nazareth, perhaps we too will be touched as they were, moved by His words, challenged by His deeds, our hearts inspired with reverence and awe. Come, let us to-

gether wend our way through this history-making week that left the world forever changed. —*THE EDITORS*

23 ***A QUESTION OF BELIEF*—PALM SUNDAY**

If I have told you earthly things, and ye believe not, how shall ye believe, if I tell you of heavenly things? —*JOHN 3:12*

As parades went, it was paltry at best. There was no brass band to signal the arrival of a dignitary, no brightly waving flags or banners, no quick-stepping marchers. In fact, it was difficult to figure out just exactly what was going on.

"Father," one boy said, puzzled by the commotion on the road where he and his friends had been playing, "if this man is so important, why is he riding a donkey?" And he pointed at Jesus.

David's father shook his head. "Some say he is the Messiah."

"The Messiah?" David asked, "Is he?"

His father continued watching the motley parade but said nothing. Some had begun shouting "Hosanna!" and throwing their coats and palm branches in the path of the oncoming throng. But for whom? Would God really send as the Messiah such an ordinary-looking man as this? And on a donkey?

"Is he?" David asked again.

"Is he what?" the father replied.

"What they say he is, the Messiah. He isn't, is he?"

David's father paused again. How he wanted to believe that this Jesus was the Messiah; how he had prayed and hoped for the coming of the Redeemer. But the Messiah was supposed to make everything right again, and yet life was still so hard.

"David, my son," he finally said, "you must decide for yourself." He knew of no other words to give the lad. "You have seen him. Do you believe he can do for you what he says?"

Do you believe? *The question ordinary people asked themselves two thousand years ago faces us just as squarely today as we enter Holy Week. In a world so fragile, where pain and heartache are all too real, can you shout "Hos-*

anna!" on this day? Can you believe that this ordinary-looking man can do for us what he says he can do?

Father, yes, how I long to believe. Help Thou my unbelief.

—JEFF JAPINGA

24 *A QUESTION OF COMMITMENT*—MONDAY

Many therefore of his disciples, when they had heard this, said, This is a hard saying; who can hear it? —*JOHN 6:60*

He looked them square in the eyes, his own flashing with a kind of anger those merchants had never before seen.

"This is a place of prayer," Jesus thundered, sweeping his hand in a wide arc as he stood inside of the temple. "You've made it a den of thieves!" Suddenly he turned to his right and threw over a table. There was a wild scramble as caged doves and coins went flying in all directions. Another table went over, then another. The entire selling area was in chaos: people lunging for clattering coins; doves flapping away from broken cages; temple officials racing to restore order.

Then, just as quickly, it was over. Their wares gone, the disconsolate merchants began to gather up what was left before starting home. "I'm ruined," moaned an old man, clutching a bag of coins considerably lighter than when he had arrived earlier in the day. "That man, he turned not only my table upside down, but my whole life."

"He's turned a lot of lives upside down," another told him. "My sister says he is the Christ, the Messiah. She has heard him preach. She said he even told that rich young fellow in Judea to sell all he had, give the money to the poor and follow him. Imagine—demanding a person give up everything he has!"

"What kind of life would that be?" the old man muttered, picking up a few more broken pieces of cage.

In the distance they heard children shouting "Hosanna to the Son of David!" A man was running through the temple shouting that he was healed. Others talked excitedly of the joy that this man called the Christ had brought them.

"What kind of life would that be?" the old man repeated slowly.

And gathering his last coins, he shuffled out of the temple where Jesus was teaching.

> **What kind of life would that be?** *Christ has always asked much from His followers; alas, sometimes not only their livelihoods, but their very lives. And if we are truly to commit ourselves to Him today, we must be just as ready to respond to His call. Is this the kind of life you want? Are you prepared to live it?*

Lord Jesus, sometimes it is difficult to be a loyal follower. I pray for the courage that will earn me the right to be called Your disciple.

—JEFF JAPINGA

25 *A QUESTION OF PRAYER*—TUESDAY

If ye shall ask any thing in my name, I will do it.

—*JOHN 14:14*

She stared intently at the crowd gathered in the corner of the town square. Had this man Jesus really said all those things her daughter had reported, she wondered. Or were the girl's words simply the product of a youthful imagination?

She pushed through the throng, trying to ignore the cruel taunts she always heard about how great her sin must have been to have brought her such a crippling illness. When she could go no farther, she cupped her hand behind her ear. Yes, it *was* he, and he was talking about prayer.

"You may ask for anything in my name," Jesus was saying, "and I will do it."

Her heart leaped...for this was just what her daughter had reported that he had said to his disciples. And now, here, right out in the open of the town square—surely it was a statement meant for all.

"You may ask for anything," she repeated to herself as she hobbled home, a fierce determination burning in her heart. Once inside, she went straight to her room and bolted the door behind her.

"Father, I have heard Your messenger. I beg You, heal my lameness so I may better serve You," she prayed, her eyes tightly shut.

But when she opened them, nothing had changed. God had not heard her prayer; God had never heard her prayer. Perhaps she *had* sinned too grievously.

Only the urgent pounding at the door forced her to cease crying. "Mother," a voice said, "you must come. Two men are here—they want to reserve our upper room."

The woman nodded. All week followers of this man Jesus had come to her house looking for shelter and protection. And she had given it to them, knowing that the authorities would not bother the house of one they considered a sinner.

"I will come," she said, "but let me alone for just one moment." And when the footsteps retreated, she prayed again. "God," she said, "I did not ask before *in Jesus' name*. Now that You have sent him to us—this one called the Christ—please, God, be patient with me while I learn to know him, to trust his words and accept his love. In the meanwhile, I will continue to serve You just as I am."

Then the woman stood up, threw back her shoulders, raised her chin and walked as straight as she could. She was being called.

> **Am I being called?** *As we move closer to our Lord Jesus in this Holy Week, as our insight into His eternal message of love and hope is deepened, what would you ask of Him? And what do you think He would ask of You?*

My prayer, O Jesus? To know my Father's will for me. —JEFF JAPINGA

26 *A QUESTION OF SERVICE*—WEDNESDAY

Verily, verily, I say unto you, The servant is not greater than his lord; neither he that is sent greater than he that sent him. —*JOHN 13:16*

"But, Papa, I tell you, I saw it with my own eyes," she wailed, pounding her small fist on the tabletop.

"Nonsense, my child," the old man said. "No rabbi would wash his guests' feet. That is servants' work!"

"Yes, Papa, that is exactly what Jesus said, that the Son of Man must be a servant. One of his followers, Peter, tried to make him stop. And Jesus told him, 'Unless I wash you, you have no part with me.' Jesus said he was our example; now we must all be servants one to another."

"He is mad, I tell you. No Messiah of ours would ask us to do such a thing. Can you imagine what it would be like if we did?" And with that, the father stormed out of the house.

For hours she waited for him to return. Finally there was a knock at the door. She rushed to answer it, only to see standing there a man she did not know.

"Will you take these?" he said, holding out two loaves of fresh, warm bread. "It is very little, I know, but it is the only way I could think of to thank your father for what he did for me and my family."

"My father?" the girl questioned.

"Yes, has he not told you?" the man replied. "I owed him some money. But tonight he came to my house and said that he considered my debt paid in full. And that he would help me find a job so I and my family might live more comfortably."

"Did he say why?" the girl asked.

"When I told him that I didn't deserve such treatment, he said, 'Tonight I met our Messiah, and neither did I deserve the treatment he gave me. He has shown me a whole new life. Now I want to try to live as he does and learn to follow his example.'"

> **Can I be a servant?** *In His words proclaiming that the first shall be last, Jesus calls upon us to revolutionize the world, to serve one another in humility. And it was in Christ's greatest act of servanthood—allowing Himself to be crucified for our sake—that He blessed us with the freedom to follow Him and gave us the incomparable example of love to lead us through all our days.*

You have shown me the greatest Love there is, O Lord. What now may I lay down for my brothers and sisters? —JEFF JAPINGA

27 *A QUESTION OF PEACE*—MAUNDY THURSDAY

Blessed are the peacemakers: for they shall be called the children of God. —*MATTHEW 5:9*

Smoke billowed from his torch, casting an eerie pall over the quiet, moonlit night. Though a mere servant of the high priest, he thought it a strange group of people he was escorting down that narrow path: members of the ruling religious body and a band of guards from the emperor's garrison. Even a slave knew how antagonistic the relationship between the two factions had always been.

Suddenly they broke into a clearing, and the young slave gasped. Now it all made sense: the swords and clubs the guards carried; the top-secret meetings that had been going on all week; the odd way in which two opposing groups had joined in working together. They were after Jesus, and this was the occasion of the long-threatened action.

Jesus stood to the side surrounded by men the boy knew were his followers. In the harsh shadows, this Jesus looked tired, his face crossed with lines of weariness. *Why doesn't he run?* the boy wondered. *He must know what is happening.*

Judas made the first move. There was some quiet talk; then the voices turned angry. Swords were raised. In an instant the lad saw a blade flashing toward him. It struck him in the ear; there was blood and searing pain.

"No more of this!" Jesus shouted. And taking the boy's shorn ear, he lifted it back in place. The pain and blood were gone. The boy stared incredulously at this man—whom he thought evil—and wondered how it could be. And how, amidst an arrest the boy knew could mean death, could Jesus show such peace and serenity? What made this man so different from all others?

> **Can I bear peace?** *On this Holy Thursday we are being called to show to the world the difference that was Jesus. To be free for the business of God—peace, justice, mercy and love—calls on us to rid ourselves of the strife and warring in the world. Today let us, as Christ did, fall to our knees and pray for peace in the hearts of all people.*

Christ, this night, in the midst of anguish, You sought the peace of prayer. Tonight, in the midst of the anguish yet in the world, I also pray. . . .

—JEFF JAPINGA

28 *A QUESTION OF LIVING*—GOOD FRIDAY

If we suffer, we shall also reign with him: if we deny him, he also will deny us. —*II TIMOTHY 2:12*

For nearly three hours darkness had covered the land. It was not just from the heavy clouds of the storm, but an uncanny midday black. Ominous rumblings had been rippling across the hill all afternoon; many of the bystanders had fled because of the tremors. Nervously the centurion shifted his weight from one foot to the other and back again.

"Something isn't right; it isn't right at all," he muttered, looking about at the three crosses on which he had been instructed to crucify three criminals. As he looked at one of the accused—Jesus of Nazareth—their eyes met. Quickly, the centurion turned away. *What is it about this man?* he thought. *What is it?*

He had heard stories about this Jesus, but that was a matter solely for the Jews, certainly not for an ordinary Roman centurion like himself. Sure, he'd nailed that mocking plaque above the man's head—"This is Jesus, King of the Jews." But there had been more self-proclaimed Jewish kings than he could count; many of them had even been crucified on this very hill. So what was it about this one? Why did he feel so strangely drawn to him?

"I am thirsty," Jesus suddenly said.

"Let's give him vinegar," snarled one soldier. The centurion wanted to stop them, but he held his tongue. What would his friends think?

Jesus drank a bit, then whispered, "It is finished."

Suddenly the ground shook with a mighty heaving, and everyone ran off in panic.

Except for the centurion. For it was all coming back to him now. This was the same Jesus who, according to the rumors, had healed the dying daughter of another centurion, in Capernaum. Yes, yes, now he recalled how people had speculated over why this so-called Jewish Messiah had done this for a Roman. Suddenly he knew; he knew why.

"Truly," the centurion said to himself, "this man was the Son of God." And he lowered his head in mute sorrow at what they had done.

What can I do for Him? *This Good Friday we too must stand at the foot of the cross. We have heard the claims of that man nailed there; we have seen His works in the lives of those around us. But do we really realize Who He is, what He had done for us? And what we have done to Him in return?*

Father, in both word and deed I have too often denied Christ. Forgive me and show me how my life can reflect You. —JEFF JAPINGA

29 *A QUESTION OF HOPE*—SATURDAY

I will not leave you comfortless: I will come to you.

—*JOHN 14:18*

The two of them had been hiding in the back room of their house for more than a day now, the door bolted, the lamp unlit. He did not even go out to the synagogue to participate in Sabbath Day services. And he would not go out this night either. He did not dare. Why, oh why, hadn't he just minded his own business? What had he done by getting mixed up with this Jesus?

"But we're just ordinary people," he heard his wife saying. "How would they know? Come, let us go out and see what is happening."

"No," he said, his voice adamant. "They'll recognize us. They'll know we were with him. If they could kill him, they could kill us also."

And so they sat, in silence and fear. They had come to Jerusalem for the Passover, thinking this might be the time for Jesus to take power, the time for the Messiah to liberate the Jews, as all had been prophesied. They had believed in him, taken him at his word and cast their lot with him. And now he was dead. And they themseves were wanted.

"How could he do this to us?" the man complained bitterly. "How could he leave us alone, with all of this trouble to face by ourselves? How could he?"

Has He left me alone? *Do you ever feel, in the midst of deep trial, in the throes of paralyzing fear, that Christ has abandoned you? Then take hope! For though it may now seem the darkest, the dawn of a new day lies just ahead.*

Behold, my fellow pilgrim, see the stone—it is rolled away!

In the dark depths of loneliness and waiting, Lord, give me the courage to look toward the dawn. —JEFF JAPINGA

30 ***THE SPIRIT ANSWERETH*—EASTER SUNDAY**

And the Word was made flesh, and dwelt among us, (and we beheld his glory, as of the only begotten of the Father,) full of grace and truth. —*JOHN 1:14*

Their voices were hushed, a quiet muddle of wonder and excitement, of fear and disbelief. What were they to do now? The body was gone—Peter had seen the grave clothes, and the women said an angel had told them that Christ was risen. Could it be...what he had told them? Or was this just a cruel hoax?

Deep in their own private worlds, the eleven wrestled with their thoughts, only occasionally speaking. "Peter, are you sure?" Andrew pleaded. "I want to believe, oh, how I want to believe. But *alive*?"

There was talk about hanging it all up, talk about going back to the fishing boats in Galilee and returning to their former lives. In their misery, no one even noticed that there was someone else in the room with them. That is, until he spoke: "Peace be with you."

And from the back of the room there came an excited cry of exultation: "HE IS RISEN!"

Only three words—and yet, each who heard them knew that at that very moment the world as they knew it had ceased to be. For brought into being for all time forward was the gift of hope, the promise of eternal life and the blessing of God's absolute love. HE IS RISEN!...and with those words, the spirit of the disciples lifted and they rose and moved into the world as one, bearing the glorious tidings of our Lord's Resurrection.

Our Lord, alive? *This week we have been on a pilgrimage like no other one, sometimes causing us confusion and doubt, sometimes filling us with joy and affirmation. And now, at the end of our journey, as we stand here before an empty tomb, we too can cry with thankfulness and gladness, "HE IS RISEN!" And we know that our lives can never* *be the same either for, yes, that humble man—that Jesus of Nazareth—has touched us as He touched those two thousand years ago.*

HE IS RISEN! Three ordinary little words... bearing the most extraordinary message in all history.

Lord, You live! Live in me... Live in me! —JEFF JAPINGA

31 **MONDAY**

Ye... justify yourselves before men; but God knoweth your hearts... —*LUKE 16:15*

My stepmother-in-law Linda is an honest, forthright person, with the most generous impulses and best intentions. I liked her from the first day we met. The wedding breakfast she put on for Carol and me was superb: fresh-buttered rolls, smoked salmon, fresh orange juice and huge bouquets of flowers. It was her expression of love.

After our honeymoon, Carol and I, preoccupied with newlywed housekeeping, let several weeks slip by before we called to thank Linda. When I finally did phone, there was an unusual silence from her end of the line. "Why didn't you call sooner?" she finally asked. "Haven't you neglected your family?"

I fumbled around for words, trying to justify my behavior. I felt her treatment unfair. After I hung up, I said to Carol, "She was so cool to me. She would hardly listen to my side of the story. We were, after all, still getting settled."

For days I was upset. Even my parents heard about it long distance. "But Rick, dear," Mom said gently. "Linda was hurt. That's all."

"*Oh*," I said sheepishly, but with sudden understanding.

In my attempt at self-justification, I had forgotten about Linda's feelings. And I'd forgotten what Christ showed us again and again—that fairness doesn't count where love is called for. I needed to speak with my heart, not my head. With that in mind, I sat right down to write Linda an apology.

Lord, when stubborn pride would like to reign,
Let humility fill my heart's domain.

—RICK HAMLIN

Praise Diary for March

1

2

3

4

5

6

7

8

9

10

11

12

13

14

15

16

17

18

19

20

21

22

23

24

25

26

27

28

29

30

31

April

S	M	T	W	T	F	S
		1	2	3	4	5
6	7	8	9	10	11	12
13	14	15	16	17	18	19
20	21	22	23	24	25	26
27	28	29	30			

The Bread of Life

Our souls are hungry, Lord,
for a peace that satisfies,
Our hearts are thirsty, Lord,
a need within us deeply lies,
We have offered You our wills,
but not our dedication,
We have offered You our lives,
but not with consecration,
Our sin is our half-heartedness,
of withholding just a slice,
Of keeping life raft at the dock,
demurring total sacrifice,
Teach us, Lord, that each must choose
at some appointed Calvary,
Else we will never know the joy
that comes from giving all to Thee.

I have food to eat
that you know not of,
Come to Me and hunger never,

I have wine to drink
that is from above,
Come, be thirst-quenched ever.

—*FRED BAUER*

1 Practicing His Presence in April

TAKING A WALK WITH GOD

...Walk humbly with thy God. —*MICAH 6:8*

Our eight-year-old spaniel, Captain Marvel, was overweight and in need of exercise. My nine-year-old daughter Ann dug out the dog leash. "Mama, let's take Captain for a walk," she said.

I didn't usually take walks, but at her urging I agreed.

The three of us struck out along our little neighborhood street. "Look, a squirrel!" squealed Ann. "Oh, Mama, see how yellow those flowers are!"..."Why is there moss on trees?"..."That cloud looks like a rhinoceros." She constantly drew my attention to the things around us. Things I knew were there, but never looked at.

Cloud, sun, tree bark, grasshoppers, the smell of jonquils, the green in the grass. Something began to well up in me—a sense of Presence of the One Who had made all this. The line of an old hymn flickered through my head. "In the rustling grass, I hear Him pass."

We walked on. I began trying to discover God through the ordinary things along the path, for I was finding that the sights and sounds in God's creation could be signposts pointing to Him, like voices speaking of His mystery and beauty. And it was not like trudging through a familiar old neighborhood at all. It was like strolling into a brand new awareness of God.

This month, try taking a "daily walk with God" in which you take time to practice His Presence by seeing Him in the small and splendid things around you. Why, it could even spill over into your walk through the rest of the year!

Dear God, let me walk with You and seek You in both the ordinary and extraordinary that dot my path. —SUE MONK KIDD

2 **WEDNESDAY**

Are not two sparrows sold for a farthing? and one of them shall not fall on the ground without your Father.

—*MATTHEW 10:29*

It was a mystery how the newborn squirrel had gotten in our backyard. There were no tree nests in the area and it was certainly too young to have come there by itself. My wife Shirley helped me feed it milk from a formula we had used previously on two baby raccoons...and then we wondered, what should we do with it?

A check with the county wildlife office revealed that the nature center no longer cared for orphaned animals. "But some inmates over at the prison care for them. Perhaps they will help you," we were told. An hour later we handed over our squealing baby squirrel to a soft-spoken young man named Jim. He accepted our present in the prison lobby and then invited us to see the animals he was tending. They included some more squirrels, birds, a skunk, rabbits, several raccoons and a beautiful wing-damaged hawk.

Jim told us how he had served as "pet nurse" for nearly a year, helping dozens of animals before they were released. Soon he would be released himself, and would turn his job over to another inmate. "I don't know how I would have ever stood the confinement without these animals," he said, tenderly stroking a baby rabbit. "They've given so much to me."

Driving home, I thought about the discovery Jim had made behind bars...that the more we give, the more we receive. To learn that truth had cost him a year of his life. But then some of us never learn it in a lifetime.

Remind us daily, Lord, that You came to serve, not to be served, and help us to find ways to do likewise. —FRED BAUER

3 **THURSDAY**

I am the way... —*JOHN 14:6*

I have just finished driving 110 miles through dense fog, over unfamiliar roads, at night. I'd taken my son John to southwest Nebraska to get his car. When we started home, the sun was down and fog had settled over the plains like a fat white hen, roosting. As I fretted, John said, "It'll be all right, Mom. Just follow my taillights." When we pulled onto the road, I could feel the muscles at the back of my neck tighten. But as I focused on John's taillights, I began to relax. He had traveled this road many times and knew it well, even the bumpy detour. I knew I could trust him to get us safely home.

There's a serious problem in my life, and right now I can see no way out of the dilemma. But just as John's taillights led me through the night, I think to myself, I don't have to be able to see the road ahead at every turn. All I have to do is to keep my eyes on God's Son and let His light guide me. He knows the way. He'll lead me safely through this darkness.

When my future is uncertain, Lord,
I'll stop squinting to where I cannot see
And focus on Your Son who's near to me.

—MARILYN MORGAN HELLEBERG

FRIDAY

All flesh is grass, and all the goodliness thereof is as the flower of the field... —*ISAIAH 40:6*

When my husband Jerry and I were building our first house, that sea of barren red clay surrounding it worried me. No grass, no shrubs, no flowers. Would anything ever grow out there? One day Jerry came out with a big bag and began sprinkling seed around. *As simple as that?* I wondered. "Will grass really come up from those seeds?" I asked him. "Of course. You'll see," he replied serenely.

Every morning I went to the window to see if we had any grass. "It's not coming up, Jerry," I would say, but he would only mumble something about giving it more time, turn over and go back to sleep. Then there came the morning when I looked out to find that some magnificent artist had crept in during the night and gently tinted our red clay with a delicate green brush. Drops of dew clung to the incredibly tiny blades, glistening in the freshness of dawn. *We had grass!*

"Jerry!" I screamed. He sat up straight in bed. "The grass is up!" He stared at me in stunned disbelief and went back to sleep. I flew outdoors and ran my hand over the fragile fuzz, joyful and amazed.

Years later I thought I saw a parallel. When God brings us into life on His earth, He watches and waits to see how we will grow. Will we dawdle, some of us never meeting His expectations, or will we steadily deepen our roots according to His plan? Wouldn't it be nice if we all worked together to make His world just as beautiful as He intended and hoped it would be?

Father, let me do Your work by sowing seeds of love and kindness in the world around me. Amen. —MARION BOND WEST

5 **SATURDAY**

The Lord... bringeth low, and lifteth up... for the pillars of the earth are the Lord's, and he hath set the world upon them. —*I SAMUEL 2:7, 8*

My husband sometimes brings home a patient's electrocardiogram to study. He was doing so one morning while I cleaned up the kitchen. We had just finished our after-breakfast reading from the Bible.

But one thing disturbed me: the assurance that anyone who is truly close to God will always be at peace. "If that's true I'm failing," I protested. "You know how happy I usually am, but how often things upset me. Being criticized. Interruptions when I'm writing. Even right now, feeling troubled about a friend. If only life could be smooth and simple. If we just didn't have these peaks and valleys!"

In reply, my husband beckoned me to examine his electrocardiogram. "Look at this," he said. "This EKG monitors one of the body's most important organs, the heart. It has the biggest job of all; it beats a hundred thousand times a day. It expands and contracts. When the auricle contracts it makes a peak on the chart, then there is a line, then a larger peak that comes down, then there is another line.

"These elevations and pulsations fluctuate," he went on. "But the heart itself has to have these peaks and valleys; it couldn't *run* in a straight line." He paused to let this sink in.

"This is a fact of life. Life *can't* be a regular, unbroken line. These peaks and valleys don't prove you're *not* in tune with the Creator—

they prove you're *alive.* Why, without them, you'd have no life at all!"

Father, grant me the patience to see the value of the valleys, as well as the joys of the mountaintops. —MARJORIE HOLMES

6 SUNDAY

Thou art my hiding place; thou shalt preserve me from trouble; thou shalt compass me about with songs of deliverance. —*PSALM 32:7*

My eighty-two-year-old mother is known in the Ozark town where she lives as a good "pray-er." Friends and family members, including me, often call her with prayer requests. Sometimes her prayer list grows to an overwhelming length. But she's found a solution. She picks a hymn that pertains to a particular person's need and, as she goes about her daily work, she sings the hymn, inserting the name of that person into the words. For instance, when I asked her to pray for her granddaughter Meghan who was discouraged at college one semester, my mother chose the religious folk song *Kum Ba Yah*:

> Here is Meghan, Lord. Be with her.
> Meghan's troubled, Lord. Stay with her.
> Meghan's crying, Lord. Comfort her.
> Oh, Lord, stay with her.

Now I've started singing prayer hymns, too, especially when I can't find the right words for someone's problem. If you have trouble praying for someone because you don't know what to say, why not sing it? Besides, there's an added benefit: even if your voice isn't on key, prayer hymns can help you stay tuned in to God.

Try it with me now....

I am singing, Lord, kum ba yah. —MADGE HARRAH

7 **MONDAY**

But they that wait upon the Lord shall renew their strength; they shall mount up with wings as eagles...

—ISAIAH 40:31

The bald eagle is our national emblem. You can see it on U.S. half dollars and quarters; it also appears on the great seal of the United States and many documents. Bald eagles aren't really bald, but get their appearance of baldness from the white feathers on their heads. They have wingspreads of as much as eight feet, and when they soar through the skies, nothing could be more regal.

But flying isn't as easy for them as it looks. They have a problem getting themselves launched. The secret of becoming airborne is that they perch on the edges of high cliffs and wait for the right wind to come along. When it does, they let go and mount up, soaring with the current, rising higher and higher as they bank. Their uncanny understanding of air currents enables them to get maximum mileage from the wind.

Just as eagles are borne aloft by waiting for the right wind current, Christians must wait to catch the direction of God's spirit, and then let go and soar according to His will.

Lord, may we be so sensitive to Your direction that we, too, will be able to soar on the spiritual currents You send our way.

—SAM JUSTICE

TUESDAY

Ye are our epistle written in our hearts... written not with ink, but with the Spirit of the living God; not in tables of stone, but in fleshy tables of the heart.

—II CORINTHIANS 3:2, 3

In China's Forbidden City, there is a single slab of marble that weighs more than two hundred tons. Our guide told us that thousands of laborers had pulled the block many miles from its quarry site during the winter so that it would slide over the icy ground. I was fascinated by the snakelike dragons, symbols of power, which were chiselled into the face of the marble. For five hundred years, only the Ming emperors were allowed to walk upon these legendary creatures, used as a ramp to one of their temples.

Everywhere pavilions were surrounded by ornate fences with serpentine posts, all made of marble. The sights aboveground were so marvelous and exciting that sometimes I failed to glance down at the intricate sidewalk mosaics that had been so painstakingly and delicately wrought by artisans in centuries far gone.

I said to my husband, "Do you realize what a privilege this is? Here we are, thoughtlessly walking on a portion of antiquity created at a tremendous cost of human life and labor."

But then we remembered reading that similarly carved works of art abounded back in Biblical times. Doubtless St. Paul's listeners were well aware, as he spoke the words in the verses above, that working with human hearts is often far more difficult than trying to carve upon stone.

O Father, make my heart soft and pliable to Your touch. Mold me in Your way. —ISABEL CHAMP

WEDNESDAY

As the servants of Christ, doing the will of God from the heart; With good will doing service, as to the Lord, and not to men. —*EPHESIANS 6:6, 7*

There was no generation gap between my fifteen-year-old granddaughter Jessica and myself when it came to any rewards, compensation or gratitude coming our way. We believed in them.

Rewards—an A+ or a letter from my editor for a job well done. *Compensation*—appropriate pay for our work. *Gratitude*—a "thank you" for good deeds. When I didn't get them I suffered from hurt feelings, resentment, even—most unbecoming at my age—a tendency to sulk. So I sympathized with my granddaughter when she did likewise. Then, in her Bible class, Jessica studied Paul's letter to the Ephesians and demonstrated for me that it isn't *Christian* nature.

Jessica is a competent and highly paid baby-sitter, in great demand on weekends. When she came home one weekday from school, her mother relayed an emergency call from a neighbor who had wrenched her back and needed Jessica to feed and bed down her three obstreperous youngsters. Despite an impending exam, Jessica went. Three hours later, exhausted, she came home and opened her books.

"How much did she pay you?" demanded her practical young brother.

"She didn't," said Jessica. "She didn't even say 'Thank you.'" Then she added quietly, "But that's all right. I didn't do it for her. I did it for Jesus."

Lord, when I go unrewarded, uncompensated, unthanked, help me to hold to my Christian nature, and rejoice that I could do it for You. —ELAINE ST. JOHNS

THURSDAY

Now faith is the substance of things hoped for, the evidence of things not seen. —*HEBREWS 11:1*

It had been a season of prayers unanswered in my life. People who were sick were not getting better, people who asked for our help were sinking deeper into their problems. One morning I found myself crying out, "Are You really there, God? Can You give me one sign of hope?"

Throughout the gray, drizzly day I waited for my sign—and received none. Later, enroute to town with my son, we passed by Stone Mountain, an immense hump of granite rising out of the Georgia plain. Our family had climbed it last fall, and every time we see it little David says, "Look, Mama, there's the mountain we climbed." That afternoon, however, he exclaimed, "It's good we aren't going to climb Stone Mountain, Mama, because it isn't there."

"Of course it's there," I muttered automatically, then turned to look. It wasn't. The fog and mist had come down so low that not one bit of that huge mountain was visible.

I remembered suddenly the Bible verse above. The *substance* of Stone Mountain was temporarily hidden by mist, but I believed it was still there and I would see it again. My evidence? I had climbed it!

As I stared at the mist where I knew the mountain had to be, I knew again that God was really there, too. I've experienced His presence in my life time and time again. Until I received the substance I hoped for, I could walk by my evidence of things not seen.

Father, thank You for past times of closeness that can guide me when You seem far away. Amen. —PATRICIA HOUCK SPRINKLE

11 **FRIDAY**
I am come a light into the world... —*JOHN 12:46*

One night under a starry sky, I talked with my husband's grampa about Halley's comet—that dazzling heavenly body that appears about every seventy-six years. I remember because Halley's comet returned late last year and is most visible to the naked eye about now, astronomers say. Gramp was still young when it streaked into view in 1910. In May of that year the head of the comet was brilliant as it passed Earth. But it was the tail of the comet that Gramp remembered best.

"Imagine a train of light trailing behind the comet, lighting the sky for miles," he said. His eyes were fixed on the black, sparkling dome cupped over our heads and my eyes followed his, searching the vast curves of space. I seemed to recall that a comet's tail was actually the luminescence of tiny particles reflecting the sun's light.

We sat quietly. Then Gramp, who was very old, seemed to speak what was in his heart. "Life goes by awful fast," he said with a smile. "Near 'bout fast as a comet."

Today as I think about my own life, it does seem my days come and go quickly, sweeping by like a comet. But I think about the comet's tail and remind myself that each of us is able to leave a trail of light behind that touches the lives of others long after we have passed by. For every act of love, every gesture of gentleness, every gift of self becomes a particle of light that shines behind us.

God, may I live my life each day in such a way that it casts light across Your world. —SUE MONK KIDD

12 SATURDAY

Father, I thank thee that thou hast heard me. —JOHN 11:41

I'd gone on the weekend retreat with a question. The monastery was located at a beautiful spot on the banks of the Hudson River, eighty miles north of New York City. Somewhere in that prayerful place, perhaps during a chapel service, perhaps in the course of a talk by the retreat leader, answers to a special concern I had would come.

And one came, too, but not as I'd expected. It happened as I walked along the narrow shale beach at the edge of the river. There I watched a giant oil tanker make its way downriver, huge and silent, then growing smaller, leaving the water's broad surface as undisturbed as though no ship had passed. I strolled on downstream.

The noise came from behind me—a roar like an approaching freight train. I whirled around to see two-foot waves rushing across the water that seconds earlier had been so placid. Curling, cresting, crashing in white foam around my feet—where had they come from! What had suddenly roiled the water on this windless morning?

Of course: the tanker, now far away.

I had wet shoes as I started back up the path to the monastery, but I had an answer, too. Our efforts at prayer are not unavailing. Our pleas are registered in heaven and God's redeeming power is in motion. The effect has just not reached the surface of things where our land-bound eyes are watching.

Thank You, Father, for answers that in their time come cresting forth.

—ELIZABETH SHERRILL

13 SUNDAY

And they heard the voice of the Lord God walking in the garden in the cool of the day... —GENESIS 3:8

I've always liked the word 'saunter.' To saunter implies a state of cheerful relaxation. No tension. No hurry. No strain. No stress. To saunter is less purposeful than to stroll, but it's not as vague as to wander or to drift. It just means ambling along, enjoying your surroundings, going no place in particular, but aware of everything.

Where did this delightful word come from? My big Webster's dictionary says, "origin uncertain." But I remember once reading that Henry Thoreau fashioned it from two French words, *sainte terre*, meaning "holy land." Thoreau wrote, "The swiftest traveler is he that goes afoot," and felt that anyone who walks slowly and peacefully through the countryside, looking and listening for beauty, is really on a pilgrimage through the holy land of God's creation.

Sainte terre. Saunter. Why not take time out to be a saunterer today? The rewards may be greater—and more pleasurable—than you think.

Slow me down, Lord, as I saunter through Your beautiful world.

—ARTHUR GORDON

MONDAY

...Forgetting those things which are behind, and reaching forth unto those things which are before...

—PHILIPPIANS 3:13

Father built his home in 1929. It was never paid for and, as a child, our family's lack of money always frightened me.

Then, on an April evening in 1965, the terror returned. At the time, I was an engineering technician on the Gemini space program, and my company expected to get the Apollo contract. Our son was an honor student at a Massachusetts college. Our apartment was comfortable, and best of all, we were debt-free!

One evening a real estate agent called to tell me that a house I had liked immensely had come on the market. But suddenly memories of my parents' ordeal silenced my enthusiasm. The very idea of assuming a mortgage filled me with fear...and my hands shook as I signed the purchase agreement. Days later, the bank loan placed me twelve thousand dollars in debt. I gasped. If my parents couldn't pay for their home—how could I expect to pay for all this?

Unexpectedly, the Gemini project terminated and the Apollo contract was awarded to a competitor. My income was reduced by half. My apprehension soared.

Then one day I chanced upon my 1937 diary. There I rediscovered my own handwritten words: "Take no thought...for what ye shall eat or what ye shall drink...you must never think of defeat!" Words, partly Scripture, I had recorded twenty-eight years earlier. Words that had guided me, a penniless lad, through college. *Why not use them again?*

I recited those words as I began keeping up my minimum payments. I established a firm budget and my bills were paid. Our son completed his education *and soon our home was paid for.* Now, in retrospect, I can see the folly of allowing boyhood fears to spoil adult opportunities.

Do you tend to let past fears control your life today? Take courage from one who knows.

Father, help me to remember that the past is gone and that each day brings a fresh chance to follow Your purpose. Amen.

—OSCAR GREENE

15 *WILDERNESS JOURNEY* (Exodus 15)

Trudging Through the Desert

And... the waters were made sweet. —*EXODUS 15:25*

I can't imagine what it would be like to go three whole days without water, can you? But that's what happened to the Israelites after they had crossed the Red Sea. Shall we join them as they trudge along in the desert of Shur on their way to the Promised Land?

There's dust in our nostrils and sand between our teeth. Our throats are raw and our lips are cracked. Our craving for water is driving us wild. *But look! Up ahead! It's a spring of water!* We begin to run toward it, bumping into one another, gasping, scooping up water in our hands and sucking it into our mouths. *Ach! Pttt! It's bitter! Bitter! We can't drink it!* In our frustration, we turn on Moses, shouting, screaming at him, blaming him. And he cries, "Help, Lord! Help!"

Poor Moses. You've tried so hard to follow God's leadings, but the people blame you for everything that goes wrong. I think I have some idea how you feel. A friend recently told me that my teenage son lacks self-discipline because I wasn't firm enough with him as he was growing up. It could be true, but oh, how that hurt! I've tried *so hard* to be a good mother. Things haven't been the same between my friend and me since she said that.

But what is Moses doing now? We watch as he finishes his prayer, then breaks off a branch of a nearby tree and throws it into the water. He drinks from the spring and holds out his cup to me. The water is pure and sweet now! Oh, could it be that my resentment toward my friend has contaminated the spring of Living Water the Lord has placed within me? I want to get rid of the bitter taste! I close my eyes now and stand next to Moses, allowing myself to feel the full sting of my friend's blaming words; then I cry to the Lord for help.

Moses breaks another branch from the tree and hands it to me. I hold back at first, not sure I'm ready to give up my grudge. Then letting go, I throw the branch into the cloudy, bitter water. Instantly it clears, and I drink from it. How sweet and cool and fresh it tastes! I'm free to love my friend again. In opening myself up to her again, I have been set free and delivered from the bondage of bitterness.

Resentments are hard to give up, Lord. Only You can sweeten the bitter waters within me. I'll reach for Your purifying branch—and then let go. —MARILYN MORGAN HELLEBERG

16 WEDNESDAY

Take heed therefore how ye hear... —*LUKE 8:18*

Jonathan and Elizabeth, two young friends of mine, accompanied my dog Kate and me into the woods one day for one of our favorite pastimes: tracking. The object is for me to go on ahead and get "lost"; then Kate leads Jonathan and Elizabeth to where I am by following my scent.

I had gone quite far and was hiding in a thick cluster of trees and brush. I stood very still, waiting for the sound of Kate's excited breathing. If you have ever been alone in the woods, then you must know it is anything but silent. I wondered how I would hear Kate with so many other sounds around me—birds high up in the trees, a bee buzzing near my head, small animals rustling in the bushes, a lively little stream gurgling nearby, the breeze playing in the leaves. I thought I heard Kate, far off, but I wasn't sure. So I listened harder, trying to catch that one familiar sound among all the others.

Then I heard her. No doubt about it. Kate was very close and soon she broke through the underbrush and came running to me.

I think of that experience now whenever I listen for God's voice in my life. Sometimes, when I can't hear Him, it's because all the other sounds of life are getting in the way. That means I have to listen harder for that one familiar Voice. And it's always there.

Lord, teach me to recognize the sound of Your Voice in the midst of a noisy world. —PHYLLIS HOBE

17 **THURSDAY**
A sower went forth to sow... —*MATTHEW 13:3*

It was hopeless. For years I had been gently trying to convince my beloved uncle that there is a God. But because he was so implacably opposed to the idea, I had to drop my little hints with exceeding care. I would casually say things like, "Isn't it wonderful God gave us such a beautiful morning?" or "I finally got a raise. God has answered my prayers!" My uncle would shake his head, mutter something under his breath about "God stuff!" and turn away, until I finally began to wonder if I shouldn't just give up and consider the matter closed between us.

Then one day I was taking him down from his mountain cabin to keep a doctor's appointment in the city. I drove carefully along the winding road, where the cliff loomed straight up on one side and the rocky canyon dropped far below us on the other. My uncle, then eighty years old, sat silent beside me, staring out at the magnificent scattering of huge old weather-worn boulders. Suddenly—and gruffly, as though pulling each word out—he spoke.

"You just may be right," he acknowledged. "It sure does seem that *Somebody* must have put all this here."

I felt a rush of warm gratitude flood through me and reached over to squeeze his big, gnarled hand. Hopeless? No. In God, nothing is ever hopeless.

Keep me tireless...my spirit ever hopeful...as I witness for You, Lord God. —DORIS HAASE

18 FRIDAY

...Grow in grace and in the knowledge of our Lord...

—II PETER 3:18

When Lynn and I got married five years ago, I made her a promise: When she finished her seminary work three years hence, she could attend the graduate school of her choice. Wherever it was in this country, that's where we'd go.

And so I encouraged Lynn to apply to the best schools, no matter where they were. She did, and was accepted by most. That's when I realized that keeping my promise meant I would have to turn down a good job here in New York so Lynn could accept a scholarship at her first-choice school in the South. I got cold feet. We stayed in New York, so I could take the job, and Lynn enrolled in graduate school—a good one, but not the one she'd wanted.

I'd broken my promise. For weeks that haunted me, until one day when a professor of Lynn's, who knew how I'd been agonizing, told me, "The two people who made that agreement no longer exist. You've changed, grown as individuals. You'd be cheating both yourselves and God if the way you dealt with life did not change, too."

Someone said that life's only certainties are death and taxes. But there's a third: *change*. It doesn't necessarily make our lives easier. With God's guidance, Lynn and I will continue to search for His Will in that change, even if it means we have to change ourselves.

Thank You, Father, for making it possible for me to change and grow in accordance with Your Will. —JEFF JAPINGA

19 **SATURDAY**

O come, let us worship and bow down: let us kneel before the Lord our maker. —PSALM 95:6

As we walked the dogs at sunset, we discovered that overnight the daffodils had bloomed! A lovely yellow spill of them poured down the hill, like a river of light. We gazed at them in wonder, for this same spot had been green only last night. Yet while we slept, the buds had opened, the satin petals opening into these yellow stars with their deep golden throats. We bent down to breathe their delicate fragrance. They smelled new and pure and all were exactly the same height.

"I wonder why these flowers don't keep right on growing?" I asked my husband. "Or get to be as tall as lilacs...or why lilacs don't keep right on growing as tall as trees?"

"Even the trees have to stop someplace," my husband said. He pointed to the horizon where oaks and elms and towering pines were silhouetted against the orange afterglow. "Trees don't just go on climbing forever into the sky. Each species of tree has a cut-off point; it gets just so tall and no more."

And so do people, we both realized. God designed the whole picture: trees, flowers, human beings. Yes, of course we vary. "Look how short you are, how tall I am," George went on. "But each of us reach our destined height." He grinned and gave me a little shake, knowing how often I've bemoaned my size. "We should no more envy anybody else than a daffodil should envy a tree!"

Thank You, Father, for creating me according to Your perfect specifications. I'm glad to be "just right"—just me. —MARJORIE HOLMES

20 **SUNDAY**

He healeth the broken in heart, and bindeth up their wounds. —PSALM 147:3

In an interview, Dr. Christiaan Barnard, the famous heart transplant surgeon, told of a patient who asked to see his old heart after the operation. "It was an unusual request," Dr. Barnard said, "but I told him I would arrange it."

Later, a glass jar containing the man's damaged heart was brought into the room. Dr. Barnard pointed out to him the areas that had caused his suffering and endangered his life.

With tears in his eyes, the patient said, "I am truly grateful that you took that old heart that was killing me and gave me a new one."

Is something troubling your heart today? Take it to the Great Physician. He will remove the problem and give you a clean, fresh start.

Father, here is my hurt. Transform me with Your healing love. Amen. —DRUE DUKE

21 MONDAY

For God hath not given us the spirit of fear; but of power, and of love, and of a sound mind. —*II TIMOTHY 1:7*

"Mom, you worry too much," my daughter Lindsay told me as we prepared to load her horse into our new trailer. She was right, of course, but today I had reason to worry. You see, I had never driven the horse trailer with horses in it before. And because my husband was out of town, *I* would have to drive Lindsay and a friend to a 4-H riding event. That meant taking not one, but *two* horses down the highway, through busy intersections, and even—God help me—around corners.

No wonder I was scared.

When the time came, we loaded the horses, climbed into the car and inched our way onto the highway. I prayed for God's comfort and strength. At the first stoplight, the horses kicked a bit. Around the first corner, they shifted their weight, making loud noises and rocking the trailer, just as I had feared. But we didn't tip over. Slowly, we moved through traffic, and slowly, I gained confidence. In fact, I was beginning to get the hang of it!

I suppose I secretly expected all the people in the dusty parking lot to cheer wildly when they saw us coming. Obviously they didn't. My reward instead was the satisfying realization that I'd stepped out in faith and conquered a fear. With God's help. That made me happy and maybe even a bit cocky as I pulled into a parking spot in a middle row. That's when a new panic gripped me.

How do I back this thing up?

Lord, let me remember I need You to conquer the things I fear the most. —CAROL KUYKENDALL

22 TUESDAY

...Be clothed with humility: for God resisteth the proud, and giveth grace to the humble. —*I PETER 5:5*

Many years ago a famous religious leader was scheduled to speak at a large convention. He had a reputation of being one of the world's most inspiring religious speakers. One of the sponsors of the convention, a man prominent in national affairs, offered to take him to dinner on his first night in town. "I'd be proud to be the host of a man like that," the dignitary said. "What a power he is!" And he reserved the best table in the finest restaurant, so everyone would be sure to see him in such illustrious company.

But when at last the dignitary met the religious leader, he was shocked to find that the speaker was a strange-looking, crooked little man, no more than five feet tall. Reading the disappointment in his host's face, the speaker smiled and said, "Isn't it wonderful what God can use?"

Truly great people are so because they have great souls. It doesn't matter to God what a person looks like; it is his or her inner greatness that counts. Whenever we feel self-doubt, we need to affirm that God, who gave this strange, little man his great ability, can also give us power and boundless blessings, whoever we are, whatever our limitations.

Lord, help me remember that the seeds of greatness are planted within everyone, and it is up to us to make them grow.

—ZONA B. DAVIS

23 WEDNESDAY

In all labor there is profit... —*PROVERBS 14:23*

The other day on a commuter train I overheard a fellow and a girl talking. They seemed to be at the getting-to-know-you stage, because she was telling him about herself and her theatrical aspirations. She had appeared in a few minor productions and was hoping for some better roles. "Are you performing right now?" he queried. She shook her head. "What are you doing?" he asked. The girl hesitated a moment. Then she said, almost sheepishly, "Oh, I'm just a secretary."

Is she ashamed of being a secretary? I wondered. She shouldn't be. Secretaries are important people. Where would thousands of harried executives be without their secretaries to organize their business lives?

I'm sure secretaries would have been useful in Biblical times, too. Think about it. Jesus probably could have used someone to keep track of His appointments and the many people who waited to talk with Him. And St. Paul could have used a secretary while he was dashing off all those letters to the Romans, the Corinthians, the Galatians, the Ephesians. Peter was fortunate. He did have a scribe—Mark—who recorded his remembrances.

"All true work is sacred," said Thomas Carlyle. If a saint can be a secretary, why not you?

Lord, whatever job You give me, help me to do it gladly for Your glory.

—ELEANOR SASS

24 THURSDAY

For ye were as sheep going astray; but are now returned unto the Shepherd and Bishop of your souls. —I PETER 2:25

It was only recently that I noted seven important words while studying my Bible: "Then were the days of unleavened bread" (Acts 12:3). It was another story of Passover and Peter.

As the prison in which Herod had put Peter (for preaching the Lord's word) grew dark and quiet for the night, did Peter ponder the recent death of James, brother of John, and wonder if he would be next to die? Or did he remember that other Passover night, that terrible evening when he followed Jesus to the high priest's palace—and denied Him? He may have recalled, too, those subsequent days of bitter self-hatred, the glorious Resurrection, Jesus' total forgiveness: "Do you love me, Peter? Feed my sheep" (John 21:15-17).

If so, he could now rejoice. On this Passover, too, he had been accused of following Jesus, and, this second time, he had accepted imprisonment rather than deny his Lord. Acts 12 shows how completely Jesus forgave Peter: He gave Peter another second chance.

Sometimes, when the going gets rough, I, too, deny Jesus by words or acts that could make those around me doubt that I know Him. These times leave me full of remorse and regret. But I don't have to stay in pain and bitterness. I can ask the risen Christ to forgive me and give me, like Peter, a second chance.

Dear Jesus, Thank You for forgiveness and second chances... and third ... and fourth... and.... —PATRICIA HOUCK SPRINKLE

25 FRIDAY

...That I might know the proof of you, whether ye be obedient in all things. —*II CORINTHIANS 2:9*

I quickly discovered after my husband died that raising two teen-aged boys alone was no picnic. The twins were fifteen when their father died. When they were nearly seventeen, rebellion openly surfaced.

"God, do something with my boys," I begged frequently.

Then we heard a minister speak about the time God told Moses to throw down the rod. The minister, because of a speech difficulty, substituted "stick" for "rod." Then he explained how some of us needed to throw down *sticks of disobedience* in our lives. He invited those wishing to do so to come forward. I waited for my rebellious sons to go.

You go, God seemed to say to me. *You have several sticks in your own life.* My boys sat perfectly still while their mother went forward. It wasn't easy, but at that moment, I threw down some sticks of rebellion that I'd been harboring for a long time. *Rebellious children,* God seemed to make clear to me, *almost always have a rebellious parent.*

Practically overnight, my relationship with my sons began to change. We talked openly about some of their concerns, their hurts and needs. We laughed and cried together. What a powerful transformation had begun.

Perhaps you'd like to throw down a stick today...and experience the release of discord and the freedom of love.

Father, I long to know the blessings that always *follow obedience.*

—MARION BOND WEST

26 SATURDAY

Yea, brother, let me have joy of thee in the Lord...

—PHILEMON 20

I'm an occasional jogger. Three times a week I take a leisurely run on the track bordering the reservoir near our apartment—nothing too strenuous, just a little exercise.

The other morning as I was running around the lake, I approached another runner who seemed to be going at the same speed. Suddenly an unexpected competitive urge to pass him overcame me. But as I quickened my steps, he quickened his. And when I slowed down, he did too.

"You keep up a nice pace," I observed, panting from my unsuccessful effort. "You too," he replied good-naturedly.

With that, he pulled alongside me and we continued our jog— together. Our sneakers pounded in rhythm and our arms pumped in unison, like synchronized oil wells. I quickly forgot my fatigue and found myself running at double the pace—and surprisingly with twice as much pleasure.

"Thanks," my running partner said when it came time to part. "I go much faster running with another." His words almost made me stop in my tracks.

Isn't that the truth? When we cooperate, when we share our part of the track with another, we travel swifter, higher and faster. What's more, the trip becomes a pleasure.

So, you might ask yourself from time to time as I do: Where am I running today? Is there anyone who might want to come along? Is there someone *I* can join?

Father, help me turn my drive to compete into a golden opportunity to share.

—RICK HAMLIN

27 SUNDAY

But by the grace of God I am what I am: and his grace which was bestowed upon me was not in vain...

—I CORINTHIANS 15:10

Samuel Morse, inventor of the telegraph, once confessed to a friend,

"Whenever I could not see my way clearly, I knelt down and prayed to God for light and understanding. And when I was given acclaim for my invention, I knew I never deserved it. I had made a valuable contribution to mankind, not because I was so brilliant, but because God needed someone to convey this gift to the world, and He was good enough to choose me."

The first message ever sent over the telegraph was: "What hath God wrought!" When I read that, I thought of times I had preened myself with pride. "I did that," I would glow. Or, "I accomplished...I helped... I created...." Suddenly I knew that it was never *I*. They were all gifts from the Father.

Is there someone who needs help? He selects one of us to be there. Is there a job to be done? Carefully He prepares another to do it. If a wrong needs to be righted, He bestows awareness and courage in a human heart. We must give credit where credit is due, so let us bow our heads and together echo those words of Samuel Morse.

"What hath God wrought!"

Praise be to You, Lord God, and blessed be Your name.

—DORIS HAASE

28 MONDAY

...For as he thinketh in his heart, so is he... —*PROVERBS 23:7*

Do you ever wake up "ho-hum" bored, "poor-me" lonely or "cruel-world" unhappy? If you do, consider the little mood-lifter passed along to me by a friend.

"You aren't alone, you know," she reminded me when I expressed being less than happy with my lot. I thought she would go on to remind me that I have good friends, a loving family and to count my blessings. Instead, she surprised me and said something quite different:

"Meaning that you have a lot of company! We all have those downers. Know what I do when one threatens? I have a card tucked away in a drawer that reads: 'I Will Postpone All Negative Feelings for Twenty-Four Hours.' I pull that card out and prop it up in plain sight so that if I start to backslide, I am bound to be reminded to shift my thinking."

Well, I never did get around to actually printing those words on a card, but I prayed about the matter. "Please, Father," I said, "don't let me be negative today." And do you know, when I began to make an active effort to turn my thoughts around, I became increasingly aware of the many blessings in my life. And at the end of those twenty-four hours, I found I had gotten along so nicely that I might just as well coast through the next twenty-four hours in the same way.

Now if you are feeling down today, try letting go of all the negatives in your life. Use a card as my friend does if you want to, but either way, be sure to shift your thoughts into a positive channel...and see if something wonderful doesn't happen inside you!

Lord, let me think *as You want me to* be! —JUNE MASTERS BACHER

29 TUESDAY

Men ought always to pray... —*LUKE 18:1*

I know a doctor who suggests to his patients and their families that they go home and pray after they have been treated. It's part of his prescription for them. One day last week he told me that two different people objected to that suggestion. One, a father whose child had strep throat, said, "Why pray? That penicillin shot will cure him." The other was the husband of a terminally ill cancer patient, who said, "What good could prayer do at this point?"

"It never occurred to these people," said my doctor friend, "that prayer could be something more than begging God for favors, that it could be a way of tapping into undreamed-of potentials within themselves; a chance to bind themselves to Something greater than their bodies or individual egos; to glimpse the wonder and mystery and hope that is God. No matter how many 'miracle drugs' there are, no matter how hopeless an illness is, there will always be a need for the belief, the wonder, the faith, that expresses itself in prayer."

Great Physician, I will daily take my dosage of prayer, for healing of spirit as well as body. —MARILYN MORGAN HELLEBERG

30 WEDNESDAY

Let all bitterness, and wrath, and anger, and clamor, and evil speaking, be put away from you, with all malice.

—EPHESIANS 4:31

When our son Eric was in grade school, he was diagnosed as "learning disabled." Although his I.Q. tested high, he still had trouble reading and handling math skills. At that time few teachers had been trained to work with children like Eric, and so my husband and I spent the next several years trying to share with the public school system all that we were learning from psychologists and parent-support groups.

There were times, however, when I got so frustrated over the situation that I became, a yelling, table-pounding hothead. After I'd related one of those scenes to a friend, she said, "Look, you can let your problems make you *bitter* or *better.* Which do you choose?"

A good question. Answering it took a number of prayers, but I had to admit that bitterness would do neither my son nor me any good. I did return to the fray with the school system, but this time I took a positive approach, strengthened by the Lord's support, which helped me reach through to those involved.

I can change bitter *to* better, *Lord, when I share my problems with You.*

—MADGE HARRAH

Praise Diary for April

1

2

3

4

5

6

7

8

9

10

11

12

13

14

15

16

17

18

19

20

21

22

23

24

25

26

27

28

29

30

May

S	M	T	W	T	F	S
				1	2	3
4	5	6	7	8	9	10
11	12	13	14	15	16	17
18	19	20	21	22	23	24
25	26	27	28	29	30	31

Lord, Make Me Kind

Most of all, Lord, make me kind—
a grateful stream o'erflowing,
One who sees in barren banks
a fertile ground for sowing,
Prompt me with a helpful word
when one will dry a tear,
Nudge me into action bold
if it will offset fear,
Fill my heart with gentleness
instill my soul with caring,
So that I reciprocate
the love You give unsparing.
Make me thoughtful every day
sensitize my mind,
But most of all I pray, dear God,
that You would make me kind.

No greater service do you Me
than when you're kind to others,
No greater joy do I feel
than when you love like brothers,
Feed My sheep, feed My lambs
give them water in My name,
And when you come to heaven's gate
Your soul I there will claim. —FRED BAUER

1 *Practicing His Presence in May*

RESPONDING TO THE NAMES OF JESUS

God hath...given him a name which is above every name, that at the name of Jesus every knee should bow...

—PHILIPPIANS 2:9, 10

When my friend Betty, who was struggling through a difficult time, began her prayer to Christ, "Dear Bright Morning Star," it made me wonder. Afterward she explained, "Today I am trying to call Him by that name because that is what I need most—a light rising in this darkness of mine."

I began to think of the names given to Christ in the New Testament. As I searched through it, I was amazed not only by the number of them (I counted over a hundred), but by the great variety. Christ was called everything from "The Last Adam" to the "True Vine." And all of them are important because they each reflect something deep and special about Him.

It seemed there was a unique communion I could share with Him by responding to these names, letting them echo through my days and speak to my needs.

Won't you join me in selecting one name each day during this month and let Christ "become" that name in our lives? Here are thirty-one I have selected. Choose one and begin today:

Author of Faith (Hebrews 12:2)
Living Bread (John 6:51)
Cornerstone (Eph. 2:20)
Deliverer (Romans 11:26)
Door (John 10:9)
First and Last (Rev. 1:17)
Gift of God (John 4:10)
Head of the Body (Col. 1:18)
True God (1 John 5:20)
Righteous Judge (2 Tim. 4:8)
King of Kings (1 Tim. 6:15)
Son of Man (Matt. 16:28)
Teacher (John 3:2)
True Vine (John 15:1)
Life (John 14:6)
Lamb of God (John 1:29)
Light of the World (John 8:12)
Lord of All (Acts 10:36)
Master (II Tim. 2:21)
Prince (Acts 5:31)
Resurrection (John 11:25)
Rock (1 Cor. 10:4)
Savior (Luke 2:11)
Servant (Matt. 12:18)
Good Shepherd (John 10:11)
Bishop (1 Peter 2:25)
Truth (John 14:6)
Word (John 1:1)

Friend of Sinners (Matt. 11:19) Way (John 14:6)
Horn of Salvation (Luke 1:69)

Your Name is above every name, Lord—especially this month.
—SUE MONK KIDD

2 **FRIDAY**

...We look not at the things which are seen, but at the things which are not seen: for the things which are seen are temporal; but the things which are not seen are eternal.
—*II CORINTHIANS 4:18*

I've always been a sucker for efficiency tips. For years, I read almost any book and attended any seminar that promised to tell me how to squeeze more meaningful moments out of my twenty-four-hour day. And for years, I got all tangled up in complicated systems about color-coded lists, A, B, and C-rated priorities, and the clarification of long- and short-term goals.

Then one morning over coffee, a friend gave me a nugget of advice that simplified all these systems. "When choosing priorities," she told me, "ask yourself, 'Will it matter in five years?' If so, go for it. If not, reconsider."

It sounded too easy, but soon I found that her formula was helping me make big and little daily decisions: whether to serve on a committee or cut back to spend more time with my family; whether to make cookies with the kids or preserve my spotless kitchen; whether to have lunch with my husband or finish a pressing project at home.

Too bad my friend was not around to give Martha the same advice when she grumpily chose to busy herself in the kitchen while her sister Mary took advantage of the precious opportunity to sit at the feet of Jesus. No doubt the question "Will it matter in five years?" would have zoomed Martha right out of the kitchen to join Jesus.

Dear Heavenly Father, today let me escape the tyranny of my "To Do" list and choose instead to fill my hours with the deeds that truly matter.
—CAROL KUYKENDALL

3

SATURDAY

The stone which the builders rejected, the same is become the head of the corner. —*LUKE 20:17*

In the Shawangunk Mountains, seventy-five miles from our home, is an unspoiled lake where I go when my spirits need recharging. Shaded and inviting in summer, scarlet and gold in fall, pristine and silent in winter, Mohonk is special in all seasons. But the most spectacular time of year there is May, when the laurel blooms, turning whole hillsides pink and white. People from all over come to gaze at nature in her most blithesome mood.

Laurel wasn't always so appreciated. In fact, it was the presence of this stubborn evergreen shrub that, in 1869, made it possible for two Quaker brothers to purchase the land around the lake. Albert and Alfred Smiley had a conviction that scenic areas ought to be preserved for coming generations to enjoy. Land costs money, however, and the Smileys didn't have a lot of it. It didn't take much, though, to buy land "infested" with laurel. "It never grows big enough for firewood," one local farmer complained as he signed the sale papers. "And cattle won't eat the good-for-nothing stuff."

Good for nothing...except refreshment of body and renewal of heart and mind.

Am I failing to value some shrub You have planted? Lord, show me the laurel in my life today. —ELIZABETH SHERRILL

SUNDAY

And God blessed them, and God said unto them, Be fruitful, and multiply, and replenish the earth, and subdue it: and have dominion over... every living thing...

—*GENESIS 1:28*

Every year on the first Sunday in May in the gardens of the Hammond Museum in North Salem, New York, a group of interfaith clergy perform a "Blessing of the Animals." The custom is an ancient one, now observed chiefly in European churches, but it was new to me and I drove up for it.

I'm glad I did. It was something to see. Men, women and children formed a long line, their pets in hand or in tow: dogs and cats, ponies and horses, birds, turtles, goldfish, ferrets, etc., etc., etc. One little boy even brought his terrarium (I did not ask what about-to-be-blessed animals lurked inside the greenery!).

It was a lovely ritual. My face must have looked like one big benign smile as I watched these dumb creatures presenting muzzle, beak and fin for the Lord's benediction. What I did not expect, however, among the mingling of the animals was their calm demeanor—not a yap or yowl or chirp from any in that long, decorous parade. It was a peaceable kingdom....

Except...

...for Renée. A peach-colored poodle, she was not in the procession. She had come with her master, a handicapped man from New Jersey who was confined to a wheelchair. Renée sat beside him in the audience and barked and howled at every animal that passed by.

"Dumb creatures," did I say? From biped to centipede, which one of us wants to be left out of the Lord's blessing?

I'd bark, too, Renée.

Father, may we—whom You have made a little lower than the angels—never be too proud to love Your lesser creatures.

—VAN VARNER

5 **MONDAY**

...Thus saith the Lord... choose the things that please me, and take hold of my covenant. —*ISAIAH 56:4*

It was just another in a long series of choices we've had to make during the last few years. But this one seemed so clearcut. If Lynn wanted to turn her full-time but temporary pastorate into a permanent position, she would have to cut back on her graduate work. If she felt graduate work was more important, then the pastorate would have to go. One or the other; simple as that.

But Lynn was having trouble making the decision. "Just decide," I told her. "Make up your mind and then do it." She couldn't seem to. It drove me crazy.

But you know, when Lynn finally did decide to concentrate on her studies in church history and to resign her pastorate, she knew it was the right decision. And, as the months unfolded, more than just a few things that happened supported the choice she'd made—a choice that might have been different if she'd made it in a hurry, the way I'd wanted her to.

Maybe that's because Lynn understood better than I that you wait for God; that only when you do something in God's time are you doing it at the right time.

God, give me the patience to wait for Your Word. —JEFF JAPINGA

TUESDAY

I am not come to destroy, but to fulfill. —*MATTHEW 5:17*

Many years ago, when Larry and I applied to adopt our two children, the caseworker asked if we were prepared for the restrictions that children might bring into our lives. At the time I was a writer, painter, and composer, and Larry was a busy scientist and astronomer. We paused to think about it, but decided to go ahead with the adoptions.

And what happened? From the start Eric and Meghan "did their own thing," bypassing my piano and Larry's telescope to get involved in sports. *Sports?* What a surprise! Larry and I found ourselves learning about football and rodeo with Eric, gymnastics with Meghan. And limits? Oh no—the children *broadened* our horizons!

Do you sometimes worry that a commitment to Christ might demand too much of you? So did I—once. But when I decided to be God's person, my life opened up like a flower opening to the sun. Instead of inhibiting my creative endeavors, God expanded them, giving me the chance to write for Him, to travel, to speak, to meet other creative Christians.

What a surprise! Commitment does not mean limitations—but fulfillment!

Father, today I commit my life to You and await the fulfillment You bring. —MADGE HARRAH

7 WEDNESDAY

Wherefore comfort yourselves together, and edify one another, even as also ye do. —*I THESSALONIANS 5:11*

It was hard asking for help after I broke my wrist and ribs. I had to ask friends to help me with the least little thing—tying my shoes, buckling my watch, zipping my jacket. It was embarrassing feeling so helpless. That is, until a friend talked to me one day.

"Good," he said. "It's about time you discovered you can't do it all alone. You don't suppose God intended us to get along by ourselves, do you?"

"No, I guess not," I said.

"*You* don't mind helping other people, do you?"

"No," I had to admit.

"Then stop minding other people helping you!" he barked good-naturedly.

Talking with him reminded me that my friends were there waiting to give me gracious and loving assistance, if I needed it—but they respected me enough not to force help on me if I didn't ask for it. And that's the way it is with Jesus—He's always there, waiting for me. But He respects me enough to wait for *me* to make the first move.

I know it's hard, but listen to one who's learned. Swallow your pride, face up to your need and take the first step. It's the only way to get the help you need.

I ask You, Lord, most humbly, for help in ____________________.

—SAMANTHA McGARRITY

THURSDAY

For God giveth to a man that is good in his sight wisdom, and knowledge... —*ECCLESIASTES 2:26*

Last year I was hospitalized twice for surgery on my eyes. Both times the worst part was sitting around for a couple of days for post-operative examinations. I wasn't allowed to read or watch television, and how many hours a day can you sleep? So I would get up, put on my robe and slippers and stroll around the hospital chatting with other patients.

One morning I was passing the chapel and decided to go in to say some prayers of thanksgiving because everything seemed to be going so well. To my surprise, there was my doctor, kneeling, head bowed, hands folded, obviously in prayer. The moment seemed so private that I said my own prayers quickly and left without approaching him.

Later that day, when he came to my room to check me out, I said to him, "I saw you in the chapel this morning. I was surprised to see you there."

Still concentrating on my eyes, he said, "I hope you don't think I can do my job all on my own, do you? Neither can you."

I thought about that a great deal. For years it had been my habit to spend my first morning hour planning my workday, with some time for spiritual meditation, but I had never put the two together. I do that now. My work doesn't come any easier, but I have more confidence in it these days—and in myself.

Father, the source of all our talents, we serve You best when we serve You first. —GLENN KITTLER

FRIDAY

But the mercy of the Lord is from everlasting to everlasting upon them that fear him, and his righteousness unto children's children; to such as keep his covenant, and to those that remember his commandments to do them.

—PSALM 103:17, 18

Following my mother's death, I found tucked in her Bible part of a handwritten note addressed to her. It was from a woman I didn't know and read in part:

> "I love you for a number of reasons, the most important of which is what you did for my husband. When he was in the hospital, you visited him, though you'd never met him, and you encouraged several men in the church to visit him. As a result of those visits, my husband said, 'When I get better, I'm going to that church and find out what it has that makes those people so kind and lovable.' What he found, my dear

friend, was Christ and Christian fellowship, neither of which he had known before."

It had been years since my mother was physically able to visit anyone, so I knew the letter was quite old. I recognized it as a treasured keepsake and I folded it carefully and placed it back in her Bible.

How she must have treasured those words! What joy she must have experienced in the knowledge that she helped to introduce that man to Jesus! What a wonderful heritage—and challenge—our Christian parents pass on to us!

Father, today I will be alert for someone I can lead to You. Amen.

—DRUE DUKE

10 **SATURDAY**

I can do all things through Christ... —*PHILIPPIANS 4:13*

During a recent career setback that made me want to give up, my friend LeAnn told me about her experience in 4-H club. A city-bred girl trying to compete with ranch kids, she'd spent all of her free time grooming, training and caring for her chestnut calf, Jessica. Still, they placed last at the Spring showing. After leading her calf out of the ring, she ran into her father's arms and sobbed. "Oh Daddy, I give up. Jessica and I just aren't good enough. I'm going to quit showing."

"Well, that's up to you," said her father. "But remember one thing. You're only a loser if you quit praying and trying."

Several months of intensive work and many prayers later, LeAnn and Jessica were named champions at the county fair, and then went on to capture two purple ribbons at the state showing, the largest 4-H show in the world.

If you're feeling discouraged today, remember that you're only a loser if you quit praying and trying.

I will not give up, Lord. With Your help, I know I can be a winner!

—MARILYN MORGAN HELLEBERG

SUNDAY

Who are those with thee? And he said, The children which God hath graciously given thy servant. —GENESIS 33:5

Mother's Day—a day set aside to honor mothers. However, have you ever thought of turning this special day around? This year, why not give honor to your children instead of expecting it from them?

Here's what I wrote my sons last Mother's Day:

> I'll never forget the first time I saw you. After the nurse brought you in, I unwrapped you and saw that every little bit of you was perfectly formed. I was awed...you truly were "fearfully and wonderfully made" by God. It was a precious moment.
>
> You have always been a warm, loving son. To me, the nicest endearment I could wish for is "Mom." Without you and your brother, I'd never have this title.
>
> I'm proud of your accomplishments and your steadfastness, and it's been a wonderful thirty-seven years having you as the sons God gave me.
>
> I love you. Mom.

I know my sons' loving response to my letters gave me my most memorable Mother's Day ever.

Maybe you could find your own way to turn this special day around— a day for celebrating others?

Today, Lord, I celebrate__________________. —ISABEL CHAMP

MONDAY

Thou hast put gladness in my heart... —PSALM 4:7

Martha works in the first-floor bank in the building where my office is located and we often confer together. I never mind waiting at her desk when she is busy because I enjoy reading the little inspirational gems she tucks under the desk's glass top.

One that I like to read sports a happy face and says:

> A smile costs nothing but gives much.
> It enriches those who receive it without making poorer those who give.

> It brings rest to the weary, cheer to the discouraged, sunshine to the sad, and it is nature's best antidote for trouble.
> Yet it cannot be bought, begged, borrowed or stolen, for it is something that is of no value to anyone until it is given away.
> Some people are too tired to give a smile. Give them one of yours, for no one needs a smile so much as he who has no more to give.

Each time I read that, I leave Martha's desk wearing a smile, with determination to give it away all day long!

I hope reading this makes you feel the same way. The world can use *lots* of smiles.

Father, today let my face smile for You that I might spread Your love to all I meet. Amen. —DRUE DUKE

13 TUESDAY

For in him we live, and move, and have our being...

—*ACTS 17:28*

One morning as I rushed about packing lunches, cooking breakfast, ironing my husband a shirt and dressing for a meeting—nearly all at the same time—Ann brought me her ribbon box so I could tie back her hair. I looked down into that multicolored tangle of ribbons and felt I was seeing a reflection of my own life. My demanding days with family, home, church and community sometimes left me feeling fragmented—a jumble of ribbons, having little connection.

Later that day while cleaning a drawer, I came across an old photograph of my mother as a young girl taking part in a May Day celebration. She was one of many who were dancing around a maypole, holding onto long ribbons. Something clicked inside me. *Ribbons...*

I stared at the picture. The ribbons weaved back and forth in all directions around the pole. But they were not disconnected and scattered like those in Ann's box. They were connected to the maypole in the center.

In that moment, something seemed to say to me, "You need a maypole rooted in the center of your life, around which all the ribbons of your day can flow like a festival." And of course, I knew that what I needed was once again to make God my center that could connect all the little pieces of my life and give it meaning and celebration.

So right then I bowed my head and prayed:

Be my center, Lord, and help me make every task, even the mundane ones, a joyous dance that revolves around You. —SUE MONK KIDD

WEDNESDAY

Give unto the Lord the glory due unto his name...

—PSALM 96:8

Our "bottom forty" had become a joke between my brother and me. Inherited from our mother, the field borders Salt Creek, which overflows nearly every other year, taking the soybean crop with it. When a new farmer took over its management, we warned him that we might get sixty bushels of beans to the acre one year and half the next. "I'll plant corn," he said. "It can take water."

This past season, the creek overflowed seven times, inundating the crop. "There won't even be enough money to pay the taxes," we predicted. Then the farmer came, with the largest check we'd ever gotten for the field. "Corn can take water," he repeated. "And I had a helper—God."

But why hadn't God helped the other times, when we'd planted soybeans? The answer is simple. When God bestows His bounty on us, He expects us to have the sense to know what to do with it. We—or the people who planted the field for us—should have found out ourselves that corn can take water and soybeans can't. The saying "God helps those who help themselves" may be old, but it's nonetheless true.

Have you been waiting for God's blessings to fall on you without considering how to make the best use of what you already have? If so, look again....

Lord, is there a resource, an angle, a plan I'm overlooking, one that is already within my reach, that will help me solve my own problem? If so, open Thou my eyes.... —ZONA B. DAVIS

15 *WILDERNESS JOURNEY* (Exodus 16)

Manna Rains from Heaven

I will rain bread from heaven for you. —EXODUS 16:4

We're in the desert still, travelling with God's people. All day, we've trudged over cracked, dry earth, stumbling on the hard rocks, sweating under the hot sun, and we're desperately hungry. But our food has run out! Children are crying, old people are moaning. Suddenly, we're seized with anger—at Moses, at God, at ourselves.

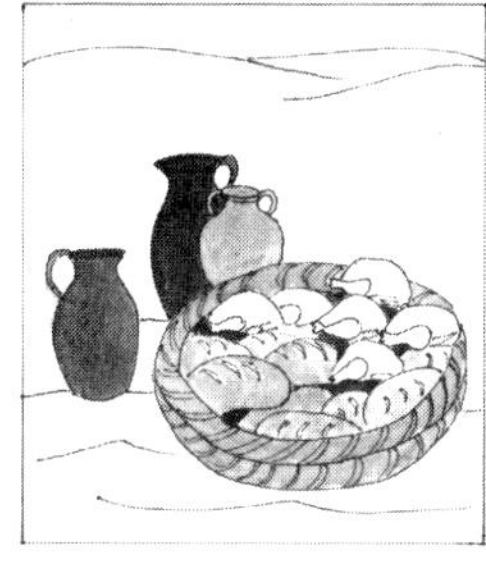

We wish we'd never come. We want to give up. We'd rather go back to being a slave in Egypt than die in this horrible place!

The men of the tribe begin to shout at Moses and Aaron. As always, those two turn to God... and the Lord promises that we will have meat to eat at night and bread in the morning. How can we believe that? We kick away the dry bones of a desert-starved animal and pitch our tent. Our hopes sink with the setting sun. Then suddenly, there's a ruckus in the sky! Out of nowhere, whole flocks of quail plop down at our feet. Meat! At least for tonight, we'll sleep without hunger pangs. *But what about tomorrow?*

We step from our tents at sunrise, still fearful about what we lack. Our food pots are empty. How will we eat today? And what are these white flakes all over the ground? "This," says Moses, "is the bread which the Lord hath given you." Hesitantly, you and I pick up some of the round white pieces. They taste like wafers made with honey, and we eat our fill. *But what about tomorrow?* Quickly now, we start gathering more so we won't run out. "Stop!" says the old man with the long white beard. "God will give us more for tomorrow. He's brought us this far. We can trust Him to supply each day's needs."

Every day for forty years, the Lord did supply his wandering people with that white food, called *manna*. I wonder if *I* could learn to trust Him for each day's needs. There are so many things I lack. Our bank account is low right now. What if the money runs out? What if time runs out before I do everything I've planned? What if illness or an unexpected emergency prevents me from working?

What about tomorrow? As I catch myself thinking, *what about tomorrow?* an old man with a long white beard steps out of my imagination to remind me, "There *will* be bread on the ground in the morning." At last, I think I can believe it.

I'll focus on Your abundance, Lord, instead of on my apparent lacks. Then I'll get my container ready!

—MARILYN MORGAN HELLEBERG

16 FRIDAY

And God said, Let the waters bring forth abundantly the moving creature that hath life... and it was good.

—GENESIS 1:20, 21

Cerro Tortuguero, it's called, the lone mountain that stands like a beacon a little way back from the beach. It is this mountain, the local legend says, that guides migrating sea turtles from all over the Caribbean to Tortuguero Beach on the east coast of Costa Rica. How a turtle can return to the same nesting place, even though several years may have elapsed, is a mystery, or a miracle, or both.

As part of a zoological society team, I had come to track and tag these endangered creatures. Since it was night work, a group of us decided to spend a free afternoon climbing Cerro Tortuguero.

It was a hot and arduous climb, but when we reached the top it was exhilarating. To the west was the rain forest, green and lush, stretching as far as the eye could see. To the east lay the Caribbean, shimmering in the sun. I tried to envision hundreds of turtles swimming down below, waiting for the cover of darkness before coming up onto the beach to lay their eggs in the sand.

"Do you really believe that legend about the turtles and the mountain?" someone asked our guide, Deanie. "Well," she replied, "some instinct, some unseen force, *something* must guide the turtle."

Yes, I thought, and that same *Someone* is a guiding Force in my life, too. He is the Light that leads me through darkness onto safe shores, and shows *me* the way home.

Thank You, dear Lord, that You are not unseen to me, but the knowable, one true Force in my life. Amen. —ELEANOR SASS

17 SATURDAY

Whoso is wise, and will observe these things, even they shall understand the loving-kindness of the Lord.

—*PSALM 107:43*

In World War II, I was stationed in the Philippines. How we rejoiced as the surrender documents were signed aboard the *USS Missouri* in Tokyo Bay, almost two thousand air miles away. I was elated. The war was over and soon I would be heading home to my wife Ruby and a son I had never seen.

But, like many others, I hadn't reckoned with adjusting to civilian life. After this long separation, my wife and seventeen-month-old son seemed like strangers. Fortunately I landed a job quickly and it afforded me Sunday overtime. But it left me little time to get reacquainted with my family, and consequently unhappiness abounded.

Then one Sunday I attended church—alone—for the first time since leaving the army. It was a strange feeling. I bowed my head when the pastor blessed us with, "The grace of our Lord Jesus Christ, the love of God, and fellowship of the Holy Ghost be with us all evermore."

Grace... love... fellowship. A spiritual family. It became clear to me now—my priorities were all wrong. Work was interfering with those dearest to me.

That evening I talked things over with Ruby, and I never again went off to work on a Sunday. Ruby joined the church choir and we enrolled our son in the church's Sunday nursery school. Day by day we grew into the close-knit, loving family I'd dreamed of back there in the Philippines.

I shall always cherish the memory of that Sunday when I became a member of His family—and became a member of *mine*, too.

Lord Jesus, on this Armed Forces Day, let me remember the many families to which I belong: my nation, my church, my immediate family. And I give thanks for each one. Amen. —OSCAR GREENE

18 SUNDAY

...If I may touch but his clothes, I shall be whole.

—*MARK 5:28*

My granddaughter Jamie is three, but she still derives enormous comfort from a piece of ragged, yellow baby blanket. When anything seems to threaten her security or she is hurt, she turns the house upside down until she finds the beloved blanket.

One day she fell and hurt her knee. Quickly I found the blanket and handed it to her. We sat together on the front steps and I watched her examine the blanket carefully, moving it about in her small hands. "Here it is, Nanny, the best place of all," she sniffed, showing me a worn corner. "This spot here. It makes everything good again."

She pressed the spot to her bruised knee, and then she began to smile. "It always works," she explained.

Like Jamie, I'm discovering a healing source. My Christian friends, God's word and prayer are woven together in *my* security blanket. And, when my pain is sudden and acute, there's a special spot in the "blanket" that I cling to—the Cross of Calvary, a place of healing. It always works.

Thank You, Jesus, that by Your wounds we are healed.

—MARION BOND WEST

19 **MONDAY**

But other fell into good ground, and brought forth fruit, some a hundredfold, some sixtyfold, some thirtyfold.

—MATTHEW 13:8

We planted seeds in Sunday school yesterday. I don't know how many of those dead-looking nasturtium seeds will sprout in their styrofoam cups, nor whether we will get any phlox in our dishpan. But I hope they all grow. Even more than that, as the Sunday school teacher I pray that the seed of God's word that I also tried to plant will take root in each child's heart.

I am a naturally impatient person, however. Waiting for seeds to sprout is very hard. Waiting to see the results of my teaching and parenting is hard. Giving love to hard-hearted people is difficult, and being sunny among grouches is hard. I want to *see* results of my actions, not *wait* for them.

That's why I keep this quote from Albert Schweitzer at my desk:

"No ray of sunshine is ever lost. But the green which it awakes into existence needs time to sprout. All work which is worth anything is done in faith."

Dear God, into Your hands I commit the work of this day, that You may bring forth its fruit in Your season. Amen.

—PATRICIA HOUCK SPRINKLE

20 TUESDAY

...Faith without works is dead. —*JAMES 2:20*

There are two kinds of Christian witnessing—verbal and active, word and deed. I was reminded of this recently while hearing a sincere and earnest streetworker proclaiming the Gospel in a large city. Though he was shouting at the top of his voice, busy passersby were for the most part doing just that, passing by.

Some Christians are guilt-stricken because they don't testify to their faith more vocally; and we all know that a willingness to tell others about the saving grace of Jesus Christ is part of our charge. Yet I have come to believe that there is often a prerequisite to telling our faith story. It is capsuled in a beautiful phrase, "earning the right to be heard."

How do we earn the right? Most often by being present at the point of need, being where the action is, being where people are lonely, hungry, suffering, dying...and doing what Christ would have done. "Before we can give a message," someone said, "we must *be* a message." Amen.

Teach us, Lord, that with our deeds,
We may say more than with our creeds.

—FRED BAUER

21 WEDNESDAY

Blessed are the meek: for they shall inherit the earth.

—*MATTHEW 5:5*

Mother and Dad flew up from Georgia to visit me last May. I'll never forget our drive to the hospital to pick up my brother John. Our father had always been a strong man, an adventurer at heart. He had hunted in the jungles of South America and traveled up the Amazon River. He was the "rock" in our family. The decision-maker, the bread-winner, the repairman, the disciplinarian, the instructor.

Now, though, as we walked into my brother's ward, Dad stooped horribly from a curved spine, a condition that has caused him to drop to barely 100 pounds and look much older than his years. My brother John's illness, on the other hand, has caused a regression over the years so that he is childlike in many ways.

Years ago there had been some harsh feelings between my father and brother. Some were the normal growing pains every family must have; some perhaps were caused by the pains and confusion that surrounded John's schizophrenia.

But now I saw my brother full of concern, his eyes meeting Dad's with compassion. He went up to Dad, put an arm around his hunched shoulder, and taking his hand, led him outdoors.

Both are meeker men than in years past, both have let go of some bitterness, and both are stronger men for that. And now, I hold great love and respect for them—and sometimes, simply *awe*.

Dear God, in our weakness teach us strength.

—SAMANTHA McGARRITY

22 THURSDAY

Be merciful unto me, O Lord: for I cry unto thee daily.

—PSALM 86:3

My trash basket fills so fast. I empty it and yet the next day it is again overflowing. Sometimes I weary of carrying old newspapers and bags of debris down my apartment stairs. Up and down. If only I could empty it out *once and for all*.

Have you noticed how life is like that, too? Each morning I kneel and pray, clearing my mind for the day. But all too soon the junk begins to pile up. My mind chatters away.

"They don't appreciate how hard I work." "What did she *mean* by that remark?" "Why does my boss give dictation at quitting time?"

When evening comes, my head is overflowing with negative debris—and it is cleaning time *once more*. Down on my knees I must go and ask God to grant me His gifts of charity, patience and peace. Up and down. Until at last my house, both inner and outer, is in order. How good that feels!

Until tomorrow....

Daily I must turn to You, Lord God... and daily You bring me peace.

—DORIS HAASE

23 **FRIDAY**

A time to get, and a time to lose; a time to keep, and a time to cast away... —*ECCLESIASTES 3:6*

One day my eighty-four-year-old friend telephoned me. "Oscar," she said, "I have a few things for your Trash and Treasure tables. Will you pick them up?" I said I'd be glad to.

For weeks I hauled for our church fair until our cellar was crammed. Then I sorted, washed, polished and priced the six thousand items. During all this hurried activity, I forgot my promise to my friend.

Then, on the day of the fair, her housekeeper appeared on the lawn, carrying two pieces of luggage. "These are Amelia Earhart's bags," she said, dropping them at my feet. I studied the faded luggage as questions raced through my mind. I wondered, because my friend is Muriel Earhart Morrissey, sister of famed aviatrix Amelia Earhart.

I hurried inside to the telephone, and Mrs. Morrissey's cheerful voice answered. "Is this really Amelia Earhart's personal luggage?" I asked her.

There was a pause. Then, "Yes, dear, they are. Amelia left them with me just before her last flight (1937). I've kept them for forty-eight years, hoping she would need them. I guess it's time now to share them with someone else."

By the time I returned to the fair, the luggage was sold. Yet in those few short minutes, I had been given an unforgettable lesson about hope, letting go and sharing.

Dear Father, give me the grace to know when I should relinquish hope to acceptance... and then the courage to move forward with renewed faith.

—OSCAR GREENE

24 SATURDAY

...Behold, I will proceed to do a marvelous work among this people, even a marvelous work and a wonder...

—*ISAIAH 29:14*

Many of you may know that every year on this day some friends and I have a party in the very middle of the Brooklyn Bridge—a birthday party for the bridge itself. Sure, it's an odd sort of celebration, but it's always amusing and it feels good for a change to honor something wonderful that man has done with his God-given talents.

Last year the Bridge was 102 years old and as usual we gathered on the footpath, put on our party hats, cut more than one birthday cake and invited joggers and bicyclists and homeward-bound pedestrians to join us. And as usual the party grew bigger and bigger and friendlier and friendlier. I talked for a long while with a seamstress from Soho and her carpenter boyfriend, then a couple visiting from Israel and a family of four from West Germany, but the guest we all enjoyed the most was a little boy named David Pollack. It was his birthday, too.

"David was born the night they had all those fireworks for the Bridge's centennial," his mother explained. "We brought him out on the Bridge for his first birthday last year. We're making a tradition of it."

As people began to drift away, I stood for a moment listening to the whirr of the traffic underneath and looked out to where the Statue of Liberty stood sheathed in her restoration scaffolding. I thought of what a gentle May day it had been, and of the party that had brought friends and friendly strangers together. I thought of how the noble old Bridge—the very symbol of connecting things—shows us not just what man can do, but what he can *be*. *Well*, I said to myself in a warm surge of hope, *it's a pretty nice world we live in, after all..*

Happy third birthday, David Pollack, and a happy lifetime. See you on the Bridge!

Show us, Lord, how right Your world can be, if we but look for it, and work for it.

—VAN VARNER

25 SUNDAY
...Exercise thyself... unto godliness. —*I TIMOTHY 4:7*

"And this," our guide said, "is the bowling alley."

Lynn and I exchanged smiles. It was true: There, in the basement of the church where she was to be pastor, right beneath the sanctuary, was a bowling alley! Two lanes, semiautomatic pinsetter, ball return, all circa 1930.

A bowling alley? In a church? I couldn't imagine anything more incongruous. Then I talked to a fellow I sit next to in our church choir who, as a teenager in the thirties, set pins for four church bowling leagues, at $1.25 per night—good money then, he said. And there was the old gentleman who came to the alley one evening with a friend to bowl and became a regular churchgoer. Now Lynn and I have started taking friends there for an evening of good times and (mostly) bad scores.

That first day I was surprised to see a bowling alley in the church. But I realize now that there's no reason why there shouldn't be a bowling alley...or ping-pong tables...or exercise equipment. Nothing is a better complement to a healthy spiritual life than a healthy physical life. That's what Lynn's church realized a long time ago and that's what I've just learned.

Father, I will show my appreciation of Your gifts by taking the best care I can of the body You gave me. —JEFF JAPINGA

26 MONDAY
Understanding is a wellspring of life unto him that hath it... —*PROVERBS 16:22*

When my mother-in-law phoned us one evening to tell us about her new dog, I was surprised. "Another dog?" I said to my wife Carol. "She already has three, McGhee, Siegfried and Butler...and now Missouri the Pug?"

But when we visited at my in-laws' home a week later and were greeted by this puppy wagging his tail so hard he almost fell over, licking our faces, his eyes shining with friendliness and trust, I soon relented.

Missouri *is* adorable. He stares at you through liquid eyes, furrowing his prematurely wrinkled brow, and when he jumps into your lap with his stumblebum paws...he's impossible to dismiss.

Since then, "Mo" and I have become great pals—going for walks, playing catch, shaking hands. But I'm especially grateful to Missouri for a very simple lesson he taught me. If ever I am uneasy about meeting strangers, I think of our first meeting. He didn't hesitate or hold back, waiting to see what I thought of him, if I wanted to be friends or not. No, he jumped right into my lap. He was instantly lovable because he instantly loved.

Lord, lead me from my self-concerns, to discovering the joys—and rewards—of being other-centered. —RICK HAMLIN

27 TUESDAY

A new commandment I give unto you, That ye love one another; as I have loved you... —*JOHN 13:34*

Our town fathers have beautified our streets with "magical" trees. Standing in proud lines up and down the avenues, these ornamental fruit trees seem to defy the seasons. In the autumn, when the other trees are decked out in crimson and russet, these trees, their leaves still green, blossom out to rival the glory of an orchard in the spring. Then when the other trees have put on their catkin wigs, these trees turn scarlet and gold.

One day I said to my husband, "Our lives are so defined by the seasons, yet look at how very special these trees are because they are *out* of season." And I began to think about the traditional rhythms we follow. Christmas cards are exciting, but they are in season. Birthday cards add to the excitement of our day, but we have grown to expect them. Even convalescent cards, welcome as they are, come only when we're ill. How about all the days in between? Those days that are the major part of our calendar? It would be such unexpected joy to receive little "love notes" out of season, beyond our normal expectations.

I thought of the boy who mowed our lawn so nicely...my neighbor, who is always here when I need her...the new couple in church on Sunday. Oh, I can think of lots more, but now it's your turn....

Father, remind me that the season of love is NOW.

—JUNE MASTERS BACHER

28 **WEDNESDAY**

To him give all the prophets witness, that through his name whosoever believeth in him shall receive remission of sins. —*ACTS 10:43*

Every year I used to make a trip south to visit old acquaintances—relatives, classmates and working associates from earlier days. I always looked forward to those visits. They not only provided me with an opportunity to talk about the good old days, but also to reflect on the present state of the world.

I still make those trips, but my itinerary has shrunk in recent years. A long-time friend in Virginia, an old classmate in North Carolina, and a cousin in Georgia have all died, and in Savannah a former business associate has developed Alzheimer's disease, so communication with him is impossible.

The visits were fun and evoked warm memories, but one thought keeps returning to me. Where are the absent ones now, and what kind of witness was I during our many contacts? Could they see Jesus in me and what steps, if any, did I take to encourage them to walk daily with the Lord? I know I tried, but did I honestly do my best?

There's no way to reclaim the past, but, from now on, each time I visit a friend I will share my love of the Lord with him and encourage him to walk daily with Him.

Lord, let me speak Your word to those I love, and be a faithful witness to You. Amen. —SAM JUSTICE

29 **THURSDAY**

In his hand are the deep places of the earth... —*PSALM 95:4*

May is strawberry-picking time in northern Alabama. Farmers open their fields to the public to "pick your own." Bob and I, with white buckets provided by the farmer, set out to gather berries for our freezer. After a time, I looked over into Bob's bucket.

"Where did you find those great big berries?" I asked. "Look how tiny mine are!"

"Are you picking off the top of the vine?" he asked.

"Of course. That's where I see the most fruit."

He reached over and lifted the leaves of several plants where I'd just picked and underneath were dozens of big, beautiful berries I had not seen.

"You can't just skim along the top," he said. "Sometimes the best berries are hidden under the leaves. It takes a bit more effort to find them, but they're worth it."

Later I thought about how "just skimming along the top" could refer to so many things in my life—learning a Bible verse but not going into the depth of its meaning, or talking with a friend but not listening with my heart, or saying, "Let me know if I can help," without doing anything positive.

How much I had been missing by doing things halfheartedly. More than that, how little I had been giving by just going through the motions. I decided I would start looking deeply into every situation, to be sure I was getting—and giving—the very best.

Lord, help me not to hold myself back under any circumstances, but to be always ready to give my best. —DRUE DUKE

30 FRIDAY

Be strong and of good courage; be not afraid, neither be thou dismayed;for the Lord thy God is with thee whithersoever thou goest. —*JOSHUA 1:9*

It was October 21, 1861. The Eagle Regiment, as the Eighth Wisconsin was called, marched into its first battle of the Civil War at Fredericktown, Missouri. Leading them was James McGinnis, carrying a staff with a perch on which their mascot, an American Bald Eagle named "Old Abe," proudly sat. The bird seemed to know his comrades faced death, for he gnawed at his rope and screeched until the victory had been won. The regiment fought thirty-five more battles. And through them all, "Old Abe" was a good soldier, never flinching under fire. He led them with remarkable courage, made them laugh when they were downhearted and whistled encouragement during the long, hard marches. At the end of the war, "Old Abe" was presented to the State of Wisconsin and the regiment captain told the Governor, "We shall never forget him!"

This is a good day for remembering both the dead *and* the living. God sends protectors. There are "Old Abes" in every life, so what better way of celebrating Memorial Day than to think of all who have brought us blessings and offer a prayer of thanks to Jesus—the greatest Blessing of all.

Thank You, Father, for the many reminders that You are always near, even during times of great travail. —DORIS HAASE

31 SATURDAY

The fear of man bringeth a snare: but whoso putteth his trust in the Lord shall be safe. —*PROVERBS 29:25*

I've been guilty in the past of envisioning all kinds of dark possibilities for my family and me, worries that greatly disturbed my peace of mind: What if our son were injured on his construction job? What if our daughter's car broke down in the middle of nowhere on her way back to college? *What if...*

Then one day I heard a story about St. Francis of Assisi that gave me pause. It seems that the saint was hoeing beans in a garden near the monastery when another monk rushed up and announced excitedly that the world would end in the next fifteen minutes.

"What are you going to do?" cried the monk.

"I'm going to finish hoeing these beans," replied St. Francis.

What a simple, commonsense way of dealing with unsubstantiated worries. Instead of distressing himself over something that might not come to pass, he put his faith in the Lord and got on with his everyday tasks.

I wonder—is there a lesson there for me?

I turn this day over to You, Lord, as I get on with the work at hand. —MADGE HARRAH

Praise Diary for May

1

2

3

4

5

6

7

8

9

10

11

12

13

14

15

16

17

18

19

20

21

22

23

24

25

26

27

28

29

30

31

June

S	M	T	W	T	F	S
1	2	3	4	5	6	7
8	9	10	11	12	13	14
15	16	17	18	19	20	21
22	23	24	25	26	27	28
29	30					

For Restless Sleepers

Trouble-filled, I toss and turn,
Like curdled milk in butter churn,
My rest is clouded by worries vague,
Faceless demons a night-long plague,
Sleep that's robbed by fears unfounded,
Cries for faith that's in God grounded,
Alas I beg His intervention,
"Release me from this unnamed tension,"
And in the grip of darkest hour,
He reassures with sovereign power...

Be not anxious in any way,
Place tomorrow with yesterday,
They're both the same, don't you see,
Equal in reality,
Only this moment really counts
And only lasting faith surmounts,
Trust is all I need to carry
You through trials of which you're wary,
Now close your eyes and go to sleep,
Your soul's securely in My keep.

—FRED BAUER

1 Practicing His Presence in June

MEETING GOD THROUGH MY NEWSPAPER

Praying always with all prayer and supplication in the Spirit... —*EPHESIANS 6:18*

At a conference I attended recently, there was a woman who radiated a kind of presence evident to almost everyone. I noticed people always gathered about her—automatically, it seemed. During a break I found myself talking with her about prayer. "Sometimes there is so little time for prayer in my life, for taking moments to be in God's Presence," I said.

"Do you read the newspaper?" she asked.

I nodded, wondering what the newspaper had to do with it.

"Well, I've been combining the two for years," she said. "Each morning I carry out a spiritual exercise with the paper, and not only does it save time, but it enriches my prayer life *and* my reading."

Her method was simple. She began on the front page, praying for each event as she read. "Sometimes my heart is torn by all the tragedy, and I feel God hurting along with me for all His children," she said. "But there is also some good news, and I try to praise God and thank Him for each occurrence."

The local news was her "listening page," where she listened for any prompting God might give her—writing a letter about a council vote or carrying clothes to a needy family.

At the wedding and birth announcements, she prayed for each marriage, each baby. At the obituaries she prayed for the families of those who had died. While smiling over the comic strips, she opened herself to God's healing through laughter and found Him present in her joy. The possibilities seemed to expand as she talked. And I began to reflect on how much impact praying in this way might have on others. "Do you think your prayers have made much difference in the world?" I asked.

"Oh, yes," she said. "But it begins with me...."

Father, this month, I will pray through the pages of my newspaper, listening for Your promptings and, I hope, making a difference in Your world. —SUE MONK KIDD

2 **MONDAY**

...They had also seen a vision of angels. —*LUKE 24:23*

We were at one of those summer festivals in Britain that they call a "tattoo," a colorful display of military bands, drill teams and so on. One of the events was to be a parachute jump by a special squad of airborne troops. I have no particular affection for parachutes, having been damaged a bit in one in a jump during World War II. So I watched indifferently as a big transport plane droned overhead just at sundown and a dozen tiny dots spilled out.

But when the chutes opened—rectangles of brightly colored nylon instead of the old-fashioned circles I remembered—I became aware that these men weren't falling, they were *flying*. Gracefully, they arranged themselves in a descending curve, following their leader in a majestic arc against the crimson sky. Down they came, lower and lower, maintaining perfect formation.

The target in front of the grandstand was a piece of cloth no bigger than a blanket. One after another, they settled on it, each man landing lightly on his feet and leaping aside to make room for the next. The contrast with my own clumsy efforts of four decades ago was almost unbelievable. Yet the laws of gravity and of aerodynamics hadn't changed.

What had made the difference? Discipline. Practice. Willingness to follow the leader. That was why they could soar with such control, such assurance, such grace. Maybe if we could achieve the same dedication of the spirit, our souls could soar in the same joyous formation.

Lord, help me to step out in faith, trusting Your spiritual laws to carry me safely to Your "mark." —ARTHUR GORDON

3 **TUESDAY**

Heaven and earth shall pass away, but my words shall not pass away. —*MATTHEW 24:35*

When Larry and I got married in June long ago, the women from my parents' church gave us a family Bible with our names stamped in gold on the cover. We started housekeeping with little furniture and few appliances, but when I placed that Bible on the table in our apartment, I felt that we were truly home.

Other things we received at showers—the toaster, the iron, the sheets—are long gone, but the Bible we still have. In the family register we've written the dates marking the progress of our marriage: our wedding, our children's births and adoption dates, the education we all received, our son's marriage, his children's births...a record of Larry's and my years together.

Now it's June again, the traditional month for weddings, and we have a shower gift to buy for a young friend of ours and her fiancé. What better present than the book of God's Word to start that couple on their way! In fact—what better gift to share at any time of the year!

Dear Father, in this world of built-in obsolescence, thank You for Your Word, which is everlasting! —MADGE HARRAH

WEDNESDAY

The Lord is my shepherd; I shall not want. He maketh me to lie down in green pastures... —*PSALM 23:1, 2*

Finally, I know what a "green pasture" is. I've been lying in one lately.

We have always served city churches and communities, enjoyed living in old houses we've restored and working with inner city people and their problems. We were surprised, therefore, when God called us out of our world of broken glass and pocket-sized lawns to start a church in a suburban, almost rural community, and gave us an almost-new house with a rolling, green lawn.

Soon after we arrived, in a small group worship service, a woman read the twenty-third Psalm. For the first time I really *heard* one word: *makes.* God *makes* me lie down in green pastures.

Not "offers-me-the-choice-to-lie-down," but "*makes*."

I remembered a day when my car broke down and "made" me stay home for an unexpected morning of leisure, a day when a minor illness "made" me go to bed for a day of reading, a day when a friend's schedule "made" me spend an hour sitting in a lovely park.

I see now that a wise and loving Shepherd knows when His sheep need a rest—and provides one.

Dear Shepherd, today I will especially look for that green pasture of blessing in which You make *me lie down.*

—PATRICIA HOUCK SPRINKLE

5 **THURSDAY**

In every thing give thanks... —*I THESSALONIANS 5:18*

While on my first mission trip abroad, we visited a slum area. The heat was as intense as the desperation of the poor, hungry people there, but the overwhelming beauty of the ragged, patient children nearly took my breath away. Many had never seen Americans before and they watched my every move in fascination. They were incredibly shy, slow to return a smile, and inevitably I would ask, "What's your name?"

I asked this question of a thin, amazingly beautiful, unforgettable child. Her black eyes danced when I singled her out. "Annabelle," she said. I had nothing to give her, but the lines of a poem came to me.

"What a beautiful name! Someone has written a poem about you." I bent down to her level and whispered,

> It was many and many a year ago,
> In a kingdom by the sea
> That a maiden there lived whom you may know
> By the name of Annabel Lee...

She absorbed the poem as though it were a marvelous gift just for her.

Later as we were driving away, Annabelle ran in the thick dust with the other children after our car. I could make out her words, "Thank you ma'am...for the poem...don't ever forget me. Thank you...."

Genuine gratitude. Such a rare quality today. I am trying to learn to give thanks with a new sincerity.

Thank You, thank You, thank You, my Lord and my God.

—MARION BOND WEST

FRIDAY

Give, and it shall be given unto you; good measure, pressed down, and shaken together, and running over...

—LUKE 6:38

My grandfather always had a big garden and a yard full of flowers. One of his greatest pleasures was providing beautiful bouquets for church every Sunday, and in carrying overflowing baskets of his tomatoes and other vegetables to families up and down the block. Mother also enjoyed providing a little extra for somebody else: baking a few extra loaves of bread for the neighbors, tucking an extra apple or a few more cookies in our lunchboxes for a hungry friend.

I tried to practice this habit in raising my own children. But their father was an even better example. Once, after he'd plowed the soil for his garden, he received unexpected orders for a transfer. We arranged to rent our house, and began the arduous job of packing and cleaning. The last exhausting night before our move, to my surprise I saw the yard light on, and a lone figure busy in the patch.

"What in the world are you *doing?*" I called out.

"Planting," he called back. "Since I'd already bought the seeds, I thought I might as well get them into the ground. By the time the new people get settled, the lettuce and radishes will be up."

It was a special thrill for all of us when, months later, the new family wrote us: "It was so generous of you to give us a head start on that garden. Our kids were so excited; they'd never gardened before and neither had my husband, but everybody pitched in to finish it—and before the summer was over, we had enough to share with the neighbors!"

Lord Jesus, this is surely what You meant when You told us to go the extra mile. And how it multiplies joyously for everybody.

—MARJORIE HOLMES

SATURDAY

Ye are the light of the world. *—MATTHEW 5:14*

On a dark, rainy night, I carried the Olympic flame through Westport, Connecticut. The symbolic fire had just arrived in the United States that morning, on its way to Los Angeles, where it would herald the beginning of the Twenty-third Olympic Games. This was the torch relay, and, for one memorable kilometer, I ran with the flame.

Conditions were not promising. It was ten o'clock at night, and the dull drizzle never let up. What's more, the Russians had just announced their withdrawal from the Games, and the loss of our most formidable rival seemed to make the Games lose a great deal of their luster. And so I didn't expect to draw much of a crowd in this small New England town.

I was wrong. As I jogged over the crest of a hill, I found the street lined with families in cars, pedestrians carrying umbrellas and shopkeepers waving from doorways. They clapped and cheered, and some even sang. It wasn't me they were applauding; it was the Olympic ideals symbolized by that flame: sportsmanship, brotherhood and peace. For that moment, as the center of their enthusiastic response, I could feel the crowd united in a bond of belief.

All it took was one flame in the dark.

Although my witness may seem small, Lord, help me remember that a spark is enough to start a fire. —RICK HAMLIN

SUNDAY

Be filled with the Spirit...singing and making melody in your heart to the Lord. —*EPHESIANS 5:18, 19*

Once I went backstage at the Metropolitan Opera to interview Jerome Hines, the internationally famous singer, who is also an active Christian layman. While he applied makeup and got ready for his role in that night's production of *Aida*, I asked him questions about his life and faith. One of the most fascinating things he told me was that he wasn't a good enough singer to make his high school glee club. He auditioned for the choir, but failed to impress the director. Did that detour his career? I wanted to know.

"No, I kept on singing," Hines said, "because I loved to sing. I never considered it anything more than an avocation." So he went off to college, studied math and sang as a hobby—which led to some amateur roles, which led to his discovery by people who had a different opinion of his talent from his high school teacher. Later that night, as I stood in the wings and listened to his rich bass voice sweep over that famous hall, I wondered how there could have ever been any question.

I am always uplifted by such stories because they tell me that God has a place in mind for each of us and a plan for our getting there. It may not always be clear and we may get sidetracked, but if we trust Him and obey Him, nothing will thwart His will. He will prevail.

Teach us to trust You, Lord, even when the results are discouraging at first. —FRED BAUER

MONDAY

...According to the power which the Lord hath given me to edification, and not to destruction. —*II CORINTHIANS 13:10*

When I was learning to play competitive tennis a few years ago, my coach was a tenacious, sixty-three-year-old former women's champion who never lost her touch for teaching. Tirelessly, she drilled me in ground strokes, volleys, serves and strategy. It wasn't until she died recently that I realized she taught me more about how to live life in those sessions than she did about tennis.

It had to do with her techniques of bringing out the best in people. On the court, she always told me what I was doing right instead of what I was doing wrong. Always—well, almost always—I puffed right up and started doing even better, because she filled me with powerful self-confidence instead of paralyzing self-doubt. "Positive thoughts make positive shots," she hammered into me.

Off the court, she sometimes explained the reasoning behind her technique. "The Good Lord gave us all gifts," she told me, "and we reach our fullest potential when we feel good about ourselves."

I've never forgotten her words. And they work every time I remember to put them into practice, with my husband, a child, a friend, or the checker at the supermarket. If I build up instead of tear down, encourage instead of discourage, seek the positive instead of the negative, the results are powerful—and satisfying.

Go ahead and try it. I dare you to feel good today.

Dear Heavenly Father, throughout this day, let me seize upon every opportunity to be an encourager. —CAROL KUYKENDALL

10 TUESDAY

Let us therefore come boldly unto the throne of grace, that we may obtain mercy, and find grace to help in time of need. —*HEBREWS 4:16*

China's Forbidden City consists of two hundred fifty acres of pagodas, statues, courtyards and magnificently carved marble bridges and balustrades. Massive pairs of snarling bronze lions guard the gates. Horrendous stone dragons skulk atop walls and rooftops.

Had I been born back in the early 1400s, I would never have been permitted inside. Only the emperor and high officials were allowed.

"Why is that called 'The Temple of Heaven'?" I asked our guide. We were looking at an especially ornate edifice that looked like three huge toy tops stacked one above the other, their golds, blues and reds glinting in the sunlight. "Because the emperor entered it once a year to pray to its gods for a good harvest," he explained.

His answer reminded me of something similar in Old Testament days. Only the priests could enter the hallowed inner courtyard, and the High Priest alone set foot in the Holy of Holies once a year when he prayed to God for the people.

I'm not sure of how or when outsiders were allowed into the Forbidden City...but I *am* sure that we outsiders are allowed access to God's Temple of Heaven, for it is written that Christ told us: "Behold, I have set before thee an open door...." (Revelation 3:8)

Yes, you and I can come confidently into our Lord's presence at any time—certain of His love, knowing we are always welcome.

O Father, I come...seeking You now. —ISABEL CHAMP

WEDNESDAY

A faithful man shall abound with blessings... —*PROVERBS 28:20*

When he reached his teens, our eldest son James suddenly became an exercise aficianado. He bought barbells, weights and various other strenuous equipment. But for his major exercise, he didn't require any gear at all. He simply ran from five to ten miles daily, building up tremendous endurance. He ran no matter what the weather—through rain, heat and cold. Sometimes he was exhausted, but he forced himself to run nevertheless.

The result? In addition to stamina, James developed beautiful rippling muscles and the strength of a young Tarzan. Sometimes I'd watch him puffing after a run on a hot day, his sweatshirt dripping wet, his face ruddy and glowing with exertion, and I'd be a bit concerned. But he'd look over and smile at me in a way that said, "You don't need to worry, Dad. I'm okay. This is something very special to me."

When I think of James' faithfulness to running, it encourages me when I become weak about my own commitments. And don't we all have those moments when good intentions get lost in the fog of discouragement, laziness or sheer loss of willpower? How do we begin again and put failure behind us? We can smile, like my son James, and remind ourselves each time we feel a slip, "This is something very special to me." And then we can pick up our heels and keep on going.

Gracious Lord, even though the course before me today be steep and arduous, I will be faithful to it . . . and You. Amen. —SAM JUSTICE

12 THURSDAY

We then that are strong ought to bear the infirmities of the weak. . . —*ROMANS 15:1*

Have you ever taken someone's strength for granted—a parent, a spouse, a friend, a minister? Not long ago, I came into my doctor's office to have the cast removed from my left arm. "I'm sorry," the nurse said, "Dr. Smith is ill. He's referring his patients to—"

I broke in angrily, "I've spent six weeks in this cast and half a day getting here, having to ask someone to drive me. I've already paid Dr. Smith's bill, and I want to get out of this cast *today*!"

The nurse looked haggard. "It was so unexpected," she apologized. "These things happen—doctors sometimes get sick, too."

I felt ashamed at having been so insensitive. Then, that afternoon, while I was waiting to have my cast removed at another doctor's office, I overheard a telephone conversation. *Dr. Smith had had a nervous breakdown.*

Going home, I thought about my doctor and how his illness stunned and saddened me. I'd never thought of a doctor as a patient. But even those people in whom we see only strength get sick, grow frail. They need cheering, kindness and support, too. Dr. Smith had done so much for me in the past; now what could I do for him?

It was small, but I posted a get-well card the next day, with this prayer in my heart:

Dear Heavenly Father, send this wish for healing along with Your comfort to ______________________________ today.

—SAMANTHA McGARRITY

13 FRIDAY

...If any man have a quarrel against any: even as Christ forgave you, so also do ye. —*COLOSSIANS 3:13*

When I was a sophomore in college, I fell head over heels in love with a young lady who didn't return my affections. I was loath to take the hints she dropped until, finally, she began to shower attention upon a young man I felt unworthy of her. My pride and jealousy surfaced then and destroyed the relationship. I was crushed.

Many young people know this experience...and most bounce back quickly from it. But not me. The rejection left an ache in my heart, shattering my confidence.

Later I taught school in the Midwest. One day a colleague confronted me. "It seems like you turn away when anyone tries to be your friend. Why? I care for you, but when I reach out, you back off."

Well, I was startled and perplexed. But after some reflection it became clear to me: I had been using that early rejection as a shield against others because I hadn't found it in my heart to forgive that young lady. Then a healing began. I knew that the rejection was not the problem—it was the way in which I had reacted to it. Our Savior asked forgiveness for those taking His very life. On a far lesser scale, why couldn't I do the same?

Do you still harbor feelings of regret and distrust over something that happened back in your past? Or maybe it's as recent as yesterday—or this morning? Then let's pray right now:

Lord, when loving fails, help me to forgive and then trust Your healing love. —OSCAR GREENE

14 SATURDAY

Lord, thou hast been favorable unto thy land...

—PSALM 85:1

I have a very special flag to present to my two grandchildren on this Flag Day. It's not a new flag; rather, one I've saved since before either of them was born. It has forty-nine stars and came into being when Alaska joined the Union. It is unique in that it is our shortest-lived flag, lasting only one year, from July 4, 1959 to July 4, 1960. On that day, we added another star for Hawaii.

How thrilling it is to trace the changes in our flag! From thirteen stars representing the original colonies along the Atlantic coast, the constellation has grown to fifty, and we have spread from sea to sea. As our banner grew, it picked up the name "Old Glory," and fortunes and lives have been sacrificed to protect what it stands for.

Let's take time today to salute our flag and to thank God for every star that shines in it.

Father, we are humbly grateful to You for this land of the free and home of the brave. Amen. —DRUE DUKE

15 SUNDAY

Son, thou art ever with me, and all that I have is thine.

—LUKE 15:31

They say Father's Day is set aside to honor dads, and that makes sense. I'm not a dad yet, just a son, but still I know all about what it means to be a dad. You see, my dad taught me.

He taught me in our backyard on 26th Street in Holland, Michigan, where the clothesline pole was third base and the sandbox first. He taught me the night I heard him say into the telephone, "I can't go; my son is sick." He taught me the night it snowed so hard I didn't get home from my date until 1:30 A.M. and found him still up waiting. (After chewing me out, he told me how concerned he had been.) He taught me about jump shots and a sense of humor and how to live as if God made a difference in your life.

So what does a kid like me who can shoot a jump shot but has never had to deal with children's headaches know about being a father? Just what my dad taught me: that being a father means having one day in your honor, but giving your child three hundred sixty-five Son's Days.

Yep, every day, I got a present from my dad: *himself.* And that's why, when I get to be a father myself, I know I'm going to be ready.

Happy Father's Day, Lord. Thank You for giving us Your Son to show us the way.

—JEFF JAPINGA

16 *WILDERNESS JOURNEY* (Exodus 17)

Moses Prays for Victory

...I will stand on the top of the hill with the rod of God in mine hand.

—*EXODUS 17:9*

This month, as we join the Israelites on their journey toward the promised land of Canaan, they are camped peacefully in the desert, weary from marching but comfortable at last. We sit with them on the ground in front of our tent, visiting quietly as the children play nearby. But in an instant, our peace is shattered as shrill war cries pierce the air. They are the Amalekites, a wild people who have come down from the mountains to rob and kill! Children cry. Mothers grab up their babies and run screaming into their tents. Fathers, brothers, husbands seize their swords to fight back. Some are killed.

Now Moses speaks to that brave young man with the flashing brown eyes named Joshua. "Choose the bravest and strongest of the young men of Israel. Tomorrow you will attack the Amalekites. I will go stand at the top of the hill, holding the rod of God in my hand. God will help you fight the battle."

In the morning, Moses and Aaron and a man named Hur stand on a hill where all can see them, and pray for God's help. While Moses' arms are raised to heaven, the Israelites prevail, but when his arms grow tired and fall down, the enemy has the advantage. Now, as we watch from below, Aaron and Hur stand on either side of Moses and hold up his arms, until the sun goes down. In this way, the battle is won!

Although I've never fought in a war, there is a certain wildness in me, not unlike the Amalekites, that wars against my higher self. I am often called to fight inner battles against bad habits, personality weaknesses, character flaws, temptations. Sometimes the battle rages when I am tempted to skip my prayer time, or eat too much, or tell a little lie, or pass on a rumor, or sleep late on Sunday morning. Like Moses, I know that when I am in close touch with God, the battle is easy. Maybe the moment an inner clash begins, I need to raise my arms to God. I can do it mentally or I may even do it physically. I'll hold them there long enough to convince my heart that God is in control of the battle and will bring the victory.

My inner battles are fought in the valleys, Lord, but I know that they are won in heaven. —MARILYN MORGAN HELLEBERG

17 TUESDAY

What is man that thou art mindful of him?...For thou hast made him a little lower then the angels, and hast crowned him with glory and honor. —*PSALM 8:4, 5*

It was among my son's school papers—an assignment to "name three things you like about yourself." Bob had written, in large fifth-grade letters: "1) I can play baseball real good. 2) I always make 100 on my spelling tests. 3) I am pretty great!"

The little list (especially that last modest self-description) made me smile. How often Bob had heard negative things about himself: "You didn't pick up your room...You aren't nice to your sister...You're forgetting your manners...." It was only right that he should be as aware of his bright side as he was of his darker one.

Then I remembered how aware I constantly was of *my* negative qualities, probing my faults and confessing things about myself that I felt needed changing. Of course that was important. But it was just as important for me to remember those things God had already changed in me—the strengths he had given, the places in me where "grace did abound." So I sat down and wrote three things I liked about myself. And it really did me good.

Why not celebrate yourself today? You're a child of God and you're pretty great!

Lord, here are three positive things You put into my life and for which I am thankful:

1. ______________________________
2. ______________________________
3. ______________________________

—SUE MONK KIDD

18 WEDNESDAY

And I will bring the blind by a way that they knew not; I will lead them in paths that they have not known...

—*ISAIAH 42:16*

On graduation day nine years ago, my classmates and I received our diplomas, our hearts filled with the highest hopes. One friend had aspirations of becoming a best-selling novelist. Another dreamed of being a successful businessman. Still another hoped to lead a world-wide church ministry, while I—well, I nursed ambitions of becoming a musical comedy star.

Today I smile at our naiveté. Oh yes, God *has* answered our prayers. There's the author who writes intriguing letters from abroad, the entrepreneur who runs a general store in a tiny village in the Adirondacks, the minister who is an English teacher at a small private school, and that musical comedy star whose best performances are in the choir loft of his neighborhood church.

But our *real* successes? They're our happy marriages, our close ties to family and friends, plus unexpected adventures—serving up a great home-cooked dinner, dreaming up bedtime stories to delight the kids or nieces or nephews, building the perfect bookshelves in a first home.

Yes, we've been richly blessed, but in ways we never could have planned. Maybe that's because we didn't do the planning ourselves—not at all. At our finest moments of decision, we left ourselves open to discover the wonders of His greater plan.

When my ambitions are aligned with Your will, Lord, so too will be my successes.
—RICK HAMLIN

19 THURSDAY

Take therefore no thought for the morrow...Sufficient unto the day is the evil thereof. —*MATTHEW 6:34*

At thirty-plus, I considered myself too old to ride a merry-go-round. But my little five-year-old niece Dina pleaded with me to ride the big, white horse. Fearing she might fall, I strapped her into the saddle, then climbed onto the big brown steed next to hers. As we whirled faster and faster to the calliope strains of "In the Good Old Summertime," I noticed Dina's radiant face. Happily, she was singing along with the music, and so I relaxed too, getting into the spirit of things.

When we passed the gold-ring pole, I called over to Dina proudly, "I'm going to try for the gold ring." She didn't respond. She was engrossed in the thrill of the ride, enjoying it all. I leaned out toward the pole and made a grab...and missed. "Next time," I said. But next time I missed again. And again. I was annoyed. "If only I can get a gold ring, Dina," I said, "you'll have a free ride!"

"I don't care about a free ride, Auntie," she answered. "I'm having fun on this one."

Well, some people might say that Dina was too complacent for her young years. But her words soothed the defeat of not getting the gold ring. And instead, like Dina, I sat back to enjoy the simple thrill of the ride.

Do you suppose that's what Longfellow was saying in his famous poem:

> Trust no Future, howe'er pleasant!
>   Let the dead Past bury its dead!
> Act, act in the living present!
>   Heart within, and God o'erhead!

And that, too, is what Jesus was saying when He suggested that we "Take no thought for the morrow...."

Lord, when the effort of reaching a goal or surmounting an obstacle wears me down, help me let go and enjoy Your gift of the marvelous here-and-now. —ELEANOR SASS

20 **FRIDAY**

...The Lord lift up his countenance upon thee, and give thee peace. —*NUMBERS 6:26*

I'm afraid of thunderstorms. It may seem silly, but that's why the little classroom incident took me completely by surprise.

We seldom have thunderstorms here in southern California. And on that day, when a zigzag of lightning clawed at the mulberry sky outside, I knew that my usual panic, born partly of fear but also of concern over how I could hide my fright from a roomful of third-graders, was about to come over me.

Calmly, I reached for a storybook and announced that I would read to them. With a silent prayer for steady hands and a serene voice, I tried to smile at the class reassuringly and then began the story. Soon the children were engrossed in the little plot and, to my relief, the storm ended at about the same time the story did.

I had forgotten the incident when a couple of days later one of the little girls came up to me on the playground and said the words that surprised me: "Know what I do when I'm afraid, teacher?"

When I bent down to listen, she said, "I squeeze my eyes shut and remember your face that day of the big storm when you read to us. I knew you weren't afraid, so I wasn't either."

I was about to correct her thinking when the beauty of it struck me. Of course! When surrounded by life's storms, all I need to do is close my eyes and see the reassuring face of God!

Father, the next time I feel frightened, anxious or afraid, I will look to You for courage. Amen. —JUNE MASTERS BACHER

21 SATURDAY

...And they shall be one flesh. —*GENESIS 2:24*

For one breathless moment, as I adjust her wedding veil and smooth her train, Karen is a newborn baby again, her small sleeping form warm against my chest. *How long have I had to reach up, Lord, to kiss her cheek?* The bells chime, the usher seats me, the wedding march begins. *How radiant she is, Lord. When did that soft, womanly smile replace her pixie grin?* I wink at the graying man with the nervous smile who walks beside her down the aisle, and I see a barefooted, pigtailed girl run squealing into her daddy's arms, her first fish dangling from a bamboo pole. I'm probably the only one who hears Rex's little sigh, as he places her hand in Dave's and steps back.

Now these two, so much in love, kneel together and everything blurs....*Thank You for lending her to me, Lord. Help me to love her enough to let go now.* "Wilt thou have this woman...for better for worse...with this ring...what God hath joined...I now pronounce you..." and they are husband and wife. Truly one in Christ. As I put my arms around my new son-in-law, I am suddenly aware of *Who* has authored this beautiful love story.

May they always acknowledge You as Author of their living book, Lord. May I do the same—in every *chapter.*

—MARILYN MORGAN HELLEBERG

22 SUNDAY

For she said within herself, If I may touch his garment, I shall be whole. —*MATTHEW 9:21*

I once read a joke about a man who owned an axe he claimed had belonged to Abraham Lincoln. Said he, "Except for two new heads and three new handles, it's the very same one."

I laughed over that story, but later I got to thinking about it and found that I could accept that axe as Lincoln's. Because there would be a mystique about it, some kind of intangible sense of Lincoln's presence passing from the original parts, a relic that would make the owner feel he was still somehow in touch with Lincoln's greatness.

I think it was a wish to keep in contact with Jesus that led early Christians to seek relics—the Holy Grail, bits of the Crucifixion robe, pieces of the "True Cross," something they could hold and say, "This was His, I am touching something He touched." Many of the objects they found were probably fake, but that doesn't matter. What counted was the faith of those who held the objects, their desire for the power of Jesus's presence in their lives.

The beautiful thing, of course, is that we don't need an object Jesus touched to keep in contact with Him. When I go to Him each day in prayer, I feel His presence with me so strongly that it's as though His hand is on my shoulder. His last words to us, as reported by Matthew, are: "Lo, I am with you always, even unto the end of the world" (Matthew 28:20).

He's with me this very moment. And with you.

Let me manifest Your presence, Lord, to those who don't yet know Your touch. —MADGE HARRAH

23 MONDAY

Of the increase of his government and peace there shall be no end... —*ISAIAH 9:7*

A long time ago a teacher listened to his students recite the Pledge of Allegiance with great monotony each morning. One day he said, "Boys and girls, I would like to recite the Pledge to you, word by word, and explain its meaning." He said:

I: me, an individual.
pledge: promise, dedicate myself.
allegiance: devotion and loyalty to our flag, a symbol of freedom.
United: standing side by side: together, like a family.
States: individual communities joined in a common purpose.
and to the Republic for which it stands: a state in which the government *is* the people, elected *by* the people.
indivisible: incapable of being divided; forever bonded.
with liberty: freedom; to live freely, without fear or threat from others.
and justice: dealing fairly with one another.
for all: you and me; or, it's as much your country as it is mine.

When the teacher had finished, he said, "And now, boys and girls, let's all recite the Pledge of Allegiance again."

The Bible verse above speaks of a world with an ideal government, one that promotes each tenet of our Pledge of Allegiance. To make that dream real, we must work for it, not just give it lip service. I like to think that the students of that long-ago class grew up to believe in those words.

Father, our country, founded with Your guidance, is still not perfect. May we continue to grow as one nation—under God. Amen.

—ZONA B. DAVIS

24 TUESDAY

And this is love, that we walk after his commandments.

—II JOHN 6

One day as I was watching the evening news, I was particularly struck by a story about a thoroughbred racehorse—a winner who loved to come from behind, they said. Then veterinarians discovered the horse had an incurable disease, one that would have cost most horses their lives. But this one had his diseased leg amputated and replaced with an artificial steel leg. And he lived. In fact, he thrived. No more races, of course, but he watched his foals grow up instead—and now they're racehorses. Thoroughbreds all.

"Not a lot different about him," his owner concluded. "Except where he used to run, now he walks. I suppose that's a good lesson for all of us."

I suppose it is—especially for those of us who try to do all, make all, be all, on our own. Because sometimes life slows us down. We have to recognize that maybe we won't always be able to run like the wind—but we'll always have a purpose if we walk with God.

Walk with me today, Lord: help me to run my life at Your pace.

—JEFF JAPINGA

25 **WEDNESDAY**

And I will make all my mountains a way, and my highways shall be exalted. —*ISAIAH 49:11*

It is a famous monastery that was erected on the hill above the Hudson River in 1902: a three-story red brick, dutch colonial building. But the interesting thing to me about it is the way it's oriented. The main entrance faces the river, the dock and the long driveway up from the water by which, it was assumed, people would arrive.

Amazing to recall that, as late as the early years of this century, the river was still the chief thoroughfare of the area. Dirt roads, dusty or muddy depending on the season, served local farmers, and there was passenger train service on the opposite shore. But on this side of the river, fast modern steamboats provided by far the most dependable transportation.

Automobiles? They were too new, too few, too experimental for most people to glimpse their potential.

Those scheduled daily steamboat runs must have seemed as dependable as the tides. The building's designers never questioned positioning the front door facing the boat landing. But that sweeping drive up from the river's edge was never actually built. Automobile highway construction was begun about the time the monastery opened, and for eight decades people have been coming in...through the back door.

It makes me think of our assumptions about God, and the avenues He chooses to use. How often we limit God with our preconceived ideas of the way He's going to enter our lives. How often He surprises us by turning up where we least expected.

Father, build a royal road into my heart today and let me trust the traveling to You. —ELIZABETH SHERRILL

26 **THURSDAY**

The Lord shall... bless all the work of thine hand... —*DEUTERONOMY 28:12*

My house is right in back of a stately brick church with magnificent stained glass windows, each nearly twenty feet high. Last summer these windows were restored. For this very intricate work, a gentle afternoon sun is best. At least the fellow who worked on these windows thought so because there he was every afternoon when I returned from work.

I made a habit of stopping to watch him. He was nimble on the scaffolding, scampering from place to place like a squirrel. He whistled or hummed as he worked, and his hands, while quick, seemed almost reverently to caress the glass. Soon he was greeting me with a friendly, "Hey, how are you?" but he'd never pause to chat. His work in the soft, waning afternoon sun was too precious to fritter away in small talk.

What an inspiration he was. At a time when many complain that craftsmanship is as dead as the work ethic, here were both flourishing in my backyard! This energetic person made me think about the way I approach my workday. And how about you? Perhaps your enthusiasm and energy for the tasks at hand today can be an inspiration to another.

As I go about my work today, Lord, let me be mindful of the genuine joy I get from a job well done. —JAMES McDERMOTT

27 FRIDAY

. . . And the third day He shall rise again. —*LUKE 18:33*

He was so old and wrinkled that it was hard to imagine he had ever been young. He sat on the same bus-stop bench every sunny day and watched people come and go, and his lined face seemed alive with an inner joy. *How could anyone so shabby and wrinkled always have such a warm smile*, I wondered. And so one day I asked him.

"You look so happy this morning," I said.

"Why not?" he answered. "Life is good."

"But don't you ever have troubles?" I asked.

"Oh, you can't live a long time and not have troubles," he agreed. "It's just that... well, you know the Easter story? How God's Son died on a cross on Good Friday? Well, just look what happened only three days later! So you see, whenever things go wrong, I've learned to wait three days. Somehow by then, things always seem much better."

The bus came and I climbed aboard. From my seat, I turned to look back at him. He was petting a poodle on a leash and smiling up at its owner.

Suddenly, for me, it felt just like Easter.

Whenever things go wrong, dear Jesus, let me always be willing to wait three days. —DORIS HAASE

28 SATURDAY

Beloved, think it not strange concerning the fiery trial which is to try you... but rejoice... that, when his glory shall be revealed, ye may be glad also with exceeding joy. —*I PETER 4:12, 13*

Every June the neighbor's cottonwood trees shed white tufts that make the entire area look as if it's been dusted with snow. These cotton balls are pretty when they skip through the air, but are a nuisance when they settle into every available nook and cranny. When I open my garage door, they swirl like a cloud of white gnats and land in my hair or flutter into the house.

Why do my neighbors have to keep these messy trees around? I was thinking darkly as I looked out of my living room window. At least there was one bright spot I could enjoy. A hummingbird was building her nest outside just below the sill, unaware that her choice of nesting site gave me a grandstand view.

On that particular day I noticed that Mama Hummer had lined her nest with the pesky cotton that was flying around the yard. During the next few days, she laid three eggs—that looked like navy beans—on top of it. Two weeks later three naked little creatures had a fine "eider-down" mattress. The cotton from the neighbor's trees, which had been such an annoyance to me, had been a blessing to them.

Do you suppose that means there's some good in everything, no matter how much of a nuisance it may seem?

Father, help me to look for the silver lining behind every storm cloud.
—ISABEL CHAMP

29 SUNDAY

This is the word of the Lord unto Zerubbabel, saying, Not by might, nor by power, but by my spirit, saith the Lord of hosts.

—*ZECHARIAH 4:6*

The story of Zerubbabel is scattered throughout the books of Ezra, Nehemiah, Haggai and Zechariah. I followed his story one day as I read in my Bible and found a layman of courage and faith.

Zerubbabel returned to Jerusalem with Nehemiah and was appointed governor of Judah. He and his close friend Jeshua, the High Priest, were told by God to rebuild the temple. But people who'd been living in Jerusalem during the Jews' absence had threatened violence against anyone who attempted to rebuild the temple. What should Zerubbabel and Jeshua do?

Ezra 3:3 (N.I.V.) tells us: "Despite their fear of the peoples around them, they built the altar on its foundation and sacrificed burnt offerings on it to the Lord...."

Sometimes I feel threatened by those around me. I *want* to speak up for truth, do what is right, but I am afraid of what others may say or think.

Those are the times when I need to remember God's promise to Zerubbabel: "Your 'strength' is not your own might nor power, but My Spirit." *His Spirit!* I want to remember that when my fear overtakes me, then perhaps like Zerubbabel I can obey God even when threatening enemies loom near.

Dear God, so fill me with Your Spirit that I hear what You would have me do—and dare do it even when I'm afraid. Amen.

—PATRICIA HOUCK SPRINKLE

30 MONDAY

Now we exhort you, brethren... comfort the feebleminded, support the weak, be patient toward all men.

—*I THESSALONIANS 5:14*

I got up earlier than usual this morning so I could begin an important—and overdue—writing project. But it turned out to be a day of one interruption after another.

First, my son, the magician, calls from Atlanta. He needs a rush shipment of raccoon puppets stored away in my basement for his act. So I wrap and cart them ten miles to the nearest carrier. When I finally get home, the phone rings again—a housebound neighbor needs some groceries. Forty minutes later, I bring them, fix a jammed window and empty her trash. Then back home again, and the minister calls. Can I drive an ailing church member to the doctor? Sure, and I'm off again. And so it went for the rest of the day....

Finally it's late night now and I'm in bed. *Well, so much for my good intentions—it was another wasted day.* As I turn over, a new thought hits me: Was it *really* wasted? Maybe I didn't accomplish what *I'd* set out to do, but wasn't I useful to others in some way? *Maybe,* I think as I relax peacefully under the covers, *those distractions were God's way of telling me what* He *set out for me to do today?*

Lord, let me fill my day with Your *business. Amen.* —SAM JUSTICE

Praise Diary for June

1

2

3

4

5

6

7

8

9

10

11

12

13

14

15

16

17

18

19

20

21

22

23

24

25

26

27

28

29

30

July

S	M	T	W	T	F	S
		1	2	3	4	5
6	7	8	9	10	11	12
13	14	15	16	17	18	19
20	21	22	23	24	25	26
27	28	29	30	31		

Walking Through Fire

No one who's born to earth
  and walks upon this plane,
Is free of life's vicissitudes
  its suffering or its pain,
Christ didn't say we wouldn't die,
  or hurt or be rejected,
He didn't say we wouldn't cry,
  or some times feel neglected,
The only promise that we have,
  one worth our walk through fire,
Is at life's end we live again
  —never to expire.

I know of your worries
  I know of your needs,
I know of your trials
  when your heart bleeds,
No harm shall befall you,
  no evil ensnare,
No peril that threatens
  shall your soul impair,
My love knows no boundaries,
  My grace is the same,
And I'll never forget one
  who honors My name. —*FRED BAUER*

1 Practicing His Presence in July

WRITING GOD A LETTER

...I have called you friends... —*JOHN 15:15*

Today I came across a letter my daughter Ann wrote at age six when she had received a box of stationery for her birthday. "Who can I write to, Mama?"

"Write one of your friends," I suggested. She sat down and filled nearly a whole page. "Whom did you write?" I asked as she signed her name. "Jesus," she said matter-of-factly, as if it were the most natural thing in the world.

Now, finding her letter and reading the tender, special things she wrote to Him, I felt a tug inside to write Him, too—as a friend. I pulled out my stationery. At first it was a little awkward, but soon the words were spilling onto the blue paper. I wrote to Him about things that lay in the cellar of my spirit, things I was not really aware of until I began to sort through the darkness and put them on paper. Though I'd "talked" to God about my life many times, taking my thoughts and making them tangible in black and white letters gave everything fresh perspective and clarity. It seemed my friendship with Him took on a sense of immediacy.

So during July, I'm planning to write to God each day—letters sharing my problems and needs, thank-You notes, invitations for Him to join me in activities and projects...correspondence full of tender and special things. Won't you join me in an "offering of letters" to God? For just as letters draw friends together across the miles, so these will bring Him near.

Dear God, You are my friend and confidant. I will correspond with You throughout this month. —SUE MONK KIDD

2 WEDNESDAY

...More bold to speak the word without fear.

—*PHILIPPIANS 1:14*

Jose Rodriquez, a nationally rated chess expert, is so highly skilled that he plays blindfolded against opponents who can see. My husband and I were surprised when our grown-up son Bryce decided to compete against him. The odds were so obviously stacked in Mr. Rodriquez's favor. Blindfolded or not, he was a champion.

But, to the amazement of everyone, Bryce won the match. People crowded around him with questions. "How did you do it?" they asked.

"With confidence in the Lord," Bryce said simply. "And in myself."

Confidence. Yes, that summed it up: a replay of David and Goliath. But was there a lesson here for all of us? Can we change ourselves and the world around us by being more confident?

God's word tells us we can!

Are you worn down by continuing conflict with difficult co-workers? Approach the problem with confidence (the Lord will supply it). Are you afraid of possible reprisals when you take a stand for the right? Move out confidently (with His strength and power). The neighbor who refuses to speak...the job that is too great...the unmended quarrel? Pray for confidence. Then step out in new faith—remembering that with God all things are possible (Mark 10:27).

Lord, I would be bold as a lion and gentle as a lamb, confident in Your word. —JUNE MASTERS BACHER

3 THURSDAY

For we walk by faith, not by sight. —*II CORINTHIANS 5:7*

I was away from my office attending a week-long workshop in interpersonal relationships. Each day found me increasingly disturbed as the riveting self-examination engaged in by the group rubbed my sensitivities raw. I wondered if I was really learning anything.

Then one morning the group leader said, "Today we're going on a Trust Walk. Choose a partner; one of you close your eyes and the other lead in a silent walk."

I closed my eyes as my partner took my arm. At first I was fearful, hesitant and untrusting, holding back, walking haltingly. Then gradually I began to relax, and as I did, I noticed the lovely fresh smell of the nearby lake. I heard the insects buzz. I felt the warming sun touch my head, then the cooling shade as we passed beneath a stand of trees. My feet sensed the pavement, the lawn and the gravel pathway. A sense of peace and order swept over me.... *This* was trusting?

At supper that evening, a waitress paused and asked, "Who were those blind people here today?"

I smiled to myself because she had spoken the truth. How blind I had been to our Lord's blessings, to His love and generosity. *Funny,* I thought, *I had to travel far away from home to find something—or Someone—Who had always been there. I just didn't see....*

Gracious Lord, with every step I take, I will trust in Thee. Amen.

—OSCAR GREENE

FRIDAY

For he loveth our nation, and he hath built us a synagogue. —*LUKE 7:5*

This past summer my Scotch-Irish husband and I spent the week of July 4 in our mountain cabin with our fair-skinned red-haired daughter, a friend who is black, another friend who is Anglo and her darling brown-eyed Spanish baby. We attended church services that Sunday in the loft of a nearby lodge, where the windows framed views of snow-rimmed mountain peaks gleaming against sapphire skies. The song leader ended the service with a patriotic hymn:

America! America!
God shed His grace on thee,
And crown thy good with brotherhood
From sea to shining sea!

As I listened to the voices rising in harmony toward the rafters, I thanked God that we live in a land where different races can worship together in freedom.

Freedom. The very word stirs the blood. Freedom to assemble, to speak out, to vote, to choose without coercion among many churches and denominations. As we approach another Day of Independence, please pray with me that these precious freedoms will not be taken for granted, but that, with God's help, we'll work together to make things even better for all our people, regardless of color or creed, "from sea to shining sea."

Dear Lord, please shed Your grace this day on our nation helping us to continue to walk in brotherhood. Amen. —MADGE HARRAH

5 **SATURDAY**

A time to weep, and a time to laugh; a time to mourn, and a time to dance... —*ECCLESIASTES 3:4*

Even misfortune has advantages. A recent illness, which kept me home from work for several weeks, also provided a special blessing.

When I moved to my garden apartment in Sherman Oaks, one of the things that delighted me was the huge expanses of green grass and tall trees. At last I could spread a blanket and lie in the sun. I *could*—but didn't. Not until I had all my work finished. And somehow that never happened.

Now, feeling too tired to do anything else, I spread that blanket under a tall tree and spent a peaceful afternoon lying on my back, reading and watching two squirrels chase each other along the branches above my head. It was delightful.

As I lay there, I thought of how guilty I always felt taking time for fun or relaxation. It seemed that God wouldn't approve unless I was being useful. I forgot what He said in Ecclesiastes...*there is a time for everything*.

Should you ever feel the way I did, why don't you read that passage? I'm sure our Lord takes pleasure in our enjoyment. Isn't that why He created for us such a beautiful world?

Let me make time, Lord, for entertainment...for fun, relaxation, enjoyment...to become as a child. —DORIS HAASE

SUNDAY

It is easier for a camel to go through the eye of a needle, than for a rich man to enter into the Kingdom of God.

—MARK 10:25

When I first came to New York from a small midwestern town, an acquaintance told me that to really "make it" here you needed to own a building and have it named after you. To prove his point, he rattled off a few examples of buildings that were named after their owners: the Woolworth Building, the Chrysler Building, Trump Tower....

More recently, someone I know who qualifies as a "Yuppie"—that's the term for "young, upwardly mobile professional"—told me I didn't need to set my sights quite so high. All I needed to "make it" was a summer home in the Hamptons, a BMW and a six-figure income.

Achieving those goals seemed almost as remote to me as owning my own skyscraper. I was feeling a little sorry for myself when one day at lunch I found that somebody (could it have been my wife?) had tucked a slip of paper in with my tunafish sandwich. On it were the following words: "'The value of a truly great man consists of his increasing the value of all mankind.' Adolf von Harnack. Great church historian. And he didn't have a car. And no buildings were named after him."

Will there ever be a Japinga Building? I doubt it. Does it matter? Not at all. I know now that most great people—like good old Adolf von Harnack—"made it" by *contributing* something good to this world, not by getting enough money to *buy* it.

Father, help me to learn that greatness of spirit means far more than material possessions. —JEFF JAPINGA

MONDAY

...That I may daily perform my vows. *—PSALM 61:8*

Every New Year's Day I make up a new set of resolutions. "I'm going to give up desserts," I say. Or, "I'll read the entire New Testament through." But by January 31, I'm having dessert—twice a day. By Groundhog Day, I've only read a chapter of Matthew. And by June, it's quite obvious to me, and everyone else, that I have no willpower at all.

But I'm going to start the second half of 1986 fresh. I think I finally see what I've been doing wrong. Instead of thinking of what I want to accomplish each day, I've been trying to "picture" the whole year ahead of me; in other words, using a wide-angle lens where a microscope is called for. Laying out a whole year's work is too broad—and too disheartening!

How about if, instead of giving up desserts for the whole year, I decide I just won't eat that piece of peach cobbler *tonight*? And what if I don't think about the entire New Testament, but take fifteen minutes reading a *chapter or two* a day in Matthew?

Starting now, I'm going to think in "bite-size" pieces, taking one "babystep" forward at a time. My intentions will be no less resolute, but each link, daily considered, will make a stronger chain along the road to successful completion.

Lord, let me think of what I have to do before me on this *day and trust Your provision for tomorrow.* —JUNE MASTERS BACHER

TUESDAY

But seek ye first the kingdom of God... —*MATTHEW 6:33*

I found a dime as I walked along my street this afternoon. I picked it up from the road and slipped it into my pocket. A few yards away something else glinted in the bright sun. I detoured to look—only a bottle cap. Nearby was the ring-tab from a soft-drink can; several steps farther on, a foil gum wrapper glistening like silver.

I'd allowed myself half an hour away from my typewriter, and so after fifteen minutes I turned around. In that quarter-hour I'd seen a lot of litter and developed a crick in my neck from staring at the ground. One thing was sure, anyhow: there were no more coins beside the road.

So I raised my eyes as I headed back...and saw a rose bush with wine-colored blooms, a nuthatch spiraling down a tree, maple leaves brightly red against a gray trunk.

July 1986

What we look for determines what we find. I knew that. But also, I thought, what we find at the start of a walk, a job, a new acquaintance, is what we're apt to go on looking for the rest of that walk, that work, that relationship.

Lord, help me to see a shining token of Your love at the outset of each new adventure. —ELIZABETH SHERRILL

WEDNESDAY

. . . Thou shalt love thy neighbor as thyself. —*LEVITICUS 19:18*

One summer not long after I bought a house in Brooklyn, my neighbors decided to update the electrical access cables to our block of row houses. It was decided that we'd share the costs. But my house was considerably narrower than the rest, so I reasoned I should pay less. "Not fair, we all get the same electricity," said my neighbors. After our inability to resolve our rather heated differences, we decided to meet in the fall to try again.

That September our family took a trip to northern California to visit the magnificent redwood forests. The majesty of those forests took my breath away. Talking to a ranger about the trees, I said, "I'll bet their roots must go down deep into the earth."

"Actually, they don't," he said. "Their roots are really quite shallow so they can pick up as much surface moisture as possible."

"What about high winds and storms, then?" I asked. "How do they keep from being toppled over?"

"Their roots are all twined together," the ranger said. "The trees have a kind of lock on one another that supports them. I guess that's why you always see them in groves. Just one couldn't stand alone."

"Kind of like a close-knit neighborhood," I said.

"Yes, exactly."

I sent a postcard with a picture of redwoods on it to my neighbors and apologized for being so "narrow-minded about my narrow house." When I got home, the electrical cable work was already under way. "Pay whatever you think is fair," my neighbors told me. Then, that winter, when my furnace broke down, my immediate neighbors took us in. "Stay as long as it takes to get it fixed," they told us.

When I told them about the redwood forests, they just smiled.

Father God, help me to be the kind of neighbor that I'd like to live next to... and help me to act *it too.* —JAMES McDERMOTT

10 **THURSDAY**

He hath made the earth by his power, he hath established the world by his wisdom, and hath stretched out the heavens by his discretion. —*JEREMIAH 10:12*

It was my first day at the beach and I was out early, walking in the wet sand and playing games with the outgoing tide. Sometimes a wave caught me unaware and swirled around my ankles.

Returning, I walked farther inland where the sand was still warm from yesterday's sun. As I looked down, I was amazed at all the footprints. Some had been made by bare feet, some by shoes, some by children, some by strong, active young people and some by those with a lighter, more halting step. There were the thick herringbone marks of sand-tractor tires, and the shallower dents of dune buggy wheels. Bicycle tires, too. I could make out the rhythmic-patterned prints of dogs' paws and the fragile tripod markings of seagulls foraging washed-up shells.

So many of God's creatures going in so many different directions—each one a different story, each one crossing the trails of other lives and forming new chapters. Yet God keeps track of us all. He not only created us; He knows each of us as His own. We may not know where our various paths will lead, but He does. And we can trust His guidance.

Make me more sensitive to Your direction, Jesus, and keep me from going around in circles. —PHYLLIS HOBE

11 **FRIDAY**

If the Son therefore shall make you free, ye shall be free indeed. —*JOHN 8:36*

Not long ago, after my mother died, I had the unwelcome task of selling our family home. Sorting through sixty years of accumulation, deciding what to keep and what to throw away, was emotionally exhausting. Then came the hassles with real estate brokers and prospective buyers. As weeks dragged by with no sale in sight, I began to think of the house as an albatross around my neck. I was afraid I would never be free of it.

Finally, a buyer appeared and we set the closing date. After the papers were signed, I went back to the quietness of my own apartment. There, suddenly, a feeling of enormous relief washed over me. I raised my arms and shouted joyously to the ceiling, "I'm free! I'm free!"

Even as I said it, something nagged my memory. I remembered back to when I was twenty-one and had attended my church's young adults retreat. In contemplating my life, I knew I had to get rid of old sins and ask Jesus Christ to be Lord of my life. So one afternoon in a garden at the retreat center, I prayed, "Lord, take my life." And, as I surrendered myself to Him, I felt that soaring, exhilarating feeling of freedom.

Then I was completely sure. But now I was older, wiser....Was I *still* free? Was my life *still* surrendered to Him? I knelt down and repeated the same words I had said so long ago: "Lord, take my life."

And, once again, I knew that wonderful, indescribable sense of freedom.

Dear Lord, with You in my life I shall always be free. —ELEANOR SASS

12 SATURDAY

And the rain descended, and the floods came, and
winds blew and beat upon that house; and it fell nc
was founded upon a rock. —*MATT*

After attending the Winter Olympics at Lake Placid in 1980 anc ing the great camaraderie of peoples from all over the world, our family couldn't wait for the 1984 summer games in Los Angeles, and we sent off for tickets with high enthusiasm. But when reports of all the possible problems began to surface, we wondered if it might be wiser to forego the event.

The smog at that time would be horrible, it was predicted. The freeways would never be able to handle the traffic, announced a newspaper story. Shortage of housing, exorbitant prices, the possibility of international terrorism...the list went on and on.

"They couldn't pay me to go to Los Angeles," one person exclaimed. "Don't you think you're asking for trouble?"

"We may be," I told him, "but we have an old family saying: *Only rain cancels our picnics, not the forecast of it.* So we're going." In the end, none of the bogeymen that people feared materialized and the color and excitement was a never-to-be-forgotten experience. The rain clouds passed and the picnic was a success.

Show us, Lord, when to take care and when to take risks.

—FRED BAUER

13 **SUNDAY**

Casting all your care upon him; for he careth for you.

—I PETER 5:7

In a church where I once belonged, there was a particular stained glass window that always drew my attention. It simply portrayed Christ standing with His hands held opened together in front of Him.

One Sunday I was sitting in church feeling the heaviness of a problem that had been troubling me for some time. My eyes were suddenly drawn to the stained glass window once more. The figure of Christ holding His hands cupped in front of Him seemed to be waiting for someone to drop something into them. I stared at the opened hands. *Give it to me*, the scene seemed to whisper. And in that moment I imagined myself reaching inside me and lifting a heavy stone of worry from within and placing it into the hands in the window—the waiting hands of Christ. And as I did, I felt a sudden lightness...the freedom that comes only from releasing our burdens and trusting them to God. The problem I had been carrying did not immediately resolve itself, but my heart was now freed to find a creative solution I had not seen before.

From that day forward, that window became my "worry window"—a place to which I could come and leave my fears and anxieties every week...a place that helped me learn to let go and trust.

Maybe there is an open pair of Hands waiting to receive your burden right now.

Jesus, take this concern from me—and grant me Your peace.

—SUE MONK KIDD

MONDAY

For now we see through a glass, darkly; but then face to face: now I know in part; but then shall I know even as also I am known. —*I CORINTHIANS 13:12*

We are talking a lot about "Blessed Ambiguity" in our home and church these days. Blessed Ambiguity is that span between the time you hear a call from God or put in a petition to God, and the time God acts.

My husband Bob has been called to start a church in Atlanta. So far he's knocked on a lot of doors and talked to a lot of people, but few of them come to worship yet. Meanwhile each week he labors faithfully over his Bible and prepares his Sunday sermons.

Donna and Carl feel called by God to move from New York to Atlanta to help us start the church. So far he has not found a job, their house has not been sold and they have not found a new one here.

Glen and Lucille have a daughter who was severely brain-damaged in an automobile accident. So far their prayers for her healing have not been answered.

Blessed Ambiguity. *Ambiguity*, because it's a time of many unanswered questions, a time of stress in spite of faith and hope. *Blessed*, because it's a time of deep (though sometimes anguished) prayer, a time of puzzled but trusting "waiting upon the Lord."

During these times, all of us can look back to past times of Blessed Ambiguity and trace exactly how God brought everything together in His perfect timing. We can pray. But mostly, we wait. As we wait, we discover that God has no better school to teach trust, patience and hope. May He also give us each grace to graduate with honors!

Lord, walk with us through times of Blessed Ambiguity and keep reminding us that You will do all things in Your perfect timing. Amen.

—PATRICIA HOUCK SPRINKLE

15 WILDERNESS JOURNEY *(Exodus 18)*

Moses Is Tired

Thou wilt surely wear away... this thing is too heavy for thee. —EXODUS 18:18

Moses leans on his staff, shoulders stooped, deep lines furrowing his brow, dark circles under his deep-set eyes. All day, people have been bringing their problems to him—great throngs clamoring for his attention, demanding that he settle their disputes, help with their decisions, answer their questions about God. This man who was specially called by the Lord, who has led us these many miles, brought us victory in battle, interceded for us with our God, is starting to show signs of fatigue. *Oh Moses, haven't you learned that only God can be a flame that burns without being consumed?*

But who am I to talk? I often take on too much, overspending my emotional and physical reserves. It seems that there are so many demands on my time and energy, and it's so hard for me to say no. So many people are counting on me.

Now who is this pushing his way forward in the crowd? It's Jethro, Moses' father-in-law. He talks quietly with our leader, and we hear the older man say that what Moses is doing is not good. "You'll wear yourself out," says the older man. "Then you won't be any good to the people. Appoint some others to help you."

Taking advice from a father-in-law may not be easy, but Moses does it, and perhaps the first case of potential burnout is averted! As I watch him choosing others to share his burden, I sense a great humility in this act—and a great freedom. In a flash of insight, I see that it's *pride* that makes me think I have to do it all. I'll do a better job of the things that matter most, if I can learn to say no to the rest. I sense an important inner turning as I resolve to simplify my life.

O Lord, help me to sense what You *want me to do. Then give me the humility to pass the rest to others.* —MARILYN MORGAN HELLEBERG

16 WEDNESDAY

Then was our mouth filled with laughter... —*PSALM 126:2*

It was one of those days: In the middle of rush-hour traffic there was road construction equipment ahead. In the distance I saw a beast-like machine laying steaming-hot macadam atop criss-crossed iron bars in the bedrock. As I glanced at the tightly packed cars around me, I sensed all the other drivers fuming and just as frustrated at the delay.

We finally approached the bottleneck as three lanes compressed into one. There, none of us could miss the huge sign that the burly, tar-spattered operator had attached on the machine's side:

The road to happiness is always under construction.

Our frustration dissolved into smiles, of course, and we waved friendly, appreciative acknowledgment as we passed.

It's sort of like the Christian life, isn't it? God's Word doesn't claim our path in life will always be smooth—because we all need repairs and construction along the way from time to time. But maybe our way will be smoother if we accept the rough spots for the small irritations they really are?

Lord, when diversions appear in my path, give me a smile, some humor—and the generosity of spirit to share with others.

—ISABEL CHAMP

17 THURSDAY

But covet earnestly the best gifts... —*I CORINTHIANS 12:31*

When I was four, my Southern family of grandmother, aunts, uncles, cousins, along with my immediate family, gathered for a reunion in a large beach house on St. Simons Island off the coast of Georgia. When it came time to leave, no one seemed to realize my misery at not being able to stay at the ocean. I felt so happy there.

But, I discovered, someone did realize. As Mother, Dad, John and I headed for our car to drive the two hundred miles inland to Macon, Rosa took me aside. Rosa, the large, quiet, smiling cook, handed me a bag, saying, "Here, missy, this is for you."

The little brown bag was full of beautiful shells—memories to take home with me. She'd awakened early to gather them before making breakfast. Somehow in my childish mind I knew her act of kindness to be much more important than anything I'd yet received.

Rosa wasn't wealthy, but she was perceptive and willing to go out of her way to make someone else happy—and those, I realized as I gazed at the shells throughout my childhood, were the essentials for gift-giving.

Dear Heavenly Father, give me a sensitive spirit, one that looks outward to fill another's need. —SAMANTHA McGARRITY

18 **FRIDAY**

...And when thou hast shut thy door, pray to thy Father which is in secret... —*MATTHEW 6:6*

Today is my birthday, the only day of the year that is totally mine to spend exactly as I please. And probably I'll do exactly what I did last year because it seemed a perfect way to savor my special day.

I spent the whole day, from sunup till sundown, alone at a favorite spot in the Colorado mountains near our home. With my Bible, paper and pencil, but no watch, it was a timeless gift to have a day to think, read, pray, reflect, and be alone with God.

At home, I seek little pockets of quiet time early in the morning, or even in the car, tucked between errands and appointments. But the setting is hardly inspiring, and I am always interrupted by other responsibilities: a cat meowing at the door, the buzzer on the dryer, or a child's footsteps in the hallway.

My day alone in the mountains was different. I had time to get down through the layer of details that usually clutter my mind, into the deeper level of more meaningful thoughts. I had time to talk to God. And most of all, to listen to Him. I had time to refill my cup to overflowing.

And as I came back down the mountain at sunset to join my family for an ice cream-and-cake celebration, I knew I had received a blessed, precious gift—hidden in my *giving up* a day to Him.

Lord, in the midst of all my attention-grabbing responsibilities, help me find time away and alone with You. —CAROL KUYKENDALL

19 SATURDAY

By faith Abraham, when he was called... obeyed; and he went out, not knowing whither he went. —HEBREWS 11:8

Fear of the Unknown. I can feel it creeping up on me when I'm about to enter a room full of strangers, or when I have to attempt something new. In those times that I manage to resist the urge to run, it's partly because of a lesson my mother taught me.

I was a timid six-year-old at the time, and we were vacationing at the beach. Now that I was a good swimmer, Mom felt it was time I learned how to handle a boat. Every morning that summer, when the winds were slight, she would take me out in a small dinghy and teach me to sail. I learned how to find the wind, maneuver the boat, adjust the sail and come into shore.

Then one day I stepped into the dinghy...and Mom pushed me out into the bay all by myself. (Little did I know she was ready to rescue me if I needed it.) I was petrified. Clinging to the tiller and mainsheet, I sailed from buoy to buoy, never venturing far from shore. Gradually, with each tack, my confidence grew. By the time I landed, I was radiant. "I can sail!" I announced to Mom.

"All by yourself," she agreed, smiling at my victory.

That day I learned that overcoming fear is a matter of taking risks and trusting what you know. *Facing the unknown requires faith.* It's the courage God lends us; for, when we push off into unfamiliar waters, there's no telling what wonders we'll find!

Be with me, Lord, when I am afraid, and give me the courage to face the unknown... tack by tack. —RICK HAMLIN

20 SUNDAY

The Lord thy God shall bless thee in all thy works, and in all that thou puttest thine hand unto. —DEUTERONOMY 15:10

As I drove down the highway, I noticed a young man carrying a large plastic sack walking beside the road. Ahead of him the roadside was littered with trash. Each time he saw a piece of paper, he would stab it with a long stick and put it into the sack.

I looked back at the area through which I had just driven. How much more attractive the roadside was where he had done his work. I thought of other people who enrich my life through the everyday tasks they perform: the boy who tosses the newspaper on the lawn in time for me to enjoy it with breakfast; the postman who brings me letters from friends; the telephone engineers who keep my line in good condition; the people who keep our parks green and beautiful....

Think about it. How many people are there whose work you take for granted, yet who help to make your life a little better? Why not make a point of giving each one a smile and a special "Thank you" the next time you see them?

Lord, this day I ask your blessing on ________________, who adds so much to my life. Amen. —DRUE DUKE

21 MONDAY

Happy is the man that findeth wisdom, and the man that getteth understanding. —*PROVERBS 3:13*

I don't know about you, but I'm a pack rat. I'm always clipping inspiring or memorable sayings from magazines, or making notes of wise things people have said to me. I used to drop these clippings into drawers where they piled up like little snowdrifts. But unlike snow, they didn't melt, they just got increasingly dog-eared and lost in the shuffle.

Now I have a better method. I mount these items in a scrapbook. Just as I sometimes need a cup of coffee between meals for a quick pick-me-up, so do I sometimes feel the need for a spiritual pick-me-up during the day or when a problem threatens to overtax me. I thumb through my scrapbook and choose one item at random to think about. For instance, I'm going to open it now....

Okay, here's one clipped recently from a magazine:

Looking back—Thank Him.
Looking ahead—Trust Him.
Looking around—Serve Him.
Looking up—Expect Him.

How's that for a pick-me-up! (And it contains no calories or caffeine.)

Now I find that my scrapbook of sayings, which once only cluttered up my drawers, speaks out to me and enriches my life every day. Maybe starting one could enrich yours, too?

Wherever I find inspiration, Lord, I'll hold it close to my heart.

—MADGE HARRAH

22 **TUESDAY**

I write not these things to shame you, but as my beloved sons I warn you. —*I CORINTHIANS 4:14*

Last year my son and I spent a vacation at Banff Springs in the Canadian Rockies. On our last day there, we visited the construction site of a new hot springs resort area. The air was crisp as we strolled along the wooden walkway that led among large pines and down toward the steaming springs.

Suddenly Gary stopped. "Mom. Bear tracks!"

I stopped and stared at the mud along the path. There, frighteningly fresh, were the imprints of a large bear.

"Wow," I said. "I *did* want to see the springs—but I think maybe we'd better go back."

"I agree," Gary said. "Let's get out of here." We hurried back to our car, grateful for the warning that spoke so clearly: "*Danger*. Bear nearby!"

I am not always so grateful for warnings. Sometimes, like Lot's wife, I would rather ignore them and do just as I please. But God never allows me to be headstrong like that. He has clearly expressed His love and concern for all His children through His gift of the Ten Commandments. "Thou shalt not..." He warns us. What He is really saying to His beloved children in such warnings is this: "*Danger*. Devil nearby!"

Today I will be obedient, Father God, alert for Your signposts to guide my path.

—DORIS HAASE

23 WEDNESDAY

The Lord God hath given me the tongue of the learned, that I should know how to speak a word in season to him that is weary... —ISAIAH 50:4

The other day, out of the blue, I received a postcard from a young man who had been one of my charges when I was a camp counselor years before. "Maybe you never knew," he wrote, "how you reached me that week."

The letter reminded me of a kid I knew back in elementary school. Twice a week, a tall woman in black-rimmed glasses would knock on the classroom door to ask him and two others to come with her to speech therapy class. But for all he knew she could have been the Grim Reaper. The shy kid had always talked a little funny—when he talked at all. You could tell those sessions embarrassed him by the way he tried to sneak back. And he never, ever would talk about what went on there.

But something good must have happened, because, by the time he was in high school, he was doing radio announcing; and, after he graduated from college, he ended up as announcer for the local college's football and basketball games.

One other thing about that kid: he was—is—I. And what that card reminded me of is the lesson I learned back in fourth grade from the speech therapist whose name I no longer remember: that we never realize how much good we do by just showing love for someone. For that caring woman turned things around for me. A smile, a touch, a gracious word—that's how a speech teacher in Holland, Michigan, made a difference to a kid with a bad "s" and no self-confidence. And that's how you and I can make a difference to other people, too—even though we may never know it.

Help us to give life, O Lord, by giving love. —JEFF JAPINGA

24 THURSDAY

...Ye shall say unto this mountain, Remove hence to yonder place; and it shall remove; and nothing shall be impossible unto you. —*MATTHEW 17:20*

My daughter Ann loves the cartoon characters Garfield, the wise-cracking cat, and Odie, the ever-cheerful dog. One morning she wanted to share a Garfield comic strip with me, but I was busy at my desk and waved her away. "Later, honey," I said, concentrating hard on the papers before me. The project I'd taken on was giving me trouble, and I was beginning to wonder if maybe I had bitten off more than I could chew.

Later that evening, Ann returned with her comic strip. It showed Garfield and Odie racing up the side of a tree and sitting on a limb, while below their owner protested, "Odie! Dogs don't climb trees."

"It's amazing what you can accomplish when you don't know what you can do," observed Garfield.

The cartoon's message really got to me: Refuse to accept your limitations and those things you think you can't do, and you will be amazed at what you can accomplish.

No, I wouldn't drop the project that seemed so mountainous. With God's help, I would finish it in the spirit of Odie, who climbed a tree because no one ever told him he couldn't.

Lord, help me believe that with You all things are possible.

—SUE MONK KIDD

25 FRIDAY

...In quietness and in confidence shall be your strength... —*ISAIAH 30:15*

When my mother was a little girl, the family homestead burned down and almost everything they owned was lost. To this day she has a vivid memory of the family riding away from the gutted house in a horse-drawn wagon, her father, silent and inscrutable, sitting up front. The children looked to their mother to understand better how to deal with their loss.

My mother says she will never forget the expression on her mother's face. She was serene, smiling slightly and appeared totally unshaken. In one hand she held a single rose—the only thing she had been able to rescue.

As the children watched, she lifted the flower to inhale its lovely fragrance and then she allowed each child in turn to sniff at the delicate petals. No words were spoken. While the wagon carried the family away from their devastated home, the children and their mother focused their attention on the pristine beauty of a single red rose.

When as a child I heard that story, I thought my grandmother had made a poor selection as she ran out of the burning house. But now that I am a mother and grandmother myself, I am astonished by the wisdom and faith of this remarkable woman. Very simply she gave her children the essence of faith: to let go and let God in, for if we are but quiet and listen, we will find Him speaking to us in even our most difficult moments.

O Father, in the midst of misfortune, even loss, show me Your peace.

—MARION BOND WEST

26 **SATURDAY**

And the Lord shall guide thee continually... —*ISAIAH 58:11*

As I walked along the path leading to the beach, I heard a sharp whistle behind me. Turning, I saw a bicycle approaching rapidly and as it neared, I realized it was a tandem with two youthful riders. Noting their perfectly synchronized pumping, I called out as they raced past me, "Nice teamwork, fellows!"

As they headed up the ramp to the beach area, one of the riders turned and momentarily stopped pumping as he sought the source of the compliment. Then the lad in the front seat yelled, "Hey, Steve, keep pumping or we won't make it up the ramp!"

I watched, fascinated, as they reestablished their teamwork and easily took the incline. As they disappeared down the other side, it occurred to me that their pulling together was somehow like a proper walk with the Lord. He's up front in the driver's seat and He'll steer us safely. But we must yield our will to His and we can't let up, turn away or coast. As long as we're part of God's team, we have to be right there pumping, even when the going gets the steepest—or perhaps, *especially* when the going gets the steepest.

Father God, no matter how difficult the task You set before me, I know that together we can do it. You lead, and I will follow.

—SAM JUSTICE

27 SUNDAY

And we know that all things work together for good to them that love God, to them who are the called according to his purpose. —*ROMANS 8:28*

My mother had many virtues, but she had one persistent if amusing fault. She took the weather personally. A beautiful day she regarded almost as her individual benediction. She sang and accomplished things and loved the world. On gloomy days, she would stew and fuss.

My grandfather, on the other hand, had two dear homilies that he applied to life's situations, including the weather: "It's all for the best, but we can't see it," and "No great damage without some small good." When hailstones battered the crops, or droughts parched the fields, or a cyclone took the roof off the barn, "We'll be strengthened by this adversity," he'd solemnly declare. "The barn needs a new roof anyway, and look how many neighbors want to help. Why, except for this we'd never appreciate how good people really are."

Grandpa's philosophy has helped me through many of life's storms. There are some conflicts and problems simply beyond our control. But there is one thing we *can* do about them: Trust in God to use them for some great good we can't see yet, and meanwhile find some immediate good in them for which to thank Him.

Father, help me to look for the blessings that are always scattered through the storms. —MARJORIE HOLMES

28 MONDAY

Bear ye one another's burdens, and so fulfill the law of Christ. —GALATIANS 6:2

All through church yesterday, I choked back tears. A personal concern lay heavy on my heart. There's a time in our service when we are asked to greet one another, passing the peace of the Lord to our neighbors in the pews. As I squeezed the hand of my friend and prayer partner, Carolmae, I noticed that there were tears in her eyes.

After church, as we sipped coffee in the fellowship hall, I asked her if something was wrong. "Do you have a minute?" she asked, and we went back into the sanctuary. Her problem tumbled out in a stream of tears and we prayed together until her eyes were dry and there was a smile on her face. As I drove away from the church, it occurred to me that I no longer felt like crying. My problem was still there, as serious as before, but inwardly I felt renewed.

Are you hurting today? Do you know anyone who is lonely or ill or grieving? Perhaps visiting and praying with someone else who is in pain will bring healing to you both.

Lord, send me to someone who needs my love and prayers. Then wrap us both in Your loving arms. —MARILYN MORGAN HELLEBERG

29 TUESDAY

For where your treasure is, there will your heart be also. —MATTHEW 6:21

I once clipped a newspaper interview with Chicago art dealer Ruth Volid. "I have taken care," she said, "not to clutter up my space with things that have no meaning in my life."

I found that clipping one evening when I was knee-deep in boxes of books, clothes and dishes, packing to move from Chicago to Atlanta. I can't say that it made me immediately toss out everything, but it did give me a new criterion for keeping things. Not merely "Do I need this?" but *"Does this have meaning in my life?"*

As we've settled into a new house, a new routine and new friendships, I find myself asking, "Does this have meaning in my life?" I see that if I don't clutter my mind with certain books or television shows, my time with certain meetings, my daily schedule with certain errands, I have more time to enjoy my family *and* more time to do the things that really matter.

For most of us, clearing away clutter will be a lifelong task. But it's a task worth doing again and again. For it's only when we get rid of the true unessentials in our lives that we can find the most lasting "treasures."

Dear Lord, my greatest Treasure, help me to let go of things that don't matter so that I can better treasure the things that do. Amen.

—PATRICIA HOUCK SPRINKLE

30 WEDNESDAY

Give us this day our daily bread. —*MATTHEW 6:11*

"There's so much to get done!" I cried in frustration to my friend Bob. "I feel like I'm just spinning my wheels!" Renovating a cottage I had recently bought in the country was more demanding than I had anticipated.

"Look, Sam," he said, "when my wife died and left me with a baby daughter, I used to worry myself sick about the future. How would I be able to raise her *and* hold down a job? How could I send her to college? Then I realized that all I really needed to get us through was one day—*one day at a time*. So each morning I made a list of the things that had to get done, and concentrated only on it.

"You don't have to worry about the next decade, or next year, or even tomorrow, Sam. Just today. So, make a list of things that have to get done today and work on only that."

Later I thought about Bob's wise advice. And I thought about lists that God had made for us: the Ten Commandments, the traits of a virtuous woman in Proverbs, the guidelines in the Sermon on the Mount, even those things He accomplished each day of Creation in Genesis. Maybe Bob had something there, after all.

Later that day, I made a list. Slowly, one day at a time, that little cottage is becoming a home.

Heavenly Father, these are my goals. Please bless my attempts today, while I leave tomorrow in Your care. —SAMANTHA McGARRITY

31 THURSDAY

Now unto him that is able to keep you from falling, and to present you faultless... —JUDE 1:24

I'm embarrassed to confess: I'm terrible at sports. As a schoolboy I was usually picked last when teams where being chosen. The basketball bounced off my toes, footballs never stayed in my hands, my volleyball serve was shaky. And yet, then as now, I enjoy taking part: inexpertly whacking a tennis ball across the net, trying to dribble a soccer ball down the field, maneuvering a dinghy without tipping over.

Why keep trying? When I taught English, a colleague of mine once advised me: "Rick, if you're a teacher, you should also be a student once in a while *at something you don't do well.* It'll help you understand what your students are going through, give you patience...and a sense of awe."

For me that something is sports. Maybe there's something for you—cooking, sewing, singing, riding a horse? I just know that whenever I do something athletic, I come away awestruck at the effortless grace of an ice skater, the brute strength of a weight lifter, the endurance of a long-distance runner.

Oh yes...and awe brings me closer to God.

Oh Lord, give me the daring to face new challenges, the strength to keep on trying when I fail, and the awe of Your presence in all my achievements. —RICK HAMLIN

Praise Diary for July

1

2

3

4

5

6

7

8

9

10

11

12

13

14

15

16

17

18

19

20

21

22

23

24

25

26

27

28

29

30

31

August

S	M	T	W	T	F	S
					1	2
3	4	5	6	7	8	9
10	11	12	13	14	15	16
17	18	19	20	21	22	23
24	25	26	27	28	29	30
31						

Keeping Score

Time and again I tell myself
and time and again I fail,
To remember that in serving You
the goal's not to prevail,
I need not always lead the pack
or win at every game,
I need not prove each day my worth
or seek to raise my name,
It's not how much we're worth
or for a spell possess,
It's not how big our house
or how richly we are dressed,
But rather how we serve and give
how we love and share,
The gifts that You have lavished
beyond all earth's compare.

Blessed are the poor,
blessed are the meek,
Blessed are the merciful
and those who My will seek,
Blessed are the lonely,
blessed are the last,
In heav'n they'll be exalted;
it all shall come to pass. —*FRED BAUER*

1 *Practicing His Presence in August*

CARRYING AN IMAGE OF PRESENCE

...Keep this for ever in the imagination of the thoughts of the heart of thy people, and prepare their heart unto thee.
—*I CHRONICLES 29:18*

One evening while my husband was on a trip and the children were visiting Grandma, I grew lonely, feeling deserted in the quietness of the house. To take my mind off my aloneness, I began to prepare for a Bible study I was to teach. I sat on the floor, leaning against the foot rest of a wing chair, and read some verses in Luke. Then I came upon a phrase—"Mary...sat at the Lord's feet." Suddenly the scene seemed very real to me.

I closed my eyes and began to imagine it, letting it unfold in my thoughts. I saw Christ sitting on a large winged throne, His face turned down toward a woman who was sitting on the floor beside His feet. I could only see her back, but I watched her—Mary—as she stared up at Him. I sensed how warm and precious being with Christ was for the woman. She seemed totally absorbed, as if a curtain of love had shut away every distraction and care. Then as my imagined scene was about to end, I saw the woman turn her head. And it was not Mary at all. *The woman was I!*

The image of myself at Christ's feet gave me a vivid sense of nearness to Him. The picture lingered in my mind, and I found myself returning to it throughout the evening. The lonely feelings I'd had earlier were forgotten.

During August, join me in making time for some imaginative adventures. Select some of your favorite Gospel stories in which a person encounters the presence of Christ. Then, closing your eyes, try to "image" the event, letting the face of the person become your own...and let the healing power of Christ touch you in a special way.

Lord, fill me with bright images of Your Son that will travel with me through this month.
—SUE MONK KIDD

2 SATURDAY

Thou art that God that doest wonders: thou hast declared thy strength among the people. —*PSALM 77:14*

It wasn't an ordinary sandcastle. I've seen many of those in my lifetime and they are sweet endeavors, with scooped-out moats and sandpail-shaped mounds all smooth and round and tapering toward a blunt top. But this one was a work of art that must have taken hours to complete.

Graceful pinnacles rose like stalagmites, some of them two and a half feet high. None of the pinnacles was smooth; each was delicately sculpted with not a sign of a finger mark. It was as if some great power had caused the castle to rise from the beach in joy. Moats and pools of ocean water threaded their way around the pinnacles, reflecting their loveliness. At one point water flowed under a delicate bridge of sculpted sand connecting two spires that rose into needle-fine points.

The scene was so magnificent that it stopped me in my early morning walk. Yet the castle was made of ordinary sand. Just as we are ordinary sand—until God touches our lives and turns us into masterpieces.

Oh, yes, He can do that—if we let Him!

Bring out the best in me, Lord, so that I can serve You better.

—PHYLLIS HOBE

3 SUNDAY

Sir, we would see Jesus. —*JOHN 12:21*

I picked it up from among the soiled clothes and toiletries in my husband John's suitcase: a rectangular block maybe five inches long, redwood from its color, rough beneath my fingers, with short pieces of a lighter wood glued on one side. The blond pieces formed no pattern or image—they didn't even make a pleasing design.

"I admired it on someone's desk," John said, "and he gave it to me."

"I don't get it. What's it for?"

"To look at." He set it on the dresser.

"I don't get it," I repeated. It was just a few scraps of wood, crudely stuck together. I hoped he wasn't planning to keep it on the dresser beside my Spode pin dish. I carried the laundry down to the basement and when I came back upstairs I saw it. From the dressertop the word leapt across the room at me:

JESUS

That seemingly meaningless arrangement of light-colored sticks in fact formed the negative spaces of a name-plaque that speaks to me about God's presence in my life.

Sometimes I see the name so clearly I wonder how I could ever miss it; other times I can't see it at all. Then I remember: It's not the raised parts that spell His name, but the hollows and valleys in between.

When things don't make sense, Lord, it's only because I haven't caught sight of You in the midst of them. —ELIZABETH SHERRILL

MONDAY

Let us come before his presence with thanksgiving...

—PSALM 95:2

Last year I had the opportunity to meet a very successful Christian businessman. When I asked what had contributed to his success, he answered, "Gratitude." He explained that during his early days of struggle, his pastor had counseled him.

"Never forget to be grateful, Carl. Be generous in showing appreciation. Remember, none of us can succeed alone."

Carl listened. Each day as he traveled to work on the bus during those early years, he busied himself writing thank-you notes to *every person* who had offered any kind of help or encouragement. He follows that practice today and owns one of the most successful public relations firms in Los Angeles.

How wise that pastor was. It takes only sixty seconds to say "thank you," and yet you may reap friendships that last a lifetime. And so I, too, have decided to follow that advice. Therefore I would like to thank *you* right now for spending these few moments with me...and I pray that God's blessings may rest upon you throughout the day.

Thank You, Father God, for... everything. —DORIS HAASE

5 **TUESDAY**

And he said unto them, Full well ye reject the commandment of God, that ye may keep your own tradition.

—*MARK 7:9*

Whenever I read about the Pharisees and Sadducees in the New Testament, I find I can easily become smug about my ways and indignant about theirs. How could they have been so blind to Jesus in their midst? Even after He had risen from the dead and had witnesses to the fact, we are told in Acts 4 that the Sadducees arrested Peter and John for preaching His resurrection because they "did not believe in the resurrection of the body." They preferred their traditions to God's continuing revelation.

Then I read again Jesus' summary of the commandments: "*Thou shalt love the Lord thy God with all thy heart, and with all thy soul, and with all thy mind...(And) love thy neighbor as thyself.*" (Matthew 22:37, 39). I begin to get uneasy. I remember a day when a friend suggested I change a habit, and I countered with an angry, "That's the way I *am*." I remember a meal to which I could have invited a lonely old man, but decided instead to preserve a tradition of "Christmas is for family." I remember insisting on my own way in a family argument. *Do I, too, sometimes prefer my own traditions to God's commands*?

It's easy to criticize the Pharisees. It's also easy sometimes for me to hold tightly to my own way of doing things—unless I keep my own eyes and heart open to Jesus.

Dear Jesus, forgive me for preferring my own ways to Yours, and teach me truly to love God and others. —PATRICIA HOUCK SPRINKLE

WEDNESDAY

Judge not, that ye be not judged. —*MATTHEW 7:1*

Last year my wife Shirley and I made a long-anticipated first visit to Japan. On a long list of sight-seeing highlights is the Daibutsu *Buddha*, the famous bronze statue of Kamakura, south of Tokyo. Cast in 1252, the sitting figure is not the country's largest statue of Buddha, but is considered priceless as a work of art.

One of the features of this monument, which is forty feet high, is that for a few yen tourists can enter its base and stand inside the figure's body and head—which we did. As we studied the inner construction of the statue, Shirley commented that the experience made her wonder what it would be like to be "inside God's head" and see and think as He does.

Her thought came back to me the other day when I started to pass judgment on the motives of a businessman I know. "First, look upon his heart," an inner voice prompted. "But I can't see his heart," was my response. The thunderous silence that followed told me to "stop right there, and to judge not."

Instead of giving others the benefit of the doubt, Lord, help me give them the benefit of my faith. —FRED BAUER

7 **THURSDAY**

For God is not unrighteous to forget your work and labor of love, which ye have showed toward his name, in that ye have ministered... and do minister. —*HEBREWS 6:10*

When I was a child, our house was so small that my parents didn't often entertain. But when they did, for days ahead we children lived in excited anticipation of "company for dinner."

While my mother prepared the meal, my father would widen the the round oak table. Its six leaves had to be cleaned, then their pegs fit into matching holes. This wasn't easy because the leaves had warped. After that the auxiliary legs had to be lowered so the table ends wouldn't collapse when they were loaded with dishes.

Then there was the matter of chairs. Two were stored in the attic and had to be let down through the ceiling. Another two were stacked with books in the bedroom (we had no bookcases). Several more were brought in from the barn. Often Daddy glued in a leg or wired a broken seat. And patiently, he wiped each one of them clean.

At mealtime, the company would always compliment Mother's cooking. "You did a marvelous job!" they'd say. I can't tell my father—he's been gone for years—how it was only recently that it came to my mind that nobody ever said to him, "Thanks for rustling up chairs for us," or "You did a great job in setting everything up."

I recalled this childhood scene recently when I noticed someone in our church working at a task so ordinary it normally drew no comment. On impulse, I stopped and said, "May God bless you. He knows how truly important your work is to tonight's meeting."

Father, help me to be sensitive to the quiet, self-effacing ministries of others. May their small labors of love show me the Way.

—ISABEL CHAMP

FRIDAY

Cast away from you all your transgressions, whereby ye have transgressed; and make you a new heart and a new spirit... —*EZEKIEL 18:31*

When I first started using my word processor, where information is stored on a floppy disk, a writer friend warned me, "If you give a wrong command, you can wipe out a disk, so always keep a backup copy of everything you write."

But I got so confused just learning the basic procedures that I put off learning how to make backups. Then one day I accidentally gave my computer two commands at once and it scrambled forty-seven pages of my novel into garbage. What a shock! For hours afterwards I just wandered around, depressed, discouraged, mad at myself. God had to listen to a lot of prayers that day. It wasn't until the next morning that I found the courage to start over.

Starting over after messing up—boy, is that hard! If it hadn't been for God's support, I don't think I could have done it. But He says over and over in the Bible that He knows we're not perfect and He's standing ready to help whenever we need Him. In fact, no matter *what* we've done, he says that we can, through Him, cast away our transgressions and receive a new heart and a new spirit.

To start over with new spirit—what a wonderful gift from God!

Thank You, Father, for new beginnings. Send me one today.

—MADGE HARRAH

SATURDAY

And God blessed them, saying, Be fruitful and multiply, and fill the waters in the seas... —*GENESIS 1:22*

I do enjoy unusual vacations. So, when I was invited to spend two weeks studying the migratory habits of turtles at a remote station in the Caribbean, I accepted. But, now that I was sitting all alone on a darkened beach in Costa Rica, I had misgivings. *Did I really care all that much about turtles?*

Then, in the bright moonlight, I saw a huge green turtle coming in with the breakers, rising with a wave, then bumping back softly on the sand. She peered into the night, as if making up her mind. Then, slowly but doggedly she made her way to the beach vegetation. I watched, mesmerized, as, flinging sand in every direction, she used her long front flippers to dig a pit for her body. With one of her rear flippers, she dug out an egg chamber. Soon she began to give great heaves, and the chamber filled with a hundred eggs, each about the size of a golfball.

After a while, she stopped and seemed to be resting. Then her front flippers started up again as she covered the pit with sand to conceal it from predators. Finally, her mission accomplished, she turned and headed back to the sea.

I watched as she disappeared. *Dear God, how endearing she is*, I thought. *And how uncanny her ability to find her natal beach, so she, in turn, can continue her species.*

As I trudged back to the station, I said a prayer for the safety of the turtle hatchlings who would emerge from the shells in about sixty days. *God's creatures. He cares about them. And He wants me to care about them, too.*

Oh, Lord, when You consider all that You have made, including me, I'm in awe. Thank You for the blessing of life. —ELEANOR SASS

10 SUNDAY

Let your light so shine before men, that they may see your good works, and glorify your Father which is in heaven.

—MATTHEW 5:16

Ray Palmer was teaching school for girls in New York. One day he took out his pocket notepad and wrote down four verses he felt had come straight from his heart and soul. He tore off the little sheet of paper, folded it and slipped it into his vest pocket, carrying the verses with him wherever he went.

Then, two years later, Dr. Lowell Mason, the celebrated Boston composer, asked Palmer if he would provide him with a new hymn for his next volume of *Spiritual Songs for Social Worship*. Ray Palmer removed the four verses from his pocket and handed them to Dr. Mason, who subsequently composed the tune "Olivet" to accompany the words.

Years later the composer told Ray Palmer, "Sir, you may live many years and do many things, but you will always be known as the author of those words."

So today when you or I hear or sing the following words:

My faith looks up to thee,
Thou Lamb of Calvary,
Savior divine!

why don't we pause and give thanks for the gift God gave Ray Palmer a hundred and fifty-six years ago?

Dear Lord, help me share my light with another today—whether in thought, word or deed. Amen. —OSCAR GREENE

MONDAY

He shall call upon me, and I will answer him: I will be with him in trouble... *—PSALM 91:15*

I was moving from my Brooklyn apartment, storing all my things in a warehouse, then going to live in a country cottage. At three in the morning, after I had just packed box seventy-four, I sank exhausted on the sofa. "Am I doing the right thing?" I wondered aloud.

My new home, lacking in modern conveniences, was in dire need of renovation. I was feeling afraid and alone—this change would be quite a responsibility. "God," I said, "I need some reassurance, right now."

The radio had been playing softly, keeping me company during the drudgery of packing. At that moment a song came on: "Like a bridge over troubled water, I will lay me down...." It seemed to hold a special message for me.

That's how God is, I reminded myself. He's going to help me through the rough times—be that bridge over difficulties. With that I pulled myself up from the sofa and began packing box seventy-five.

Isn't it comforting? *We* may change, but *He* never does.

Heavenly Father, sustain me through all my comings and goings.

—SAMANTHA McGARRITY

12 **TUESDAY**

That the trial of your faith, being much more precious than of gold that perisheth, though it be tried with fire, might be found unto praise and honor and glory at the appearing of Jesus Christ. —*I PETER 1:7*

At those times when I feel things are getting to be too much for me, I remind myself of the bristlecone pine.

I first saw these trees last summer with Mom and Dad, when we were driving through California's Sierra Nevada. At 10,500 feet, we met our first bristlecones. What astonished me most was the terrain in which they thrive. Perched on steep cliffs, clinging to boulders, these gnarled and twisted trees live in the most inhospitable environment. Although the bitterly cold winter winds buffet them, they flourish in the high dry air of a mountainous home, which gets only ten inches of rainfall a year. Yet many of them are as much as four thousand years old.

"One of the enigmas of the bristlecone pine," the guide told us, "is that the oldest trees are found in the worst terrain. Trees on the protected side of the mountain rarely survive for more than a couple of hundred years. But these slow-growing granddaddies have weathered so many storms and suffered so many trials, it gives them mettle."

It gives them the sort of perseverance and strength that I try to emulate, because that, I believe, is what God intends for us to learn from their example.

Our Father, give me a "bristlecone" faith—one that remembers that today's struggles will bring me the strength to face tomorrow.

—RICK HAMLIN

13 **WEDNESDAY**

Let the words of my mouth... be acceptable... —*PSALM 19:14*

"Did you know you spoke forty-eight hundred words yesterday?"

The question from my son lifted my eyes from the typewriter to his face, ready to add to the total. I relaxed when I saw there was nothing personal in the question. He had lifted it from the Trivial Pursuits game.

But did I? I wondered. Did I *really* give voice to nearly five thousand words? I turned back to my work. But my thoughts were not with the written words but the spoken ones. Had they been kind? Had they been loving? Had I comforted someone...reassured another...or even coaxed a smile? Or had I been too busy getting my thoughts on paper to share them with another?

God's Word has a lot to say about ours! "A word spoken in due season, how good is it!" (Proverbs 15:23). "Comfort one another with... words" (I Thessalonians 4:18). "Every idle word that man shall speak, they shall give account thereof..." (Matthew 12:36).

There's more! But these references were good for starters. And it was not in "trivial pursuit" that I decided it would be wise to bridle my tongue in the days to come. Those five thousand words I am going to speak today should count for something in God's Kingdom.

Can my speech pattern be changed? Yes, I believe it can—not magically but by prayer:

Let the words in my mouth be acceptable, O Lord, I pray!

—JUNE MASTERS BACHER

14 **THURSDAY**

Therefore choose life... —*DEUTERONOMY 30:19*

Have you seen that Xerox commercial on television? The one showing the laughing little monk who proclaims, "It's a miracle!"? Well, I met him in person the day he visited our office. I was in the middle of filing when I heard a loud, cheerful voice. I looked up and saw a round little man wearing the flowing brown garb of a monk. He sauntered over to each desk, joked and pasted something on the front of every typewriter. When he reached my desk, I saw that "something" was a small disk imprinted with his smiling face and the words, "It's a miracle!"

It was only an innovative advertising idea, and yet to me it has come to mean much more. Daily that colorful label catches my eye and I am freshly reminded that being alive *is* a miracle! The reminder has brightened many dark moments.

If you would like such a memento, draw a happy face on a piece of paper with those words and put it where you see it every morning. It will lift your spirits and open your eyes to seldom noticed miracles. Miracles like sunlight, bacon and eggs, family and, yes, life itself!

Let me catch sight of today's miracles, Father God—large and small.

—DORIS HAASE

15 *WILDERNESS JOURNEY* (*Exodus 19, 20*)

Moses Receives the Ten Commandments

And the Lord called Moses up to the top of the mount...

—*EXODUS 19:20*

We're standing at the foot of Mt. Sinai. What a bedraggled, travel-weary band we are, as we hover together looking toward the peak where Moses, our leader, has gone to meet God. What is it about these mountaintop experiences? It was on Mt. Horeb that God first spoke to our leader. Again and again, Moses has gone back to the

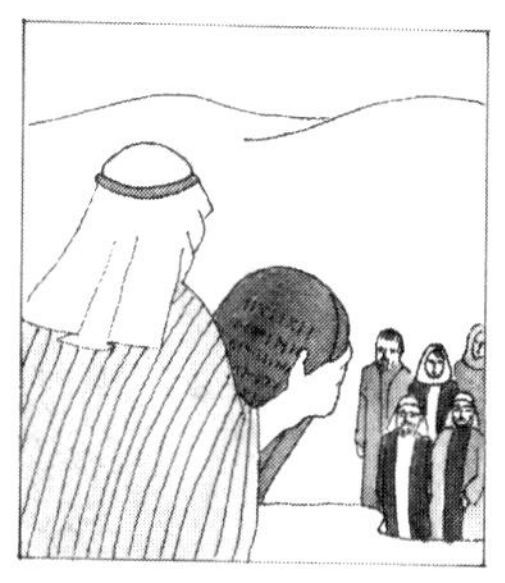

peaks, seeking his God, speaking with Him there, listening for His powerful voice.

Early this morning, we were startled out of sleep by ear-splitting thunder and blazing lightning. Then Moses gathered us at the foot of the mount and God called him to go up. Now we draw back and cling together, trembling, as fire gushes forth from the peak and smoke billows out. Was that the sound of a trumpet? After what seems like an eternity, our leader descends again, through clouds of smoke. A hush falls over the crowd as Moses begins to speak. "God has given us ten fundamental rules for our lives that we must never break...." After Moses speaks the Ten Commandments, he continues with God's message, which includes instructions for *every detail of our daily lives!*

This knowledge makes me catch my breath and fall on my knees. This God, this mighty, exalted Being Who holds the whole universe in His hands, truly cares about every little detail of my daily life. He's just as ready to help me find my car keys as He is to advise me on that big career decision. And I can climb the mountain with all of my little and big concerns and speak directly to Him! He may not answer in thunder and lightning and trumpets, but if I sit very still upon the mountaintop, I know that I will hear His quiet voice within. And I will know what to do.

If you need some good advice today, close your eyes, climb Mt. Sinai with Moses, and place your concern before the Lord. Then listen for His answer.

Almighty God, I know that whatever matters to me, matters to You. Thank You for being so infinitely caring.

—MARILYN MORGAN HELLEBERG

16 SATURDAY

Return unto me, and I will return unto you, saith the Lord of hosts.

—*MALACHI 3:7*

Where Tiger came from we'll never know, nor where he goes when we're away. He simply sprang from the hedges one night when we drove in from our weekend at the lake; a striped gray tom, howling as if to demand, "Where have you *been*?" and entwining our legs like long-lost kin.

Because of our two dogs, we fed and bedded Tiger on the porch, thinking he'd surely be gone in a few days. But he promptly took up residence. And our instant, abiding love for him has never waned.

So it was with some regret that we left for a month's vacation that first year. Margaret, our next-door neighbor, agreed to feed him, but we doubted Tiger would be there when we returned. And sure enough it was true.

"Sorry about your cat," Margaret told us. "He hung around for a couple of days, then disappeared."

But as we stood mourning the inevitable, who should come scurrying up to purr his forgiveness and welcome but Tiger?

And thus it has been for almost four years. Mysteriously, he senses our homecoming. And no matter how long we may have stayed away, he comes rushing upon us, showing in every way he can how much he loves us, and how glad he is to have us home.

Lord, no matter how far from You we stray, or how long we may be gone, You always faithfully await our return. —MARJORIE HOLMES

17 SUNDAY

O Lord my God, I cried unto thee, and thou has healed me.
—*PSALM 30:2*

My sixteen-year-old son Jeremy and I had such a bitter argument that morning he wouldn't have come to church with me, except that Glenn, the visiting minister, was an old friend. After the service, Glenn greeted us both warmly. He spoke first to me, then turned to Jeremy. "Hi, Jeremy, how're you doing, boy?" he said, looking directly into Jeremy's eyes as he shook his hand.

It had been a rough year for Jeremy. His father had died the year before, he wasn't doing too well in school, he hadn't been able to find a part-time job and he didn't like being under my sole authority.

"Fine, just fine," he said, with a forced smile.

Glenn shook his head. "No, you aren't, son. You're hurting. You're hurting real bad, aren't you?"

Jeremy blinked and swallowed. "Yes, sir, I guess I am."

"I can see it, boy. I've learned to spot pain a mile away." He grabbed Jeremy in a bear hug and held him. When they separated, Glenn said, "I love you, son. God has told me to spend some time with you and your brother. I'll call you, okay?"

"That would be fine, sir. Thank you."

But Glenn hadn't finished. "Jeremy, there's nothing wrong with hurting. We all hurt. I've hurt so bad I thought I'd die. God healed me. And He'll heal you, too, if you let Him. But He can't do it unless you're honest about your hurts, get them out in the open where they can be seen."

Driving home, I thought: When we're sick we tell the doctor where it hurts, we show the dentist the painful tooth, we even tell the man in the shoe store where the shoe pinches. Why is it we try to hide our emotional pains from God, the greatest Healer of all?

Father God, help me deal honestly and openly with this wound of the spirit: ______________________________. —MARION BOND WEST

IN CELEBRATION OF SMALL TOWNS

My Savannah: Behold Its Goodness

Last year we were treated to glimpses of the big city. Many readers wrote us in surprise, "Hey, city people aren't so different from country folk!" Exactly.

We're a diverse nation, of diverse people, interests, customs and environments. But we're one nation under God. *That means our diversity can bring us together and enrich us.*

That's what our southern writer, Arthur Gordon, hopes to do. This week he shares his historical seaport hometown of Savannah, Georgia, with us. (Actually Savannah isn't exactly a town, it's a small city, but it has the feeling that we think of as "small-town," in the warm, positive sense of the words.)

This town is America. It is you. It is me. It belongs to us all. Together, let us behold its goodness. —*The Editors*

18 DAY ONE—*A Small Town's Uniqueness*

And God saw everything that he had made, and behold, it was very good. —*GENESIS 1:31*

Whenever I visit up North or travel to cities in the West, people sometimes ask me what life is like in a small town in Georgia, in the deep South. "It must be very different," they conjecture.

Yes, it is different, I tell them. No one would ever mistake the colonial charm of my hometown of Savannah for the super steel-and-glass structures that tower over New York City. Likewise, no one would ever mistake the forested mountain slopes of New England for the Everglades of Florida, or the plains of Montana for the bluegrass of Kentucky, or the Arizona desert for the lakes of Minnesota. Diversity, that's one of the glories of our country.

We're all Americans, and we all speak the same language and watch the same TV programs and elect the same president, but the regional differences persist—different foods, traditions, values, life-styles. And it's out of this wonderful "differentness" that the color and texture and character of this marvelous country are created.

It's just one more example of the infinite variety and inventiveness of the Creator of all things. No two individuals alike, no two snowflakes alike, no two sets of fingerprints alike. And no two towns alike, either. No matter what, yours is going to be different.

So is mine. Let me tell you something about it.

Lord, give us the capacity to be amazed and delighted and enchanted by the diversity around us. —ARTHUR GORDON

19 DAY TWO—*A Small Town's Founding*

Be thou my strong habitation... —*PSALM 71:3*

Did you ever stop to wonder how a town came to be? Something must have drawn the first people to that place. Perhaps someone built a hut where two trails crossed. Maybe a spring of clear water attracted a pioneer. Or some government built a fort, and cabins sprang up around it. Whatever the reason, I think the first basic purpose often lingers and subtly colors everything.

That's true of this old coastal town in Georgia where I live. More than 250 years ago a group of 130 settlers sailed across the ocean from England and landed here. They traveled slowly up a winding river looking for a deep anchorage and a high bluff on which to pitch their tents, and found both about sixteen miles from the sea. There they started the thirteenth—and last—of His Britannic Majesty's colonies; and they named it after him, *Georgia*.

Those first settlers came with high hopes and high ideals. Since the colony was supposed to be a military outpost protecting the Carolinas from the Spaniards in Florida, their leader was a soldier, General James Oglethorpe. When the Spaniards did march north, his tough Scottish Highlanders drove them back.

Oglethorpe brought a city plan that survives to this day. It called for straight, wide streets and evenly spaced squares, some containing wells, others fenced to allow cattle to graze. Today they are green oases where squirrels scamper and people drowse on benches on hot summer days, and bronze statues of long-dead heroes look down endless aisles of tree-lined streets.

The original settlers were constantly invoking God's blessing on their labors and on their struggling little settlement. And their prayers must have been heard, because, despite wars and pirates, plagues and pestilence, Savannah did survive. Now people come from all over the nation to visit and sometimes make it their home.

Father, we take comfort in the knowledge that, wherever we make our dwelling place, You are and always will be with us.

—ARTHUR GORDON

20 DAY THREE—*A Small Town's Heritage*

...And after the fire a still small voice. —*I KINGS 19:12*

A poet once observed: "The trouble with a kitten is that it grows up to be a cat." Sometimes I wonder if that isn't also true of small towns. If you don't watch out, they grow to be large towns, and then big cities.

There's nothing wrong with big cities—or with cats, for that matter. But there's something comfortable about smallness. In a smaller town or city, you can see the threads in the tapestry more plainly.

Take the spiritual heritage of my hometown, Savannah, for instance. The Church of England came ashore with the first settlers in 1733. Within three years John Wesley was holding services in a crude wooden building and George Whitefield had started the first Sunday school in the world. Moravians came from what is now Czechoslovakia—strong, pious Christians who refused to bear arms even when the Spaniards were threatening the colony. Irish settlers fleeing from the potato famines brought the Roman faith and built the tall cathedral on Abercorn Street. The Greeks came with their Orthodox forms of worship. Jews from Portugal sought freedom from persecution here. And always the blacks, whose yearnings for a better life found a haunting voice in their spirituals, songs that wring the soul with their beauty and their sadness.

Many forms of worship, each like a single silvery note of music, combine to make a mighty chord of worship that reverberates in the air we all breathe. You can hear it all over America, of course, but perhaps it's a bit more audible in small towns. I like to think so, anyway.

Thank You, Lord, for providing a safe harbor for all, no matter how we worship You. —ARTHUR GORDON

21 DAY FOUR—***A Small Town's Spirit***

Simon Peter saith unto them, I go a fishing. They say unto him, We also go with thee. —*JOHN 21:3*

When you live in a seaport, you're never far from the spell of the ocean. My hometown lies on the edge of vast coastal marshes, laced with silver creeks. I always feel most myself when I'm on salt water—or in it. So this morning I got up early to go shrimping. I put my cast-net in a tin bucket and went out in my little boat.

I usually go at low tide, when the chocolate-brown mud flats are exposed. In summer the shrimp feed along the edges of these flats in water one or two feet deep and you can catch them by throwing out a circular net, one with weights around the edges and a retrieving rope fastened to your wrist. You seldom actually see the shrimp; you just keep casting the net until some are caught. Sometimes the net brings up other creatures, too: an angry blue crab, or maybe a terrapin—the small tidewater turtle once considered a great delicacy. I always let the terrapins go, though. It's funny—as you grow older, releasing things you've caught is just as much fun as catching them.

The design of a cast-net hasn't changed much in four thousand years—I've seen similar nets in Egyptian wall paintings dating from 1300 B.C. The one I use is ten feet wide and is quite heavy when wet. But when I swing it with a kind of lazy rhythm, I don't get tired because I feel I'm part of a pattern that goes back to the dawn of time.

Sometimes when I grip the edge of the muddy net with my teeth, as a cast-netter must, and spin it out over the quiet water, I feel a certain kinship with the disciples of Jesus. Some were fishermen too, and good ones, I'm sure. Remember that last scene in the Gospel of St. John where they fish all night with no luck? Then a stranger standing on the shore tells them to cast their nets on the right side of the boat, and they haul in 153 fishes (not just "a great many" or "a lot," but exactly 153—a fisherman always knows precisely what he has caught!). Then John cries, "It is the Lord!" And Peter gets so excited that he jumps into the sea and flounders like a big shaggy retriever to the shore....

I like to think that whenever I fish, I'm casting my net...on the *right* side.

Lord, guide us as we cast our nets into the sea of life, so that when we draw them out they will be full. —ARTHUR GORDON

22 DAY FIVE—*A Small Town's Legacy*

...Your old men shall dream dreams, your young men shall see visions. —*JOEL 2:28*

Sometimes here in this old town an early morning mist steals in from the marshes, blurring trees, muffling sounds, softening the outlines of buildings, touching everything with mystery. If you go out and walk along the paths that wind through the old colonial cemetery, you can feel all around you the presence of departed ones who walked this way before you were born, but have left something of themselves behind.

They are there in the shadows...General Oglethorpe and John Wesley and his brother Charles...the Count D'Estaing whose French fleet tried to take the town, and the fiery one-armed British Major Maitland who fought so hard to stop him—and succeeded ...the elegant Lafayette who spoke to the populace from a balcony...Eli Whitney, the young Yankee from Connecticut who invented his cotton gin on a plantation nearby...Robert E. Lee, who was stationed here as a young officer...William Makepeace Thackeray, who wrote part of his novel *Henry Esmond* during a visit here ...Jefferson Davis, who was received with quiet dignity after the Confederacy collapsed...and General Sherman, whose bluecoats staged a victory march down Bay Street at Christmastime in 1864.

And with them, perhaps, the seven-foot figure of Tomochichi, the Indian chief who was such a good friend of the first settlers that Oglethorpe took him all the way to London to visit the King. The old warrior once told Oglethorpe that, instead of praying to God, he left it to Him to do what was best for everybody. He thought that "God of Himself would do for everyone what was consistent with the good of the whole, and that our duty to Him was to be content with whatever happened in general, and thankful for all the good that happened in particular."

Gratitude. The old Indian believed that it was one of the sincerest forms of worship. I believe that he was right.

Lord, we thank You for answering not just the prayers of each one of us but the prayers of every one *of us.* —ARTHUR GORDON

23 DAY SIX—*A Small Town's Flavor*

...And let them keep food in the cities. —*GENESIS 41:35*

Every section of the country prides itself on certain dishes, certain recipes handed down for generations. Clam chowder in New England. Spoon bread in Virginia. Mountain trout in Colorado. Chili in Texas....

Which area offers the best when it comes to really good eating? I'm prejudiced, of course, so I'm inclined to agree with Ogden Nash, who wrote something to the effect that "every region thinks its food is gorgeous, but I like Georgia's."

Down here in Georgia we can offer you a marvelous she-crab soup. And marinated shrimp you spear with toothpicks. And Creole kisses—feather-light little meringues full of chopped nuts. And, of course, grits, much maligned by ignorant outlanders, but still a noble food, especially when combined with shad roe. I always liked the story of the English visitor who, when offered some grits, asked apprehensively, "Could I try just one?"

Soul food is a phrase that's used to describe things like spareribs and collard greens and black-eyed peas. But isn't all food soul food when you come right down to it? What about the five barley loaves and two small fishes that fed the five thousand (Matthew 14:17)—wasn't that soul food?

The subject of food appears all through the New Testament. When Jesus raised the young girl from the dead, He told her parents to give her something to eat (Luke 8:55). And, in His last appearance to the disciples, He said to Peter, "Feed my lambs...feed my sheep" (John 21:15, 16).

Feed them with what? *With food for the soul, of course.* Because the soul needs nourishment just as much as the body. Maybe more.

Father, we thank You for giving us our daily bread, and for giving us spiritual sustenance along with it. —ARTHUR GORDON

24 DAY SEVEN—*A Small Town's People*

Remember now thy Creator in the days of thy youth...

—ECCLESIASTES 12:1

There's a lovely old house downtown where two stone lions guard the doorway, just the right size for a small boy to sit upon and imagine he is monarch of his own private kingdom. The house once belonged to my aunt, Juliette Gordon Low, who was founder of the Girl Scouts of America. And I was the little boy with the big imagination. The lions were British lions because my aunt's father-in-law, who built the house around 1840, was an English merchant.

The lions are still there but, now that the Colonial Dames own the house, nobody rides them anymore.

Aunt Daisy, as we called her, died when I was fifteen but I remember her vividly—warm, spirited, very deaf, but full of the joy of living. Aunt Daisy had no children of her own, but she had millions of adopted ones—Girl Scouts. They were always around in their (then) khaki-colored uniforms and Aunt Daisy was always busy with them: feeding them, teaching them, showing the girls ways to be their best.

She was an animal lover too. She had a mockingbird that would sit on her shoulder and nibble at her pen as she wrote letters. And a huge, somewhat malevolent parrot called Polly Poons. And countless dogs. All living things were loved and cared for by Aunt Daisy's generosity and goodness.

I recall Aunt Daisy once gave me a strong-willed pony named Buster Brown. When Buster Brown didn't get his way, he would rear up and fall on his back. I learned to be very good at jumping off in a hurry. When I complained to Aunt Daisy about this, she explained solemnly that ponies and horses see everything eight times its normal size, and so they're easily frightened. This struck me as a reasonable explanation, and I forgave Buster Brown everything...which was what she wanted me to do, of course.

At night, when I used to conclude my small-boy prayers with a list of people I wanted God to bless, Aunt Daisy was always on my list. Her love and zest for life lives on in my small-boy's heart and her legacy to all American girls has touched countless generations, from town to town, city to city, across our land.

Lord, let my life be an inspiration to others. —ARTHUR GORDON

25 DAY EIGHT—*A Small Town's Melody*

...And his sound shall be heard when he goeth in unto the holy place... —*EXODUS 28:35*

Different sections of the country have their own sounds. The hollow roar of an approaching subway train in New York, the muffled clang of a fog-shrouded sea buoy off the Maine coast, the mournful wail of a locomotive echoing through the hills of West Virginia, the brassy jangle of a San Francisco cable car...these are signature sounds that give a place its own distinct character and personality.

Sometimes such sounds fade away and are gone forever. When I was a very small boy, I used to wake up early and listen to the cries of the street vendors, marvelous old women with baskets on their heads, calling out, "Oyster buy-ah-h-h; ay-y-y shrimp, ay-y-y crab...." Each word a musical note, lovely and lonely, in a plaintive minor key, ringing out over the quiet town.

Still, not all has been lost. The songwriter Johnny Mercer grew up around here, and he caught the languid, indolent Savannah mood in many of his songs—"Lazybones," "Blues in the Night," "Moon River." Southern sounds all of them. And, when the chimes of St. John's Church ring out the old familiar hymns on a sleepy Sunday afternoon, the echoes swirl around the tall spire of the First Presbyterian Church where organist Lowell Mason wrote such hymns as "Nearer My God to Thee" and "From Greenland's Icy Mountains." Then I know that most of our signature sounds are safe and will be heard and loved long after I am here to listen to them.

Lord, we thank You for allowing us to dance to the music of life.

—ARTHUR GORDON

26 DAY NINE—***A Small Town's Faith***

In my Father's house are many mansions... —*JOHN 14:2*

This morning I went out to Laurel Grove to visit the graves of my parents and grandparents. It's a very old burial ground, quiet and peaceful, with gray streamers of Spanish moss swaying in the soft breeze and golden dust motes hanging in the shafts of sunlight that sift through the live oaks and the tall magnolia trees. There's a sense of timelessness about Laurel Grove.

In the South we accept death as a part of life. I've seen cars stop on country roads and the occupants get out and stand respectfully, hat in hand, while the funeral cortege of some total stranger winds slowly by. I remember, too, when I was a small child, being petrified about funerals. I remember apprehensively asking one of my aunts after a funeral, "Is Cousin Nelson really in that big box?"

And Aunt Harriet smiled and said, "What he used to wear is in that box."

She was right, of course. When you have that kind of faith, you know that death is only of the body, not of the soul. That's why you can hear a note of hope, and even joy, in the words of the old spiritual:

> I looked over Jordan, an' what did I see,
>   Comin' for to carry me home?
> A band of angels, a-comin' after me;
>   Comin' for to carry me home.
> Swing low, sweet chariot,
>   Comin' for to carry me home....

Nothing gloomy there, just a song about a chariot picking us up for the happy ride back to our Eternal Home.

At the end of the road, Lord, grant us a safe lodging, a holy rest and peace at last. —ARTHUR GORDON

27 WEDNESDAY

But as for me, I will come into thy house in the multitude of thy mercy: and in thy fear will I worship toward thy holy temple. —*PSALM 5:7*

Early one evening a young girl was stopped by the police in a Communist country and asked where she was going. She dared not admit that she was en route to a secret Christian meeting. Thinking quickly, she said:

"My Brother has died. Some of us are going to meet and talk about His life and how we love Him. He left each of us something wonderful, and I want to be present when His Will is read so that we can all celebrate together."

Can you think of a more beautiful way to describe going to church or a more wonderful Gift for which to be thankful today?

Lord Jesus, Brother of us all, never let us lose sight of the Gift of Your life for us. Amen. —DRUE DUKE

28 THURSDAY

And in the fourth watch of the night Jesus went unto them, walking on the sea. —MATTHEW 14:25

One summer on a visit South, a friend and I drove along the New Orleans Gulf Coast. Gazing out over the huge body of shimmering water, I observed three people in the distance—*walking on its surface!*

"But that's impossible!" I gasped.

My companion laughed. "The water barely covers a mile-wide shelf of rock in this area and is only a foot or two deep. So you see," he said, "there's a logical explanation for everything."

Not quite everything, I thought, remembering Jesus' disciples. They had no logical explanation on the day they saw Jesus walking toward them on the Sea of Galilee. They had only faith. In fact, Peter's faith was so strong he was able to get out of the boat and walk on the water toward his Master—until fear caused him to sink (Matthew 14:30).

Sometimes fear attacks my faith too, and I think I might sink into despair, but when I follow Peter's example and cry "Lord, save me!" Jesus *never* fails me—He always stretches out His hand. What can be more logical than to believe in a Lord like that?

Your hand is all I need, Lord Jesus. —DORIS HAASE

29 FRIDAY

Charity never faileth... —I CORINTHIANS 13:8

A large withering plant stood abandoned in our office. "Sam, why don't you take this one on?" a co-worker suggested, poking his head into my office. "Maybe you could save it."

"I don't know," I mumbled, looking around at the abundant pots of cheerful greenery that already surrounded me. I gave the plant an unsympathetic glance. It looked hopeless.

"Come on, Sam, he'd fit right into that sunny corner."

"Oh all right," I reluctantly agreed. We tied up its stem with an old brown tie, watered it, and dusted off its lackluster leaves. Then we named it! Leif.

The plant became a center of attention. People dropped by to give it words of encouragement. "Come on Leif, you can do it!" "Good morning, Leif, have a good day!" In just two weeks, Leif completely revived and towered proudly over my desk. I was amazed—and elated.

"Would you look at this plant!" I called out to everyone that day. "What a transformation."

"Plants aren't too different from people," Ginger, one of the secretaries replied. "A little love, and they perk right up." We all nodded glowing in our triumph.

Looking at Leif today, I am reminded of the importance of giving love—so very essential to *all* of life. And then I taped this message to one of the plant's gigantic leaves: "People need perking up, too."

Keep me ever alert, God, to the opportunities You give for passing along love—for "perking" up a friend, for helping a stranger to stand taller. —SAMANTHA McGARRITY

30 SATURDAY

...Help thou mine unbelief.

—MARK 9:24

I'm not what you would call a *real* golfer, but I like the game. One day I was playing in a foursome—and swinging badly. My partner took her shot, a nice high arc right over the lily pond and up close to the green. Then it was my turn. "Watch me botch it," I called out. My club connected and sent the ball hurtling. Down it came, on the wrong side of the lily pond and in the rough. Out loud I laughed; inside I cried.

As we walked to the pond, Hal, one of the other players, came up alongside me. "Eleanor, your attitude is wrong. You're tense. Relax and try to imagine where you want the shot to go before you take it."

We reached the rough. "Remember," Hal said, "think positive." I looked down at the little yellow ball, then up at the red flag waving in the breeze over the ninth hole. *That's where I want it to go.* Concentrating on this picture I swung. Up out of the rough shot the ball. It soared high in a perfect arc, landed on the green, bounced and rolled toward the hole. "You did it! It's in the cup!" Hal screamed.

Needless to say, my golf game isn't always that good. I still swing wildly at times and land in the rough a lot. But when I do, I try to remember to relax and imagine how I want the shot to go.

Golf is a lot like life. Keeping out of the rough isn't the object; getting out after we are in it is. Jesus gave us the formula when He said, "If ye have faith...nothing shall be impossible unto you" (Matthew 17:20).

Oh, Lord, whenever I'm in a rough spot, let me visualize, with Your help, a way out of it. —ELEANOR SASS

31

SUNDAY

The inhabitants of Zidon and Arvad were thy mariners: thy wise men, O Tyrus... were thy pilots. —*EZEKIEL 27:8*

It was six A.M. and I was out on deck watching our ship approach Moorea, near Tahiti. There it sat in a vast circle of ocean, biting a saw-tooth outline out of a rosy morning haze.

Long before we reached the island, a small boat had delivered a ship's pilot who would guide us into the harbor. I wondered why we needed a pilot—the sea looked like miles of pure, pink silk. There was barely a ripple in it anywhere.

But, as we drew closer and the sky turned from rose to bright light, I saw the bay was rimmed and studded by jagged rocks jutting through the surface of what had resembled pure silk. Waves smashed themselves against the boulders, breaking into spume, which then dissipated like smoke as it blew away.

It wasn't the rocks that one could see that were the most dangerous—it was the ones we couldn't see. But our pilot knew where they were. He took us in safely.

In that moment standing alone on deck, I bowed my head and said, "Jesus, pilot ME over my own life's dangerous seas."

Dear Pilot-of-my-life, guide me through the known and unknown dangers that surround me every day. —ISABEL CHAMP

Praise Diary for August

1

2

3

4

5

6

7

8

9

10

11

12

13

14

15

16

17

18

19

20

21

22

23

24

25

26

27

28

29

30

31

S	M	T	W	T	F	S
	1	2	3	4	5	6
7	8	9	10	11	12	13
14	15	16	17	18	19	20
21	22	23	24	25	26	27
28	29	30				

When We Can't Forgive Ourselves

He slipped and fell, this friend of mine
  Yielded to ambition's wine,
Made mistakes and paid the price
  Though paying back did not suffice,
Did not remove the guilty stain
  Or rid him of his gnawing pain,
Speak to us, Lord, of sin that's past
  When memory's arms still hold us fast...

Your life, My son, is only beginning
  Look up to Me, not to your sinning,
When you asked Me to erase
  Transgressions that your life debased,
I wiped the slate of every wrong
  You still are Mine, you still belong,
Now lift your head and walk in peace
  My pardon's final, My love won't cease.

—FRED BAUER

1 **MONDAY**

...Serve him... with a willing mind... —*I CHRONICLES 28:9*

On Labor Day I am reminded of my grandmother. "What we do on this day we will do for the rest of the year!" she said, her words carrying a word of warning.

Consequently, there was a flurry of activity. Rugs were beaten, windows washed. Lawns were mowed, hedges trimmed. Labor Day was literally translated.

I was thinking of my grandmother's philosophy this Labor Day as I (yes, from childhood training) cleaned a neglected closet. When I picked up a book to dust beneath it, this item fell out:

> Are you willing to consider the needs and the desires of children; to remember the weakness and loneliness of people who are growing old; to stop asking how much your friends love you, but rather to ask yourself whether you love them enough; to bear in mind the things that other people bear in their hearts; to trim your lamp so it will give more light and less smoke; to make a grave for your ugly thoughts, and a garden for your kindly feelings, with gate wide open—these, *even for one day*? —*Henry van Dyke*

It was almost as if Gran were standing there with me...reminding me that today set the pattern for the days to come. Well, doesn't it? Each day is a new start. Don't I owe the best I have to the Heavenly One by Whose grace I have this day?

Lord, let my labor be for Your Cause today, tomorrow... and the rest of my life. —JUNE MASTERS BACHER

2 *Practicing His Presence in September*

MAKING CHAIN LINKS OF LOVE

And whosoever shall give... a cup of cold water only in the name of a disciple, verily... he shall in no wise lose his reward. —*MATTHEW 10:42*

Before communion the choir stood to sing "In Remembrance," a song that begins with Jesus offering the cup and bread to His disciples and asking them to eat and drink in remembrance of Him. But as the rest of the song unfolds, the vision of remembering Christ begins to expand..."In remembrance of Me, heal the sick...In remembrance of Me, feed the poor...In remembrance of Me, open the door and let your brother in."

As I listened to these words, I thought how receiving the cup and bread brought Christ's Presence to me. Now the song was suggesting there was another special way to bring Christ's Presence to my life, not just on communion Sunday, but *every day*. Go out and do some tangible act of love in remembrance of Him, it suggests. Feed a hungry person, open a door, send a card, cry with a friend, give a gift, say "I'm sorry," visit a shut-in, hug a child...*but pause and do them deliberately in the name of Jesus.*

That is what sets them apart and makes them special; that is what transforms them into acts of remembrance. There in church I began to see that all the small expressions of love done in His memory could be chain links of Presence that run through my days, binding them with a sense of communion.

After the service, I saw an elderly woman drop keys that had spilled from her purse and rolled under her car. "Let me get them for you," I called. *It's a small thing, Lord, but I do this in memory of You*, I thought. And somehow the mere act of kneeling down and groping for keys became an experience of Presence, as full as bread and wine.

Each day in September help me keep Your Presence ever with me, by doing some loving act every day, especially in Your name and memory.

—SUE MONK KIDD

WEDNESDAY

...Paul called unto him the disciples, and embraced them...

—*ACTS 20:1*

On Wednesday nights when I come home from work I go directly to church where I help with the youth group—Chris, Jenny, Judy, Paul, Chance, Claire, Shawn, Sandra. We get anywhere from six to twenty junior high school kids at our weekly meetings. Together we sing songs, put on plays, do skits, play charades, study the Bible, delve into our problems and, most important, pray.

Sometimes, though, I'm too tired to attend these meetings and wish I were going home instead; but once I'm there, the kids always pick me up. Such as the other night when I was pouring juice during our refreshment hour. Suddenly I felt a tentative tap on my shoulder. I turned around to see Chris—our break-dancing champ—beside me. "Yes?" I asked. "Did you want something?"

"Nothing," he said shyly, shrugging his shoulders. "That was just a pat on the back." Then he paused, "We all need one every once in a while."

That's what my Wednesday night friends give me, a thousand pats on the back. In more formal language it's called Christian fellowship, but I like Chris' way of seeing it.

Once in a while, when it's least expected, it's nice to *show* our care, one for another.

There's unbroken fellowship, Lord, when I reach out to all who cross my path today.
—RICK HAMLIN

THURSDAY

Therefore be ye also ready: for... blessed is that servant whom his lord when he cometh shall find so doing.

—*MATTHEW 24:44, 46*

Recently, when we were talking about faith, a friend said, "I read a lot about getting through tough times or times of joy, but what about those weeks when you go to work and come home, do the same things over and over, waiting for something to happen?"

"Yeah," her husband chimed in, "you think you are where God wants you, but nothing's going on."

I thought of that the next Saturday, when my son Barnabas played his first "Under-Eight" soccer game. A brand-new player, he was assigned to stand back near the goal to keep the other team from scoring. Since his team's opponents were inexperienced six-year-olds, the ball came his way only twice during the entire game.

I was prepared to console him afterwards for missing the action. Instead, he ran up with shining eyes. "Did you see that, Mama? I was ready for them all the time!"

I can still see the way he stood during the game—alert, watching what others were doing, sometimes jumping up and down in his excitement. And I wonder, can I approach my own routine tasks with that same kind of anticipation and readiness?

Dear Lord, as I serve You, help me to value the humdrum periods of my life, which prepare me for action when it comes.

—PATRICIA HOUCK SPRINKLE

5 **FRIDAY**

Fear thou not... I will help thee; yea, I will uphold thee with the right hand of my righteousness. —*ISAIAH 41:10*

One day when I was fourteen, I visited my friend Laura, who lived in a small town where the local youngsters used an old mill pond as a swimming hole. On the right the pond was shallow; to the left the water dropped sharply to vast depths. There was a marker, but no one thought to warn me of the danger it indicated.

Being a poor swimmer, I hovered close to shore and inadvertently bobbed over into the depths beyond the marker. I went down, down into what seemed like a bottomless abyss. It was an eternity before I could struggle to the surface, and there I floundered about, paralyzed with fear, too self-conscious to scream.

Suddenly a hand pushed against my back, shoving me into the shallows. Laura's older brother threw an arm around my shoulders, steadying me. "Hey, I'm sorry I hit you," he apologized. "I thought you were Laura, and I was just kidding around."

It wasn't until years later that he told me he had known all along it wasn't Laura he shoved. He had pretended to think it was his sister rather than embarrass me in front of the others by openly "rescuing" me.

Many times since that day I have gotten in over my head because of a problem, and I have learned not to be ashamed to cry out for help.

Stubborn pride, I've found, can only get me into deeper waters, while an honest appeal for help could have saved me from the start.

When I want to hide my needs from others, Lord, give me the humility to speak up and the grace to accept their support. Amen.

— DRUE DUKE

SATURDAY

Incline your ear, and come unto me: hear, and your soul shall live; and I will make an everlasting covenant with you... —*ISAIAH 55:3*

The paint is peeling on Grandmother's old house in rural Danbury, Nebraska. The gate I used to swing on has disappeared, the roof sags, and the upstairs sleeping porch has been enclosed. Gone are the summer nights when I slept on that porch, watching the gaslight swinging over the dirt street below, soothed to sleep by croaking frogs and singing katydids, and wakened at dawn by crowing roosters.

But now my son and his family live in Danbury, in a cozy little bungalow at the edge of town. Paul tells me that, when the windows are open on a summer night, the frogs and katydids can still be heard. And old George Henderson, who sickled hay with my dad when they were boys in the early 1900s, now brings pumpkins to my son's children.

This morning, I walked by the old place and noticed the chunk of concrete that Granddaddy had poured one summer day, so that we kids wouldn't wear away all the dirt under the swinging gate. It was in the ditch at the edge of the road, but our handprints were still on it—and so was the crooked cross I'd drawn in the fresh cement with an old stick. Something inside of me knelt.

Help me to sense echoes of eternity, Lord, in the rhythmic patterns of time.

—MARILYN MORGAN HELLEBERG

7 **SUNDAY**

The kingdom of heaven is like unto treasure hid in a field... —*MATTHEW 13:44*

There is a story in the vein of the Good Samaritan parable, which tells about a young game trapper who took ill and lay dying in his remote log cabin. And he would have died had it not been for the efforts of an old miner who found him and painstakingly nursed him back to health.

When it came time for the miner to leave, he wanted to give the trapper something of value to help him get back on his feet. He decided upon his most valuable possession, an uncut emerald. But the young man had a reputation as a heavy drinker and the miner feared his friend would quickly waste the treasure. Then one night inspiration came. The miner secretly sewed the gem into the lining of the young man's sheepskin coat, praying that some day his friend would find it and put it to good advantage.

Many years later the miner again found the young man, nearly starving, begging on the street. Though dirty and tattered, the man wore the same sheepskin coat. "What did you do with the emerald?" he asked. "What emerald?" the younger man asked. And then the old miner showed him the hidden gem that had been in the coat's lining all those years.

In the Bible Jesus compares the kingdom of heaven to the "one pearl of great price," and His gift of salvation is like the gem sewn into the lining of the man's coat. There is no need for any of us to go spiritually hungry when life's most precious jewel is so close at hand.

Lord, give me eyes to see Your rich blessings—by first looking within. Amen. —FRED BAUER

 **MONDAY**

...I have a message from God unto thee. —*JUDGES 3:20*

Fall had arrived and with it, the first day of school. As I drove down the street, I was aware of all the children getting picked up by shiny yellow school buses.

At a red light, my attention was caught by one particular child. He had on new clothes and clutched tightly at his mother's hand. She stood next to him in her bathrobe and his father, dressed in a business suit, waited with them also.

His very first day in school, I thought, and at that my heart went out to him and I waved. His face lit up and he waved back. As I drove away I saw him grin and look up at his parents as if to say, "Hey, maybe it's going to be okay out there in the world after all."

Driving on, I thought, *Which one of us hasn't had our share of fear and anxiety and apprehension in new or unfamiliar circumstances? And who among us hasn't needed some small reassurance—a smile, a word of encouragement, a hug?* I believe those small, fleeting gestures are gifts from God, His way of telling us He is near and we are not alone, no matter how frightened we are.

Father, let my presence minister to others, offering comfort, that I too might be comforted. —MARION BOND WEST

TUESDAY

And God hath chosen the weak things of the world to confound the things which are mighty. —*I CORINTHIANS 1:27*

During a recent trip to Hawaii, Ruby and I took a tour of a tea garden located in a lush, tropical valley outside of Honolulu. While the others visited the little chapel on the grounds, I strolled down the hill past the exotic greenery to the small grass shack where the writer and poet, Robert Louis Stevenson, had once lived.

Alone, I stood and studied the little room that contained only a cot, a bookcase, a table and five wooden chairs. My eyes glistened with tears.

I could picture Stevenson, suffering from illness and pain, as he wrote, sometimes by candlelight, in that barren room, giving the world his great gifts. Then I thought of the countless others who have left us priceless legacies despite deprivation and hardship: Abraham Lincoln, studying late at night by candlelight in his log cabin; the artist Van Gogh, poverty-stricken, painting feverishly in his small studio; and Beethoven, who composed beautiful symphonies although deaf.

Inexplicably, I found myself offering a prayer of thanksgiving for the multitude of courageous people who have gone before, sharing their gifts with the world...with you and me.

Dear Father, though my own gift be small, help me use it generously and unselfishly in Your world. Amen. —OCSAR GREENE

10 **WEDNESDAY**

...Having heard the word, keep it, and bring forth fruit with patience. —*LUKE 8:15*

I used to think that making a copy was a simple mechanical process. Not so. The people who operate copying machines have a lot to do with the results. I learned that from Warren, who works at my local copy store. Since his arrival, my manuscript copies are crisp and clean. That's because Warren is fussy. If he doesn't like a page, he adjusts the machine and does it over. Naturally, that takes time. But I don't mind waiting. If Warren is fussing over a single page for the customer ahead of me, then I know he'll do the same with my entire manuscript.

One day as I was leaving the store, I stopped to tell the manager how much I appreciated Warren's work. "Funny," the manager said, "when he started working here, all I heard were complaints. People said he was too slow. Now all I hear is praise for his work."

Then I remembered how impatient I was when Warren began doing my copies. All I wanted was speed—until Warren taught me there is such a thing as quality.

Fortunately, God knows it takes time and patience for us to do our best. So if we're a little slow developing our talents, He doesn't give up on us. He's more interested in quality. And maybe we should be, too.

Lord, You created us with great care, and when we do our best we're honoring You. Help us to resist the temptation to do less because we're in a hurry. —PHYLLIS HOBE

11 THURSDAY

...Be careful to maintain good works. —*TITUS 3:8*

There were five of us at the table, finishing our dessert—me, three business associates, and Bishop Desmond Tutu, winner of the 1984 Nobel Peace Prize. One of the others asked the bishop what we could do in the world to promote peace. He gazed into the distance for a moment then answered in a quiet voice, "You must care."

Yes, that's important, I remember thinking. *But how does one person, like me, care for a whole world?*

I might simply have forgotten the bishop's words, but for the things I began to notice, events that had certainly been there but that I just had not paid attention to: schoolchildren in New York raising $150,000—one penny at a time—for starving Ethiopians; parishioners in a local church collecting money and clothes for fire victims in Philadelphia; a group of professionals giving up their weekends to renovate a halfway house for troubled teens.

Lynn and I have started to make sure that each week we pick up some extra food at the grocery store for a local food pantry that gives food to the poor. It's only a small thing, sure but, by picking out one little corner in which to begin, it's our way of saying, "We can begin to care for this world."

Have you stopped today to think how you, too, can do something to show you care?

Lord, let there be peace on earth. And let it begin with me.

—JEFF JAPINGA

12 FRIDAY

...It is an ancient nation, a nation whose language thou knowest not, neither understandest what they say.

—*JEREMIAH 5:15*

Photography is one of the latest hobbies in China today, and young people especially seem to take particular delight in snapping pictures of one another.

I had heard about the portable darkrooms on the streets in China. Sure enough, there they were. The customer's film is threaded into developing fluids within a black box about the size and shape of a picnic basket. Soon its smiling owner is showing a long strip of black and white negatives to friends. They hold it toward the light and scrutinize it carefully. Then they walk away with it fluttering behind them like a ribbon.

One afternoon I was engrossed in taking pictures of an intricately carved dragon sculpture when I was surrounded by a group of young men curiously pointing at my camera. Never have I been so frustrated—and grieved—as I was that day at our language barrier. Obviously their eager chatter concerned my equipment. I was just as eager to explain it to them, and to ask my own questions. But there was no interpreter about, and finally we simply burst out laughing. How ludicrous it was for us to keep on repeating words neither they nor I could understand.

Yet, because of our smiles, a universally understood gesture, we parted as friends.

Father, teach us to speak the language of love: the smile that says "Friend!"... the gesture that says "I see!"... the touch that says "I like you!" —ISABEL CHAMP

13

SATURDAY

For I would that all men were even as I myself. But every man hath his proper gift of God, one after this manner, and another after that. —*I CORINTHIANS 7:7*

One recent Indian summer day, my husband and I were hiking to the top of Longs Peak, a popular but strenuous climb up one of Colorado's highest mountains. After several hours, we fell in line behind a couple of lean, athletic-looking men who introduced themselves. "Hi, I'm Rick, a stockbroker from Kansas City, and this is my buddy Allen from Texas. He's a bank vice president."

Their titles automatically came with their names, which didn't seem unusual until Rick questioned us. "And what do you do?" he asked. "I'm Lynn," my husband responded. "I'm a lawyer in Boulder, and this is my wife Carol." Period. No title. As they continued in conversation, I began to feel left out.

The uncomfortable moment soon passed. But days later I thought about it again, and the question echoed in my mind: *And what do you do?* If I had to come up with a title, apart from the definition of wife and mother, what would I choose? What *do* I do? Then I hit upon a title.

I'm an investor, I decided. I try to invest in relationships. With my husband. My children. My friends. And most of all, with Jesus. It's a full-time job and sometimes it takes a lot of effort. But it has an exciting, eternal future.

Now that's an important-sounding title and I can hardly wait to use it the next time I'm asked, "And what do you do?"

Dear Lord, in all I do, *let my best work be to invest in relationships today.* —CAROL KUYKENDALL

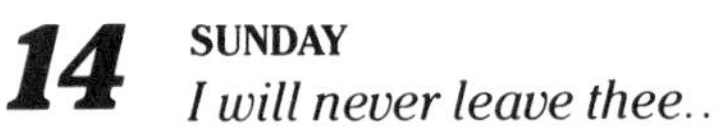

14 SUNDAY

I will never leave thee... —*HEBREWS 13:5*

I'm writing this by the light of the clown night lamp I bought one day so that my two-year-old granddaughter would feel safe. Dawn has never been away from her parents overnight before, but I've told her I won't leave her, and she's learned that she can trust me.

I'm not sleepy, so I let my thoughts wander. What kind of legacy is my generation leaving for its Dawns? Arms limitation talks go on. What will Dawn's history books say about them? There is a terrible famine in Ethiopia. Will there be enough food for Dawn's children and grandchildren? Dawn's great-great-grandfather traveled by covered wagon to homestead in Nebraska. His son, my father, flew around the world in a jet. Will Dawn's children colonize other planets?

My mind is filled with unknowns about Dawn's future and the world's. Like a child away from home, I sometimes long for the comfort of familiar ways. But One Who loves me has told me that He will never leave me, and I've learned that I can trust Him. That trust may be the best legacy I can give my little Dawn. Perhaps it will be a night light for her in the twenty-first century.

Oh Lord, thank You for being the changeless certainty in a world of unknowns. —MARILYN MORGAN HELLEBERG

15 *WILDERNESS JOURNEY* *(Exodus 25–27)*

Worshiping a Golden Calf

Let them make me a sanctuary; that I may dwell among them. —EXODUS 25:8

After receiving the Ten Commandments, Moses came down from Mount Sinai only to find God's people worshiping a golden calf! They thought they needed a god they could see. I guess the Lord understood the humanness of that because He told them to build a holy place for Him in the middle of their camp, a place about which they could say, "This is the house where God lives among His people."

Sometimes I, too, wish I had a God I could see. At times, when I pray to my invisible God, He seems so far away. His voice is drowned out by all the others... "Chemical leak kills two thousand people...Where's the beef?...Arms talks at a standstill...Earthquake shatters village... Try the real thing...Tune in at ten for details on nuclear buildup...." The dog barks, the doorbell rings, and holiness escapes me like an early morning dream. Maybe, like the Israelites, I need a holy place at the center, where I can go to reassure myself that God really does dwell with me.

The Israelites' tabernacle was portable, so that they could take it with them wherever they went. I'm going to build a tabernacle in my heart so it will always be with me. Today let's build an inner temple, too, following God's pattern. Let's imagine within our heart a holy room. Let's close our eyes and enter. See its walls of gold, its ceiling of fine linen in colors of blue and purple and scarlet. On our right, there's a golden table with twelve loaves of bread on it. When we're spiritually hungry, we can go within and be fed. To our left is a golden lampstand with seven branches. Let's light the lamps in our tabernacle now. At the far end of this room is an altar of incense. We can offer a prayer here now, knowing it will be lifted to God all day as the incense rises heavenward.

Now we enter the Holy of Holies, that hidden room where God's glory shines over the Ark of the Covenant, containing the Ten Commandments. Before the Ark are a pot of manna, Aaron's rod that blossomed and other remembrances of God's mercies. Is there something with special meaning for you that you'd like to place here, to remind you of God's mercy? I think I'll place pictures of my family, a pen to represent the writing God helps me do, and my mother's Bible, before the Ark in my Holy of Holies.

Deeply aware, now, that God really is with us at this moment in our inner tablernacle, we fall on our knees in thanksgiving.

Thank You, Lord, for helping us create a holy place within, where we can meet You, as we travel our everyday desert paths.

—MARILYN MORGAN HELLEBERG

16 TUESDAY

...Live in peace; and the God of love and peace shall be with you. —*II CORINTHIANS 13:11*

Recently I witnessed a striking demonstration in a worship service. The minister placed a metal trash can on top of the podium. Then he held up a tiny silver BB and said, "This one single BB represents all the guns, bombs and fire power of World War II." He let it drop into the trash can, causing a hollow ping to echo through the chapel.

Next he lifted up a huge jar filled with BBs. "Based on the same scale, I'd like to show you the bombs and firepower we have available in our world today." He tilted the jar and 5,995 BBs came crashing down into the metal can like peals of thunder. The entire chapel was ringing.

"Peace, like war, must be waged," he said. "Let us each do what we can for peace in God's world."

As I left I could still hear the roar of those BBs in my ears—but even more so in my heart. I thought of all the instruments of war and hatred...the call to "wage peace." But what could *I* do?

Later that day I found myself thinking of the box a friend had brought me from Europe. Inside was a picture of St. Francis and the words, "Lord, make me an instrument of thy peace." And suddenly an idea came. Why not write my concerns about peace on cards and place them in the St. Francis box? Each day I could pull out one and pray about it. Prayer was one thing I *could* do.

The box began to fill up with all kinds of peace concerns, not only for reconciliation between nations, but also for reconciliation in my own life ... and in the lives of friends and neighbors.

My box is very small and some days I forget it's there. But one thing I know. There are fewer "BBs" in my own heart because of it. And in the world? Who knows?

Lord, help me to wage peace not only in my world, but in my own heart. —SUE MONK KIDD

WEDNESDAY

The Lord thy God is with thee whithersoever thou goest.

—*JOSHUA 1:9*

Over 38 million people left Europe between 1864 and 1924 to come to the United States. They left poverty and political and religious oppression, seeking the wonderful, free new life described in letters received from friends and relatives who had gone before.

Among them were my ancestors. In Germany, my great-great-grandmother Buchholz prayed with her sons Friedrich and Wilhelm, then watched from her kitchen window as they walked away. She never saw them again. Great-grandfather George Sutton, born in England, arrived in Chicago in 1851, where he helped to build the Cleaver Soap Factory. After that he drove pilings for the bridge on Lake Street. And we still have an ice skate that he wore when he won a contest by skating a mile in two minutes and eighteen seconds.

George brought his wife to Effingham County via the new Illinois Central Railroad in 1861. They raised Jersey cattle, apple trees, and three daughters, one of whom was my grandmother, Sara Buchholz. He lived 101 honorable years, helping the people of his community, and was one of the founders of Salt Creek Church.

Most of us are descendants of immigrants who built new lives for themselves while they helped to build this great country. Let us especially remember them on this, Citizenship Day.

Wherever, I make my home, Lord, abide with me there, and with my children's children too. —ZONA B. DAVIS

18 **THURSDAY**

Lo, children are a heritage of the Lord…As arrows are in the hand of a mighty man; so are children…Happy is the man that hath his quiver full of them. —*PSALM 127:3–5*

When our thirty-year-old son Jeff announced that he was moving back in with us temporarily, my wife and I had mixed emotions. Oh, we loved our children—we'd raised five—but we were enjoying our increasing freedom and privacy. And now would we have to give it up—and become parents again?

Then Jeff arrived. To our surprise, somewhere between the time he'd left home and come back, he'd become an adult. He was neat—he kept his room clean and his clothes tidy. He was so quiet coming in at night we never heard him. He pitched in around the house and paid his share of the load. And, we were discovering, he was good company.

But it was more than that. It wasn't just fun to have Jeff home; it was *special.* I never got to see much of my father after I finished college. I moved to another city and got enmeshed in launching a career. Then, when I was twenty-four, my father suffered a stroke—and in a few days, he was gone. I never really knew him as an adult.

So we are relishing Jeff's stay. It's been a special blessing getting to know our son...uh...er...this new person. And what a likable fellow, too.

Lord, all of life's moments are the best time to get to know one another better. Let me begin now. —SAM JUSTICE

19 **FRIDAY**

Be kindly affectioned one to another with brotherly love… —*ROMANS 12:10*

Recently, while out for a stroll, I saw a boy of about ten walking toward me. Not long before, I'd encountered some mischievous teen-agers who had teased me. Not wanting to risk another incident like that, I frowned and turned my head away when this boy and I passed each other. But then I heard him exclaim, "Aren't ya gonna speak to me?"

I wheeled to find him peering toward me with a tentative smile. Immediately contrite, I said, "Hey, I'm sorry. Hello!" His freckled face split open in a wide grin. "Hi. Have a nice day."

A small incident, maybe, but later I kept thinking how I'd allowed the impolite actions of that first group to influence my behavior. That night I was re-reading the Sermon on the Mount, and the Golden Rule leapt out at me with new force: *"Therefore all things whatsoever ye would that men should do to you, do ye even so to them"* (Matthew 7:12).

I've heard that Rule all my life, but that day I'd turned it around, trying to pass along the unfriendliness I'd encountered instead of obeying God's simple but excellent advice. Thanks to one friendly boy, however, I was reminded that God's plan for sharing *positive* behavior is indeed the better way.

Lord, let all my actions toward others today say, "Have a nice day!"

—MADGE HARRAH

20 SATURDAY

Be strong and of a good courage, fear not... for the Lord thy God... will not fail thee, nor forsake thee.

—DEUTERONOMY 31:6

A long illness had me discouraged, when an old friend dropped by to cheer me up.

"As a young fellow," he said, "I attended a small prep school. In football we had a very fine first team, but since we were small in numbers, we had no substitutes. We played these big high schools and for three quarters we were nose to nose. But by the fourth quarter we began to drag. That's when the coach called us off the field, one by one.

"'If you *will* it,' he said, 'no one can beat you. Even if you lose, you will be better men because only you can control how you think and feel. I know you're hurting, but pull your socks up and do your best. God asks for nothing more.'

"Years later, whenever one of us got into difficulty, someone would call and say those magic words. So, Doris, I'm saying them to you: 'Socks up!'"

That simple phrase restored my spirits. It means, "Have courage, persevere, trust in God."

Are you ill or discouraged? If so, then remember that with God on your side, you are unbeatable. So...*socks up!*

You make winners of us all, dear Jesus. Help me be one today.

—DORIS HAASE

21 SUNDAY

I will praise thee; for I am fearfully and wonderfully made... —*PSALM 139:14*

Today I put a contribution in the mail for the public library. I do this every year on the birthdate of my Aunt Helen as a kind of thank-you to her. You see, Helen was an avid reader who loved books, and wanted me to appreciate them, too. So as soon as I was old enough to sit still for a while, we'd spend many hours together in the big rocking chair next to the dining-room window in her house. As I rested my head against her shoulder, she'd read aloud from *Heidi*, the Bobbsey Twins series, *Alice in Wonderland*, or the poetry of Robert Louis Stevenson.

Helen never seemed to tire of reading to me. What kept her going, I suspect, was her love for me and her desire to give me something lasting and meaningful. Well, she accomplished it, because I, too, love to read.

In a few days now I'll put a contribution in the collection plate at my church. I do this every Sunday—the Lord's Day—as a kind of thank-you to God. Because, like Aunt Helen, He loves me and desires to give me something meaningful and lasting. And He has, too.

It is the gift of *Life*. And He's given me a life*time* to use it well.

Lord, thank You for creating me in Your image. Let my works be worthy of Your trust. —ELEANOR SASS

22 **MONDAY**

What profit hath he that worketh in that wherein he laboreth? I have seen the travail, which God hath given to the sons of men...He hath made every thing beautiful in his time... —*ECCLESIASTES 3:9-11*

What an afternoon! I had struggled for what seemed an eternity with the extension ladder—lugging it around to the back of the cottage, then trying unsuccessfully to lift it against the house so that I could climb to the roof and remove the chimney covers.

Then it dawned on me—if I took apart the two halves of the ladder, one half would be light enough for me to handle and long enough to reach the roof. But when I scrambled to the top rung, I froze. I stayed there a long time gazing at the steeply sloping roof and worrying about sliding off.

"If you don't get up there and remove those covers," I told myself, "you won't be able to use the fireplace or the stove this winter!"

So, with shaky knees, I slowly crawled up the roof and snatched off the chimney covers, throwing them to the ground. Then I sat on the apex to catch my breath. Suddenly before my eyes, out over the Hudson River, was a beautiful, glowing orange sun setting. I'd never seen anything quite so magnificent and it was a sight I never would have seen from the ground.

It had been an afternoon of strenuous work, scary too; but there had been a beautiful reward, a glorious unexpected moment that would always be a reminder that everything, every experience, *in time* will reap its blessing.

Dear God, when I'm deep within a fiery trial, lift me to a rooftop to await Your unexpected reward. —SAMANTHA McGARRITY

23 **TUESDAY**

Seek ye out of the book of the Lord, and read...

—*ISAIAH 34:16*

"When should we *especially* read God's word?"

The question—posed to our Bible-study group—hung in the air unanswered. Should it be morning, evening, at mealtime? In time of conflict and confusion...or when we are serene and contemplative?

Aunt Mollie, our senior member, showed us that she was unconcerned with the time of day when she said, "When? *I'll* tell you when!"

With that she pulled a crumpled envelope from her knitting bag and began to read. The words, jotted down in her spidery handwriting, took up the rest of our hour together. May they bless you as they have blessed us:

- WHEN in sorrow (John 14)
- WHEN men fail you (Psalm 27)
- WHEN you have sinned (Psalm 51)
- WHEN you worry (Matthew 6:19-34)
- WHEN you are discouraged (Isaiah 40)
- WHEN you need courage for a task (Joshua 1)
- WHEN God seems far away (Psalm 139)
- WHEN you need peace or rest (Matthew 11:28-30)
- WHEN you feel bitter or critical (I Corinthians 13)
- WHEN you are lonely or fearful (Psalm 23)

As has often been said, the Bible itself is timeless. Not only does it live down through generation after generation, but it is there any hour of the day or night to comfort and inspire us. We can turn to it as we do to our Dear Father... Who is there for us twenty-four hours a day!

Lord, Your Word is a living Presence among us. Deepen my understanding so I may better serve You. —JUNE MASTERS BACHER

24 WEDNESDAY

I pray not that thou shouldest take them out of the world... —*JOHN 17:15*

I love nature shows on television, not only for what they teach about the world around us, but for what they suggest about the world within.

In David Attenborough's *The Living Planet* series, we see a blue and sunlit stretch of the North Atlantic. Situated in the peaceful "eye" of circulating ocean currents, it's an area immune from gales and hurricanes—so calm that immense floating mats of Sargasso weed form on its unruffled surface. Through the seaweed scuttle tiny crabs and...very little else. Few fish inhabit these crystal-clear waters, few birds pass overhead. This idyllic-looking region of the ocean is in fact a sterile world, lifeless and silent.

More than a thousand miles to the north is another sea. Shrouded in perpetual fog, swept by savage storms, it is the meeting place of the tropical Gulf Stream and the icy Labrador Current. Unlike the tranquil blue surface of the Sargasso Sea, these dark waters are in a constant froth of agitation. And they are so filled with fish that the ocean itself seems alive—millions upon millions of silvery shapes feeding on the plankton that thrive on nutrients churned from the sea floor by the swirling water. Mammals are there too, herds of seal and humpback whales, while the air is dense with sea birds wheeling and diving into the richest feeding ground on earth.

Conflict, disturbance, sudden change—not so different, I thought, from the world we ourselves inhabit. Both of them scenes of life, not of stagnation.

Thank You, Father, for the turmoil in my life today that tells me I am alive and a part of Your world. —ELIZABETH SHERRILL

25 **THURSDAY**
...By his light I walked through darkness.... —*JOB 29:3*

My little niece Laura died on a warm September day. She was four months old. When I arrived at my brother's home, my mother met me in the driveway, tears trickling down her face. Under the pine trees, we held each other.

"I'll be all right," she said finally, straightening up and brushing her cheeks. "Everyone says that time will heal this hurt."

"I'm sure it will," I replied, aware how feeble the words sounded.

Mother looked off into the distance. "I know it's foolish, but I find myself wishing it were next year already."

I've thought a lot about her words. When we come down into some valley of tragedy in our lives, we long for a bridge to get us to the other side, so we won't have to walk through the pain below. Many times I've been tempted to step over or skirt around the pain and problems in my life or even refuse to see them.

But to get to the healing side we must go ahead and dry the tears, experience the ache and struggle with the questions for which there are no answers. But there is also a promise to hold on to: "Though I walk through the valley of the shadow...thou art with me" (Psalm 23:4). God goes down into the pain with us and helps us reach the other side.

A year has gone by since Laura's death. With God, Mother walked through the grief. Today, at Laura's name, she can smile without tears.

Lord, with You as my Rod and my Staff, I will walk through life's pain and problems without fear. —SUE MONK KIDD

26 FRIDAY

[He]...which maketh Arcturus, Orion, and Pleiades, and the chambers of the south. —*JOB 9:9*

Autumn again. My favorite season for stargazing.

Soon Orion would begin to rise in the east. He had ruled my autumn sky in Florida, in the Highlands of Scotland, above the rooftops of Chicago. Now, in a new home, I scanned the Georgia skies. Where could he be?

For several nights I looked without success. In fact, none of the constellations looked familiar. Had something strange happened to the whole sky? Then one night, as I turned to go inside, I was surprised to see Orion's three-starred belt rising in what I was certain was the west.

I checked with my husband, who seems to have an internal compass. He assured me Orion hadn't moved—I was just turned around.

I stood for many minutes in the crisp stillness to regain my bearings. Now that I had found my old friend Orion, one by one I found other familiar constellations in their proper places. My night-sky world was oriented once more.

Sometimes my whole world gets disoriented. I am baffled by friends, family, my own behavior. That's when I need to stand in stillness and seek the One who made Orion—the One who, like Orion, never changes His place. Once I find Him, the rest of my world finds its pattern again. My life is a bright, shining galaxy of unmistakable lights.

Dear God, I turn to You this moment to get a proper orientation for the day ahead of me. —PATRICIA HOUCK SPRINKLE

27 **SATURDAY**
Uphold me according unto thy word, that I may live...
—PSALM 119:116

As part of the fall fix-up of our house, I decided to paint the ornamental iron around our front porch. "You'll need to pull the ivy off the post at the top of the steps," Bob cautioned.

"No problem," I said. Surely I could do a small job like that!

However, removing that ivy turned out to be a real challenge. The vine, trained to climb up the post, had grown there for fifteen years, weaving in and out of the open design and curling its tendrils tightly around every place it touched.

For most of the day, I pulled and snipped and hacked, but there were still a few sprigs that stubbornly resisted. Finally, I had to admit defeat. Those sprigs had become so much a part of the post that it was impossible to remove them.

I sat down in the porch swing to rest. Gliding gently back and forth, I thought back over the day. My work had opened my eyes to a real truth: *Whatever we cling to, we become a part of.* And sometimes, like the ivy, we need something—Someone—strong, to Whom we can cling and with Whose help we can grow upward daily.

We need Jesus.

Be my Support, Lord, so that I not grow wildly nor stray from Your purpose. —DRUE DUKE

28 **SUNDAY**
For if ye forgive men their trespasses, your heavenly Father will also forgive you. *—MATTHEW 6:14*

It was Sunday morning and instead of the usual spoken message, the minister put us to work. He asked the ushers to pass out paper and pencils.

"I would like you to write down the names of all the people you can think of who have done something you feel requires your forgiveness," he told us.

After some rustling in our seats, the congregation settled down in hushed concentration. I finished my brief list and waited expectantly.

"Now, add *your name* to the top of that list," the minister continued, "and then bow your head and repeat this prayer: 'Lord God, in Thy infinite mercy, please forgive me for passing judgment on these, my brothers.'" He lowered his head and we—in varying degrees of surprise and shame—followed his example. When I looked up, he was smiling.

"God asks us to forgive," he explained. "However, forgiveness can be offered only if we have *first judged*. Therefore, since Scripture clearly tells us to 'judge not,' we find that we ourselves are *also* in need of forgiveness. Now," he added, "you can tear up those lists."

We did. I was amazed at how light of heart I felt as I walked out of church into the sunlight, a new person, forgiving...and forgiven.

Before I judge another, Father, let me remember that I too need Your forgiveness.

—DORIS HAASE

29 MONDAY

Wherefore also we pray always for you...

—II THESSALONIANS 1:11

Ruth is the first to arrive each morning at the elementary school in Arizona where she teaches. After she opens her classroom and hangs up her coat, she walks from room to room in the empty building, pausing beside each door. There she prays quietly, lifting up to God the children who will study in that room that day and the teachers who will lead them.

"I don't know each one by name," she told me. "But God does. I like to think their days go a little better because someone is praying for them."

Each day as I go to work, I get off the elevator at the far end of the hall in which my office is located, and I've started following Ruth's lead. As I walk to my job, I pause briefly beside each door along the hallway. I silently ask God to bless the people who will do business behind those doors that day. I hope it makes their day go better.

Perhaps there are neighbors or friends, business associates or family members for whom you'd like to pray each day. Today is the perfect time to begin.

Father, today I would ask Your special love and care for__________ __________and____________________. Amen. —DRUE DUKE

30 TUESDAY

For we walk by faith, not by sight. —*II CORINTHIANS 5:7*

Oswald Chambers wrote: "Faith must be tested, because it can be turned into a personal possession only through conflict."

I didn't know that during all those years I prayed, "Lord, give me faith." God gave me simple tests at first, even though at the time they seemed overwhelming: unrequited love in high school, frustrations in learning to be a wife and mother, years of rejection slips, minor surgery, broken friendships, a move to a new town.... Then came the harder ones: sick children, the death of our beloved collie, a ten-year battle against fear and cancer, and finally the departure from this life of my husband of twenty-five years.

So life isn't like my first-grade reader, where Mother and Father and Dick and Jane laughed and played together in golden sunlight, and all their words to one another were of encouragement and love. Even Spot and Puff were obedient and healthy. But maybe Mother and Father and Dick and Jane never had the opportunity of having their faith tested and turned into a personal possession through conflict.

If I had to choose between the life of conflict and a life without the assurance that faith is my personal possession, I would choose the life of conflict. Whatever you are struggling with today, when you feel so tired and helpless and perplexed that you're ready to give up, remember, God has an answer for you. And there too you will find a faith that belongs solely to you.

Father, in Jesus' name, I ask You to help me understand that faith becomes real as it is tested; and that no matter what our circumstances, You will meet us in our times of need. —MARION BOND WEST

Praise Diary for September

1

2

3

4

5

6

7

8

9

10

11

12

13

14

15

16

17

18

19

20

21

22

23

24

25

26

27

28

29

30

October

S	M	T	W	T	F	S
			1	2	3	4
5	6	7	8	9	10	11
12	13	14	15	16	17	18
19	20	21	22	23	24	25
26	27	28	29	30	31	

For Empty Vessels

When the world too much presses in,
  And the noise of the crowd becomes din,
I seek a refuge for my storm-tossed soul,
  Away from shouting waves to gentler shoal,
And there at anchor undisturbed
  I try to hear God's holy Word,
Hidden gems heart-stored from youth,
  Priceless pearls, eternal truth,
But naught today I vainly quote
  Serves as tranquil antidote,
So with my lonely self I wrestle,
  Thirsty as an empty vessel...

Then He speaks to me at mooring,
  In words both healing and restoring,
He used them once to calm a sea—
  They echo through eternity,
Quietly they still fulfill,
  Three whispered words: *Peace, be still.*

—*FRED BAUER*

1 *Practicing His Presence in October*

LEARNING TO LET GO

Cleanse your hands... and purify your hearts... —*JAMES 4:8*

Today I am filled by a desire for something I don't need. A frivolous, material thing. I have thought about it for days. As the desire grows, I go to the window where the days are turning amber. Beyond I see the oak tree as if burning like a yellow fire. Its thick leaves swirl in the wind like wood smoke.

Outside I stroll under the limbs and gaze upward at the full branches. *So many things crowd the branches of my life, too.* Desires, possessions, jealousies, resentments, worries. Things that clutter my mind, blocking away God's Presence.

Suddenly wind shakes the oak and a leaf drifts down in a ragged dance. I pick it up, for somehow it has become a symbol of what God seems to be asking of me—"Let go of the thing that crowds My Presence from your life." *Oh Lord,* I pray silently. *I abandon this desire to You. I am shedding it like a leaf. Fill the space it has taken up in my life.*

As I walk inside, I bring the fallen leaf with me and drop it in a basket on the table, a reminder that my desire has been shed.

Today as October begins and leaves start to fall, I want to begin an exercise in trust and letting go. Each day I will abandon "one leaf" to God...something particular that crowds Him from my life. I want to fill the basket on the table.

Lord, make me an "empty tree" where Your Spirit can blow full and free this month.

—SUE MONK KIDD

THURSDAY

...And let us run with patience the race that is set before us.

—*HEBREWS 12:1*

Sometime early in my 39th year, I decided to run a half-marathon. I guess I needed a physical challenge, and a thirteen-mile race seemed a respectable goal for an inconsistent, fair-weather jogger like me. So I picked a mid-October race and began to train. On nice days, at least. And I prayed, "Lord, please give me the perseverance to reach this goal."

The day of the race dawned, sleet-gray and spitting frost. "That's it," I announced. "I'm staying home."

But somehow I couldn't.

So off I went to the starting line, all bundled up and reluctant. The gun went off and I moved out with the throng. The wind blew stinging ice crystals in my face. The miles began to pass by. Slowly. Four... then five. I wanted to quit.

But again I couldn't.

"Just take little baby steps," a voice inside commanded. "Keep moving forward toward the goal."

Somehow I kept going. One jogging step at a time. At Mile 10, my legs felt like jelly. My lungs screamed with every gulp of frigid air. I wanted to quit.

But I couldn't.

"Baby steps," said the voice. "Don't stop." Mile 11...12...only one more to go. I saw the finish line in the distance. With a burst of energy I crossed it, numb with cold but exhilarated, and thankful that the voice inside me hadn't let me quit.

Lord, even when I'm tired and discouraged, help me persevere toward my goal, albeit one baby step at a time. —CAROL KUYKENDALL

3 FRIDAY

Be still, and know that I am God... —*PSALM 46:10*

The bus was late and I was upset. I'd promised to be part of a prayer chain at my church that was praying for our minister, who was in the hospital undergoing bypass surgery. If things had gone according to my plans, I would have been home now, calm and relaxed, ready to begin my 7:30 to 8 P.M. prayer vigil.

Instead, because the bus was late, at 7:30 I was standing on one of the busiest corners in the city. Sirens wailed, voices babbled and the traffic swirled around me. *Oh, Lord, I'm caught. What shall I do?* I pleaded. And, immediately, a small voice responded, "Be still."

Be still? Amid all this confusion? I looked up at the sky. Fleecy clouds moved serenely across it and a few early stars twinkled. "*Turn*

your eyes upon Jesus..." that glorious old hymn came to mind and I began to hum softly. "*Look full on His wonderful face...*" I imagined Jesus standing next to our minister's bed, looking down at him. "*And the things of earth will grow strangely dim...*" Jesus was placing His healing hand on his brow. "*In the light of His glory and grace.*"

Over and over again I hummed the words, completely lost in the vision of Jesus and our minister, and the healing that I knew was taking place. After a while I stopped humming and my eyes came down from the clouds. The bus was approaching. I looked at my watch. It was exactly 8 P.M.

Thank You, Lord, for showing us that any time, anywhere is a good time and place to pray. —ELEANOR SASS

SATURDAY

I commend unto you Phoebe our sister... that ye receive her in the Lord, as becometh saints, and that ye assist her in whatsoever business she hath need of you...

—*ROMANS 16:1, 2*

Sarah is a street person. For whatever reason—a mental disorder or simply a fragile emotional state—she shirks the normal responsibilities of adult life. With a few personal belongings gathered in a well-worn plastic bag, she takes to the streets. At night she sits on empty stoops; in winter she hovers over grated steam vents; sometimes she sleeps in a city-run shelter—until even that becomes too much responsibility for her.

Recently our church adopted Sarah. We've found a safe place for her to sleep, bring her to worship services, and most of all, we provide her a community to which she can belong...so that she is rarely on the streets these days.

This was our cause for rejoicing last October 4, when we celebrated her birthday. Bringing her a cake and candles after worship, we sang; gentle Sarah smiled. "Speech, speech," we said. Sarah responded, "Thank you."

By some lovely coincidence, October 4 was also the day set aside in the church year to commemorate St. Francis of Assisi. As I listened to the lay leader describe this thirteenth-century saint's vocation to poverty in order to give of himself, I thought of Sarah. I thought of how her lost, bewildered condition had given us a chance to court St. Francis' bride of poverty. Her destitution gave us a chance to help.

There are many others in this world like Sarah, some with far less dramatic needs, but they all give us something precious: the opportunity to serve.

Dear Lord, use me to serve wherever You beckon. —RICK HAMLIN

5 **SUNDAY**

And as they were eating, Jesus took bread, and blessed it... and said, Take, eat; This is my body. —*MATTHEW 26:26*

Today is World Communion Sunday. The idea being that barriers between nations come down when people everywhere unite in spirit in a corporate act of worship. It didn't happen on World Communion Sunday, but I'll never forget the Sunday when God used a communion service to break a barrier between me and a friend.

Diane and I attended the same church and sat next to each other in choir. Although we weren't *best* friends, we had spent time together when I'd confided some of my deepest secrets to her. When I learned that she had broken confidence, I was devastated. Later she tried to apologize, but I rebuffed her. "I'll never forgive her," I said, "Never!"

Then one Sunday during a communion service, I listened as our minister read from the Book of Common Prayer. "For on the night in which He was betrayed...." *I know what betrayal is*, I thought bitterly. My attention returned to the minister. "My blood of the New Testament, which is shed for you, and for many, for the remission of sins...."

Sins. The word seemed to reach something deep inside me. *It's a sin not to forgive Diane, isn't it, Lord?* I didn't have to ask; I knew.

"Do this in remembrance of Me...." Now the minister was issuing the invitation. As I walked up, I saw Diane in line across the aisle. At that moment I knew what I had to do. As if God were helping me do it, we reached the altar at the same time. She saw me, hesitated, a questioning look on her face. I nodded and smiled.

Then, side by side, Diane and I knelt to receive communion together.

To err is human, to forgive divine. Thank You, Lord, for showing us this wisdom. —ELEANOR SASS

WAITING THROUGH AN ILLNESS:

"I Need You, Lord"

Last spring Sue Monk Kidd spent ten days in the hospital recovering from a ruptured appendix. Afterward she spent three more weeks recuperating at home. Though she'd spent time on the "other side" of the hospital bed as a registered nurse, she'd never been on the "patient side," finding herself vulnerable—facing fear, doubt, uncertainty, depression, helplessness. Yet through it all she discovered God meeting her needs in quiet and gentle ways.

Journey with her this week and discover help for those difficult times in life when we find ourselves whispering, "I need You, Lord."

—The Editors

Day One—***In Need of Presence***

Perfect love casteth out fear. —*I JOHN 4:18*

It was the night before surgery, my husband Sandy had gone home, and I lay in bed thinking how wholly unpredictable life could be. That morning I was packing school lunches for the children, planning a busy day. Now a few hours later I was in a hospital room where darkness hugged the window.

I thought back to the pain that had prompted me to visit the doctor. "Sue, there is a large mass here," he'd said. "I want you to go straight to the hospital." I was numb with shock, wondering what the next few days would bring.

The fear I'd managed to hold back all day now crowded in. I tried to sleep. Instead I found myself remembering the time when I was a little girl and woke during a rainstorm to the sound of branches scratching the window. I'd sat in bed trembly with fear until finally I called out for Mama. I remembered how she came through the darkness and sat on my bed. With her presence comforting me through the storm, my fear evaporated.

Here in the hospital I felt like that little girl all over again. "God," I whispered, calling out for Him from the deepest place of my heart. And suddenly it was as if He came through the darkness and sat on the bed beside me. I knew He was there...His presence comforting me through this whole uncertain event. "What happens to you happens to Me," He seemed to be saying. "We are in this together."

In those moments my fear was swallowed up in His love. As my eyes closed in sleep, I heard the word "together" far into my dreams.

When difficulty comes, dear Lord, may there be no "me"... only "us."

—SUE MONK KIDD

7 Day Two—***In Need of Strength***

O Lord... be thou their arm every morning, our salvation also in the time of trouble. —*ISAIAH 33:2*

The day after surgery, when the nurse asked if I wanted to sit up in the chair, I shook my head. I didn't have the energy for it. During the night my fever had climbed over 102 degrees and it seemed to take all my strength simply to turn on my side. Even my spirits seemed depleted.

As I stared at the IV bottles overhead, a hazy recollection of something I'd once heard came to mind—that each of us has an inner well of strength that we draw upon in difficult times. I wondered if coping with the uncertainty, the surgery and now a post-operative infection had taken most of my reserves. "Dear God," I struggled to pray. "I need strength. Please fill my well again."

I fell asleep only to be awakened shortly by a volunteer wheeling in a little cart of flowers and get-well cards from family and friends. "We love you...We are praying for you," the cards read.

Sandy placed the flowers and cards around the room in a small semicircle about my bed. Looking at them now, they were like a pair of arms wrapped around me.

All morning the "arms of love" lifted me up as I drew strength from God's family. I summoned the nurse. I was ready to sit in the chair. To smell my flowers. To think about tomorrow....

Thank You, Lord, for nourishing my "well" through the loving expressions of others. —SUE MONK KIDD

8 Day Three—***In Need of Faith***

Daughter, be of good comfort; thy faith hath made thee whole.
—MATTHEW 9:22

Each day I improved. Then one morning after a small setback, I found myself growing anxious about my condition. Restlessly, I worried, unable to find peace.

Later in the day there was a soft knock at the door. To my surprise, in came Rosie, the cleaning woman who worked at my husband's college. I'd often run into her when I dropped in at Sandy's office. We had made small, but friendly talk, about her granddaughter, her hobbies. But certainly I never expected her to come to the hospital.

She stood at the end of my bed and smiled down at me. "Everything's going to be fine with you," she said. "You just got to have faith." Then she did a curious thing. She made the sign of the cross over my bed. And slipped away.

Her gesture touched me deeply. It seemed she had woven a reminder over my bed, and it hung there like the sheer folds in an organdy canopy. A reminder that no matter how alone or frightened we feel in uncertain circumstances, His invisible presence sustains us.

I gazed at the space over me, draped and surrounded by the words God had spoken to me through Rosie. And once again in the silence of my heart, I handed Him my life.

Lord, my life belongs to You. Especially in my dark and lonely hours. Amen.
—SUE MONK KIDD

9 Day Four—***In Need of Gratitude***

The earth is full of the goodness of the Lord. *—PSALM 33:5*

One morning I was quietly sitting by the window in my hospital room, missing my husband, my children and work. Suddenly, there before my eyes, a tiny yellow butterfly flew up to the glass. His wings moved on the air like petals, so sheer that when he turned, for a split second it seemed as though the sun rays shot right through his wings. I could not move. He was life and joy and fragility all at once.

I bent close to the pane as he lit on the ledge and crawled up to the glass. His beauty was mesmerizing. And there we were, eye to eye, with only a sheet of glass between us.

Then, flash! he was gone, sailing off into the sun. Something within me leaped with joy. I was being reminded of how precious life is. And all at once I felt profoundly grateful to be alive...to be sitting by a window, any window, eye to eye with life.

I gazed long at the place where the butterfly had been. "Thank You, Father," I kept saying. "Thank You."

Oh Lord! There is nothing to surpass the simple joy of being alive in Your world. Let me never take it for granted. —SUE MONK KIDD

10 Day Five—*In Need of Stillness*

Commune with your own heart upon your bed and be still...

—PSALM 4:4

I came home from the hospital one morning in a blaze of eagerness to get on with my life. I wanted to pick up where I'd left off—packing school lunches and digging through the work on my desk. I was home a mere fifteen minutes when reality set in: I was so weak, I had to go straight to bed, unable to do anything requiring even the smallest effort.

The surgeon's instructions came back to me. "I want you off your feet and doing nothing for three weeks."

Time rolled by on slow, lumbering wheels. I reclined on the sofa feeling bored and useless. After a week of forced inactivity, I crept to my desk and began digging through papers, ignoring the tired feeling overtaking me. As I opened a drawer, I saw a little framed sign someone had sent me. It read: "Be still and know that I am God." I stared at the words, letting them sink down and unravel my heavy spirits. "Grow quiet deep within," I heard. "Take time to let your spirit heal and grow...to listen to the Voice that whispers when you are silent. Take time to 'be.' Only then can you find the spiritual rest and balance you need."

I made my way back to the sofa and sat still, gazing through the patio doors at a crabapple tree quietly turning pink. And I heard a sigh, like a wind, sweep through the room. It had come from me.

Slow me down, Lord, and send me a pink apple tree as a reminder to seek You every day.

—SUE MONK KIDD

11 Day Six—*In Need of Growth*

We are troubled on every side, yet not distressed; we are perplexed, but not in despair, persecuted but not forsaken, cast down but not destroyed. —*II CORINTHIANS 4:8, 9*

"You've healed well," the doctor said as he left the examining room. "You can go about your life normally again." It had been six weeks since my surgery, and as I left his office, I found myself thinking back over the whole episode.

I recalled the initial uncertainty about my condition, fear, pain, bouts of exhaustion, nights when sleep would never come, interrupted work and what had sometimes seemed like an unending, uphill road back to health.

I remembered too, the quiver in my daughter's voice when she'd called my hospital room and said, "I miss you, Mama;" the special vacation we had planned that had to be canceled; and those small, irreplaceable things like not being there to see my son score the winning soccer goal.

Six lost weeks, I thought with a small note of bitterness.

Later that day a friend dropped by bearing a little clay pot arranged with flowers. After she left I picked it up, feeling the hard clay surface. Suddenly I remembered an old saying I'd heard all my life: "The sun that melts the wax, hardens the clay."

I turned it over in my mind, knowing it was truth I needed to hear again—that difficulties come to us all. Some of us harden and become bitter because of them, like the pot. But others become soft and pliable like the wax, ready to be molded by God into new and better creatures because of the adversity.

Could I take my experience and turn it into something positive, seeing the possibility and meaning within it? Could I grow because of it? I looked at the pot.

Six lost weeks? *No*, I thought, *six weeks of growing!*

Whatever difficulty comes, Lord, together we can use it creatively.
—SUE MONK KIDD

12 SUNDAY

Behold, the Lord thy God hath set the land before thee: go up and possess it... fear not, neither be discouraged.
—DEUTERONOMY 1:21

As I raised the flag in front of my house in honor of Columbus Day, I remembered the words in our Pledge of Allegiance to that flag: "...one nation under God."

Truly, our land has been under God's care from the time of its discovery by Christopher Columbus. Columbus was a devoted Christian who was convinced that God had destined him to be His instrument for spreading the Gospel, and claimed that his name, which means "Christ-bearer," proved it. He declared many times that God granted him special understanding of His Word and of seamanship and related sciences. In 1501 he wrote, "Our Lord revealed to me that it was feasible to sail from here [Spain] to the Indies and placed in me a burning desire to carry out this plan."

I watched the flag whip proudly in the morning breeze, and my heart overflowed with gratitude to God that He had implanted that desire in Columbus and had led him to our shores.

Then I remembered something else our discoverer wrote: "No one should fear to undertake any task in the name of our Savior, if it is just and if the intention is purely for His holy service."

Many things remain to be done by me and my fellow countrymen to keep our land safe and free. My prayer is that we may undertake these tasks without fear and in the name of our Lord.

Father, let all of our works be done fearlessly for Your glory and be pleasing in Your sight. Amen.
—DRUE DUKE

13 MONDAY

Behold, I make all things new. —*REVELATION 21:5*

I feel I have reached one of those points in my life of letting go. You know, that time when God calls you to a new place—whether in a job or place or relationship—and to leave the old behind. But friends have nurtured me through; and what began as a question mark is becoming a different, interesting avenue that is leading to new opportunities.

One friend gave me a prayer by Thomas Merton that helped me through my uncertainty. I pass it on that it might give you strength and courage today should you be facing a new turn in the road ahead:

> *MY LORD GOD,*
> *I have no idea where I am going. I do not see the road ahead of me. I cannot know for certain where it will end. Nor do I really know myself, and the fact that I think that I am following Your will does not mean that I am actually doing so.*
>
> *But I believe that the desire to please You does in fact please You. And I hope I have that desire in all that I am doing. I hope that I will never do anything apart from that desire. And I know that if I do this You will lead me by the right road though I may know nothing about it.*
>
> *Therefore will I trust You always though I may seem to be lost and in the shadow of death. I will not fear, for You are ever with me, and You will never leave me to face my perils alone.*

—SAMANTHA McGARRITY

14 TUESDAY

And if any man sin, we have an advocate with the Father, Jesus Christ the righteous. —*I JOHN 2:1*

Our neighbor's little boy is a handful. Bright, friendly, funny, and always into mischief.

"For his own sake," his mother told me, "he has to be disciplined." She went on to say that when this job fell to her, "Billy will start yelling, 'I want Daddy!' When it's his father's turn to deal with him, 'I want Mommy!' And when he's in trouble with both of us, 'I want Jesus!'"

Though I laughed at the time, the more I thought about it, the more I realized how Billy feels. When I too have been "bad"—failed other people and myself, said or done things I shouldn't—then I too want Jesus.

He already knows all about me, as He did the Samaritan woman at the well. He will forgive me. He will refresh me with living water. He will lift me up, as He did the woman about to be stoned for adultery. "Neither do I condemn you," He will say. "Go, and sin no more."

Father, when my conscience is heavy, I will bring its burdens to You.

—MARJORIE HOLMES

15 *WILDERNESS JOURNEY* *(Leviticus 16:5–22)*

Offering a Sacrifice

And he shall let go the goat in the wilderness.

—LEVITICUS 16:22

There's a deep red ache in my heart today. My oldest son told me something in private, and I betrayed his confidence by discussing the matter with another. I'd like to excuse myself by saying that I was feeling so burdened for my son that I felt I *had* to talk to someone, but that's no excuse. There's always been a special bond between this tall blond son, with the soft green eyes and sensitive heart, and his feet-of-clay mother. Have I burned that golden thread? I've asked Paul's forgiveness and God's, and I know I've received it, but I can't seem to *forgive myself.*

Do you ever have trouble accepting forgiveness and letting go of your guilt burdens? Thirty-two centuries ago, God gave the Israelites a way to deal with that. Maybe if you and I could take our stubborn guilt feelings into that scene with us, we could learn to let go of them, too.

We are standing in front of the tabernacle. A man has brought two goats to Aaron, the priest. The first one is offered as a sacrifice for the sins of the people. But what's this? Aaron is placing his hands upon the other goat's head and mumbling something. Moving closer now, we discover that he's naming the offenses of the children of Israel, symbolically placing those sins on the goat. Now a swarthy man leads the animal away. He will take the scapegoat into the wilderness and release it, far enough away from God's people that it will never find its way back. I run after them and place my hands on the goat's head. In a sudden rush of feeling, my guilt and brokenness about my son comes pouring out onto the goat. Then I let go and watch until the man and the goat have passed completely out of sight.

There's a great commotion in the crowd now. Finally trusting that our sins are gone, we begin to embrace one another. Tears of relief flow. Shouts of praise rise up. As I walk back toward the tabernacle, I am suddenly face to face with a tall young man, his blond hair shining like gold in the afternoon sun, his soft green eyes telling me that my offense is gone... and that I'm still loved.

Oh Paul, my son, my son!

Thank You, Lord, for removing my offenses so far from me that they can never find their way back. —MARILYN MORGAN HELLEBERG

16 **THURSDAY**

Being confident of this very thing, that he which hath begun a good work in you will perform it until the day of Jesus Christ. —*PHILIPPIANS 1:6*

The autumn after my mother died I went for a walk alone in the woods. Mother had loved botany, especially the study of trees. She could identify any leaf, and often gave me spontaneous lessons whenever we passed a woodland area.

As I walked, towering pines, maples and walnut trees formed a colorful canopy of green, scarlet and gold above my head. I spotted an acorn on the ground and stopped to pick it up. As I held it in the palm of my hand I could hear my mother speaking: "In this tiny shell is everything needed for all that the tree will be. The acorns sink down under the soft mulch of old leaves. Then they wait for the sunlight and rain to send the tap root down deep into the earth to start the long process to maturity."

As I stood there, looking at the acorn, a thought struck me: The faith I have today, the faith that strengthens me in moments of crisis, had been cultivated in me from early childhood by my mother. It was her faith and guidance at the crossroads of decision in my life that helped me become who I am today, just as Mother Nature cultivates an acorn to full growth. This was the most important thing my mother ever did for me. And it is something we all can do—every day—for each other.

Dear God, help me be a builder of faith in the budding life of ______________________ today. —ELEANOR SASS

FRIDAY

I press toward the mark for the prize of the high calling of God in Christ Jesus. —*PHILIPPIANS 3:14*

Although I am not much of a sports fan, I rooted for my home team, the San Diego Padres, during the summer of 1984. It was an exciting moment when they won the championship playoff over the Cubs. I stood up and cheered as loudly as my husband and son.

Although they didn't win the World Series—the Tigers outplayed them—I was proud of them all the same. To us in the San Diego area, they were winners simply by virture of the spunky and determined way they played the game.

Moreover, I learned something from them that day. I jotted it down in the car on our way home after the game:

- Be aggressive.
- Come out swinging, whether you are ahead or behind.
- Walk tall in time of defeat.

I am convinced it is not the score that impresses our Lord. It is how well we play the game for Him—if we are aggressive in our devotion to His cause, if we keep our faith when the odds seem to be against us, if we can suffer loss and still hold firm to our belief. God's true "winner" is the team member who never gives up.

Though I "strike out" now and then, Father, I won't give up, since there's no such thing as "defeat" with You. —JUNE MASTERS BACHER

18 SATURDAY

...He hath spread a net for my feet, he hath turned me back... —LAMENTATIONS 1:13

Jogging grows daily in popularity as millions seek to get in shape. To my son-in-law Jeff, though, jogging turned out to be less than a blessing.

He had finished his first day on a new job in Sarasota, Florida, and decided to tone up with a daily run. He suited up and struck out along the back streets of his neighborhood. He was feeling exhilarated as he jogged along the palm-lined streets when he heard someone yell a greeting. As Jeff turned to see who it was, he stepped on a palm frond, lost his balance and fell flat on his face. Damage: a broken tibia in his ankle, a bloody knee, a scraped hand and—worst of all—a bruised ego.

Later, with his foot in a cast, he moaned, "If only I hadn't allowed myself to be distracted by that fellow calling to me."

His experience reminded me that when there is an objective to be gained—no matter what it is—it's vital to concentrate on your goal. And isn't it also true that as many of us jog through life, we're continually diverted from our relationship with the Lord by the attractions and enticements of the world?

Lord, help me to keep my eye on You as I run the race of life.

—SAM JUSTICE

19 SUNDAY

Freely ye have received, freely give. —MATTHEW 10:8

Between the countries of Israel and Syria four streams merge to form the Jordan River, where John baptized our Lord. From there, the river tumbles into the northern end of the Sea of Galilee where the water is sweet and transparent, abounding in fish. It was here that Peter and his brother Andrew were fishing when Jesus called them to follow Him.

The Jordan splashes on through the Sea of Galilee and continues until it empties into the Dead Sea. What a contrast this body of water is to the Sea of Galilee! No fish swim in its salty brine. No trees grow on its banks. Some even claim that birds will not fly over the Salt Sea (as it is called in Genesis 14:3 and Deuteronomy 3:17).

What makes the difference? Surely, the Jordan carries the same water into each sea. The diversity lies in the way each uses the water it receives. The Sea of Galilee welcomes the river, making it a part of itself and then passing it on. The Dead Sea, with no outlet, greedily holds all of the water that comes into it. Through years of evaporation, the minerals of the river have accumulated, making its water useless.

The love of Jesus Christ flows freely into each of us. What will I do with it? What will you? Will we receive it, make it a part of ourselves and then pass it on to others? Or will we hold it, provide no outlet for it, and become spiritually useless?

We have a choice.

Lord, let me be a vital living witness—a Sea of Galilee—for You. Amen. —DRUE DUKE

20 **MONDAY**

The water that I shall give him shall be in him a well of water springing up into everlasting life. —*JOHN 4:14*

On the grounds of a famous Japanese shrine in Kyoto, there is a spot that features three waterfalls. They represent wisdom, health and longevity. Visitors who pass beneath the falls are invited to choose which of the blessings they want for their lives by placing their hands under the appropriate falls. But according to tradition, a person can choose only one.

Which of the three would you choose? After some soul-searching, I selected health, reasoning that without it longevity would be tainted and with it one would be free to pursue wisdom. But my wife Shirley picked wisdom because she believed longevity would be hollow without it and that wisdom would prompt more healthful habits.

Fortunately, if you know Christ, you don't have to make the choice. He came to ensure our longevity—eternal life; through His Holy Spirit we have at our disposal divine wisdom; and as far as health, we can rest in His promise: "I came that you might have life and have it more abundantly" (John 10:10). Drink at His waterfall and never thirst again.

Restless we are until we rest in Thee,
Empty, O Lord, 'til You gain victory. —FRED BAUER

21 TUESDAY

For we are laborers together with God...

—*I CORINTHIANS 3:9*

Budget reductions had cut funds to our public library. To help keep the library going, I sponsored an ongoing book sale. Friends of the library donated books from their attics, cellars, and libraries. Daily I put out new hardcovers and paperbacks, which were quickly snapped up. Our sales soared.

One day a woman I knew came up to me and said, "I never knew you were a librarian. Or don't you have to be a librarian to hold this job?"

I smiled and explained that I was a volunteer worker. The woman gaped. "You mean you work here for nothing? But you should be paid for your time."

Paid? Then I recalled that George Washington Carver received only twenty-eight dollars per week at the height of his fame. He sought no patents for his two hundred discoveries. He felt these were gifts from God.

And then there was our Savior, trudging from place to place, healing and ministering. His counsel has comforted us for two thousand years, yet He never sought a salary.

I was paid. Had not God given me good health, a love for books, and the opportunity to serve? I was very well paid, indeed.

Gracious Lord, I will serve You with gladness. —OSCAR GREENE

22 WEDNESDAY

If we love one another, God dwelleth in us... —*I JOHN 4:12*

Around the time some people began saying, "God is dead," a friend told me a story. He had gone for a walk, when he noticed an elderly man lying on the lawn in the park. The man's head rested on an outstretched arm and his right hand grasped the left, which was trembling as though in a spasm.

"At first I kept going," my friend said. "Then I thought, 'He's in trouble!' I hurried back and touched his shoulder. 'Are you all right?' I asked. The man looked up at me and smiled. 'Sure thing,' he said. 'Just getting a bit of sun and doing a few isometric exercises. But thank you. You're the third person to ask.'

"As I walked away, suddenly I felt good. Three of us had cared enough to offer help if it were needed—and that, after all the talk about God being dead! Before I reached the end of the block, I noticed two more strangers had stopped to inquire. I can't tell you how proud I was of all *five* of us!"

What if it were twenty-five? I thought. *Or a hundred? God dead? Never!* In fact, if each of us were to do only *one* thing today in the name of the Lord, then that five might be multiplied—perhaps by a million. Who knows what our world could really be like....

May our actions today, Lord God, tell the world that You live!

—DORIS HAASE

23 **THURSDAY**

But God hath chosen the foolish things of the world to confound the wise... —*I CORINTHIANS 1:27*

We live in an apparently well-cared-for area. I say "apparently," because you never see the litter there that you'll find in other neighborhoods. That's only because of my neighbor John. Wherever he walks—to the shopping center, the corner mailbox, the bus stop—he constantly picks up debris and deposits it in the nearest trash container.

Most people consider John's behavior foolish. "You don't have to go around picking up stuff others drop," we tell him. "Let someone else do it." But he just shakes his head smiling and goes along the street picking up trash.

As I watched John one afternoon from my front window, I wondered, *What if everyone had said, "Let someone else do it?"* Well, the hungry five thousand wouldn't have had loaves and fishes...the lame wouldn't have walked or the blind regained their sight...the Bible wouldn't have been written...and our neighborhood wouldn't be litter-free!

John seems to know what the Lord has given him the capacity to

do. And he's happy because he's doing his job to the best of his ability.
I just hope I can learn to follow his example.

Lord, give me the backbone to serve You—even at those times when others call it foolishness. —ISABEL CHAMP

24 FRIDAY

For he is our peace, who hath made both one, and hath broken down the middle wall of partition between us.

—*EPHESIANS 2:14*

Even though I had seen pictures of China's Great Wall ever since childhood days, I was awestruck upon actually viewing it. Its cement-colored body stretched thousands of miles, and it is earth's only man-made object that has been seen from the moon. Made of enormous, hand-hewn blocks of granite, it is as high as a two-story building, as broad as thirty feet at the base, and its width on top was designed to accommodate five horses galloping abreast. Although its first sections were built in the fifth century B.C., it is said that three hundred thousand men worked from 221-210 B.C. to link it together and complete this barrier against northern invaders. Although thousands of years have passed, the Great Wall of China still stands today.

In Jesus' day, the people had not heard of the Great Wall in China, but nevertheless they were familiar with walls. Paul wrote about a similar wall in the verse above, one that divided man from man, and man from God.

Unlike the impenetrable Great Wall, the good news the Bible has to tell us is that the wall between God and man has been surmounted for all eternity, giving us hope for dissolving the walls that stand between man and man.

O Father, on this United Nations Day, help us to break through the walls that separate us one from another, nations from nations, and help us seek peace. Amen.

—ISABEL CHAMP

25 SATURDAY

...Sleepest thou? Couldest not thou watch one hour?

—MARK 14:37

"Don't forget to turn back your clocks, folks. You have an extra hour to sleep tonight." Every October we hear those words by newscasters on the last Saturday of the month and think to ourselves, "We've gained an hour!" And many of us can think of no better thing to do with it than to "get some extra z-z-z's."

But time is relative. Suppose we *have* gained an hour; suppose the Lord has said, "Here is an extra hour—yours to do with as you please."

Would you choose to sleep? Now, suppose He cautions, "It will be your last." Then how would you respond?

Sobering, isn't it? Yet the "when" of it is not so important when we realize that *every* hour is a gift to us from our Lord. One of the sayings my grandmother favored came back to me this year when it came time to turn the clock back: "Live each hour as though it were your last."

Would I then want to spend my last hour sleeping? Would you?

Awaken me to Your precious gift of time, Lord Jesus, and let me use it wisely. —JUNE MASTERS BACHER

26 SUNDAY

For the Lord God will help me; therefore shall I not be confounded...

—ISAIAH 50:7

I have a friend who sometimes climbs alone in the rugged Rocky Mountains of Colorado. Recently he told me about a time when he ascended a steep mountainside and then couldn't get back down. At first he panicked. The afternoon light was waning and soon darkness would fall, that sudden overwhelming darkness one experiences in high country where the air is too thin to hold the twilight. He could no longer see exactly how he had climbed to that particular spot. No one knew where he was; no one would come searching.

As he sent a prayer of distress winging into the air, he wasn't sure what he expected in return—maybe a rescue helicopter appearing miraculously over the hilltop, or a party of climbers with ropes and pitons.

What he saw, instead, was a narrow trail off to one side, very faint, zigzagging down the cliff. He made his way to the spot and found that it was a deer trail, almost indiscernible and still precarious. But by picking his way down that trail, he arrived safely in the valley below.

He concluded, "You know, I've found it's okay to pray, but you also have to *look for* and *recognize* the solutions God provides."

Father, keep me alert to Your answers. —MADGE HARRAH

27 MONDAY

Hold up my goings in thy paths, that my footsteps slip not.
—*PSALM 17:5*

I once saw a magazine cartoon that showed a driver looking under the hood of his car to find out why steam was spewing out. What he didn't see was that the anchor from the boat he was pulling had dropped onto the road with the anchor's chain still attached. As a result, a furrow was being plowed in the highway behind.

I smiled and leafed on in my magazine. Then I paged back and looked at the cartoon again.

That anchor reminded me that hindrances—failures, mistakes, disappointments—often drop into my daily routine. At the moment, I don't perceive they're pulling me back. No wonder my mental attitude heats up!

If something is holding you back, maybe you're dragging an anchor behind you. And if you're anything like me, it'll be too heavy to pull up by yourself—but no anchor is too heavy for the Lord.

Father, take this burden and lighten my spirits today. Amen.
—ISABEL CHAMP

28 TUESDAY
Teach me to do thy will; for thou art my God...
—PSALM 143:10

Dinner was over, the table cleared, and I was helping my ten-year-old daughter Lindsay with her homework. Tonight it was math and common denominators. We needed to add long columns of mismatched fractions. Some were obvious. Others seemed hopeless and frustrating. In each case, however, the key to solving the problem was discovering the pesky denominator they had in common. Once found, the unlikely combinations added together easily.

I was surprised how rusty my skills had become, but as we chewed our way through each problem, the answers came easier. As we put away the books, I thought about the importance of common denominators in relationships, especially as we seek to make conversation with strangers or new acquaintances.

I want to greet the stranger sitting next to me in church, or the tired-looking customer standing behind me in the grocery line, but I hesitate because I'm afraid we won't have anything to talk about—no common denominator. I fumble around for a conversational opener, but I usually quit and turn away. Or worse yet, I avoid the problem altogether by simply saying nothing!

Surely God gives us common denominators in every relationship. We just have to search for them. And surely we please Him when we don't let our skills get rusty!

Lord, help me to be persistent in seeking the common denominators of every relationship I encounter today. —CAROL KUYKENDALL

29 WEDNESDAY
Come ye yourselves apart... and rest a while. *—MARK 6:31*

I am writing this in a little one-room cottage nestled among the tall pines at a monastery in Hastings, Nebraska. I've come here to be alone with God for twenty-four hours. There will be no interruptions, no meetings, no preparing of family meals. I have a freshly baked loaf of nutritious bread and a thermos of water, my Bible, this notebook, cushions and my sleeping bag.

I love the reflection of the lantern glow on the wash basin and pitcher, the crackling fire in the little wood-burning stove, the open Bible on the desk in front of the window. In the distance, I can hear a train whistle, children playing, leaves blowing. Yet there is a silence greater than any of the sounds. It's an inner silence. Already the emotional load I carried in with me is beginning to slip away in awareness of Christ's healing presence. I will spend my time praying, pondering His words in the Bible, and just being present with Him in love. And I know that I will go away refreshed, nourished, made whole.

You don't need a monastery to take a twenty-four-hour sabbatical. A camper will do, or a motel room, or a tent in nice weather. I hope you'll try it and discover what a healing grace twenty-four hours alone with God can be for you!

I will seek Your whispers in the silence, Lord.

—MARILYN MORGAN HELLEBERG

30 THURSDAY

For I will restore health unto thee, and I will heal thee of thy wounds, saith the Lord... —*JEREMIAH 30:17*

When Larry and I revisited the big island of Hawaii, after an absence of twenty years, one of the places I wanted to see again was Devastation Trail. I remembered how the boardwalk then had meandered through a moonscape of gray and red ash, land laid waste by a violent volcano eruption. Now I found, to my amazement, that a forest had grown up in only two decades to heal the land. Tall *ohia* trees nodded their fringed red blossoms above the ferny undergrowth while birds twittered overhead.

As I followed the trail through the green solitude, I thought of my brother Bill and his wife Marilyn, who had walked there with us that first time. Later they had died in a car wreck, a tragedy that had devastated our families. But gradually, helped by the loving prayers of friends, our grief had healed, and now I could walk the trail without pain, remembering the happy times with Bill and Marilyn.

Time *does* heal, a grief, a hurt, a disappointment. Time—and prayers shared with our Lord who restores not only wounded land to new life and beauty, but wounded souls as well.

Father, I accept the gift of Your healing love. Thank You.

—MADGE HARRAH

31 **FRIDAY**

Turn thee unto me, and have mercy upon me, for I am desolate... —*PSALM 25:16*

During a particularly painful time, loneliness invaded my heart with such unexpected intensity that I thought I might never recover. It was as though I had fallen off a cliff and landed on a tiny ledge about halfway down. There I sat, alone, helpless, with no one ever knowing I was marooned there. Often, with comfort lacking and healing slow, I would pray. Mornings, sometimes noon and night. One day a startling thought came to me.

God, the Alpha and the Omega, the great I AM, the God of the entire universe Who spoke a whole world into existence, must have been...lonely. He must have longed to have fellowship with man... with me! God...lonely, for even my friendship, my love. This was the thought that began to heal me, infusing me with new life. I've never forgotten how much I hurt and that many others are hurting and isolated in loneliness today.

William Wordsworth described loneliness beautifully—and it is my gift to you who are lonely today:

> I wandered lonely as a cloud
> That floats on high o'er vales and hills,
> When all at once I saw a crowd,
> A host of golden daffodils...

Father, help me be a golden daffodil to someone wandering lonely as a cloud today. —MARION BOND WEST

Praise Diary for October

1

2

3

4

5

6

7

8

9

10

11

12

13

14

15

16

17

18

19

20

21

22

23

24

25

26

27

28

29

30

31

November

S	M	T	W	T	F	S
						1
2	3	4	5	6	7	8
9	10	11	12	13	14	15
16	17	18	19	20	21	22
23	24	25	26	27	28	29
30						

Lord, Show Me How

If on my journey, I can impart
Some simple kindness from the heart,
Lord, show me how.

If on the job, I can employ
Christ-like virtues of peace and joy,
Lord, in me be Thou.

If in my family I can heal
Wounds that rob life of its zeal,
Lord, help me be heard.

If some lost soul I can guide
To salvation at Your precious side,
Lord, prompt me with Your word.

If one in need I can relieve
And his self-love thereby retrieve,
Lord, lead me now.

My child, you pass but once this way
Do all the good you can today,
Seize each day, each hour, each minute
And all the love that life has in it—
I've shown you how. —*FRED BAUER*

1 **SATURDAY**

To all the saints in Christ Jesus... Grace be unto you, and peace, from God our Father, and from the Lord Jesus Christ. —*PHILIPPIANS 1:1, 2*

Doubtless you've received letters from strangers beginning with "Dear Sister" or "Dear Brother" and signed, "Your Christian Sister (or Brother)"—terms that immediately acknowledge that you both belong to the family of God.

But if I were to be seriously addressed as "Saint Isabel," I would be dumbfounded.

I think of "Saint" as being reserved for such spiritual figures as Peter, Matthew, Mark, Luke and John, whose exemplary lives seemingly have nothing to do with everyday people like me. Yet in Biblical times, "saint" was simply another name for..."believers." Ordinary people *just* like me.

Today is All Saints' Day, a day set aside to honor God's saints, both living and dead. This year on All Saints' Day I'll look up to the spiritual leaders around me with reverence. But I'm also going to take a long look at the vast community of Christian brothers and sisters who make up my family of *everyday saints*. And I'm going to thank each and every one of you this day.

Father, I'm going to treat my fellow believers with the respect they deserve—as saints. —ISABEL CHAMP

2 *Practicing His Presence in November*

PERCEIVING EVERYDAY GIFTS

...Thou openest thine hand, they are filled with good. —*PSALM 104:28*

We ambled along a dusty path through the woods, toting our fishing poles to the pond. My husband and children threw themselves into the delight of fishing. But I'm ashamed to say I grumbled about the weather, which was muggy; the fish, which weren't biting; the children, who were sliding down a mud bank; and of course the fishing worms, which were...well, worms.

As my misery grew, I escaped to the pine trees beside the path and quietly sulked out of sight. Soon my attention was drawn to the footprints etched in the dust along the trail. My husband's tennis shoe, my sandal, the children's bare feet. I gazed at them attentively, and found myself thinking how good it was simply to be able to walk. How wondrous to move through the woods! I studied how big the children's feet are today compared to the tiny feet blotted in their baby books. Then I noticed how our footprints mingled together in the dirt, feeling anew how precious my family is to me, how glad I am for their steps intersecting through my life.

Gratitude poured over me—for woods, health, growing feet and the sharing of lives together. But most of all I felt freshly aware of the Giver. His Presence, which had been so remote all day, seemed to spill out of those moments of praise.

I learned something that day. Buried under our busyness and grumbling are gifts we no longer see. Small, everyday blessings we take for granted.

In November let's take time to praise God during each day for those positive, but overlooked blessings in our lives—music, laughter, memories, books, friends, second chances, orange pumpkins, warm fireplaces and, of course, all the footprints scattered through our days.

This month I will be alert, O Lord, to the myriads of blessings for which I can be thankful, but sometimes overlook. Like ________ and ________________ and ________________. —SUE MONK KIDD

3 **MONDAY**

But whoso looketh into the perfect law of liberty, and continueth therein ... this man shall be blessed in his deed. —*JAMES 1:25*

Kurt Mazur, conductor of the symphony orchestra of Leipzig, has a reputation for bringing fresh insights to the music he conducts. Even though he has led the orchestra for fifteen years and is himself nearing sixty, critics rave that he continues to evoke a dramatic crispness from compositions, particularly those of Beethoven.

He dislikes mechanical playing and has a knack for introducing surprise—particularly in the form of an altered tempo—in his rehearsals. But Mazur has another practice that is perhaps even more effective in bringing life to his music. He often pores over Beethoven's original manuscripts, believing that, by going to the source, he can capture the music as Beethoven himself intended it to sound. He reports that his library trips have been invaluable for clarifying Beethoven's "dynamic markings."

How right to go to the source! For as meaningful and valuable as all my spiritual exercises can be, there is no substitute for going to the Source, the Bible. It is only here I can find the "dynamic markings" that can unerringly lead to meaning in my daily life.

Dear Jesus, may Your Word help me to approach my day with the originality and uniqueness with which You have endowed me. Amen.

—JAMES McDERMOTT

TUESDAY

...I pray, that your love may abound yet more and more in knowledge and in all judgment; That ye may approve things that are excellent... —*PHILIPPIANS 1:9, 10*

On Election Day, I often think of a story my grandfather used to tell. First, though, I have to admit to a strictly partisan heritage. My grandfather was a dyed-in-the-wool, unswervingly loyal Republican. And he had a thing about one of our great Democratic presidents, a man whose policies my grandfather could not abide, with a wife whose altruistic motives he heartily mistrusted. In short, my grandfather's politics interfered with his appraisal of the people behind them.

Then once, back in the late forties, on a flight going from Washington, D.C., to Pittsburgh, Grandfather was seated by chance next to this president's widow. I don't doubt that he considered it an unpleasant surprise, but—always a gentleman—he made polite conversation, and gradually their chitchat turned into a stimulating discussion that lasted the whole flight. Not surprisingly, my grandfather came away from the flight absolutely entranced with Eleanor Roosevelt.

I don't know if my grandfather's voting habits changed after that eventful meeting, but I *do* know his thinking changed.

"Don't let politics get in your way of appreciating people," he used to remind his grandchildren. "Republican, Democrat, liberal or conservative, always vote for the *best* man or woman...and get out to vote!"

Guide my decisions and actions today, Lord, to make the best choice for our nation. —RICK HAMLIN

5 **WEDNESDAY**

Neither do men light a candle, and put it under a bushel, but on a candlestick; and it giveth light unto all that are in the house. —*MATTHEW 5:15*

Many years ago my Grandma Parry knitted for me a beautiful afghan of rosy-pink and gray stripes. She wrote a little note to accompany it: "For Eleanor, there's love in every stitch. I hope you put this to good use."

But I was afraid to use it. I worried that it would get soiled, or that the wool might catch on something and rip, so I carefully wrapped it in tissue paper and stored it in the back of my closet. Then one night when the temperature outside hovered around zero and I couldn't find any more blankets, I remembered the afghan. After taking it out and spreading it across my bed, I began reminiscing about Grandma. At one time she had lived with my parents and me. Every evening before retiring, she would read a passage from her Bible. Now, suddenly, I seemed to hear Grandma's voice: *How long has it been since you read your Bible, Eleanor?*

What a gentle reminder! I had put my grandma's afghan in the closet and my Bible in the bookcase, and had forgotten that both were meant for good use. So I went to the bookcase and found my Bible. Snuggled down under Grandma's blanket, I began to read. And I had the feeling that somewhere, watching me, Grandma smiled.

Lord, keep reminding me that Your gifts are meant to be used, and Your Word meant to be lived. —ELEANOR SASS

THURSDAY

In the day of my trouble I will call upon thee...

—PSALM 86:7

I am seated before my new computer keyboard—with the trembling and apprehension that habitually seizes those of us who became grandparents before the age of electronics. The motor hums with unexplored capabilities, while a blinking yellow dash-line impatiently taps its electronic finger beneath the next blank space on the screen. Blank space is the most daunting thing a writer faces—and *this* blank is alive and watching me....

But, oh comfort in the midst of this technological wilderness! At the bottom of the screen is the message:

F1-Help

And in the upper lefthand corner of the keyboard, sure enough, there's a key labeled, "F1." By pushing it I am instantly shown a whole screenful of explanations, reminders and instructions.

A key for help...what reassurance! Even though I can't yet grasp and utilize all the help available, I know that it is there.

And wonders of wonders...that message can apply to my spiritual life, too: "F" for *Father*, "1" for *first*—whenever the complexities of life are great.

Father. First. —ELIZABETH SHERRILL

FRIDAY

Walk worthy of the vocation wherewith ye are called.

—EPHESIANS 4:1

One day a man asked my friend's two sons, "What are you going to be when you grow up?"

"A veterinarian," the older replied promptly.

The younger looked at him in disgust. "He didn't ask what you are going to *do*, Jesse. He asked what you are going to *be*."

As an adult, I often make that same mistake—confusing what I do with what I am. Rushing to meet a deadline, I snap at my children. Working on an important project, I neglect the personal needs of those who work with me. Harried by the demands of parenthood, I forget to take time for my own spirit.

In Ephesians 4, Paul offers a good contrast between what we Christians do, and what we are to *be*. He lists several callings we may have—apostle, preacher, teacher, shepherd—but he prefaces that list with this: "Walk *worthy* of the vocation wherewith ye are called." He urges us to be lowly, meek, long-suffering, tolerant, seekers of unity and peace. And he concludes his chapter: "Be ye kind one to another, tenderhearted, forgiving...."

What I *do* for Christ is important. But far more important is what I *am*.

Dear Christ, as I strive to "grow up into you in all things," nudge me when I let what I do *get in the way of what I should* be.

—PATRICIA HOUCK SPRINKLE

SATURDAY

There is that speaketh like the piercings of a sword: but the tongue of the wise is health. —*PROVERBS 12:18*

Few recent films have moved me as much as *On Golden Pond*, the story of a retired professor and his wife, Norman and Ethel Thayer, and their struggle for dignity and meaning in old age. A ritualistic summer migration to a lakeside retreat serves as a metaphor for their waning lives.

One of the most touching moments in the story is in a conversation between Ethel and Billy, the teenaged boy who has come to spend the summer with them. The independent boy and Ethel's traditionalist husband have a two-generation gap in their understanding of each other, one that is compounded by Norman's fear of dying.

How can he be so lovable on one hand and so disagreeable on the other? Billy wants to know of Norman. Ethel understands. Love's resourceful fingers can unknot any contradiction.

"Sometimes," she tells Billy, "you have to look hard at a person and remember he's doing the best he can."

Maybe we could all remember such wisdom before we unload our verbal shotguns the next time someone offends us or our sensibilities. It might keep all our ponds more golden.

Before we in anger lift our voices,
Show us, Lord, we have other choices. —FRED BAUER

SUNDAY

And, behold, God himself is with us for our captain...

—II CHRONICLES 13:12

Once there was a young ensign who had done so well he was given the opportunity to take full responsibility for getting his ship under way. Excited and eager to show off his skill, he barked out commands and in record time had the ship steaming out to sea.

He knew he had done well, and so when a seaman approached and told him that the captain had a few words for him, he was not surprised. He *was* surprised, however, to learn that the captain's words were in a radio message.

"You did a fine job," he read, "completing your underway preparation exercises with amazing speed. There is one thing, however, that you overlooked. You failed to make sure that your captain was aboard before getting under way."

How easy it is for me, engrossed in some challenging new project, to take off without first consulting my Captain through prayer. It's full steam ahead...until suddenly I find myself floundering. Then I, like that ensign, get the message:

"You failed to make sure that your Master was aboard, before getting under way."

Pilot my ship, Lord God, that I may safely reach that other shore.

—DORIS HAASE

MONDAY

If I say, I will forget my complaint, I will leave off my heaviness...

—JOB 9:27

There was a time in my life, several years ago, when I suffered from rheumatoid arthritis, a painful condition that causes joints to swell and inflame. When some parents in our community asked me to head up a Christian youth group, I responded with inner feelings of rebellion.

It's not fair, Lord, I thought. *These other parents can move about without pain. Why should I be the one to take on this job?*

Then I read a story about French Impressionist painter Auguste

Renoir. He had arthritis far worse than mine, and he ended up in a wheelchair. When people asked him how he could paint with his twisted hands, he replied, "For the true painter, there are no obstacles." With brushes strapped to his wrists, he finished a beautiful still life on the very day he died.

That got me to thinking about how I was allowing my arthritis to stand in my way. Could it be that Renoir's statement could be reworded? How about: "For the true *Christian*, there are no obstacles."

After much prayer, I accepted leadership of the group. Caught up in the excitement of working with those wonderful young people, I forgot about myself, and you know what? That proved to be the best pain remedy of all!

I like Your prescription, Father: two tablets of unselfishness, and a call to You each morning. —MADGE HARRAH

TUESDAY

The... prayer of a righteous man availeth much.

—*JAMES 5:16*

The preacher was talking about the importance of honesty in prayer. It was important, he said, to tell God exactly what you were feeling or thinking. An honest prayer was one based on truth, and, since God Himself was Perfect Truth, that was the only way to approach Him.

Leaving the church afterward, I found myself remembering something that happened during World War II. One day, a B-17 from one of our bomber squadrons limped home from a raid over Germany, badly shot up. Just behind the tail-gunner's position was a hole the size of a washtub, where a shell fragment had slammed through, inches from the gunner's head.

The tough little sergeant who was a tail-gunner stood on the ground in his flying clothes looking up at the hole. Then he shook his head slowly. "I never thought I believed in You," he said, "and I'm still not sure that I do. But thanks, anyway."

I think God must have smiled when He heard that prayer.

Lord, I know You understand that moments of gratitude are unspoken prayers. —ARTHUR GORDON

12 WEDNESDAY

And, lo, I am with you always... —*MATTHEW 28:20*

You've heard about the Chinese man named *Lo*, haven't you? He's the one, remember, who listened intently as missionaries expounded the word of God during a meeting he and others attended.

However, Lo became extremely excited when the pastor read Jesus' words from Matthew 28:20: "And, *lo*, I am with you always, even unto the end of the world."

Lo's heart leapt. *Just think... the Lord knows my name and makes me a promise like that!*

I think the problem some of us have—at least I do—is failing to remember that the Lord has made this same promise to each one of us.

Whatever your need today, take this moment to claim His promise. And like *Lo*, enter your name....

And, ______________________, I am with you
always, even unto the end of the world.

For Your presence always, *Lord, thank You.* —ISABEL CHAMP

13 THURSDAY

Not because I desire a gift: but I desire fruit that may abound to your account. —*PHILIPPIANS 4:17*

I remember an afternoon several years ago when my friend Corona and I sat in Norma Millay's kitchen, eating strawberries and listening to her reciting poems written by her famous sister, Edna St. Vincent Millay. I don't pretend to be a poet, but our visit made me recall a poem I'd written on assignment in high school years before—a poem that taught me something important.

My poem was about Plaz Wimbish, an elderly man who lived with his family about a mile from us. Plaz had a beautiful, quiet spirit. Whether plowing or sitting astride his horse, he was so dignified that I stood in awe of him. When I wrote the poem, he was ill. Somehow his wife got a copy of it, probably from Mother.

Later, Mother wrote to me at college that Plaz had died. "And do you know, at the funeral they read your poem about him."

I felt overwhelmed. There had been a purpose for that poem, something more than the grade the teacher had given me. And I learned from this experience that writing was more than a skill. It was a way of giving to people.

And that's something we all have in common, something to give. Maybe we all can't be writers, or composers or painters, but we all have something creative inside of us to discover and use.

Lord, help me share my gift with others. —SAMANTHA McGARRITY

FRIDAY

There came a woman having an alabaster box of ointment of spikenard very precious; and she...poured it on his head...And they murmured against her. And Jesus said, "Let her alone...she hath wrought a good work on me." —*MARK 14:3, 5, 6*

"Daddy, build a doghouse for Scruffy," asked our five-year-old daughter Ann. She squeezed the brown puppy in her arms and looked at my husband with that look I suppose no father can resist, because to my amazement he said, "Okay, we'll build a house for Scruffy."

"Will it have a roof and a door and his name on the front and everything?" she asked.

"It will have everything," he assured her.

So one day Sandy hammered and sawed a pile of new lumber into something resembling a doghouse. And when it was finished it had a roof and a door and a name on the front and everything.

I looked at it and smiled at Sandy's tired face. "You are probably the only father in the history of the world who has made a doghouse for a *stuffed* animal," I said. Yes, Scruffy was only a stuffed toy, with one ear coming unstitched and his fur wearing thin from all the hugging.

"Well," Sandy grinned, "I guess it was pretty impractical. But sometimes you need to do something extravagant like that for someone you love."

That doghouse drew them together in a special way. And it taught me something. Now and then I need to pause and do something extravagant for Someone I love. My Lord.

Lord, today I will express my devotion to You in generous ways.

—SUE MONK KIDD

15 *WILDERNESS JOURNEY* (*Deuteronomy 34*)

Moses Says Good-bye

And Moses went up... to the top of Pisgah... and the Lord showed him all the land... —DEUTERONOMY 34:1

So, Moses, once again you are on the mountaintop with God. Your work is almost finished now. You have passed your mantle to Joshua, said good-bye to your people and climbed alone to this high peak. Even though you are very old, the Lord has kept your strength. Your eyes are bright, your mind is clear, your arm and heart are still strong. Alone upon the height with your God, you look on the Land of Promise spread out before you. To the north, you can see the white crown of Mt. Hermon. Far below, the River Jordan winds its way down to the Dead Sea. Across the river, at the foot of the mountains, stands the city of Jericho, surrounded by a high wall. As you gaze across to Mt. Hebron, where Abraham and Isaac and Jacob are buried, what must your thoughts be? Do you have any inkling, as you look down upon Jerusalem, of the world-transfiguring events that will one day grace that city? How does it feel, after years of thirsting, to see afar in the west the gleaming waters of the Great Sea?

I have come to love you, old man. You have led your people well. I know that the tears in your eyes are not for yourself, even though you will never enter the land you have given your life for. Your only mistake was that you hit the rock to get water, instead of obeying God and speaking to it. I guess that's always what separates us from our dreams—trying to force them to happen on our own, instead of relinquishing them to God.

I have some long-standing dreams, too, Moses. As I stand here on this peak, my visions for the coming years parade before me. I picture each one in specific detail and see it as clearly as possible, as if it had already happened. What did you say? Oh yes, I will remember to speak to the rock instead of striking it. After I've seen my dream, I will relinquish it totally to my God, knowing that His will for me is always a land of glorious promise.

And now, good and faithful leader, the time has come to say good-bye. In an act of total surrender, you lie down before God and offer up your spirit. I have no doubt that the land you now enter flows with milk and honey beyond your wildest dreams.

Take my dreams, Lord. Mold them, shape them; then grace them with Your holy promise. —MARILYN MORGAN HELLEBERG

16 SUNDAY

Most gladly therefore will I rather glory in my infirmities, that the power of Christ may rest upon me.

—II CORINTHIANS 12:9

In a little town in the French Pyrenees, there is a shrine celebrated for miracles of healing. As the story goes, an amputee veteran of World War II appeared there one day. As he hobbled up to the shrine, a bystander remarked, "Poor man. Does he think God will give him back his leg?" The veteran overheard him. "No, sir," he replied, "I don't expect God to give me back my leg. I am going to pray to Him to help me live without it."

I've always liked this story because when I was eight years old, after a bad case of measles, I began to lose my hearing. My loss was a gradual one. Today, I have about twenty percent hearing in each ear. Well-meaning Christian friends often suggest that I ask God to give me back my hearing. When they do, I tell them that I have done this but, so far, my hearing has not been restored.

Still, when I pray to Him to help me live with my loss, I believe He answers. Lipreading skills and two hearing aids enable me to hold a job and get along just fine in the world. What's more important, through my dependence on the power of Christ, I find I am able to hear *Him* a whole lot better. And that, I have found, is the greatest gift of all.

Dear Lord Jesus, help me each day to accept my infirmities and lean on You. —ELEANOR SASS

17 MONDAY

I... beseech you that ye walk... with longsuffering, forbearing one another in love... —EPHESIANS 4:1, 2

Several years ago, as an actor in California making very little money, I took a temporary office job. It was one of those insufferable offices where the employees gossiped and jealousy and mistrust abounded. Consequently, I was miserable and frequently made mistakes.

One evening after work, I had to drive to a costume shop in an unfamiliar part of town to pick out my wardrobe for a play I would be acting in. It had been a bad day at work—nothing had gone right and my boss made sure I knew about it—and my spirits were low. As a result, I got hopelessly lost driving. Furthermore, every time I stopped to call the costume shop for help, I'd get more lost. "That receptionist keeps giving me wrong directions," I muttered. By the time I got there I was ready to give her a piece of my mind.

And then I asked myself, *What kind of pressure is she under? Maybe she's as insecure about her job as I am about mine? Why can't I forgive her... and forgive myself as well?* As I spoke to her, I tried to give her peace of mind instead. Perhaps I did, for soon she was smiling and my own tensions had eased.

There's an old psychological twist to forgiveness: When God calls us to forgive others, much of the work we do is to forgive ourselves. *I forgive you*, we say. In the process *He forgives us*.

Lord, help me to be slow to anger, quick to forgive and eager to give peace a chance today. —RICK HAMLIN

18 TUESDAY

And as we have borne the image of the earthy, we shall also bear the image of the heavenly. —I CORINTHIANS 15:49

Recently in our community, a patient was sent home from a renowned cancer treatment center with the word that "There is nothing more we can do for you." It was obvious to all of us that the man, only fifty years old, had been sent home to die—a tragic and depressing situation that sends most people into unconsolable despair. But this man was an unusually committed Christian and his tremendous faith did not allow him to view his condition as tragic. Instead, he encouraged visits and enjoyed long, happy hours with his family and friends.

One Sunday morning a few months later, he asked his wife to summon some twenty close friends and arrange for the pastor of his church to bring them communion after the morning service. Then, with all gathered, communion was held, prayers were said and favorite passages from the Bible were read. The atmosphere was that of a quiet celebration, with no tears, no sadness, no mourning...just friends acknowledging that the time had come to bid farewell to a beloved companion.

Then, as the Twenty-third Psalm was being read, he slipped away... gently...from the bosom of those who loved him here on earth to the waiting arms of the One who loved him through all eternity.

Lord, help us to understand our last moments not as the end of our earthly dreams and hopes, but as a glorious new beginning, entering into the joy and peace of Your abiding presence. Amen.

—SAM JUSTICE

19 WEDNESDAY

He maketh the barren woman... to be a joyful mother...

—PSALM 113:9

"And what are *you* doing for Thanksgiving, Mrs. Bellwood?"

The slight emphasis on "you" tattled. The rest of us who were busily filling baskets for the needy had great plans for a family day. But poor Mrs. Bellwood was less fortunate than those for whom we were preparing the baskets. She had no family.

But then the grandmother-age lady surprised us all. Bending to stuff a can of nuts into one of the baskets, she spoke. "I plan for as many members of my family as I'm able to pack into my house this year!"

Mrs. Bellwood must have seen the unasked questions on our faces. "You see," she smiled as she rose and put a hand to her back, "I have such a large spiritual family...my Sunday school class, my Brownie troop—and this year I've included a dozen children from the orphanage."

Yes, Mrs. Bellwood *was* having a family Thanksgiving. And she reminded me to reach out to the needy kinfolk *beyond* my front door and make Thanksgiving *truly* a family holiday.

Lord, let me thank You by reaching out to others of my brothers and sisters in love. —JUNE MASTERS BACHER

20 THURSDAY

Know ye not that ye are the temple of God, and that the Spirit of God dwelleth in you? —*I CORINTHIANS 3:16*

This may surprise you, but I started smoking cigarettes when I was fifteen years old. It was during a high school outing where the seniors had great fun teaching the sophomores how to light up and inhale. I was a sophomore, and I thought it was cool and grown-up. Smoking made me feel glamorous, like the film stars I'd read about in the movie magazines. Unknown to my parents, I sometimes smoked a pack of cigarettes a day.

Later, as a young career woman, I became active in a youth group at my church. One day, late for a service, I raced up the church steps. When I reached the vestibule I was huffing and puffing like a steam engine. When I sat down in my pew, I began asking myself some questions: Did the smoking have anything to do with my lack of stamina? Wasn't my body supposed to be a temple of God? If God was going to use me to do His work, shouldn't my body be in the very best shape?

That night I made a decision to try to break the bad habit. It wasn't easy. I backslid often. But whenever I was tempted, I'd pray and remind myself that it wasn't my body, but God's. And, praise the Lord, the day finally came when I stopped smoking...for good!

And you know what? That was over twenty-five years ago!

Lord, keep me mindful that I am just a caretaker of Your temple.

—ELEANOR SASS

21 **FRIDAY**

For thy loving-kindness is before mine eyes: and I have walked in thy truth. —*PSALM 26:3*

I was at the counter of a fast-food restaurant, waiting to place my order, when a woman in the next line began to speak quietly to the employee behind the counter.

"I believe we failed to pay all of our charges," she said. "At the table, we added the bill and found we were not charged for one hamburger. How much more do I owe you?"

The employee hesitated, then told her the cost of a hamburger, and the woman paid her. As she turned to leave, I touched her arm.

"That was a nice thing you did," I told her.

She smiled. "Only what Jesus would have done," she said.

Her words rang in my ears all day. When I became impatient with the driver ahead of me, the words stopped my hand before it reached the horn. The third time the telephone interrupted me with the same wrong number, I heard them again and patiently helped the caller locate the number he needed. When fatigue tempted me to skip bedtime prayers, I remembered Jesus in Gethsemane and I slipped from under the covers and to my knees.

What a good day it was! What a goal for each day of my life: striving to do "only what Jesus would have done."

Dear Lord, the aim is impossible to reach without You. Give me strength as I try. Amen. —DRUE DUKE

22 **SATURDAY**

The Lord is my strength and my shield; my heart trusted in him, and I am helped: therefore my heart greatly rejoiceth; and with my song will I praise him. —*PSALM 28:7*

Each year, in November, a charming section of Philadelphia called Chestnut Hill celebrates the genius of Johann Sebastian Bach with a series of concerts. Recently my friends and I attended one where we marveled at the youth of the musicians and singers. They played and sang with the magnificence that comes not only from talent but also from years of practice.

As we were leaving, I found myself walking alongside one of the singers. I held out my hand and thanked her for the performance she and the others had given. "I'm so glad you enjoyed us," she said. Then added, "You know, I love to sing—and I'm so grateful to God for giving me a chance to sing. It's just wonderful to know people also enjoy listening!"

I won't forget the unintentional lesson she taught me. I realized that Bach wrote his music to please God. I had heard his works performed in a sanctuary designed to be sensitive to sounds of praise. And I had listened to young performers who attained incredible achievements in their efforts to thank God for His gifts to them.

It makes me think, if we were to try to please God in everything we try to do—who knows what we might accomplish?

Dear Lord, instead of trying to succeed, I'd like to make You happy.

—PHYLLIS HOBE

23 SUNDAY

Thy hands have made me and fashioned me: give me understanding, that I may learn thy commandments.

—PSALM 119:73

I am learning a new way of reading Scripture.

Since God made me and knows me better than I know myself, it seems logical that He also knows *what* I need to know and *when* I need to know it. Therefore, I now approach Scripture with this prayer: *What book would You have me study now?*

I wait in silence until one book comes to mind. After that, I read the book through—even if it takes several weeks. As I approach each chapter, I ask, *What would You show me in this today?*

Last year I reread the Gospel of John, and found myself asking questions about what it means to be an "everyday disciple" of Jesus. This fall I reread Job, discovered Zerubbabel, and am now following the works of the Holy Spirit through the Book of Acts. Every day of reading becomes an adventure, for I have a "private Tutor" who suggests the curriculum!

If you, like me, sometimes find Scripture—let's be honest—unexciting, incomprehensible, or overwhelming, try asking the One who made you and Who wrote them to help you find in them what He would teach you now.

Thank You, Lord, for giving us Your Word. Today on this National Bible Sunday, guide my Scripture reading and show me what You have to teach me. —PATRICIA HOUCK SPRINKLE

24 **MONDAY**

And he commanded the multitude to sit down on the ground. And he took the seven loaves and the fishes, and gave thanks... —*MATTHEW 15:35, 36*

It had been ten years since I'd seen Pat, my best friend from high school. She and her husband Marc and their four children had moved from Georgia to Lake Placid, New York, where they had built a log cabin on twenty-six acres bordered by a fast-flowing stream.

I felt so much at home in that log cabin in the woods with my long-time friends. They were happy people; there was a lot of laughter, and there was tenderness and a caring spirit.

But the moment that I recall most vividly was when we sat down at the table. We all held hands and together we sang: "The Lord is good to me, and so I thank the Lord—for giving me the things I need: the sun, the rain and the apple seed. The Lord is good to me."

It had been a while since I had eaten a meal in someone else's home where a blessing was given beforehand. From then on, I decided, I would make an effort to say the Lord's blessings whenever I share food. It's another way too, I find, to share a thanksgiving for my friends.

Dear Lord, I am blessed when I give thanks to You, but I am doubly blessed when I offer thanks with others. —SAMANTHA McGARRITY

25 **TUESDAY**

Blessed are they that have not seen, and yet have believed. —*JOHN 20:29*

Every Thanksgiving during my childhood, we had wonderful family reunions at Uncle Al's farm. One afternoon, after we younger children had tired of playing hide-and-seek, we wandered over to watch the horses drink from the stock tank.

Cousin Clem, who was fourteen then, told us, "If you put a horse's hair in the tank, it'll turn into a snake!" Then he tweaked a hair from Old Sal's mane and tossed it into the water. I don't know about the rest of the kids, but I was only four years old and I believed him.

Later that day, my compelling curiosity overrode my fear of snakes. Alone, I crept back to the stock tank and peeked over the edge into the water's depths.

Sure enough...there really WAS a snake in the bottom!

My mother heard my squalling long before I burst into the house and clung to her knees. She carried me back to the scene, smiling indulgently. Well, of course it turned out that the "snake" was only a long, ordinary fishworm—not an unusual find in the stock tank—magnified by the water's depth, or simply by childish imagination.

It's really rather ludicrous, isn't it, that when I was a child, I could believe in the superstitious magic of a horse's hair turning into a snake, yet as an adult, I sometimes have trouble in fully believing the miraculous Word of God telling me that His Presence is always with me?

Forgive me, Father, for those times when I doubt and question You. Reach out then and draw me close...for that is when I need to believe most. —ISABEL CHAMP

26 WEDNESDAY

And the man of God sent unto the king of Israel, saying, Beware that thou pass not such a place... —*II KINGS 6:9*

It seemed like the last straw when the railroad-crossing barrier lowered just as I approached the tracks. All week there'd been delays that interfered with a writing deadline. Now I barely had time to get to the post office before it closed. No train was in sight, but with flashing lights and ringing bells the barrier blocked my car from passing. I sat fuming. I had now missed the last mail pick-up.

The train roared into view around the bend, looming suddenly a foot from my front fender. *Big!* I thought, gaping up at it. Overpowering—all metal and speed and noise. I wouldn't want to be on the track when that thing arrived!

What had stopped me from such a disaster? It was that barrier blocking the crossing, the one I'd fumed at when it interrupted my all-important schedule.

Then I thought of other impediments this week....

What if—just suppose—all these delays had not been casual bits of bad luck, but barriers lowered by a merciful, all-seeing God, to prevent my rushing into a situation I was unaware of?

The final car of the train flashed past. The lights ceased blinking, the barrier lifted. I drove, slowly, across the tracks, found a place to turn around, determined to take a closer look at that manuscript. Was there something about it that I needed to change? (In fact, a call came the following day that postponed the publishing of that story for two years.)

I had no answers as I headed home, only a question that had not occurred to me before. One I hoped to ask any time my plans were once again "thwarted."

Father, is this frustration today a set-back—or a signal?

—ELIZABETH SHERRILL

27 THURSDAY

One man esteemeth one day above another: another esteemeth every day alike. —*ROMANS 14:5*

Autumn-colored leaves. Tangy scent of wood smoke. Happy faces around the dining room table. The linking of hands in prayer. That's Thanksgiving.

"It would be nice," we say somewhat wistfully, "if every day could be like this."

There was a time, however, when there were those who declared we had "one Thanksgiving too many!" In 1939, during the Great Depression, President Franklin Delano Roosevelt made an unpopular decision. Wouldn't it be wise, he reasoned, to inch up the holiday from the last Thursday in November to the next-to-last Thursday? This, he hoped, would boost Christmas sales in a failing economy.

Perhaps. But Americans, steeped in tradition, were unhappy. They took up their pens in protest and deluged the White House with letters. The President held firm. So did the public.

"We will celebrate two Thanksgivings," they declared.

"One Thanksgiving too many," Congress decided. In 1941, a Congressional act decided on the fourth Thursday of November.

While I am truly thankful for this special day, I would not have minded two of them. After all, thanksgiving is a heartfelt attitude that can only multiply in blessing—both for the giver and receiver. And it's something to remember *every* day, *all* year, too.

Father, You are the Giver of all good things—this day and all others. We are grateful for our daily bread. —JUNE MASTERS BACHER

28 FRIDAY

...And his hands shall restore their goods. —*JOB 20:10*

While paying for a key I'd just had made, I noticed a display box filled with rings holding three shiny keys each. A sign read: Play Keys for Children. 25¢ a Set.

When I asked the key maker about them, she said, "I used to throw away the keys I damaged. Then one day I saw a mother hand her keys to her child to play with, and this idea came to me for putting my mistakes to good use."

On the way home I thought about that...and about mistakes I've made that felt like permanent defeats. Sometimes a remark to a friend came out differently than I had intended, or the cake for the church's potluck supper fell into a crumbled mess. Humbling, right? Yes, but there's something positive, too.

In the mending, or "putting to good use," of those mistakes—the apology to the friend, which leads to greater understanding, or the decision to pour pudding over the cake, which to my surprise proves very popular—I grow, and learn, and become (I hope) a better person.

I don't mean to err, Father, but when I do, help me put it to good use.
—MADGE HARRAH

29 **SATURDAY**
. . . While I was musing the fire burned. . . —PSALM 39:3

On brisk fall days at our cottage on Lake Erie, we enjoy gathering wood for the fire. Twigs and broken branches for starters, and heavy, clean-split logs. How cozy and cheerful it is to sit before the fireplace as the flames begin to leap, drinking coffee and reading.

"Did you ever notice that a fire never burns downward, always upward?" my husband George remarked one day.

"Maybe that's because it remembers the tree that made it possible."

"That's right. Trees reach up, toward the sun. A tree always lifts its limbs and leaves toward the light." George pondered for a moment, and continued. "Then the tree dies or is cut down. It is nothing but a log or dried branches and twigs. But, when we bring its parts into the house and light them, they once again have a form of life!"

Before us was a merry demonstration. Our sticks and logs and branches had come alive in the fire; they were leaping and dancing and crackling, laughing as if in enjoyment of this bright resurrection, stretching out their rosy arms of orange and gold, giving off warmth and light and sound. They were speaking to us in tongues of flame.

And they seemed to be speaking to us of eternity, too. Death doesn't have to be the end, they seem to say, but a glorious new beginning.

Lord, thank You for the lessons of the fire. In its warmth and beauty we feel and see Your promise of eternal life. —MARJORIE HOLMES

WHAT IS CHRISTMAS ALL ABOUT?

The God Who Gives

PERHAPS the most familiar verse in Scripture is John 3:16: "For God so loved the world, that he gave his only begotten Son, that whosoever believeth in him should not perish, but have everlasting life." But have you ever thought of that as a Christmas *verse?*

Discover with Patti Houck Sprinkle how that tiny—but powerful—verse contains the meaning behind the whole Christmas story! This year during Advent, let's take this verse apart, and look at it phrase

by phrase, and marvel at how it can help us unravel the meaning of Christmas. Perhaps it can help us begin to live *Christmas, too.*

—THE EDITORS

30 *FIRST SUNDAY OF ADVENT*

For God so loved *the world, that he gave his only begotten Son, that whosoever believeth in him should not perish, but have everlasting life.* —*JOHN 3:16*

Four weeks until Christmas: already stores are crammed with hurried shoppers. Babies cry, children clamor and adults rush from store to store looking for the right gifts. At home we pull out boxes of ornaments, bake cookies in record time, scrawl messages on cards, try to remember to water the tree. Caught in the rush of Christmas, we find ourselves cross and grumpy. *We're so busy.*

One night, pausing from my own busyness, I stepped outside to catch my breath. There under a clear sky, I stood and "considered the heavens" just as the Psalmist had done years ago. For weeks I had been watching two planets moving into conjunction while the moon waxed and waned. Tonight the planets seemed inches apart, and the moon was only a sliver.

What magnificent movement—and magic—are the heavens, I thought. Then it occurred to me—*God is busy, too*. After all, He has a vast *universe* to run!

In the middle of God's busyness, He took time to love. That's what Christmas is all about.

This Advent, in the middle of my busyness, I want to take time to love: show cheerful courtesy to other shoppers, spend unhurried moments with my family, express concern for those who are alone.

God so *loved* that He gave—Himself. Can we do the same?

Dear God, Who took time to love, help me to follow Your example this season. Amen. —PATRICIA HOUCK SPRINKLE

Praise Diary for November

1

2

3

4

5

6

7

8

9

10

11

12

13

14

15

16

17

18

19

20

21

22

23

24

25

26

27

28

29

30

December

S	M	T	W	T	F	S
	1	2	3	4	5	6
7	8	9	10	11	12	13
14	15	16	17	18	19	20
21	22	23	24	25	26	27
28	29	30	31			

Some Day in the Clearing

Our faith extols a life beyond,
 a resurrection day,
When saints in trust believing
 find surcease from the fray,
But death is such a chilling word,
 so cold it goes unspoken,
A sleep farewell we deeply fear
 from which we're not awoken,
Each time a loved one passes
 we question why the pain,
The reason for this earthly hell,
 enroute to Heaven's reign...

Let not your heart be troubled
 your doubts and fears let go,
Believe and put your trust in Me
 for some day you will know,
That after a siege of darkness
 you'll reach a clearing bright,
Black storms that left you trembling
 will yield to rainbow bright,
And after long confusion,
 My grand design you'll see,
And view at last the pattern
 of My finished tapestry. —*FRED BAUER*

1 Practicing His Presence in December

CELEBRATING GOD'S COMING

Blessed is the King of Israel that cometh... —JOHN 12:13

Early in December, the children and I pulled the Christmas wreath out of the box of decorations. Made of pinecones, acorns and magnolia pods, it was nothing fancy. Just a simple wreath fashioned out of ordinary, familiar things from our backyard.

"Why do we have wreaths?" my son Bob asked.

I wasn't sure how the tradition started, but I remembered something I'd once heard. "The circle of the wreath symbolizes God's never-ending Presence in our lives," I said.

Later that evening, I noticed the wreath sitting on the table, waiting to be hung. And suddenly a question came to me. How encompassing had God's Presence been around my day? How...when had I experienced His Presence *today*? I reached for a pen and paper, thinking back, unable to come up with anything very grand.

Then I remembered the man who sold me a Christmas tree, calling out "God bless you" as I drove away. I remembered the child who sang "Silent Night" in the check-out line as it was piped into the grocery store. I wrote faster, recalling the Christmas card verse that had touched me so deeply, the sun breaking through the silvered trees while I'd emptied the garbage. Hadn't God's Presence come to me in all those moments, through ordinary people and events? Like the wreath, His coming was not fancy. He'd simply come through the familiar, everyday stuff of my life.

Then I had an inspiration. I placed my list in the center of the little brown wreath, celebrating all those quiet and gentle comings. I would not hang the wreath this year. I would leave it right here. It would be a "Wreath of Presence" where all the family could celebrate God's "advents" in their daily lives, writing them down and leaving them in the wreath. Perhaps it would remind us of the most extraordinary miracle of all. *God comes!* He comes every plain ordinary day. Without end.

This month, O God, I will celebrate Your glorious Presence that comes and comes and comes... in all the simple and ordinary moments of my day. May I carry them into my New Year, too.

—SUE MONK KIDD

2 TUESDAY

I will hear what God the Lord will speak. —*PSALM 85:8*

The woman at the next table was obviously expecting someone—whirling around each time the door to the coffee shop opened. At last her friend joined her.

"I hated to call you at work, Ethel," the first woman plunged in, "but I need your advice."

Soon I, and everyone for several tables around, were learning—willing or not—about the woman's problems. Clearly she had more of them than anyone could be expected to handle alone. An alcoholic husband, a delinquent son, a diabetic condition of her own...the doleful list went on.

I found myself covertly studying Ethel's face. Warm, responsive, intelligent. I listened quietly for the wisdom and comfort I was confident Ethel would supply.

"I wonder, Sue," she began, "if you've thought about—"

But Sue had launched into another melancholy story of failed help. And so it went, the older woman unable to break into the stream of words. I could almost see Ethel give up the effort to communicate. The last thing I heard as I rose to leave was Sue's distressed voice, "I just wish I knew what to do!"

I stepped outside thinking of the last time *I* had said those words—this very morning—not to an older wiser friend, but to God. "Why don't You give me an answer!"

Maybe, I thought, *He's been trying*. Maybe my own voice has been so loud and insistent that the two-way conversation called prayer never had a chance.

Father, it's Your turn now. —ELIZABETH SHERRILL

WEDNESDAY

And I will give them one heart, and I will put a new spirit within you... —*EZEKIEL 11:19*

During the Great Depression many people lost their fortunes overnight. One of these was my mother's Uncle Thomas. We were living in Florida then, and he came to live with us at my mother's invitation. It was an awkward time for us all—he, embarrassed by his fallen state; and my brothers and I, three strapping, scrapping boys who didn't know quite what to make of this aging widower with quiet ways.

But, as Christmas of 1934 approached, the three of us boys made a special effort and chipped in to buy him a present (a pith sun helmet to cover his bald head!). In the noise and disorder of Christmas morning, however, we suddenly noticed that Uncle Thomas was not among us. He was nowhere to be found. Although we continued with the Christmas morning festivities, our concern mounted. Finally, we heard the front door slam, and there stood Uncle Thomas in the doorway.

"I have something to say to you," he said. "For hours I've been out walking. I've been praying that our good God in Heaven would help me find a way to give you a gift that is worthy of you, my beautiful family." We stared at him dumbly. "I think I've found it," he said, "it's these." And with that he thrust out his arms and turned the palms of his hands outward. "My present is these empty hands."

At that moment even the youngest of us understood that the real gift he was giving us was his own false pride.

From that day on, Uncle Thomas was at ease with us. He helped us kids with our homework, went fishing with us (wearing his inseparable pith helmet) and we even found him fun to be with.

That Christmas, Uncle Thomas became a member of our family...as he let us love him for who he was rather than who he thought he should be.

Keep me mindful, Lord, that the most valuable gift we can give is the gift of our true selves. —VAN VARNER

THURSDAY

For the preaching of the cross is to them that perish foolishness; but unto us which are saved it is the power of God.

—I CORINTHIANS 1:18

Often, when I send a letter or greeting card to a special friend or relative, I add three *XXX*'s after my name, meaning "I love you." Lots of other people do this. Recently, I read an explanation for it.

It seems that hundreds of years ago a cross mark meant the cross of Calvary. So revered was the cross of Jesus that this mark had all the force of a sworn oath. If a person didn't know how to write his name, he'd make a cross. It was accepted as his signature. This practice continues today. But back then, a person always kissed the cross after he made it, a sign of complete sincerity. And it was this practice of kissing the *X* that led to its becoming a symbol of the kiss.

During World War II, both the American and British governments forbade military personnel to put *XXX*'s on their letters. They feared that spies might use such marks as coded messages.

Today, for me, these *XXX*'s *are* a code. What better way to express love than the symbol of the cross of Calvary, where true and perfect love was wrought forever?

Dear Lord, may my love for You always be as true and as perfect as Your love for me. —ELEANOR SASS *XXX*

5 **FRIDAY**

The Lord is merciful and gracious, slow to anger, and plenteous in mercy. —*PSALM 103:8*

How do you deal with friends who disappoint you, children who disobey you, spouses who forget their promises and co-workers who fail to follow through on important assignments? Are you, like me, usually concerned with what is "fair" or "right"?

One night as I read Psalm 103:8-10, I seemed to hear God say, "This is the way *I* deal with those who let *Me* down."

The Lord is merciful and gracious... How often do I show mercy—compassion, clemency—to my children? My spouse? My friends or co-workers? Do I speak and act graciously to them when they have disappointed me?

Slow to anger and abounding in steadfast love... "Abounding" means "filled to the very brim with." Can I be filled with enough love to give up my own quick temper and respond to others with steadfast, "never-failing" love?

He will not always chide... God is no nag ...*Nor will He keep His anger forever.* God holds no grudges. Can I give up my own nagging and grudges?

He does not deal with us according to our sins, nor requite us ac-

cording to our iniquities. Can I learn to be gentle in dealing with the faults of others? Can I overlook their flaws?

God's way of dealing with people who disappoint Him is not my most natural way. It's harder. But it's also a far better way.

Dear Father, teach me also to be merciful, gracious, slow to anger, and abounding in love, that I may be more like You. Amen.

—PATRICIA HOUCK SPRINKLE

SATURDAY

That your faith should not stand in the wisdom of men, but in the power of God. —*I CORINTHIANS 2:5*

Lately I've been hearing the word *empowerment*, without fully understanding what it means. The last place I heard it was at a conference of women interested in reaching out to women in third-world countries. Many empowerments were discussed—political, financial, educational—but the one that drew applause was "spiritual empowerment," and how it could be used to achieve the goal of helping others.

To be empowered, I thought, was to open myself up to a greater energy. This began to make sense to me one morning in the physical therapy unit of my local hospital. The young therapist was bending my wrist, which had recently come out from a cast, in a series of exercises to make it more flexible. I winced in pain, and, to keep from fainting, tried to focus on something—anything—else.

The schedule chart! There I was: McGarrity, 9 A.M.; Davis, 10; Mendel, 10:30; GOD IS LOVE, 11." *What*? Yes, someone had carefuly printed the words "GOD IS LOVE" between 11 A.M. and 1 P.M.

I focused on that thought, opening my mind to the most powerful Love imaginable. I held it concentrated there for a moment, and then I sent it out, first to the therapist, then to others in the room, then beyond. And I found in that moment of empowerment that the pain in my wrist had been transcended.

Lord, I need to be empowered. I open myself up now to Your spirit.

—SAMANTHA McGARRITY

7 ***SECOND SUNDAY OF ADVENT*** 

For God so loved **the world,** *that he gave his only begotten Son, that whosoever believeth in him should not perish, but have everlasting life.* —*JOHN 3:16*

At this time of year I love to send Christmas cards. Some travel to friends in places I've never seen: Singapore, Taiwan, South Africa, Brazil, Japan and Equador. Some travel to distant places I have known—Scotland, England, Canada and California. Making friends in each of those places has expanded my world and enriched my life.

Jesus was born because God loved the *whole* world. That love made angels sing and drew foreigners across deserts to worship. Through the centuries awareness of that love has encircled the globe, until brothers and sisters in every nation now celebrate His birth.

God's love condenses the world and stretches us,
so that we can reach across it in love.
That's what Christmas is all about.

Much of the year it is easy for "the world" to seem a large impersonal planet—distant foreign lands full of wars, famine, overpopulation and uncertain politics. Jesus' birth gives us a chance at least once every year to remember that this world is also *people*—individuals, like you and me. They have families, they marry and give birth, they work and play. Their societies have the poor, the homeless and destitute, even as here in our own country. In truth, despite the miles, we are not so different from one another. And so God sent His Son, that we might not forget.

Dear Jesus, this Christmas I offer thanksgiving and prayers of peace and goodwill for all of Your children who live in Your world. Amen.

—PATRICIA HOUCK SPRINKLE

MONDAY

And be ye kind one to another, tenderhearted, forgiving one another, even as God for Christ's sake hath forgiven you.
—EPHESIANS 4:32

I had just hung up the phone on my cousin Diane. I'd never done that to anyone. But it had been an exhausting day; she had seemed unreasonable; and my patience had worn thin.

I felt badly for having been rude, but after all, I reasoned, Diane was at fault and she should be the one to apologize. A month went by. Neither of us made an attempt to get in touch.

Then one evening late at the office, the cleaning lady asked me to type a poem for her on some special blue paper so that she could frame it for her home. This was the poem:

GIVING AND FORGIVING

What makes life worth the living
Is our giving and forgiving;
Giving tiny bits of kindness
That will leave a joy behind us.
And forgiving bitter trifles
That the right word often stifles.
For the little things are bigger
Than we often stop to figure.
What makes life worth the living
Is our giving and forgiving.
—Thomas Grant Springer

After typing the poem, I made a copy for myself, then wrote a note of apology to Diane. Soon after, she called.

How good it was to be unburdened from bitterness!

Dear Lord, give me the strength to forgive today, and release my haughty pride.
—SAMANTHA McGARRITY

9 TUESDAY

For the Son of man is come to save that which was lost.
—MATTHEW 18:11

Larry and I belong to the Civil Air Patrol, a volunteer search-and-rescue organization that helps look for people who have gotten lost in our rugged New Mexico mountains. Sometimes when the phone rings at three A.M., I groan at the thought of leaving my warm, cozy bed. But Larry reminds me, "Hey, if we were lost out there in the night, we'd want to know someone was up and looking for us." *That* gets me up and into my clothes in a hurry. Then when one of the search parties finds that missing person *alive*! What an incredible feeling of joy.

I've never been lost in the woods, but sometimes I have lost my sense of direction in life and felt separated from God. When that happens, does Jesus leave me to wander along in the cold? No way. Working through His own search-and-rescue volunteers—such as my friend Martha who listens to me and gives me advice, and my husband, who hugs me and tells me he loves me no matter what—Jesus finds me and leads me home.

He'll do the same for you.

When I'm lost, Lord, lead me safely home. —MADGE HARRAH

10 WEDNESDAY

Our help is in the name of the Lord... —PSALM 124:8

Recently I heard someone say that maybe God engineers our failures because when we are unhappy and defeated, we may be more receptive to Him and His plans for our lives. Frances J. Roberts writes in his book, *Make Haste, My Beloved*, that God uses many vessels while they are still struggling with their failures.

Fascinated, I began to think of some of the people in the Bible who must have felt like failures: Joseph in that deep well; Daniel in the lion's den; Jonah in the belly of the fish; Moses, who stuttered and yet was called upon to make speeches.

I looked up "failure" in my thesaurus: "A deficiency or lack, unsuccessful, unfulfillment, forlorn hope, collapse, dud, washout...." Not much to cheer about, is it? I don't like the idea of being a failure—who does? And yet there is a magnificent opportunity in failure: it is simply our Lord's way of teaching us to listen to Him, to open ourselves to His help, to do things His way.

If you have been feeling defeated and are ready to give up, pray. Then wait, quietly and receptively, for a Father...who helps those who *cannot* help themselves.

Dear Father, I have failed miserably again, and today I desperately need Your guidance... Your Word is my command.

—MARION BOND WEST

11 THURSDAY

Eye hath not seen, nor ear heard... the things which God hath prepared for them that love Him. —*I CORINTHIANS 2:9*

My friend had such bad cataracts on his right eye that the vision was practically gone from it. Following corrective surgery, I met him on the street.

"It's amazing how much more beautiful the world is when you look at it with both eyes," he told me.

I can appreciate the meaning of his words. On some days, the congenital condition in my own eyes has me looking through the equivalent of a London fog. But on my good days, when my vision is clearer, God's universe is so glorious to behold! On these special days, as I enjoy the beauty I see, I am reminded of Paul's words, "For now we see through a glass, darkly; but then face to face" (I Corinthians 13:12).

How I look forward to the day when *all* restrictions are removed and I see God in His full glory!

Lord, what greater joy could Heaven hold than meeting You face to face? Amen.

—DRUE DUKE

12 FRIDAY

Well done, good and faithful servant; thou hast been faithful over a few things, I will make thee ruler over many things. —*MATTHEW 25:23*

There is a legend that St. Anthony once suffered a strong attack by evil spirits in the desert. When he had won the battle, Jesus appeared and

St. Anthony rebuked Him: "Where were you when I needed you?" Jesus is said to have replied with a smile: "I was right here, but I hid myself. I wanted the pleasure of seeing what a staunch fellow you are."

That story always struck me as unfair—after all, we need all the help we can get in this world! Until one day last week. I was in the living room and heard the boys in the kitchen. First a crash, then a dismayed, "Oh, no!" I was about to rush in to clean up, when something stopped me—the sound of hurried scrambling, then the whoosh of sweeping, and the tinkling sounds of a spoon stirring water in a pitcher, until finally the boys appeared with their new glasses of lemonade.

"You heard us drop the sugar!" Barnabas pouted at me. "Why didn't you come help us?"

"I wanted to see if you knew how to clean it up," I replied. "And I'm proud to see you did." Finally I understood about Jesus and St. Anthony.

Jesus will be with us always. He has promised that, and we can accept it as true. Sometimes He may seem absent. But He hasn't gone anywhere. He just wants to see how much we can do for ourselves!

Jesus, I am glad You help me to grow up—by letting me do for myself.

—PATRICIA HOUCK SPRINKLE

13 SATURDAY

...I am with thee, and will bless thee... —*GENESIS 26:24*

Several years ago, my two best friends had a serious falling out and I was caught in the middle. Each one wanted me to take her side against the other, but I wanted to keep my friendship with both. I was really torn, so I went to talk to Mrs. Moore, the wise old lady who was our baby-sitter. She startled me by asking, "Have you tried blessing the situation?"

"Blessing it! What do you mean by that?"

"To bless something means to recognize that God is in it," said Mrs. Moore. After that, whenever the problem came to mind, instead of begging God to do something, I simply said, "I ask Your blessing upon it." Then I let go of it. There wasn't a sudden, dramatic turnaround, but gradually my friends gave up their hurt feelings and negative emotions. I really think their friendship was stronger and more solid afterward than before.

If there's a problem you've been praying about without noticeable results, why not try blessing it?

Lord, I know that You are in this situation, and I ask Your blessing upon it.

—MARILYN MORGAN HELLEBERG

14 *THIRD SUNDAY OF ADVENT*

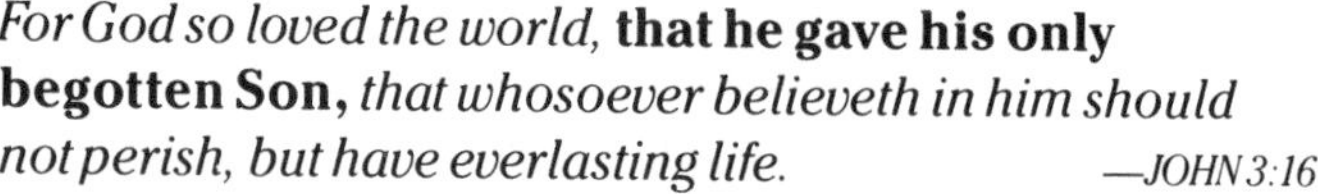

For God so loved the world, **that he gave his only begotten Son,** *that whosoever believeth in him should not perish, but have everlasting life.* —*JOHN 3:16*

Three-year-old Barnabas, two-month-old David and I were traveling alone at Christmas. As we left the plane, a helpful stewardess said, "I think I'll take that cute baby home with me."

Barnabas stared up at her in horror. "You can't! He's *our* baby, and very precious to us!" I have treasured that remark through the many brotherly battles over the years.

I also think of it at Christmas and realize, in a small way, just what it was that God *gave*. What would it take for me to send one of my precious sons to a faraway land to grow up among strangers, suffer their scorn and die from their hostility?

God gave a precious Gift out of an incredible love. That's what Christmas is all about.

Thinking of what God gave can help me choose appropriate Christmas gifts for others. My choice is not between a book and a record, or between a gift that's too extravagant and one that's cheap. My choice is between a gift that says, "Here's a present" and a carefully chosen one that says, "I love you."

Dear Father, mindful of the kind of love that inspired Your first Christmas Gift, help me choose this year's gifts with love. Amen.

—PATRICIA HOUCK SPRINKLE

15 WILDERNESS JOURNEY (Joshua 3)

Following the Ark of the Covenant

And all the Israelites passed over on dry ground... clean over Jordan. —JOSHUA 3:17

With the Israelites, we have been camped at the edge of the River Jordan for three days. Joshua has told us that, when we see the Ark of the Covenant moving into the river, we are to follow it. The sun has melted the mountain snows, swelling the Jordan to twice its usual size. How will we get to the other side? Yet if we've learned only one thing from our long journey, it's that God is faithful, even in the face of our disbelief.

Now Joshua gives the word. Up goes the sacred Ark onto the shoulders of the priests. "Go!" shouts our leader, and a shiver of excitement passes through the crowd. As soon as the feet of the priests touch the water, the Jordan stops flowing—just as the Red Sea did, all those long years ago. Out into the middle of the river go the priests with the Ark. And now it's our turn! Tribe by tribe, some six-hundred-thousand men, plus women and children, cross the last barrier and enter the land God promised to our fathers more than five hundred years ago.

The Israelites left Egypt, still in bondage to their own doubts, resentments and fears. They emerged a free people, confident in their God. They simply had to relinquish everything along the way, especially their proud self-reliance, and accept their dependence upon the Lord. As you and I have traveled with our faith-fathers, we have taken our stand to move out toward freedom. We have discovered that our real slavery is to the negative things within ourselves. Along the way, we have tried to leave behind some everyday kinds of bondage. How often we have been less than the people God calls us to be! Yet His surprising mercy has never left us.

We have laughed together, cried together, hungered and thirsted together. Let us now cross over the River Jordan as God's people, letting go of that last barrier to freedom, our own fierce self-sufficiency. Only then will we be ready to receive the young Carpenter who will one day stand in this very river, as the heavens proclaim His Sonship.

Lord, I offer You myself, with all of my enslavements. I know that the only true freedom is to be found in giving myself up so that You can claim me. —MARILYN MORGAN HELLEBERG

16 **TUESDAY**
...My expectation is from him. —*PSALM 62:5*

My mother is eighty-seven, going on one-hundred. She will reach the century mark, she "expects."

I say "expects," because that's the way Mama expresses herself. She *expects* it is going to be a bright day. She *expects* to hear from a family member. She *expects* her lima beans are sprouting...it will be an early spring...the revival meetings are going to make a change in the community. *She expects the best.*

Such a way of thinking gave me a delightful childhood. Other girls' mothers *thought* this or that; Mama *expected* it. Thinking had to do with the brain, she said; it had its place in school. But expecting? That was of the spirit.

I think of my mother today as I address greeting cards and look forward to Christmas. I am following Mama's star just as the shepherds followed the Star of Bethlehem 2,000 years ago. Because we expect wonderful things to happen.

How about you? Will you join us in the great expectations of this holy season and expect a miracle?

Come into our hearts, Lord Jesus. We are expecting You.
—JUNE MASTERS BACHER

WEDNESDAY
He staggered not... but was strong in faith... persuaded that, what he had promised, he was able also to perform.
—*ROMANS 4:20, 21*

A few years ago I received a commission to illustrate a children's book. But, fearful about pleasing the publisher, I procrastinated. Then one day I served as a judge for a girls' gymnastic meet in which my then teen-aged daughter Meghan participated. I was assigned to judge the beam event, where girls perform on a wood beam sixteen feet long by four inches wide, supported four feet above the floor. It was the event Meghan dreaded most. When she approached for her turn, her face paled and her knees trembled. She closed her eyes for a moment, breathed deeply—then leapt up on the beam. She performed beautifully and finished with a big smile and a salute for the judges.

Later, I asked her, "How did you ever get up the nerve to mount the beam?"

"It was simple, Mom," she replied. "I figured I sure can't do the routine unless I get up there, so I just say a prayer and go for it!"

Later, at my desk and facing blank drawing boards, I prayed (again) that God would inspire me to start. I suddenly understood Meghan's point. The prayer wasn't supposed to ask God to help me *begin*—it was to tell him I was *on the way.* It was up to *me* to start—to "go for it"—trusting that God would help me once I'd taken that first perilous leap. A leap that some call faith.

Lord, I'm on my way to that first step today. —MADGE HARRAH

18 THURSDAY

And it came to pass... that the soul of Jonathan was knit with the soul of David... And Jonathan stripped himself of the robe that was upon him, and gave it to David...

—*I SAMUEL 18:1,4*

There was absolutely no money for Christmas gifts. And so many people to remember! I racked my brain to find a way to show my love for those who meant so much to me.

Thinking back to memorable gifts, I remembered how Grandmother, now ninety-five, always used to give me something that had belonged to her, something she had treasured and enjoyed through the years and then passed on to me to enjoy. *Those* gifts were the most prized!

And, so I looked about me, and discovered "gifts" among my possessions—a book of Longfellow's poems for Tim, a wind chime for

Mother, cloisonné barrettes for Hillary, and so on down my list of family and friends. These gifts had been a part of my home for years, and all had given me some special joy. Now, giving them to people I loved helped me share that joy.

I felt thankful for a grandmother who taught me how to truly give—with gifts from the heart.

Dear Heavenly Father, You gave Your most cherished Possession at Christmas. Help me to remember Your example this Holy Season.

—SAMANTHA McGARRITY

19 FRIDAY

...Let not thy left hand know what thy right hand doeth.

—MATTHEW 6:3

Painted in large Gothic letters over the door of a hospital ward in Belgium are these words: "Do good, and disappear." The idea of doing something worthwhile without waiting around to collect payment in gratitude is not new. Yet I find it hard advice to follow.

Last December, I resolved to give something away anonymously. As I was walking out of the shopping mall, I heard the Salvation Army bell. The man who was ringing it didn't know me, so I reached into my billfold and pulled out a fairly large bill. But as I reached over to put it into the container, I found myself deliberately turning it so that the man could see the number in the corner! As I walked away, he thanked me profusely, and then...I was *so ashamed*! I had cheated myself out of the true joy of giving without chance of reward. I had had my reward.

Maybe I'll do better this year. I'm going to try very hard to "do good, and disappear." But I won't tell you whether or not I have succeeded!

Lord, help me to give without chance of reward. Then my gift will truly be a gift to Your Son. —MARILYN MORGAN HELLEBERG

20 SATURDAY

And the angel said unto them, Fear not: for, behold, I bring you good tidings of great joy, which shall be to all people. —LUKE 2:10

"Security measures have been intensified during this Christmas week in the Holy Land," the newscaster reported. "Special efforts are being made to protect tourists from any outbreak of hostilities among opposing groups here."

I turned off the radio. It made me sad to think of celebrating the birth of our Lord in the midst of hatred and in the shadow of war.

And yet, maybe that was part of the miracle. When Christ came among us we were not at all ready to receive Him. We were a world of conquerors and prisoners, obsessed with power and struggling to seize it from each other. The threat of war was on all sides of us and we thought only fools spoke of peace. *Survival*—that was what mattered. *Salvation*—impossible!

Into such a world He came. Into such turmoil He introduced His peace. Part of the miracle of His birth is that love is stronger than hate. Peace can survive war. Salvation is not only possible, it is here—if only we will take it.

Thank You, Lord, for the most important news of all—that the Light of the world can overcome the darkness. —PHYLLIS HOBE

21 *FOURTH SUNDAY OF ADVENT* 

For God so loved the world, that he gave his only begotten Son, that **whosoever believeth** *in him should not perish, but have everlasting life.* —JOHN 3:16

An earnest young man approached a wise old professor and asked, "Will the human experiment end in war, or love?"

The professor smiled. "Ah, but that *is* the experiment."

That story makes me uncomfortable, especially at Christmas. Angels, shepherds and a Baby remind me of God's great "experiment" of Love, but I have to admit there are some people I have a hard time warming up to. Their politics, values, lifestyles or sometimes their differing church traditions strain my tolerance.

Then I hear again the angel song: "Peace, goodwill to *all* people on earth." I remember Jesus' patience with people who were very different from each other, and from Him.

Jesus came to show that God accepts everybody. That's what Christmas is all about.

Will my relationships end in rejection, or love? It often depends on me. This Christmas, as I hear the angels sing, I pray for acceptance and tolerance for *all* God's children, especially "whosoever believes."

Dear Lord, during this Christmas time, teach me Your love and acceptance for those who are most unlike myself. Amen.

—PATRICIA HOUCK SPRINKLE

22 MONDAY

At the name of Jesus every knee should bow... every tongue should confess that Jesus Christ is Lord...

—PHILIPPIANS 2:10, 11

For a moment there was silence in the room. Each of us was staring at the picture up on the screen, amazed at the paradox brilliantly illuminated in front of us. In the great wall that forms the back of the famous Church of the Nativity in Bethlehem—the church built over the cave where many believe Christ was born—is an entrance so small that only a child can enter without stooping.

The friends who had brought the slide back from Bethlehem pointed out that in the Middle Ages the Crusaders had reduced the size of the entrance to prevent Arab horsemen from galloping into the interior, and no one had ever restored the original arched entryway.

I've seen this slide fifteen or twenty times since then, and each time I wonder if those Crusaders realized the deep symbolism of their act of protection. How fitting it is that the church where His birthplace is has a door so low all must enter on bended knee.

This Christmas, the anniversary of His birth, wherever you are, won't you join me in approaching the infant Jesus on bended knee?

O, Lord, today we kneel before You, as the King comes into our world to save us.

—JEFF JAPINGA

23 TUESDAY

Whatsoever thy hand findeth to do, do it with thy might...

—*ECCLESIASTES 9:10*

A cloud of uncertainty hovered over our Christmas Eve. Ruby was worried about the church choir, which had dwindled to eight members. For almost a lifetime, church music had been Ruby's ministry. Now she said anxiously, "I always want to do a good job, but especially tonight—the eve of the birth of our Savior."

Not only was the choir too small, but the organist had resigned, and a temporary organist would not be available until the last minute. The pastor was on sabbatical, and his replacement had served for only four months. Things looked bleak.

By the time the service began, I was sharing Ruby's anxiety. The choir entered, a hush fell over the congregation, and then...something odd happened. The music seemed suddenly to lift, reverberating through the church as never before. The voices, still only eight in number, sounded like many—a heavenly host.

Later we learned that the replacement pastor, also a musician who loved hymns, had selected the most joyous ones for the choir to sing. And these same hymns also happened to be the organist's favorites.

During the service, the words, "When two or three are gathered in my name..." floated out to me. I glanced over at the choir loft, where Ruby stood proudly, the most beatific (that's the only word for it!) smile on her face. Two or three had gathered—the choir, the replacement pastor, the temporary organist—to lead the overflow congregation in the most joyous Christmas Eve celebration we had ever known!

Are you finding roadblocks to fulfilling your ministry? If so, remember our church's problems on that most important night of nights. Then take heart and join me in saying:

Yes, Lord Jesus, Your servants can indeed do wonders in Your name.

—OSCAR GREENE

24 *CHRISTMAS EVE*

For God so loved the world, that he gave his only begotten Son, that whosoever believeth in him **should not perish**, *but have everlasting life.* —*JOHN 3:16*

On a beautiful December Saturday we piled the boys and a friend into our car and headed for a small park on the Chattahoochee River. While the three boys ran along the riverbank and tossed stones into the current, Bob and I strolled on the beach to enjoy the beauty. We found ourselves discussing litter instead. Broken bottles, discarded cans, tossed hamburger bags were everywhere. The beach's beauty had perished beneath litter.

Idly we began to pick up cans to recycle. When the boys saw what we were doing they came to help, and with all those workers we picked up paper, too. Soon we cleaned a wide stretch of the beach.

God so loved you and me that He sent Someone to "clean us up." That's what Christmas is all about.

Beneath the litter of our broken lives, discarded friendships, forgotten promises, God still sees the beauty of who we are meant to be. He sends Jesus with love and the power to "clean us up"—not merely to discard our "litter," but to restore us to whole lives, healed friendships and new intent. All He asks from us is our belief that He *can* work a mighty work in our lives.

On this Christmas Eve I don't want to think so exclusively about others that I forget to take time to present the litter of my life to the One who still comes to clean, restore and save.

Holy Jesus, today I ask You to clean and restore my life. Amen.

—PATRICIA HOUCK SPRINKLE

25 *CHRISTMAS DAY*

For God so loved the world, that he gave his only begotten Son, that whosoever believeth in him should not perish, **but have everlasting life**. —*JOHN 3:16*

Having heard the Christmas story all my life, I was surprised to find something new in it this year. For the first time I noticed how many of the people in this story are looking up: Zechariah worshiping at the altar; Joseph lying on his bed receiving a vision; shepherds gazing up at angels; wise men guided by a star.

Bounded by my daily calendar, my monthly schedule, the demands of the here and now, I seldom take time to look upward and outward, into infinity and eternity. This year, Christmas stars and angels, tapering trees and bells in the air call me to stop—to look beyond this world into everlasting life.

God has promised that we who believe in His Son live forever. THAT'S what Christmas is all about!

This Christmas Day I want to spend time reflecting on what eternity means for ME. Eternity means I can't use that old excuse, "I don't have time." I have *more* than "all the time in the world." Eternity means I need to look carefully at a relationship gone sour—after all, I may have to relate to that person forever! Eternity means I have time to grow—time to try new things. Eternity means I have time to rest, time to pray, time to sit quietly and just enjoy God's presence. *I am an eternal person!*

Lord, on this Christmas Day I'm especially thankful for Your gift of Eternity. Let all my days be mindful of how in this life I may serve You and honor Your Gift. Amen. —PATRICIA HOUCK SPRINKLE

26 FRIDAY

And he comforted them, and spake kindly unto them.

—*GENESIS 50:21*

It was the day after Christmas. I hadn't been feeling well and went to see my doctor. Three hours later I found myself in the hospital. I have always disliked hospitals and the next five days were frightening. How I longed for home!

Then, on the fourth day when my spirits were lowest, a new morning nurse came into my room. Her smile was bright.

"What you need," she announced cheerfully, "is a good rubdown!"

In a matter of moments her strong, gentle fingers were massaging me with a sweet-smelling lotion while my body, and mood, relaxed. Tingling and grateful, I thanked her as she left to complete her rounds.

Home at last, I found a friend had cleaned my apartment. A beautiful flower arrangement greeted me from the dining room table. My refrigerator had been filled. And my cup overflowed. Never before had I more fully realized how much expressions of love can cheer a heavy heart. And never again do I want to forget. In fact, I have resolved to pass this love along every chance I get.

May I begin by giving some of it to you?

Thank You, Father God, for loving hearts. —DORIS HAASE

27 SATURDAY

And she brought forth her firstborn son, and wrapped him in swaddling clothes, and laid him in a manger; because there was no room for them in the inn. —*LUKE 2:7*

During my childhood, our family lived on a farm near the edge of McPherson, Kansas. When I recently revisited the area, I drove out to see the house. *How on earth did the four of us ever fit into those three tiny rooms?* I wondered. Suddenly it was easy to understand why my parents had hidden our Christmas presents in the huge white barn way out back.

One year my mother's mother spent the holiday with us. I can't remember to this day where Grandma slept—or even sat down—but I do remember that my parents decided to put up our Christmas tree in the barn to save space.

Well, that Christmas morning, with the wind howling and snow drifting high, we bundled up and dashed out to the barn. Daddy said the cows could wait to be milked until we'd opened our gifts. They didn't seem to mind—they stuck their heads through the stanchions and, along with our pet pony, Trixie, contentedly began chewing hay. Their bodies made the barn warm, their nostrils puffing small jets of steam into the icy air.

It wasn't until years later that I realized just why that Christmas, celebrated in the drafty old barn with the animals looking on, was so very memorable....

After all, it was in a lowly stable that Christmas began, wasn't it?

Whatever my circumstances, Father, let my heart lodge Your Son through all my days. —ISABEL CHAMP

28 SUNDAY

A soft answer turneth away wrath: but grievous words stir up anger. —*PROVERBS 15:1*

We were at a friend's country house in Pennsylvania's Delaware Valley, celebrating the New Year. Just before midnight we all decided to take a late-night hike down a country road to listen to the silence, and under nature's black, starry canopy welcome the approaching year.

But the pleasantness of our walk was periodically interrupted by one in our group who was doing nothing but complaining. "Why can't we go back now?...I don't like it out here....It's too dark." It was little Jimmy, age seven, and nothing seemed to quiet him.

The rest of us strained to listen to the nearby trout stream over Jimmy's whines. He was getting to me, and I was on the verge of snapping at him when Ken, one of the men, reached down and picked Jimmy up, swinging him up on his shoulders. "Here," he said, patting the boy's leg. "Come on, I'll give you a ride back."

Jimmy became quiet, holding tightly to Ken. The walk became a treat for everyone, and Ken's understanding soothed the little boy's fears.

Maybe a dose of *compassion* is the best *companion* for a troubled heart. Next time I would practice patience and a soft answer, too.

Dear Heavenly Father, teach me tolerance and gentle ways to soften spirits. —SAMANTHA McGARRITY

29 MONDAY

Now there are diversities of gifts, but the same Spirit. —*I CORINTHIANS 12:4*

My friend June ministers to unwed mothers. Although she has eight children of her own, the whole family always seems to have room for one more young woman in need of their help.

One day in church I happened to sit next to June, listening to the inspiring performance of a famous pianist. Thunderous applause followed when the final chords faded into silence. June leaned over to me and said, "What I wouldn't give to be able to play like that!" Someone near us overheard, leaned over and whispered to her, "You do! You minister through a symphony of love."

When I got home, I thought about that perceptive remark. How dull it would be if everyone could play the piano. Someone has to plant the flowers, someone has to put color on canvas, bake the homemade bread. I have a friend who spent an entire summer constructing a ten-foot-high giraffe to illustrate a Bible story for her students! The children will always have that loving memory, to recall and share with their own children some day.

The Lord gave each of us special gifts. But perhaps more important than our talent, is whether we *contribute* our gifts...willingly, wholly and with love.

Father, show me how to use my gifts in a symphony of service for others. Amen. —MARION BOND WEST

30 TUESDAY

But now they desire a better country, that is, a heavenly: wherefore God... hath prepared for them a city.

—*HEBREWS 11:16*

Ever since my grade-school days, I've loved to read about travel and learn about exotic places I longed one day to see: China's Great Wall, Egypt's pyramids, Norway's fjords, Greece's Parthenon or Mexico's Mayan ruins. But of all the wonders of the world, one place sticks out in my mind. How about this description?

- One enters this land not through a single wooden, iron or stone gate, but via any of twelve gates—each gate being "one pearl."
- The city streets are not macadam, concrete, brick or stone—but "of pure gold, like transparent glass."
- Those inside this land have "no more death, neither sorrow, nor crying," for "God shall wipe away all tears."
- There is no darkness—God's glory lightens this land.

In fact, turn to Revelation 21 and read about it yourself. No eye has seen, no ear has heard, nor can anyone even imagine such glories as are in this land. Better yet—being there is not for the few who can afford the trip or hold the winning ticket in a drawing.

This is the heavenly country, "which God hath prepared for them that love him" (I Corinthians 2:9).

Father, day by day, let me not lose sight of—or hope for—the most glorious wonder of all that You have prepared for me.

—ISABEL CHAMP

31 **WEDNESDAY**

For what is a man profited if he shall gain the whole world, and lose his own soul? —*MATTHEW 16:26*

On a radio interview yesterday I heard a famous popular music composer retracing his career, recalling that once during a fifteen-year period he worked seven days a week, writing as many as three original scores for albums, TV shows and movies in a single day. "I don't know about the quality, but I sure got quantity," he recalled.

How did he evaluate those hectic years? "One thing for sure," he said in retrospect, "I didn't have time for anything else...but now the time is gone, the money has been spent and my children have grown up. I wonder what I was chasing?"

His lament is one with which many of us can identify. But then, of course, so could our forebears. Chasing only the tinsel and finding life's shadow instead of its substance is not a new phenomenon. In fact, there was a man who came to Christ who had this problem. Outwardly, he seemed to have it all. He was young, he was very rich and he had kept all the commandments, yet there was one thing that stood between him and God. That was his wealth. Perceiving this, Jesus told him to sell all that he had and give it to the poor. But the price was too great and the man went away very sad (Matthew 19:22).

Maybe later he came to his senses and accepted Christ. If he did, you can be sure he looked back on his wandering days and asked, "I wonder what I was chasing?"

Lord, as we face this New Year, save us from chasing rainbows that promise a pot of gold, but deliver only fool's gold. Help us instead to seek You *always.* —FRED BAUER

Praise Diary for December

1

2

3

4

5

6

7

8

9

10

11

12

13

14

15

16

17

18

19

20

21

22

23

24

25

26

27

28

29

30

31

 The Meeting Place 

Pull up a chair, stay awhile and get acquainted. Twenty-eight friends await your fellowship throughout the new year. Then return, again and again, won't you?

June Masters Bacher

This year marks June Masters Bacher's 46th wedding anniversary. Although George Bacher proposed to then-high-school-senior June on their first date, she didn't say "yes" until their *third* date! The Bachers live in Escondido, California, where they raised their son Bryce, now married and a vice principal of an elementary school. The Bachers belong to the Community Reformed Church, one of the few walk-in/drive-in churches in the country. This past year June had two major operations. After having spent six months in a cast from her second operation, on her spine, she didn't lose faith: "I'm determined to do even more with my writing—there's so much I want to share!"

Fred Bauer

"Shirley and I have gone from four nestlings to one," reports Fred Bauer. But the whole Bauer clan gathers regularly at the Princeton, New Jersey, homestead. Daughter Laraine, 30, made Fred a grandfather for a second time when daughter Ashley was born last summer. (Jessica is 2). Son Steve, 28, is a newswriter, and Christopher, 20, is attending West Virginia University, which leaves only Daniel, 16, at home. While Fred is busy working as a writer and publisher, wife Shirley is working toward her master's in library science. The Bauers manage to find time to travel in their motor home and vacation at their beach house in Florida. Finding a balance between work and play is important to Fred. He plans to spend much more time "taking my eyes off clocks and calendars and putting them on God and others."

Isabel Champ

On their second wedding anniversary, Isabel Champ Wolseley's husband Roland placed an envelope on her breakfast plate. Inside were plane tickets—for "Tulip Time in Holland." The Wolseleys have traveled far and wide this year, also visiting the United Kingdom, Norway, China, South Korea, Japan, Australia and the South Pacific. "No matter where we are," Isabel says, "we're amazed and impressed at God's imaginative creation. Like fingerprints, each place is different and special." When on solid ground, the Wolseleys make their home in Syracuse, New York.

Zona B. Davis

The *B* is for Zona's maiden name, Buchholz. In fact, she was once in *Ripley's Believe It or Not* for having a name that began and ended with *Z*—back before she married Plaford Davis. The Davises have one son, Paul, 47, a news director in Chicago, and three grandchildren: Paul Mark, 22, who just graduated from Northern Illinois University; Stokes, 20, who drove 70 miles from college to the Davises' Effingham, Illinois home "just to have a meal in Grandma's kitchen"; and Marinell, 18, who won a first in the American Open Karate Championship at Madison Square Garden recently. In her long and varied career, Zona hosted her own hour-long radio talk show five days a week for 30 years. This past year, she had cataract surgery on her one good eye and is delighted to have the gift of vision back again. She and Plaford spontaneously join hands in prayer whenever the urge strikes them. "What power there is in prayer!" she says.

Drue Duke

Last year, Drue Duke's retirement from the United Way lasted a month before she changed her mind. Now Drue and husband Bob both work two-and-a-half days a week, enjoying both "staying in the mainstream" and the long weekends that allow camping trips, visits to friends and family and a myriad of senior citizen activities. This year will mark their 42nd wedding anniversary, which they'll probably celebrate with a favorite pastime—jitterbugging. The Dukes have a daughter, Emily, 37, and two grandchildren—Bob, 14, and Christy, 11. The Dukes enjoy the "freedom and ease" of life in Sheffield, Alabama. When they take trips in their little motor home, they first join hands and then ask God's blessings on every car they pass. "It's exciting to send to those other drivers and their passengers a 'special blessing' in secret," says Drue.

Arthur Gordon

Our southern writer Arthur Gordon and his wife Pam left their home in Savannah, Georgia to visit Italy and Pam's homeland, England, as well as the "marvelous" Mayan ruins in the Yucatan and the fields of bluebonnets and Indian-paintbrush in Texas. When they passed through Austin, the state legislature passed a resolution of appreciation honoring Arthur as a writer. "A unique bit of Texas-sized hospitality," Arthur remarked. He's taken up the words of the great psychiatrist Karl Menninger as his personal yardstick this year. When asked what the purpose of one's life should be, Dr. Menninger said, "to try to dilute the misery in the world." Vowing to try to do just that in 1986, Arthur feels any small act can count, "whether it's helping a hurting soul or caring for a stray dog." The latter, by the way, was something the Gordons did last winter. Now "Andy" is a permanent member of the household!

Oscar Greene

Oscar Greene is no stranger to *Guideposts*—he was a 1977 attendee of its Writers' Workshop. Oscar, who worked on the Gemini space program as an engineering technician back in the early '60s, retired four years ago from General Electric, having obtained the prestigious position of Senior Technical Writer. A World War II veteran who served in the South Pacific, Oscar lives with his wife Ruby, a full-time tax examiner, in West Medford, Massachusetts. The Greenes have one son, Oscar Jr., a dentist with a thriving practice in Londonderry, New Hampshire, and two grandsons. Oscar spends his active days writing, giving talks, serving on bank boards and enjoying "falling in love with my wife all over again!"

Doris Haase

Every morning at 5 A.M., Doris Haase's alarm clock rings and she starts her day by reading her devotional and praying. Then, her writer's day begins with pen and notebook. At 8 she starts her day as a senior secretary in the Los Angeles school system. Doris, who lives in Sherman Oaks, California, enjoys the cultural advantages—concerts, movies, plays, the beach and, two years ago, even the Olympics—of Los Angeles. To keep fit, she bikes, swims, walks and works out at a health spa with older son Neil, 32. Her younger son Gary, 30, lives in Las Vegas. Doris feels she's growing closer to God all the time—"He's my Boss and Friend and *everything* to me!"

Rick Hamlin

Last year Rick Hamlin gave up his career as a freelancer when he became a full-time assistant editor at *Guideposts*. Rick keeps busy *after* a full day of office work, too. At church, he sings in the choir and works with the junior high youth group every Wednesday night. He sang and acted in two Gilbert and Sullivan productions, *Cox and Box* and *The Sorcerer*. In the fall he made a brief trip to London to visit his writer-wife Carol who was doing research for her latest book. And at Christmas, the Hamlins left their Upper West Side Manhattan apartment to visit Rick's family in California. Rick's morning devotions consist of Bible reading, a ritual which keeps him from being homesick for friends and family out west—"for, each morning, my devotion takes me to my *spiritual* home."

Madge Harrah

Multi-talented Madge Harrah is not only a full-time writer, but has worked as a commercial artist. She is also a professional tape-maker for gymnastic schools in both New Mexico and Illinois—she splices together pieces of music to go with gymnasts' routines. Daughter Meghan, 22, is a parole officer, and son Eric, 24, is the father of three children. Husband Larry is a physical chemist and spectroscopist at a renowned research center in their hometown of Albuquerque, New Mexico. This year, Madge and Larry got away at Christmastime to go to a scientific convention in Hawaii. "These days," Madge says happily, "I seem to be getting more and more informal with God—it's like being with an old friend and companion Who's always with me."

Marilyn Morgan Helleberg

Marilyn Morgan Helleberg became the "mother of the bride" last year when her artist daughter Karen, 28, married David Gerlach, 35. Marilyn and husband Rex have two other children: Paul, 26, who works on an oil-drilling crew and is father to three; and John, 17, a high school student. Rex sold his architecture firm this year and started working for the state. Now that he has regular hours, the Hellebergs have time for weekend camping trips. They also leave their Kearney, Nebraska, home for their Rocky Mountain cabin every year, which has been in Marilyn's family since 1912. Marilyn lectures as well as writes, but plans to concentrate more on writing this year. She reports, "My friend Carolmae and I pray together at 11 A.M. each day, signalling each other by one ring on the phone when it's prayer time."

Phyllis Hobe

For Phyllis Hobe, last year was one of quiet growth. "I'm becoming more aware of God's presence in my life—particularly of His great patience in creating things. It reminds me that I, too, must take my time putting ideas into words." She's joined a new congregation of caring members at St. Martin-in-the-Fields in Chestnut Hills, not far from her home in Flourtown, Pennsylvania. Phyllis enjoys trips to Philadelphia, where she holds season subscriptions both to the symphony and the ballet. To inspire her writing, Phyllis takes long walks every day with her faithful and energetic German shepherd Kate.

Marjorie Holmes

This past year, inspirational writer Marjorie Holmes and her husband, Dr. George Schmieler, celebrated their "fifth honeymoon." Marjorie is a *Guideposts* contributing editor and the author of such classics as *I've Got to Talk to Somebody, God* and *Two From Galilee*. Last year, the Schmielers left their McMurray, Pennsylvania, home to spend a winter month soaking up sunshine in Florida, and summer weekends they spent at their Lake Erie vacation house. Each morning and evening, Marjorie draws the sign of the cross on her forehead with fragrant oil, and prays. "This ritual," she says, "drives away the swarm of worries and frettings that forever hover about." She also takes that time to add prayers for all her loved ones.

Jeff Japinga

Last year, Jeff Japinga was the "pastor's husband" as his wife Lynn served as interim minister at Woodcliff Community Reformed Church in North Bergen, New Jersey. "This new role was rewarding," says Jeff, "even if Hallmark doesn't make cards 'For a Pastor and Her Husband/Family.'" Now the Japingas have settled into their new home in Somerset, New Jersey. Jeff, after a five-year hiatus, decided to continue studying for his master of divinity degree and started night classes at New Brunswick Theological Seminary. Lynn went back to full-time doctoral studies at the Union Theological Seminary in New York. And, for the first time in their five-year marriage, the Japingas returned to Michigan for Christmas. Jeff's long commute to his office at Guideposts, where he is editor of Book Clubs, is a rich part of his spiritual life—he reads devotional books on his train ride into the city and thinks about his reading on his way home.

Sam Justice

Last April, Sam Justice headed west to Columbia, Missouri, to attend his 50th anniversary class reunion at the University of Missouri. "It was a joyous experience of seeing old friends," he reports. Sam and his wife Ginny, who for the past 20 years has been a nursery school director, have five children—whose names all start with *J*. James, 37, is a Hollywood screenwriter; John, 35, is studying physical education; Jeff, 33, is a magician; Jennifer, 29, is a mother of two and a social worker; and Jackie, 25, is an office designer. Sam hopes to bring the whole family together this year, either in the Yonkers, New York, homestead or in their vacation house on the Jersey shore. Enjoying this time of his life, Sam writes, golfs, fixes up his home and drives elderly parishioners to church. He keeps up his daily practice of reading a Bible verse each morning, meditating on it as he goes about his day and again at night.

Sue Monk Kidd

Last year Sue Monk Kidd and her husband Sandy built a deluxe-size treehouse for their children in a huge oak in their Anderson, South Carolina, backyard. The family has stargazed through its branches, watched nesting birds and had tea parties. Sue writes every day at her big rolltop desk and has been working on her first book for Guideposts. She gets in five workouts a week, doing jogging and jazzercize. The Kidds are all Atlanta Braves fans (and even named their new beagle, Murphy, after Dale Murphy). Last year son Bob, now 12, hit his first Little League home run. Ann, 9, had her first piano recital, playing Beethoven's "Ode to Joy" perfectly. This year Sue is trying to "weave prayer into the common stuff of life—as I cook, car pool or wash the dog."

Glenn Kittler

Every morning Glenn Kittler wakes up in his Upper West Side Manhattan apartment, makes coffee, feeds his cat Louie and sits down to twelve hours of writing. Glenn, whose books have been translated into many languages, has traveled all over the world on writing assignments. His latest project is a collaborative effort on the history of religion in the United States. After cataract surgery, Glenn's baby-blue eyes are as good as new. "Dropping in" at his neighborhood cathedral church to visit with God and think awhile is a treasured pastime of this devoted writer.

Carol Kuykendall

Carol Kuykendall is a brand new contributor to *Daily Guideposts* and was a winner of the 1984 Writers Workshop. She writes in her beloved Boulder, Colorado. Carol and her lawyer husband Lynn live in a country home they built themselves, adjacent to the house in which she grew up. They have three children: Derek, 13; Lindsay, 12; and Kendall, 8. "Our children are not yet tugged by the diversions of adolescence, but are old enough to do things together and we are making the most of it!" Last year the Kuykendalls traveled to Mexico, Montana and Los Angeles, where they visited Carol's sister, actress Joan Van Ark. One of Carol's favorite devotional activities is keeping her "Book of Eternal Truths," a blank book she fills with inspirational quotes. "I reach for it whenever I feel discouraged or dry, and never fail to be renewed."

James McDermott

Jim McDermott spent long summer days swimming and sailing at his Long Island shore vacation house with wife Judy and four-year-old son David. And he headed both west and east on exotic skiing trips—to Alta, Utah, and Davos, Switzerland. While *Guideposts* celebrated its 40th anniversary last year, Jim celebrated ("enthusiastically!") his tenth anniversary with *Guideposts*. "I get the warm feeling that *Guideposts* and I are good for one another. The feeling gets even warmer when I realize that the same is true for my fellow editors, too." Jim believes *devotion* should be synonymous with *service*, but also plain likes attending Plymouth Church of the Pilgrims in his hometown, Brooklyn, New York—and belting out old, beloved hymns. "It's best not to sit in front of me!" he jokes.

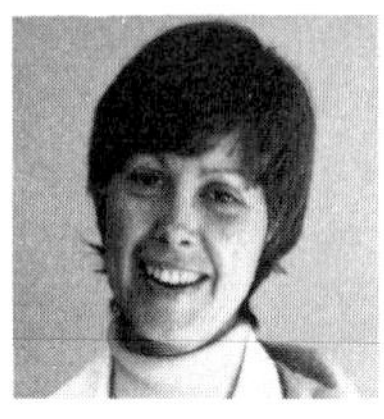

Samantha McGarrity

For Sam McGarrity, 1985 was a year of picking up pieces. "Plenty of things fell apart," she says. "Even me—I fell off a ladder and broke three ribs and a wrist." Among the changes in Sam's life this year have been trading in her ancient Volkswagen for a new Honda, selling the river cottage she'd been trying to fix up for several years (and buying a home that *works* instead) and letting go of anxieties she'd been feeling about relationships and health—leaving some things to time and God. She *has* had special moments this year: visits with North Carolinians who remember Halley's Comet's 1910 appearance; searching for shells on Ediste Island, S.C., a vacation place with family roots; lunching with writer E.B. White at his Maine farm. Sam's devotional times are practiced as she travels or is "on the move"—in the car, walking, or even on the subway. She images in her mind God's light filling her being, pushing out any darkness. "After all," she says, "we are called 'the light of the world.'"

Norman Vincent Peale

Dr. Norman Vincent Peale, son of a Methodist minister, was born in 1898 in Bowersville, Ohio. In 1924 Dr. Peale was himself ordained. Eight years later he became minister of New York's Marble Collegiate Church, where he served for fifty-two years. Dr. Peale has written 32 books—his latest being *The True Joy of Positive Living,* an autobiography. His is the longest heard voice over the radio, daily and weekly, and he writes a syndicated column, "Positive Thinking." He keeps up a rigorous schedule of about 80 speeches a year at national conventions. He and his wife, Ruth Stafford Peale, live in New York and at their farm in Pawling, New York, where they direct the Foundation for Christian Living, which they started in 1940. The Peales are co-editors and publishers of *Guideposts*.

Ruth Stafford Peale

Like her husband, Ruth Stafford Peale has had a long and distinguished career in public service as a writer, publisher, speaker and administrator. She also serves on many boards of important religious organizations. Born in Fonda, Ohio, a graduate of Syracuse University, Mrs. Peale is the author of the recently published *Secrets of Staying in Love*, her helpful guide to getting the most out of marriage. She is the recipient of many outstanding awards such as Churchwoman of the Year, New York State Mother of the Year and the Horatio Alger Award. She is an honorary Doctor of both Laws and Letters, and is listed in *Who's Who in America* and *Who's Who of American Women*. Dr. and Mrs. Peale have three children, Margaret, John and Elizabeth, and eight grandchildren.

Elaine St. Johns

Elaine St. Johns had her year slowed down when her kneecap was broken. But slowly and steadily she's been improving, from cast to walker to an eagleheaded cane, and soon, she says, "plans to be doing the rhumba." Elaine's next-door neighbors are her daughter Kristen, son-in-law Fred Wolf and their five children—Jessica, 16; Bogart, 12; Robin, 4; Hilary, 3; and Meredith, 6 months. In fact, one of Elaine's first excursions out with her cane was to a joint baptism for Meredith and confirmation for Bogart. Elaine's businessman son George and wife Pam are also parents—of Colin, 16, and Ashley, 14. Elaine, like her 91-year-old mother Adela Rogers St. Johns, has had a long career in writing. She has ghostwritten for a hotel magnate, a teenage idol, a movie star, a professor and many other unusual people. She's enjoying life at 67, with more time to read and think. She also finds her daily discipline of Christian meditation enriches her spiritual walk.

Eleanor Sass

Since her trip to Costa Rica in 1984, Eleanor Sass continues to be concerned about saving endangered sea turtles. This year she joined the board of the Caribbean Conservation Corporation, a group responsible for monitoring turtles in the Caribbean. She also speaks to groups about her turtle-tagging experience. This year, Ellie's also spending some of her precious vacation time playing golf, a new hobby. "As a New York City apartment dweller, finding a quiet place to have devotions is difficult," she says. "Even if I'm alone inside my apartment, sanitation trucks grind, car horns honk, sirens wail outside. But when I remove my two hearing aids from my ears, I can read my Bible and pray—without interruptions. My disadvantage becomes an asset!" Ellie has a four-legged roommate—her lovable 12½-year-old dachshund, Heidi.

Elizabeth Sherrill

Elizabeth Sherrill has been thankful for her husband John's patience and help as she finished yet another major book this year. "There ought to be a gold-coffeepot award for people married to writers," she says. "As I think back over 1985, heroic marriage partners seem to be the common thread." Son John Scott, 35, a Nashville musician, recently won his third gold record—aided by the smiling patience of wife Meg, who mothers their 7-year-old Kerlin. Son Donn, 32, has a commuter marriage—in his job, he travels all over South America, while wife Lorraine cares for Lindsay, 3, and newborn Andrew in Miami. Alan Flint, husband of daughter Liz, 29, turned down a promotion so she could continue her social work in Massachusetts. Elizabeth and John live in Chappaqua, New York, and spent a glorious month on Nantucket last summer. She especially treasures the devotional times she shares with others, at church and in her prayer groups: "God is speaking clearest to me when friends won't allow me to keep my fingers in my ears."

Patricia Houck Sprinkle

Patti Sprinkle has had a busy year—she and her minister husband Bob just started the Evangelical Covenant Church of the Resurrection in Atlanta, Georgia. The Sprinkles are among the first to introduce this denomination (which was brought over from Sweden 100 years ago) to Georgia. Their fledgling church is growing and has moved from their living room to a rented clubhouse facility. Patti also watched over her two active boys, David, 5, and Barnabas, 8, who, she says, "want to be involved in gymnastics, soccer and every mudhole in the county." Since 1975, the bulk of Patti's writing has been on the subject of world hunger, and moving back to Atlanta has meant a renewed involvement in anti-hunger organizations. She and her family are also tending their first-ever garden. Patti has had to learn to pray on her feet—"I have a running conversation with God."

Van Varner

This past year for Van Varner has been a special one of prayer. All year long, some of his best thinking—and praying—is done on his way to the Guideposts office in the morning. There are always stories to ponder, and as he walks through Central Park and the city's streets (and through hotel lobbies, as some of you *Daily Guideposts'* readers may recall), he is urged to pray by the sights he sees along the way—the homeless man taking advantage of the heat rising from a sidewalk grill; the headlines on the papers at the corner newsstand; the ambulance snaking through snarled traffic. But what a year it's been for our *Guideposts* Editor. To mark the 40th anniversary of the magazine, the year of "I Believe in Prayer" was inaugurated. "Every month on a designated day, at an agreed upon minute, all of us here have gathered to pray with our readers for you, for us, for a worldful of people in need" Van says. "It's a habit I don't intend to break."

Marion Bond West

This year, Marion Bond West is spending more time with her twins, Jon and Jeremy, who, she says, "need her attention more than ever at age 17." Her many speaking engagements keep her active, but she tries not to stray too far from her Lilburn, Georgia, home. She also has two married daughters: Jennifer, 23, who is studying elementary education, and Julie, 25, a mother of two daughters. This year, Marion finished a book explaining her principle of "Nevertheless Living"—helping people face despair. Marion belongs to the 14,000-member First Baptist Church of Atlanta and says she is learning not to "stew" over problems, but trusting herself instead to the care of her loving Father.

The Reader's Guide

I. SCRIPTURE REFERENCE INDEX . . . . *page 360*
II. FIRST FEW WORDS INDEX . . . . *page 363*
III. AUTHORS, TITLES AND SUBJECTS INDEX . . . *page 369*

A three-part index to all the selections in Daily Guideposts, 1986—*to help you find it quickly: Your favorite authors... memorable passages... cherished Scripture verses... timely lessons to read and reread... subjects that speak to times of special need... to help you as you prepare sermons or study lessons, or simply to locate a devotional to share with another. A treasure-filled Guide, just for you.*

 Scripture Reference Index 

An alphabetical index of Scripture references to verses appearing either at the tops of devotionals or, on occasion, within the text. Chapter and verse numbers are in bold type on the left. Numbers in regular type, on the right, refer to the Daily Guideposts page(s) on which the complete verse can be located.

Acts
10:43, 143
12:3, 112
17:28, 131
20:1, 237

Chronicles I
28:9, 236
29:18, 206
Chronicles II
5:13, 11
13:12, 298
Colossians
1:17, 18, 22
3:13, 159
Corinthians I
1:18, 320
1:27, 242, 281
2:5, 322
2:9, 326, 341
3:6, 39
3:9, 280
3:16, 306
4:14, 194
7:7, 245
12:4, 339
12:27, 57
12:31, 190
13:8, 229
13:12, 188, 326
15:10, 114
15:49, 304
Corinthians II
2:9, 113
3:2,3, 99
4:8,9, 272
4:18, 123
5:7, 179, 259
12:9, 303
13:10, 156
13:11, 248

Deuteronomy
1:21, 273
8:11, 62
15:10, 192
28:12, 169
30:19, 216
31:6, 251
34:1, 302

Ecclesiastes
2:24, 24
2:26, 127
3:1,2, 65
3:3, 51
3:4, 181
3:6, 139
3:9-11, 253
9:10, 335
12:1, 225
Ephesians
2:14, 282
3:16, 68
4:1, 296
4:1,2, 304
4:23, 72
4:31, 117
4:32, 324
5:18,19, 155
6:6,7, 100
6:18, 150
Exodus
3:17, 21
13:22, 49
14:27, 72
15:2, 73
15:25, 105
16:4, 133
17:9, 161
18:18, 189
19:20, 216
25:8, 247
28:35, 226
34:7, 58
Ezekiel
7:10, 11
11:19, 319
18:31, 211
20:47, 37
27:8, 231
34:26, 32
37:15, 52
Ezra
3:3, 172

Galatians
5:13, 44
5:25,26, 41
6:2, 45, 199
Genesis
1:20,21, 134
1:22, 212
1:28, 124
1:31, 220
2:24, 166
3:8, 103
15:1, 13
18:8, 25
26:24, 327
33:5, 130
41:35, 224
50:21, 337
Gospels *see* Matthew, Mark, Luke, John

Habakkuk
3:19, 20
Hebrews
4:16, 157
6:10, 210
11:1, 101
11:8, 192
11:16, 340
12:1, 264
13:5, 246

Isaiah
6:8, 17
9:7, 167
26:3, 10
28:16, 38

Isaiah, cont.
28:23, 27
29:14, 140
30:15, 196
33:2, 269
34:16, 253
40:6, 96
40:29, 29
40:31, 99
41:6, 28
41:10, 239
42:16, 163
49:11, 169
50:4, 195
50:7, 283
55:3, 240
56:4, 125
58:11, 197

James
1:17, 71
1:25, 293
2:20, 137
4:8, 264
5:16, 36, 299
Jeremiah
5:15, 244
10:12, 185
30:17, 286
31:13, 54
Job
2:3-6, 20
2:6, 20
9:9, 256
9:27, 298
20:10, 312
29:3, 255
38:22, 12
Joel
2:28, 223
John
1:14, 88
1:41, 78
1:43, 63
3:3, 24
3:12, 80
3:16, 314, 323, 328, 333, 336
4:14, 279
6:60, 81
8:36, 185
10:10, 279
11:41, 56, 103
12:13, 318
12:21, 207
12:46, 102
13:4,5, 76
13:13,14, 76
13:16, 83
13:34, 142
14:2, 227
14:6, 95
14:14, 82
14:18, 87
15:14, 46
15:15, 178
17:15, 254
20:29, 309
21:3, 222
21:15,16, 225
21:15-17, 112
John I
1:9, 29
2:1, 274
4:12, 280
4:18, 268
John II
6, 168
Joshua
1:9, 144, 249
3:17, 329
Jude
1:24, 201
1:2, 14
Judges
3:20, 241

Kings I
19:12, 221
Kings II
6:9, 310

Lamentations
1:13, 278
Leviticus
16:22, 275
19:18, 184
Luke
2:7, 338
2:10, 333
6:38, 154
7:5, 180
7:48, 63
8:15, 243
8:18, 106
9:62, 42
15:20, 18
15:31, 160
16:15, 89
17:4, 29
17:21, 78
18:1, 116
18:33, 170
20:17, 124
21:4, 66
22:20, 63
23:34, 74
24:23, 151

Malachi
3:7, 217
Mark
1:35, 69
3:25, 54
3:35, 23
4:39, 14, 63
5:28, 135
6:31, 285
7:9, 209
9:24, 230
10:15, 67
10:25, 182
10:27, 179
10:44, 76
14:3,5,6, 301
14:37, 283
Matthew
5:5, 137
5:9, 84
5:14, 154
5:15, 295
5:16, 213
5:17, 126
6:3, 332
6:6, 191
6:9, 56
6:11, 200
6:14, 257
6:21, 199
6:33, 183
6:34, 164
7:1, 209
7:12, 251
7:25, 186
9:21, 166
9:22, 270
10:8, 278
10:29, 95
10:42, 236
11:28, 55
12:36, 215
13:3, 107
13:8, 136
13:44, 241
14:25, 229

Matthew, cont.
15:35, 36, 309
16:26, 341
17:20, 63, 77, 196, 231
18:11, 325
18:22, 44
19:26, 51
21:22, 56
22:37,39, 209
23:12, 76
24:35, 151
24:44, 46, 238
25:23, 326
26:26, 267
28:20, 167, 300
Micah
6:8, 94

Nehemiah
4:14, 62
Numbers
6:26, 165

Peter I
1:7, 214
2:9, 26
2:25, 112
3:8, 30
4:12,13, 171
5:5, 111
5:7, 187
Peter II
3:18, 108
Philemon
20, 114
Philippians
1:1,2, 292
1:6, 276
1:9,10, 294
1:14, 178
2:9,10, 122
2:10,11, 334
2:13, 70
3:13, 104
3:14, 277
4:13, 129
4:17, 300
Proverbs
3:5, 31
3:6, 64
3:13, 193
6:6, 26
8:6, 16
12:18, 297
14:23, 111
15:1, 339
15:15, 47
15:23, 215
16:22, 141
17:17, 19
22:6, 57
23:7, 115
28:20, 157
29:25, 145
31:11, 43
Psalms
4:4, 271
4:7, 130
5:7, 228
8:4,5, 162
9:10, 50
17:5, 284
19:14, 215
22:4, 15
23:1,2, 152
23:4, 256
23:5, 74
25:4, 22
25:16, 287
26:3, 307
28:7, 307
30:2, 218
32:7, 98
33:5, 270
39:3, 313
46:10, 265
47:6, 39
61:8, 182
62:5, 330
71:3, 220
73:28, 31
76:8, 67
77:14, 207
85:1, 160
85:8, 319
86:3, 138
86:7, 296
91:15, 213
95:2, 208
95:4, 143
95:6, 109
96:8, 132
100:1, 17
103:8, 321
103:17,18, 128
104:28, 292
107:43, 135
113:9, 305
119:73, 308
119:116, 257
124:8, 325
126:2, 64, 190
127:1, 48
127:3-5, 250
133:1, 40
139:14, 252
143:10, 285
144:14, 64
147:3, 109

Revelation
3:8, 157
21:5, 274
Romans
4:20,21, 330
8:5, 36
8:28, 198
12:10, 250
12:12, 70
14:5, 311
15:1, 158
16:1,2, 266

Samuel I
2:7,8, 97
18:1,4, 331

Thessalonians I
4:18, 215
5:11, 127
5:14, 172
5:18, 153
Thessalonians II
1:11, 258
Timothy I
4:7, 141
4:14, 75
Timothy II
1:7, 110
2:3, 53
2:12, 86
Titus
3:8, 244

Zechariah
4:6, 172

First Few Words Index 

An alphabetical index to the first few words of Scripture verses appearing either at the top *of the devotionals or as full verses* within *the text, as well as the first few words of poetry, prose and songs appearing within the text. Numbers given refer to the* Daily Guideposts *page(s) on which these can be located.*

A faithful man shall abound with blessings..., 157
A friend is one to whom one may pour out all the contents..., 19
A friend loveth..., 19
A new commandment I give unto you..., 142
A smile costs nothing..., 130
A soft answer..., 339
A sower went forth to sow..., 107
A time to break down..., 51
A time to get..., 139
A time to weep..., 181
A word spoken in due season..., 215
According to the power which the Lord hath given me..., 156
All flesh is grass..., 96
America! America!..., 180
And after the fire a still small voice..., 221
And all the Israelites passed over..., 329
And all things, whatsoever ye shall ask..., 56
And as they were eating, Jesus took bread..., 267
And as we have borne the image of the earthy..., 304
And be renewed in the spirit..., 72
And be ye kind one to another..., 324
And, behold, God himself is with us..., 298
And God blessed them..., 124, 212
And God hath chosen the weak things..., 242
And God said, Let the waters bring forth..., 134
And God saw everything that he had made..., 220
And he comforted them, and spake kindly unto them..., 337
And he commanded the multitude to sit down..., 309
And he is before all things..., 22
And he said unto them, Full well ye reject the commandment..., 209
And he shall let go the goat in the wilderness..., 275
And he took butter..., 25
And his hands shall restore their goods..., 312
And his sound shall be heard..., 226
And I will bring the blind by a way that they knew not..., 163
And I will give them one heart..., 319
And I will make all my mountains a way..., 169
And if any man sin..., 274
And if he trespass against thee..., 29
And in the fourth watch of the night Jesus went unto them..., 229
And in the morning, rising up..., 69
And it came to pass...that the soul of Jonathan..., 331
And let them keep food in the cities..., 224
And let us run with patience..., 264
And, lo, I am with you always..., 300
And Moses went up...to the top of Pisgah..., 302
And peace, and love..., 14
And she brought forth her firstborn son..., 338
And the angel said unto them, Fear not..., 333
And the Lord called Moses up to the top of the mount..., 216
And the Lord overthrew the Egyptians..., 72
And the Lord said unto Satan, Behold, he is in thine hand..., 20
And the Lord shall guide thee continually..., 197
And the man of God sent unto the king of Israel..., 310
And the rain descended..., 186
And the third day He shall rise again..., 170
And...the waters were made sweet..., 105

And the Word was made flesh..., 88
And they heard the voice of the Lord God..., 103
And they shall be one flesh..., 166
And they that know thy name..., 50
And this is love..., 168
And we know that all things work together for good..., 198
And when thou hast shut thy door..., 191
And whosoever shall give...a cup of cold water only..., 236
Are not two sparrows..., 95
Are you willing to consider the needs and the desires of children..., 236
As the servants of Christ..., 100
At the Name of Jesus every knee should bow..., 334

Be careful to maintain good works..., 244
Be clothed with humility..., 111
Be filled with the Spirit..., 155
Be kindly affectioned one to another..., 250
Be merciful unto me, O Lord..., 138
Be still, and know that I am God...,67, 265
Be strengthened with might..., 67,68
Be strong and of a good courage, fear not..., 251
Be strong and of good courage; be not afraid..., 144
Be thou my strong habitation..., 220
Be ye all of one mind..., 30
Be ye kind one to another..., 297
Bear ye one another's burdens..., 45, 199
Behold, how good and how pleasant..., 40
Behold, I have set before thee an open door..., 157
Behold, I make all things new..., 274
Behold, I will proceed to do a marvelous work..., 140
Behold the day..., 11
Behold, the Lord thy God hath set the land before thee..., 273
Being confident of this very thing..., 276
Beloved, think it not strange concerning the fiery trial..., 171
Beware that thou forget not the Lord..., 62
Blessed are the meek..., 137
Blessed are the peacemakers..., 84
Blessed are they that have not seen..., 309
Blessed is the King of Israel that cometh..., 318
But as for me, I will come into thy house..., 228
But by the grace of God I am what I am..., 114
But covet earnestly the best gifts..., 190
But God hath chosen the foolish things of the world..., 281
But it is good for me to draw near God..., 31
But now they desire a better country..., 340
But other fell into good ground..., 136
But seek ye first the kingdom of God..., 183
But the mercy of the Lord is from everlasting to everlasting..., 128
But they that wait upon the Lord..., 99
But whoso lookest into the perfect law of liberty..., 293
But ye are a chosen generation..., 26
By faith Abraham, when he was called... obeyed..., 192
By his light I walked through darkness..., 255
By love serve one another..., 44

Cast away from you all your transgressions..., 211
Casting all your care upon him..., 187
Charity never faileth..., 229
Cleanse your hands..., 264
Come unto me, all ye that labor..., 55
Come ye yourselves apart..., 285
Comfort one another..., 215
Commune with your own heart..., 271

Daughter, be of good comfort..., 270
Despite their fear of the peoples..., 172
Do you love me, Peter?..., 112

Every good gift..., 71
Every idle word that man shall speak..., 215
Except a man be born again..., 24
Except the Lord build the house..., 48
Exercise thyself...unto godliness..., 141
Eye hath not seen, nor ear heard..., 326

Faith without works is dead..., 137
Father, forgive them..., 74
Father, I thank thee..., 56, 103
Fear not...I am thy shield..., 13
Fear thou not...I will help thee..., 239
Feed my lambs..., 225

Follow me..., 63
For all these have of their abundance..., 66
For as he thinketh in his heart..., 115
For God giveth to a man that is good..., 127
For God hath not given us the spirit of fear..., 110
For God is not unrighteous to forget your work..., 210
For God so loved the world..., 313, 314, 323, 328, 333, 336
For he is our peace, who hath made both one..., 282
For he loveth our nation..., 180
For I will turn their mourning into joy..., 54
For I will restore health unto thee..., 286
For I would that all men were even as I myself..., 245
For if ye forgive men their trespasses..., 257
For in him we live, and move..., 131
For it is God which worketh in you..., 70
For now we see through a glass, darkly..., 188, 326
For she said within herself, If I may touch his garment..., 166
For the Lord God will help me..., 283
For the preaching of the cross is to them that perish foolishness..., 320
For the Son of man is come..., 325
For they that are after the flesh..., 36
For thy loving-kindness is before mine eyes..., 307
For we are laborers together with God..., 280
For we walk by faith..., 179, 259
For what is a man profited if he gain the whole world..., 341
For where your treasure is..., 199
For whosoever shall do the will of God..., 23
For ye were as sheep..., 112
Forgetting those things which are behind..., 104
Freely ye have received..., 278
Friendship is the oak of life..., 9

Give, and it shall be given unto you..., 154
Give unto the Lord the glory..., 132
Give us this day our daily bread..., 200
Give ye ear, and hear my voice..., 27
Go to the ant..., 26
God hath...given him a name..., 122
Grow in grace..., 108

Happy is the man that findeth wisdom..., 193
Hast thou entered into the treasures..., 12
Having heard the word, keep it..., 243
He first findeth his own brother Simon..., 78
He giveth power to the faint..., 29
He hath made the earth by his power..., 185
He hath spread a net for my feet..., 278
He healeth the broken in heart..., 109
He maketh the barren woman...to be a joyful mother..., 305
He riseth from supper, and laid aside his garments..., 76
He shall call upon me, and I will answer him..., 213
He slipped and fell this friend of mine..., 235
He staggered not...but was strong in faith..., 330
He that shall humble himself..., 76
He that is of a merry heart..., 47
He took not away the pillar of the cloud..., 49
[He]...which maketh Arcturus, Orion, and Pleiades..., 256
Heaven and earth shall pass away..., 151
Help thou mine unbelief..., 230
Here is Meghan, Lord. Be with her..., 98
Hold up my goings in thy paths..., 284

I am come a light into the world..., 102
I am not come to destroy..., 126
I am the way..., 95
I am with thee, and will bless thee..., 327
I...beseech you that ye walk...with long-suffering..., 304
I came that you might have life..., 279
I can do all things through Christ..., 129
I commend unto you Phoebe our sister...266
I have a message from God unto thee..., 241
I have called you friends..., 178
I have planted, Apollos watered..., 39
I heard the voice of the Lord..., 17
I looked over Jordan, an' what did I see..., 228
I love you for a number of reasons..., 128

First Few Words Index

I pledge allegiance..., 167-68
I pray not that thou shouldest take them out of the world..., 254
I pray, that your love may abound..., 294
I press toward the mark for the prize..., 277
I say not unto thee, Until seven times..., 44
I wandered lonely as a cloud..., 287
I will bring you up out of the affliction..., 21
I will hear what God the Lord will speak..., 319
I will never leave thee..., 246
I will not leave you comfortless..., 87
I will praise thee..., 252
I will rain bread from heaven..., 133
I will speak of excellent things..., 16
I will stand on the top of the hill..., 161
I write not these things to shame you..., 194
If a house be divided against itself..., 54
If any man have a quarrel against any..., 159
If I have told you earthly things..., 80
If I may touch but his clothes..., 135
If I say, I will forget my complaint..., 298
If on my journey, I can impart..., 291
If the Son therefore shall make you free..., 185
If we confess our sins..., 29
If we live in the Spirit..., 41
If we love one another..., 280
If we suffer, we shall also reign with him..., 86
If we have faith..., 77, 231
If ye shall ask any thing in my name..., 82
I'll never forget the first time I saw you..., 130
In all labor there is profit..., 111
In all thy ways acknowledge him..., 64
In every thing give thanks..., 153
In his hand are the deep places of the earth..., 143
In my Father's house are many mansions..., 227
In quietness and in confidence shall be your strength..., 196
In remembrance of Me, heal the sick..., 237
In the day of my trouble I will call upon thee..., 296
Incline your ear, and come unto me..., 240
It is an ancient nation..., 244
It is easier for a camel to go through the eye of a needle..., 182
It was many and many a year ago..., 153

Judge not, that ye be not judged..., 209

Keep it up, you two lovebirds..., 28
Keep this forever in the imagination..., 206
Know ye not that ye are the temple of God..., 306

Let all bitterness, and wrath..., 117
Let not thy left hand..., 332
Let the words of my mouth...be acceptable..., 215
Let them make me a santuary..., 247
Let us come before his presence with thanksgiving..., 208
Let us therefore come boldly unto the throne of grace..., 157
Let your light so shine..., 213
Live in peace..., 248
Lo, children are a heritage of the Lord..., 250
Lo, I am with you always..., 167
Looking back—Thank Him..., 193
Lord, thou hast been favorable unto thy land..., 160
Lord, you are my strength and my song..., 73
Love one another..., 142

Make a joyful noise..., 17
Many therefore of his disciples..., 81
Men ought always to pray..., 116
More bold to speak the word..., 178
Most gladly therefore will I rather glory in my infirmities..., 303
Most of all, Lord, make me kind..., 121
My expectation is from him..., 330
My faith looks up to thee..., 213
My Father which *is* in Heaven..., 56–57
My Lord God, I have no idea where I am going..., 274

Neglect not the gift..., 75
Neither do men light a candle..., 295
No man, having put his hand to the plough..., 42
No one who's born to earth..., 177
No ray of sunshine is ever lost..., 136
Not because I desire a gift..., 300
Nothing shall be impossible..., 63

Now faith is the substance of things hoped for..., 101
Now there are diversities of gifts..., 339
Now unto him that is able to keep you from falling..., 201
Now we exhort you, brethren..., 172
Now ye are the body of Christ..., 57

O come, let us worship..., 109
O Lord...be thou their arm every morning..., 269
O Lord my God, I cried unto thee..., 218
Of the increase of his government and peace..., 167
One man esteemeth one day above another..., 311
Our faith extols a life beyond..., 317
Our Father which art in heaven..., 56
Our help is in the name of the Lord..., 325
Our souls are hungry, Lord..., 93

Paul called unto him the disciples..., 237
Peace, be still..., 14, 63
Perfect love casteth out fear..., 268
Pray one for another..., 36
Praying always with all prayer..., 150

Rejoicing in hope..., 70
Remember now thy Creator..., 225
Remember the Lord, which is great..., 62
Return unto me..., 217

Seek ye out of the book of the Lord..., 253
Serve him...with a willing mind..., 236
Show me thy ways, O Lord..., 22
Simon Peter saith unto them, I go a fishing..., 222
Sing praises to God..., 39
Sir, we would see Jesus..., 207
Sleepest thou? Couldest not thou watch..., 283
Son, thou art ever with me..., 160

Take heed therefore how ye hear..., 67
Take therefore no thought for the morrow..., 164
Teach me to do thy will..., 285
Temptation comes in many forms..., 61
That I may daily perform my vows..., 182
That I might know the proof of you..., 113
That the trial of your faith..., 214
That there be no complaining..., 64
That your faith should not stand in the wisdom of men..., 322
The earth is full of the goodness of the Lord..., 270
The fear of man bringeth a snare..., 145
The heart of her husband..., 43
The inhabitants of Zidon and Arvad were thy mariners..., 231
The kingdom of God is within you..., 78
The kingdom of heaven is like unto treasure..., 241
The Lord...bringeth low..., 97
The Lord God hath given me the tongue of the learned..., 195
The Lord God is my strength..., 20
The Lord is merciful and gracious..., 321
The Lord is my shepherd..., 152
The Lord is my strength and my shield..., 307
The Lord lift up his countenance upon thee..., 165
The Lord shall...bless all the work..., 169
The Lord thy God is with thee..., 249
The Lord thy God shall bless thee in all thy works..., 192
The New Year lies before you..., 13
The...prayer of a righteous man availeth much..., 299
The road to happiness..., 190
The stone which the builders rejected..., 124
The water that I shall give him..., 279
The word of the Lord came again..., 52
Then was our mouth filled with laughter..., 64, 190
Then were the days of unleavened bread..., 112
There came a woman having an alabaster box..., 301
There is nothing better for a man..., 24
There is that speaketh like the piercings of a sword..., 297
There shall be showers of blessing..., 32
Therefore all things whatsoever ye would that men should..., 251
Therefore be ye also ready..., 238
Therefore choose life..., 216
They had also seen a vision of angels..., 151
They helped every one..., 28
They lifted up their voice..., 11
They trusted, and thou didst deliver them..., 15

First Few Words Index

This cup is the new testament..., 63
This is the word of the Lord unto Zerubbabel..., 172
Thou art my hiding place..., 98
Thou art that God that doest wonders..., 207
Thou didst cause judgment to be heard from heaven..., 67
Thou hast put gladness in my heart..., 130
Thou openest thine hand..., 292
Thou preparest a table..., 74
Thou shalt love the Lord thy God..., 209
Thou shalt love thy neighbor as thyself..., 184
Thou therefore endure hardness..., 53
Thou wilt keep him in perfect peace..., 10
Thou wilt surely wear away..., 189
Though I walk through the valley..., 256
Thus saith the Lord God, Behold, I lay in Zion..., 38
Thus saith the Lord God; Behold, I will kindle..., 37
Thus saith the Lord...choose the things that please me..., 125
Thy hands have made me and fashioned me..., 308
Thy sins are forgiven..., 63
Time and again I tell myself..., 205
'Tis a month of reddened nose..., 35
To all the saints in Christ Jesus..., 292
To every thing there is a season..., 65
To him give all the prophets witness..., 143
Train up a child..., 57
Trouble-filled, I toss and turn..., 149
Trust in the Lord with all thine heart..., 31
Trust no Future, howe'er pleasant!..., 164
Turn thee unto me, and have mercy..., 287
Understanding is a wellspring of life..., 141
Uphold me according unto thy word..., 257

Verily, verily, I say unto you..., 83

Walk humbly with thy God..., 94
Walk worthy of the vocation..., 296, 297
We are troubled on every side..., 272
We look not at the things which are seen..., 123
We then that are strong ought to bear..., 158
Well done, good and faithful servant..., 326
What is man that thou art mindful..., 162
What makes life worth the living..., 324
What profit hath he that worketh..., 253
Whatsoever thy hand findeth to do..., 335
When he was yet a great way off..., 18
When the world too much presses in..., 263
Wherefore also we pray always for you..., 258
Wherefore comfort yourselves together..., 127
Which God hath prepared for them..., 341
While I was musing the fire burned..., 313
Who are those with thee?..., 130
Whoso is wise, and will observe these things..., 135
Whosoever of you will be the chiefest..., 76
Whosoever shall not receive the kingdom of God..., 67
With men this is impossible..., 51

Ye are my friends..., 46
Ye are our epistle..., 99
Ye are the light of the world..., 154
Ye call me Master and Lord..., 76
Ye...justify yourselves before men..., 89
Ye shall say unto this mountain..., 196
Yea, brother, let me have joy..., 114
Your old men shall dream dreams..., 223
Your 'strength' is not your own might..., 172

Authors, Titles and Subjects Index

An alphabetical index to devotional authors; titles of special series, poems and songs; proper names of people, places and things; holidays and holy days; Biblical persons and events appearing in the text; and subjects with sub-heading breakdowns that will help you find a devotional to meet that special need. Numbers refer to the Daily Guideposts *page(s) on which these can be located.*

Aaron (Biblical character)
  Moses and, 161-62
  story of the scapegoat and, 275
Ability
  doing one's best, 143-44, 243, 281-82
  finding it in oneself, 75-76
  inner greatness, 111
  learning to do for oneself, 326-27
  perseverance and, 201
  pleasing God through one's ability, 307-08
  prayer and, 330-31
  sharing one's, 213, 242-43
  using talents to serve others, 339
Acceptance
  see also Tolerance
  forgiveness and, 275-76
  limitations and, 196
  of a slower pace in life, 168
  of Christian life, 341
  of God's Will, 163-64
  of infirmities, 303
  relinquishing hope to, 139
Accomplishments
  after retirement, 70
  limitations and, 196
  taking one day at a time, 200
Advent series by Patricia Houck Sprinkle
  First Sunday, 314
  Second Sunday, 323
  Third Sunday, 328
  Fourth Sunday, 333-34
  Christmas Eve, 336
  Christmas Day, 336-37
Adventures in Loving, series about marriage by the Peales, 42-48
  on forgiveness, 44-45
  on friendship, 46-47
  on God's Presence, 48
  on laughter, 47-48
  on sacrifice, 44
  on supportiveness, 45-46
  on trust, 43
Adversity, 53
  finding good in, 40-41, 171, 190, 196-97, 198, 272-73
  patience in, 170-71
  prayer and, 283-84
  using it creatively, 272-73
Aging, 37-38
Aida (opera), 155-56
All Saints' Day, 292
Allegiance, Pledge of, 167-68, 273
Amalekites (Biblical people), 161, 162
America (song), 180
American way of life
  contributions of immigrants, 249
  Pledge of Allegiance, meaning of, 167-68
  thanking God for, 160
Amish, 17-18
Ancestors
  remembering, on Citizenship Day, 249
Anger
  bitterness and, 117
  compassion and, 339
  forgiveness and, 304
  love and, 321-22
Animals
  "Blessing of the Animals" ritual, 124-25
Annabel Lee (Poe), 153
Anthony, Saint, 326-27
Anticipation
  of God in one's life, 18
Anxiety *see* Fear
Apology, 89
Appreciation
  see also Thankfulness
  expressing, for small things given, 28
  for God's gift of life, 252
  of a brand new day, 11-12
  of beauty, 103-04, 326
  of eternity, 336-37
  of God's Blessings, 32
  of God's Creation, 38, 124, 212, 270-71, 326
  of small labors of love, 210-11

Authors, Titles and Subjects Index

Arabian proverb, 19
Ark of the Covenant, 247-48, 329
Armed Forces Day, 135
Arthritis, 298-99
Asking for help
  see also God's Help; Helping others
  humility and, 127
  pride and, 239-40
Asking the Questions of Holy Week, series by Jeff Japinga, 79-89
  on belief, 80-81
  on commitment, 81-82
  on Christ's death, 86-87
  on Christ's Resurrection, 88-89
  on hope, 87-88
  on peace, 84-85
  on prayer, 82-83
  on service, 83-84
Attenborough, David
  Living Planet series of, 254-55

Bach, Johann Sebastian, 307
Bacher, June Masters
  selections by, 28, 53, 115-16, 142-43, 165, 178-79, 182-83, 215, 236, 253-54, 277, 283, 305-06, 311-12, 330
Bald eagle
  as national emblem, 99
Barnard, Dr. Christiaan, 109-10
Baseball, 277
Bauer, Fred
  poems by:
    on communion, 93
    on death and dying, 317
    on forgiveness, 235
    on friendship, 9
    on giving, 205
    on good deeds, 291
    on worry, 149
    on kindness, 121
    on life after death, 177
    on peace, 263
    on temptation, 61
    on winter, 35
  selections by, 24-25, 57-58, 65-66, 95, 137, 155-56, 186-87, 209-10, 241, 279, 297, 341
Beauty
  appreciating, 103-04, 326
Beecher, Henry Ward, 41
Beethoven, Ludwig van, 242, 293-94
Bible
  as source of spiritual life, 293-94
  giving, as a gift, 151-52
  quote about, 50
  red-letter edition, 63
  timelessness of, 253-54
Bible reading, 295
  asking God for guidance, 308-09
  when to read the Bible, 253-54
Bible stories
  see also Exodus, The
  God tells Moses to throw down the rod, 113
  Jesus' arrest in Garden of Gethsemane, 84-85
  Jesus drives out the moneychangers, 81-82
  Jesus in the Upper Room, 88-89
  Jesus tells His disciples to cast their nets, 222-23
  Jesus walks on water, 229
  Jesus washes the feet of His disciples, 83-84
  Mary at Jesus' feet, 206
  Peter jailed for preaching, 112-13
  Roman centurion at the Cross, 86
  the widow's mite, 66
Bitterness
  handling, 117
  letting go of, 105-06
"Blessed Ambiguity," 188
"Blessing of the Animals" ritual, 124-25
Blizzard of '78, 40-41
"Bottom forty," 132
Bread of Life, The (Bauer), 93
Bristlecone pine, 214-15
Brooklyn Bridge, 140
Brotherhood
  see also World community
  between family members, 135
  celebrating, on Thanksgiving Day, 305-06
  Christian fellowship, 237-38
  gestures of, 244-45
  Olympic games and, 154-55
  passing along positive behavior, 250-51
  prayer and, 23
  treating others as "saints," 292

Camino Real, El (highway), 77
Caring for others
  peace and, 244
  showing it, 237-38
Carlyle, Thomas
  quote by, 112
Carver, George Washington, 280
Casting nets, 222-23
Cerro Tortuguero (mountain), 134

Chambers, Oswald
quote by, 259
Champ, Isabel
selections by, 12-13, 39-40, 63, 99-100, 130, 157, 171, 190, 210-11, 231, 244-45, 281-82, 284, 292, 300, 309-10, 338, 340-41
Change
see also Letting go
accepting, 254-55
God's Will and, 108
Child rearing and training
passing on virtues, 57-58
Children
see also Parent and child; School children
celebrating, on Mother's Day, 130
compassion for, 218-19, 339
getting to know them better, 250
Nature of God and, 67-68
reading to, 252
China
Christianity in, 40
expressing friendship in, 244-45
Forbidden City of, 99-100, 157
Great Wall of, 282
Shanghai, 39-40
Christian life
accepting, 341
communism and, 228
freedom and, 185-86
going without reward and, 100-01
witnessing and, 128-29
Christianity
in China, 40
Christmas
see also Advent series
celebrating, 333, 338
"expecting" Christ, 330
giving gifts with love, 328
giving of one's own possessions, 331-32
kneeling before the newborn Christ, 334
letting go of, 14
Christmas Day, 336-37
Christmas Eve, 336
Christmas wreaths, 318
Church activities, 335
Church of the Nativity (Bethlehem), 334
Churches
robbing, 74
Cigarette smoking
appreciating one's body and, 306
Citizenship Day, 249
Civil War, American, 144-45
Columbus, Christopher, 273
Columbus Day, 273
Comforting others
comforting the strong, 158-59
gift of telling stories and, 52-53
small reassurances, 241-42
speaking words of comfort, 215
Commitment
fulfillment and, 126
loyalty and, 157-58
to Jesus Christ, 81-82
to practicing the Presence of God, 10
Communion (religious rite)
forgiveness and, 267
poem about, 93
Communism
Christian life and, 228
Community life
adversity and, 40-41
sharing, cooperation, and, 184-85
Compassion
for children, 339
Competition
cooperation and, 114
Complaining
about small things, 64
Confidence
in God's Words, 178-79
Confidence, betraying a, 275-76
Constructive criticism
ability and, 156
Cooperation
community life and, 184-85
competition and, 114
Courage
faith and, 139
fear and, 165, 172
of Abraham Lincoln, 50
success and, 29-30
to face a new direction in life, 274
to face the unknown, 192
winning and, 251-52
Covenant, Ark of the, 247-48, 329
Creativity
discovering, within us, 300-01
Cross
"X" as symbol for, 320-21

Daibutsu *Buddha*, 209
Davis, Jefferson, 224
Davis, Zona B.
selections by, 19, 52-53, 76-77, 111, 132, 167-68, 249
Dead Sea, 278-79

Death and dying
  accepting, 227-28
  as a beginning, 304-05
  poem about, 317
Decision-making
  formula for, 123
  patience in, 125-26
  retirement and, 31-32
Denial
  of Christ, 112-13
Devastation Trail (Hawaii), 286
Dieting *see* Weight control
Disappointment
  handling, 54-55, 284
Discouragement
  trust in God and, 155-56
Disobedience, 113
Distractions
  relationship with God and, 173, 278
Doubt
  faith and, 309-10
Duke, Drue
  selections by, 26-27, 70, 109-10, 128-29, 130-31, 143-44, 160, 192-93, 228, 239-40, 257, 258-59, 273, 278-79, 307, 326

Earhart, Amelia, 139
Easter Sunday, 88-89
Efficiency
  formula for, 123
Election Day, 294-95
Emancipation Proclamation, 50
Emerson, Ralph Waldo
  quotes by, 13, 46
Emotional problems
  dealing with, 218-19
Empowerment, 322
Encouragement, 16-17, 156, 241-42
Eternal life *see* Life after death
Eternity
  appreciating, 336-37
Evangelism, 99-100
Exodus, The
  Israelites follow pillar-shaped cloud, 21-22
  Israelites follow Pillar of Fire, 49-50
  crossing the Red Sea, 72-73
  trudging through the desert, 105-06
  Manna rains from Heaven, 133-34
  Moses prays for victory, 161-62
  Moses is tired, 189
  Moses receives the Ten Commandments, 216-17
  worshiping a Golden Calf, 247-48
  offering a sacrifice, 275-76
  Moses says goodbye, 302-03
  following the Ark of the Covenant, 329-30
Failure
  see also Success
  opportunity in, 325-26
Faith
  see also Trust in God
  ability and, 330-31
  bristlecone pine and, 214-15
  children and, 276-77
  courage and, 139
  doubt and, 309-10
  during illness, 270
  fear and, 229
  God's help and, 230-31
  good deeds and, 137
  in Jesus as Messiah, 80-81
  in life after death, 227-28
  in Presence of God, 101-02
  let go and let God in, 196-97
  pain and, 20
  patience and, 15-16
  planting seeds of, 136-37
  story of the wild mustard, 77-78
  tests of, 259
  trust in God and, 151
Faithfulness *see* Loyalty
Families
  importance of, 25-26
  spiritual families, 57
  working together in love, 30-31
Family life
  accepting a new member, 319-20
  God's Presence in, 48
  sense of humor in, 47-48
  veterans and, 135
Father's Day, 160-61
Fear
  courage and, 165
  handling, 13, 72-73, 110-11
  of the past, 104-05
  of the unknown, 192
  of thunderstorms, 165, 13
Fellowship *see* Brotherhood
Fishing, 222-23
Flag, American
  49-star, 160
Flag Day, 160
Food, of Georgia, 224-25
For Empty Vessels (Bauer), 263
For Restless Sleepers (Bauer), 149
Ford, Henry
  quote by, 46

Forgetfulness
presence of God and, 62
Forgiveness
accepting, 275-76
anger and, 304
asking God for, 29
forgiving oneself, 29, 304
forgiving those who trespass against us, 74
God's Love and, 159
in marriage, 44-45
Jesus Christ and, 112-13, 274-75
judging others and, 257-58
poems about, 235, 324
pride and, 324
receiving communion and, 267
Fourth of July, 180-81
"Fox and Geese" (game), 12
Francis of Assisi, Saint, 145, 266-67
Freedom *see* Liberty
Freedom of religion, 180-81
Freedom of worship *see* Freedom of religion
Friendship, 140
gestures of, 244-45
God as Friend, 178
in marriage, 46-47
listening and, 27-28
making and being a friend, 19, 141-42
poem about, 9
prayer and, 199
Fulfillment
commitment and, 126
Future, the
trust in God and, 246
uncertainty about, 49-50

Galilee, Sea of, 278-79
Garfield (cartoon character), 196
Generosity
love and, 301
Gentleness
in voice and actions, 14-15
Georgia *see* Savannah (Ga.)
Gestures
of friendship and brotherhood, 244-45
Giving
the Bible as a gift, 151-52
Christmas gifts with love, 328
a gift to a friend, 23
in secret, 332
learning from God's lessons of, 66
love to others, 195
mementos of a visit, 190-91
more than we receive, 95
of one's own possessions, 331-32
of oneself, 102, 319-20
our gift of creativity to others, 300-01
poems about, 205, 324
without reward, 332
Giving and Forgiving (Springer), 324
Goal-setting, 200
God
see also Praising God, Presence of God, Serving God, Thanking God, Trust in God *and subsequent headings below beginning with* God's...
as a Source of healing, 135-36
as Creator, 256-57
as Pilot of one's life *see* God's Providence
as the center of one's life, 131-32
as the Great Physician, 109-10
relationship with, 278
writing a letter to, 178
God Who Gives, The, series by Patricia Houck Sprinkle, 313-14, 323, 328, 333-34, 336-37
God's Blessings, 132
accepting, 32
asking, upon a problem, 327-28
"Blessing of the Animals" ritual, 124-25
gratitude for, 292-93
looking within for, 241
God's Creation, 96-97, 109, 134, 207
appreciating, 38, 212, 326
diversity of, 220
Presence of God and, 94
"sauntering" through, 103-04
seeing God in His Creation, 24-25
witnessing and, 107-08
God's Hand, 17-18, 24-25
God's Help, 296, 283-84
God's Love, 31
acceptance of all peoples and, 333-34
giving His Only Begotten Son, 328
grief and, 286-87
growing in, 96-97
healing nature of, 109-10
learning forgiveness and, 159
opening up to, 322
remembering, during Advent, 314
setting out on a new venture and, 183-84
sharing, 278-79
smiles and, 130-31
spiritual renewal and, 336
world community and, 323

God's Mercy, 321-22
God's Name, 207-08
  doing good deeds in, 236-37
  names of Christ in Bible, 122-23
God's Nature, 67-78
God's Plan
  appreciating, 163-64
  discouragement and, 155-56
  fears of the past and, 104-05
  fulfilling, every day, 172-73
  learning, 74-75
  one's life and, 26-27, 109
  pain and, 20
God's Providence, 134, 222-23, 256-57, 298, 325
  adversity and, 95-96, 231
  keeping resolutions and, 182-83
  poem about, 35
  trusting in, 185
  working with God as a team, 197-98
God's Sustenance, 213-14
God's Voice
  listening for, 106-07
God's Will
  accepting, 163-64
  change and, 108
  happiness and, 70
  learning, 74-75
  spiritual life and, 99
God's Word *see* Bible
Golden Rule, 250-51
Good deeds
  faith and, 137
  poem about, 291
  Presence of God and, 236-37
  undertaking, in God's Name, 273
Good Friday, 86
Good manners, 57-58
Gordon, Arthur
  My Savannah: Behold Its Goodness series by, 219-28
  selections by, 26, 103-04, 151, 299
Grace (prayer at meals), 309
Graduation
  looking back, 163
Gratitude *see* Thankfulness
Great Wall of China, 282
Greed
  thankfulness and, 71
Greene, Oscar
  selections by, 31-32, 40-41, 50, 104-05, 135, 139, 159, 179-80, 213, 242-43, 280, 335
Grief
  coping with, 255-56
  God's Love and, 286-87
  time and, 286-87
Groundhog Day, 36-37
Group prayer, 26

Haase, Doris
  selections by, 27-28, 51-52, 64, 107-08, 114-15, 138-39, 144-45, 170-71, 181, 194, 208, 216, 229, 251-52, 257-58, 280-81, 298, 337-38
Habits and faults
  cigarette smoking, 306
Halley's comet, 102
Hamlin, Rick
  selections by, 11, 57, 89, 114, 141-42, 154-55, 163-64, 192, 201, 214-15, 237-38, 266-67, 304
Handbook of Positive Prayer (Hasbrouck), 56
Handicapped
  accepting infirmities, 303
Harnack, Adolf von
  quote by, 182
Harrah, Madge
  selections by, 20-21, 39, 70, 98, 117, 126, 145, 151-52, 166-67, 180-81, 193-94, 211, 250-51, 283-84, 286-87, 298-99, 312, 325, 330-31
Hasbrouck, Hypatia, 56
Hawaii, Devastation Trail, 286
Healing
  emotional difficulties, 218-19
  God as a Source of, 135-36
  prayer and, 82-83, 116
Health
  see also Illness
  promise of, 279
Hearing, loss of
  living with, 303
Heaven, 340-41
  salvation and, 241
  within us, 78-79
Helleberg, Marilyn Morgan
  selections by, 56-57, 95-96, 116, 129, 166, 199, 240, 246, 285-86, 327-28, 332
  Wilderness Journey series by, 21-22, 49-50, 72-73, 105-06, 133-34, 161-62, 189, 216-17, 247-48, 275-76, 302-03, 329-30
Help, asking for
  asking God, 296
  humility and, 127
  pride and, 239-40
Helping others, 28
  being God's Helping Hand, 17-18

doing God's Will and, 70
humility and, 114-15
witnessing and, 280-81
Hines, Jerome, 155
His Healing Word *see* Bauer, Fred, poems by
Hobe, Phyllis
selections by, 30-31, 36-37, 106-07, 185, 207, 243, 307-08, 333
Holmes, Marjorie
selections by, 15-16, 78-79, 97-98, 109, 154, 198, 217-18, 274-75, 313
Holy of Holies, 247-48
Holy Thursday, 84-85
Holy Week series by Jeff Japinga, 79-89
on belief, 80-81
on Christ's death on the Cross, 86-87
on Christ's Resurrection, 88-89
on commitment, 81-82
on hope, 87-88
on peace, 84-85
on prayer, 82-83
on service, 83-84
Homeless
helping them, 266-67
Homesickness, 57
Honesty, 307
in prayer, 299
Hope
in Christ's Resurrection, 87-88
relinquishing, to acceptance, 139
Human behavior
see also Friendship; Love; Marriage
passing along positive behavior, 250-51
Human relationships
meeting strangers, 285
role of wife/mother and, 245-46
Humility, 127, 137-38
see also Pride
asking others to share a burden, 189
the greatness within and, 111
of Christ, 76-77
pride and, 89
Hur (Biblical character), 161-62
Husband and wife
wife's role, 245-46
Hymns
see also Songs
"In Remembrance," 236-37
of Lowell Mason, 227
"Olivet," 213
prayer hymns, 98
"Swing Low, Sweet Chariot," 228
"Turn your eyes upon Jesus...," 265-66
"I Need You, Lord," series by Sue Monk Kidd, 268-73
Illness
being still in, 106-07, 271-72
blessings from, 181
handling, 268-73
overcoming negativity about, 289-99
taking strength for granted, 158-59
Imaging
adversity and, 230-31
Immigrants, 249
Impatience, 136-37
In Celebration of Small Towns, series by Arthur Gordon, 219-28
"In Remembrance" (song), 236-37
Independence Day, 180-81
Insomnia
poem about, 149
Inspirational sayings
collecting, 193-94
International Forgiveness Week, 29
Israelites, exodus of *see* Exodus, The

Japan, 209-10
Japinga, Jeff
Holy Week series by, 79-89
selections by, 16-17, 54, 108, 125-26, 141, 160-61, 168, 182, 195, 244, 334
Jesus Christ
and the moneychangers, 81-82
appears in the Upper Room, 88-89
as advocate with the Father, 274-75
as Brother, 228
as Messiah, 80-81
as the Way, the Truth, and the Light, 95-96
becoming disciples/soldiers of, 53, 81-82
dies on the Cross, 86
humility of, 76-77
is arrested in Garden of Gethsemane, 84-85
Martha, Mary, and, 123
Mary at the feet of, 206
Names of, in the Bible, 122-23
strength of, 257
walks on water, 229
washes the feet of His disciples, 83-84
words of, in the Bible, 63
Jethro (Biblical character), 189
Job
story of, 20
Jogging, 114, 264-65, 278
Jones, Paul
quote by, 62

Jordan River, 278-79
Joshua (Biblical character), 302-03, 329
Joy
  gift of telling stories and, 52-53
Judging others
  faith and, 209-10
  forgiveness and, 257-58
  understanding and, 297
Justice, Sam
  selections by, 17, 37-38, 74, 99, 143, 157-58, 172-73, 197-98, 250, 278, 304-05

Keeping Score (Bauer), 205
Kidd, Sue Monk
  Practicing His Presence... series by, 10, 36, 62, 94, 122-23, 150, 178, 206, 236-37, 264, 292-93, 318
  selections by, 42, 72, 102, 131-32, 162-63, 187-88, 196, 248-49, 255-56, 301
  Waiting Through An Illness series by, 268-73
Kindness, 307
  giving mementos of a visit, 190-91
  poem about, 121
King, Martin Luther, Jr., 20-21
Kittler, Glenn
  selections by, 22-23, 69, 127-28
Klopsch, Louis
  his red-letter edition of the Bible, 63
Kneeling
  before the newborn Christ, 334
Kum Ba Yah (folk song), 98
Kuykendall, Carol
  selections by, 14-15, 32, 55-56, 68-69, 110-11, 123, 156, 191, 245-46, 264-65, 285

Labor Day, 236
Lafayette, Marquis de, 224
Laubach, Frank, 10
Laughter
  healing nature of, 64
  in marriage, 47-48
Lawrence, Brother, 10
Lawrence, T. E. ("Lawrence of Arabia"), 43
Learning disabilities, 117
Lee, Robert E., 224
Leisure, 152-53
  taking time for, 181
Lent
  keeping Lenten promises, 42
  spiritual renewal in, 51
Letting go, 196-97, 274
  of guilt, 275-76
  of Christmas, 14
Liberty
  Christian life and, 185-86
  found in giving oneself up to God, 329-30
Life after death, 78-79, 227-28
  see also Heaven
  God's promise of, 313
  poem about, 177
  promise of, 279
Lincoln, Abraham, 41, 166, 242
  courage of, 50
Listening
  for God's Voice, 106-07, 316
  friendship and, 27-28
Living Planet (TV series), 254-55
Loneliness, 287
  hope of Christ and, 87-88
Longfellow, Henry Wadsworth
  quote by, 164
Lord, Make Me Kind (Bauer), 121
Lord, Show Me How (Bauer), 291
Lord's Prayer, 56-57
Love
  see also Marriage
  anger and, 321-22
  families and, 30-31
  generosity and, 301
  giving Christmas gifts with, 328
  little acts of love, 337-38
  of God, 209
  of Jesus Christ, witnessing to, 278-79
  of others, 195, 209, 229-30
  showing, in unexpected ways, 142-43
  speaking words of, 215
  taking time to, during Advent, 314
  unrequited, 159
  "XXX" as expression of, 320-21
Low, Juliette Gordon, 225-26
Loyalty, 157-58
Make Haste, My Beloved, (Roberts), 325
Manna, 133
Marriage
  customs and rites, 151-52, 166
  forgiveness in, 44-45
  friendship in, 46-47
  God's Presence in, 48
  laughter in, 47-48
  sacrifice in, 44
  supportiveness in, 45-46, 54
  trust in, 43
  wife's role in, 245-46
Martha (Biblical character), 123

Mary (Martha's sister), 123
Mary, mother of Christ, 206
Mason, Dr. Lowell (composer), 213, 227
Materialism and material possessions, 182, 264
Maundy Thursday, 84-85
Mazur, Kurt (conductor), 293-94
McDermott, James
  selections by, 41, 67-68, 169-70, 184-85, 293-94
McGarrity, Samantha
  selections by, 25-26, 67, 74-75, 127, 137-38, 158-59, 190-91, 200, 213-14, 229-30, 253, 274, 300-01, 309, 322, 324, 331-32, 339
Mealtime
  offering thanks at, 309
Memorial Day, 144-45
Memories, 240
Mercer, Johnny, 227
Mercy, 321-22
Merton, Thomas
  prayer by, 274
Messiah, The, 80-81
Millay, Edna St. Vincent, 300
Mind and body *see also* Physical fitness
  body as God's temple, 306
Miracles, 216
Missionary work, 39-40
Missouri, USS (battleship), 135
Mistakes
  handling, 284
  putting them to good use, 312
  starting over and, 211
Mohonk, Lake, 124
Morning
  beginning the day with God, 69
  waking up in, 11-12
Morrissey, Muriel Earhart (Amelia Earhart's sister), 139
Morse, Samuel, 114-15
Moses, 113
  leads the Israelites through the desert, 105-06
  receives the Ten Commandments, 216-17
  says goodbye to his people, 302-03
  tires on his desert journey, 189
Mothers
  cultivating faith in children, 276-77
  role of, 245-46
Mother's Day, 130
Mountain laurel, 124
Music, 307-08
  of Savannah, Ga., 226-27
Mustard, wild, 77
My Savannah: Behold Its Goodness, series by Arthur Gordon, 219-28

Nash, Ogden
  quote by, 225
National Bible Sunday, 308-09
Nature, 254-55
  see also Beauty; God's Creation
  appreciation of, 270-71
Negative thoughts, 55-56, 115-16, 138-39
New Year
  extending Christ's peace into, 14
  poem about, 13
New Year's resolutions, 51-52, 182-83

Obedience, 194
  throwing down "sticks of disobedience," 113
Obesity *see* Weight control
Oglethorpe, James, 221, 224
"Old Abe" (bald eagle), 144-45
"Olivet" (hymn), 213
Olympic games, 154-55, 186-87
On Golden Pond (movie), 297
Orion (constellation), 256

Padres (baseball team), 277
Pain and suffering
  faith in God and, 20
  remedy for, 298-99
Palm Sunday, 80-81
Parent and child
  see also Children; Mothers
  handling rebellious children, 113
Passover, Feast of
  John's account, 76
Patience, 125-26, 136-37, 307
  after illness, 271-72
  in adversity, 170-71
  to do one's best, 243
  with children, 339
  with life's ups and downs, 97-98
Patrick, Saint, 74-75
Paul, Saint, 112
  work of, 99-100
Peace, 304
  caring and, 244
  poem about, 263
  prayer and, 84-85, 248-49
  taking time out for, 67
Peale, Norman Vincent
  Adventures in Loving series by Ruth Stafford Peale and, 43, 44-47
  selections by, 13

Peale, Ruth Stafford
Adventures in Loving series by Norman Vincent Peale and, 44, 47-48
selections by, 17-18
Perseverance
reaching one's goal, 264-65
success and, 29-30, 251-52
Peter the Apostle
imprisonment of, 112-13
Mark and, 112
Pharisees, 209
Physical fitness, 141
Plants
mountain laurel, 124
mustard, wild, 77
Pledge of Allegiance, 167-68, 273
Poe, Edgar Allan
poem by, 153
Politics, 294-95
Positive thinking, 162-63, 230-31
Poverty
serving the poor, 266-67
Practicing His Presence..., series by Sue Monk Kidd
on celebrating His Presence, 318
on commitment, 10
on forgetfulness, 62
on God's Creation, 94
on good deeds, 236-37
on imaging Christ's Presence, 206
on learning to let go, 264
on meeting God through the newspaper, 150
on prayer, 36
on thankfulness, 292-93
on the Names of Jesus, 122-23
on writing God a letter, 178
Praise Diary
using, 39
Praising God, 39, 114-15
Prayer
see also Lord's Prayer
ability and, 330-31
adversity and, 283-84
as two-way conversation, 319
at anytime, anyplace, 265-66
beginning the day with God, 69
brotherhood and, 23
failure and, 325-26
God's Presence and, 36
peace and, 84-85
praying for healing, 82-83, 116
praying for others, 258-59
praying for peace, 248-49
praying through hymns, 98
praying through the newspaper, 150
praying with another, 199
serving God and, 20-21
solitude and, 191
success through, 129
temptation and, 161-62
thankfulness in, 103, 299
time of "Blessed Ambiguity," 188
Prayer groups, 26
Presence of God, 11, 144-45, 169, 207-08, 300
see also Practicing His Presence..., series
coming into the Lord's Presence, 157
coping with grief and, 255-56
creating a holy place within, 247-48
facing surgery and, 268-69
faith in, 101-02, 309-10
forgetfulness and, 62
God's Creation and, 94
"imaging" Bible events, 206
in marriage and family life, 48
in ordinary moments of the day, 318
prayer and, 150
relics and, 166-67
Present, the
see also Future, the
living in, and enjoying, 164-65
Pride, 209
see also Humility
asking for help and, 239-40
false, 319-20
forgiveness and, 324
humility and, 89
Prison life, 95
Problems and problem-solving, 132, 239-40
asking God's Blessing for, 327-28
praying with another and, 199
Public worship
see also Freedom of religion
diversity of, in small towns, 221-22

Rationing, 66
Reading
the Bible, 295, 308-09
love of, 252
to children, 252
Recreation *see* Leisure
Red Sea, 72-73
Redwood forests (California), 184-85
Regret, 159
Relaxation
taking time for, 181
Relics, 166-67

Relinquishment to the Lord, 22-23
Renoir, Auguste, 298-99
Repentance
  during Lent, 51
Resentment
  letting go of, 105-06
Resurrection, The, 87-89
Retirement, 70
  deciding to retire, 31-32
Retreats, 103, 285-86
  see also Solitude
Risks, taking, 186-87
Rites and ceremonies
  see also Marriage
  "Blessing of the Animals" ritual, 124-25
  receiving communion, 267
Roberts, Frances J.
  Make Haste, My Beloved, book by, 325
Rodriquez, Jose (chess expert), 178-79
Roman centurion
  at the foot of the Cross, 86
Roosevelt, Eleanor, 294
Roosevelt, Franklin Delano, 311-12
Rose Parade (Pasadena, Cal.), 11

Sacrifice
  in marriage, 44
Sadducees, 209
St. Francis of Assisi Day, 266-67
St. Johns, Elaine
  selections by, 24, 77-78, 100-01
Saints, 292
Salvation, 241
Sanctuary
  creating a holy place within, 247-48
Sargasso Sea, 254-55
Sass, Eleanor
  selections by, 23, 66, 111-12, 134, 164-65, 185-186, 212, 230-31, 252, 265-66, 267, 276-77, 295, 303, 306, 320-21
Savannah (Ga.)
  faith of, 227-28
  flavor of, 224-25
  founding of, 220-21
  legacy of, 223-24
  music of, 226-27
  people of, 225-26
  spirit of, 222-23
  spiritual heritage of, 221-22
  uniqueness of, 220
Scapegoat, the
  story of, 275-76
School children, 241-42
School work
  succeeding at, 75-76
Schweitzer, Albert
  quote by, 136-37
Sea turtles, 134
Seasons, 142-43
  see also Spring; Winter
Secretaries
  pride in being one, 111-12
Self-centeredness
  vs. other-centeredness, 68-69, 141-42
Self-confidence
  encouraging, 156
Self-identity, 26-27
Self-justification, 89
Senior citizens, 24
Sense of humor
  see also Laughter; Smiles and smiling
  in marriage and family life, 47-48
Sensitivity, 190-91
Serra, Fr. Junipero, 77
Serving God, 78, 207, 335
  one's work and, 127-28
  poem about, 205
  routine tasks and, 238-39
  strength for, 20-21
  to best of one's ability, 281-82
  working without salary, 280
Serving others, 76-77, 339
  Jesus' example, 83-84
  serving the poor, 266-67
Setting priorities, 123
Seven Pillars of Marriage, series by Norman Vincent and Ruth Stafford Peale, 42-48
Seven Pillars of Wisdom, The (T. E. Lawrence), 43
Shanghai (China), 39-40
Sharing, 154
  at mealtime, 309
  burdens of life, 189
  community life and, 184-85
  competition and, 114
  dreams and goals, 54
  one's abilities, 213, 242-43
  passing along positive behavior, 250-51
  words of encouragement, 16-17
Sherman, General William T., 224
Sherrill, Elizabeth
  selections by, 18, 51, 71, 103, 124, 169, 183-84, 207-08, 254-55, 296, 310-11, 319
Shinn, George, 69
Sickness *see* Illness

Signs and symbols
  bald eagle, 99
  Christmas wreath, 318
  "X" for the Cross of Calvary, 320-21
Silence, 286
Simplifying one's life, 189
Singing, 307-08
  prayer hymns, 98
Sleeplessness *see* Insomnia
Small towns *see My Savannah: Behold Its Goodness,* series
Smiles and smiling, 17, 190, 244-45
  inspirational saying about, 130-31
  spreading God's Love by, 130-31
Smiley, Albert and Alfred, (Quakers), 124
Smoking *see* Cigarette smoking
Snow, 12-13
  appreciating, 32
Snowstorms, 40-41
Solitude, 286
  spiritual renewal and, 191
Some Day In the Clearing (Bauer), 317
Songs
  see also Hymns
  Kum Ba Yah, 98
Soul food, 224-25
Speaking
  quietly, 14-15
  words of love or comfort, 215
Spiritual development, 65-66
Spiritual life, 151
  asking God for help, 296
  becoming soldiers of Christ, 53
  Bible as the source of, 293-94
  God's Will and, 99
  keeping Lenten promises, 42
  uncluttering our lives and, 199-200
Spiritual renewal, 217-18, 336
  during Lent, 51
  in Spring, 72
  solitude and, 191
Spiritual Songs for Social Worship (Mason), 213
Spiritual sustenance, 224-25
Sports
  persevering at, 201
Sportsmanship, 154-55
Spring, 36-37
  spiritual renewal and, 72
Springer, Thomas Grant
  poem by, 324
Sprinkle, Patricia Houck
  Advent series by, 313-14, 323, 328, 333-34, 336-37
  selections by, 20, 31, 38, 78, 101-02, 112-13, 136-37, 152-53, 172, 188, 199-200, 209, 238-39, 256-57, 296-97, 308-09, 321-22, 326-27
Stargazing, 256-57
Stevenson, Robert Louis, 242
Stories, telling, 52-53
Strangers
  meeting, 285
Strength
  for serving God, 20-21
  nourishing one's, 269
  of Jesus Christ, 257
  taking one's strength for granted, 158-59
  to face a new direction in life, 274
  to make decisions, 31-32
  weakness and, 137-38
Success, 201
  at school work, 75-76
  courage and, 251-52
  defeat and, 277
  God's Will and, 163-64
  measurement for, 182
  persevering toward, 29-30
  through prayer and trying, 129
  with God's Help, 51-52
  work and, 24
Suffering *see* Pain and suffering
Summer Olympics, 186-87
Surgery
  facing, and God's Presence, 268-69
"Swing Low, Sweet Chariot" (spiritual), 228

Taking risks, 186-87
Talent *see* Ability
Teamwork
  with God, 197-98
Temptation
  poem about, 61
  praying against, 161-62
Temptation's Tender Trap (Bauer), 61
Ten Commandments, 216-17
  as God's warnings, 194
Thackeray, William Makepeace, 224
Thankfulness, 153
  see also Thanking God; Thanksgiving Day
  as sincerest form of worship, 223-24
  expressing, 208
  for God's Blessings, 292-93
  for joy of being alive, 270-71
  for work of others, 192-93
  greed and, 71
  in prayer, 299

Thanking God, 208
at mealtime, 309
for America, 160
for answers to prayers, 103
Thanksgiving Day, 311-12
celebrating brotherhood on, 305-06
Thoreau, Henry David, 104
Thought and thinking
handling negative thoughts, 55-56, 115-16, 138-39
Tiffany, Louis Comfort, 23
Tiffany glass, 23
Time, 240
as a gift from God, 283
grief and, 286-87
taking, for leisure, 181
using it wisely, 283
Tolerance, 41
of *all* God's children, 333-34
of children, 339
of other's shortcomings, 297
Tomochichi (Indian chief), 224
Tree of Friendship, The (Bauer), 9
Trust
in marriage, 43
Trust in God, 15-16, 36-37, 169, 229, 274
adversity and, 198
discouragement and, 155-56
for tomorrow's needs, 133-34
the future and, 49-50, 246
keeping resolutions and, 182-83
learning to let go of material possessions, 264
"Trust Walk," 179-80
winning and, 251-52
worry and, 145
"Turn your eyes upon Jesus..." (hymn), 265-66
Tutu, Bishop Desmond, 244
Twelfth Night (Jan. 6), 14
Twenty-third Psalm, 152-53
2,000-Year Odyssey in Faith, A, series, 79-89
USS Missouri (battleship), 135
United Nations Day, 282

Valentine's Day, 46-47
Van Dyke, Henry
quote by, 236
van Gogh, Vincent, 242
Varner, Van
selections by, 11-12, 54-55, 124-25, 140, 319-20
Veterans
adjusting to family life, 135
Virtuousness
passing on, 57-58
Vocations, 296-97
Voice, tone of
peace, quiet, and, 14-15
Volid, Ruth (art dealer), 199
Volunteer work
pain of arthritis and, 298-99
working without salary, 280
Voting
for the *best* person, 294-95
Waiting
for results, 136-37
for God's time, 188
Waiting Through An Illness, series by Sue Monk Kidd
on faith, 270
on gratitude, 270-71
on growth, 272-73
on Presence of God, 268-69
on stillness, 271-72
on strength, 269
Walking Through Fire (Bauer), 177
Washington, George, 53
Weakness
strength and, 137-38
Weddings, 166
Bible as gift for, 152-53
Weight control, 51-52
Wesley, John, 222, 224
West, Marion Bond
selections by, 29-30, 64-65, 96-97, 113, 135-36, 153, 196-97, 218-19, 241-42, 259, 287, 325-26, 339-40
What Is Christmas All About? series by Patricia Houck Sprinkle
First Sunday of Advent, 314
Second Sunday of Advent, 323
Third Sunday of Advent, 328
Fourth Sunday of Advent, 333-34
Christmas Eve, 336
Christmas Day, 336-37
When We Can't Forgive Ourselves (Bauer), 235
Whitefield, George, 222
Whitney, Eli, 224
Widow's mite
story of, 66
Wilderness Journey, series by Marilyn Morgan Helleberg
Following the Pillar-Shaped Cloud, 21-22
Led by a Pillar of Fire, 49-50
Crossing the Red Sea, 72-73
Trudging Through the Desert, 105-06

Wilderness Journey, series, cont.
  Manna Rains From Heaven, 133-34
  Moses Prays for Victory, 161-62
  Moses Is Tired, 189
  Moses Receives the Ten Commandments, 216-17
  Worshiping a Golden Calf, 247-48
  Offering a Sacrifice, 275-76
  Moses Says Good-bye, 302-03
  Following the Ark of the Covenant, 329-30
Winter
  of '34, 75
  poem about, 35
Winter of the Soul (Bauer), 35
Wisdom
  promise of, 279
Wit and humor, 190
Witnessing, 64-65, 128-29, 280-81
  God's Creation and, 107-08
  through good deeds, 137
  to love of Christ, 278-79
Word of God *see* Bible
Wordsworth, William
  poem by, 287
Work *see also* School work
  and success, 24
  appreciating small labors of love, 210-11
  appreciating work of others, 192-93
  being God's Helping Hand, 17-18
  doing God's Will, 70, 236
  doing it well and proudly, 111-12, 169-70
  pain of arthritis and, 298-99
  reward for, 253
  routine tasks, 238-39
  serving God and, 127-28
  set-backs and, 310-11
  working without salary, 280
World Communion Sunday, 267
World community, 282
  see also Brotherhood
  God's Love and, 323
World War II, 66, 135, 299
Worry, 187-88
  conquering, 110-11
  trust in God and, 145
  poem about, 149

Youth, 24

Zerubbabel (Biblical character), 174

A Note From The Editors

This devotional book was created by the same staff that prepares *Guideposts*, a monthly magazine filled with true stories of people's adventures in faith.

If you have found enjoyment in *Daily Guideposts*, we think you'll find monthly enjoyment—and inspiration—in the exciting and faith-filling stories that appear in our magazine.

Guideposts is not sold on the newsstand. It's available by subscription only. And subscribing is easy. All you have to do is write Guideposts Associates, Inc.; Carmel, New York 10512. A year's subscription costs only $6.95 in the United States, $8.95 in Canada and overseas. Our Big Print edition, for those with special reading needs, is only $6.95 in the United States, Canada and abroad.

When you subscribe, each month you can count on receiving exciting new evidence of God's presence, His guidance and His limitless love for all of us.